AF595091

Manufacturing Metrology

Manufacturing Metrology

Editors

Kuang-Chao Fan
Peter Kinnell

MDPI • Basel • Beijing • Wuhan • Barcelona • Belgrade • Manchester • Tokyo • Cluj • Tianjin

Editors
Kuang-Chao Fan
Dalian University of Technology
China
Peter Kinnell
Loughborough University
UK

Editorial Office
MDPI
St. Alban-Anlage 66
4052 Basel, Switzerland

This is a reprint of articles from the Special Issue published online in the open access journal *Applied Sciences* (ISSN 2076-3417) (available at: https://www.mdpi.com/journal/applsci/special_issues/Manufacturing_Metrology_Manufacturing).

For citation purposes, cite each article independently as indicated on the article page online and as indicated below:

LastName, A.A.; LastName, B.B.; LastName, C.C. Article Title. *Journal Name* **Year**, *Volume Number*, Page Range.

ISBN 978-3-0365-2986-8 (Hbk)
ISBN 978-3-0365-2987-5 (PDF)

Cover image courtesy of Kuang-Chao Fan

Contents

About the Editors

Kuang-Chao Fan, Professor, received his Ph.D. degree in mechanical engineering from the University of Manchester Institute of Science and Technology (UMIST) in 1984 and the DEng of University of Manchester in 2014. He was a professor of mechanical engineering at National Taiwan University from 1989 to 2015 and has been the Cheng Kong Scholar at Hefei University of Technology since 2001. He is a Fellow of SME, ISNM, CSME, and CIAE. His current research interests include manufacturing metrology, machine tool metrology, robot calibration, and micro/nano measurement. He has published more than 170 journal papers and 250 conference papers. He holds more than 40 national and international patents. He has received many academic honors and awards. Currently, he is working with Dalian University of Technology. He is the editor of four SCI journals.

Peter Kinnell, reader in metrology, obtained a degree in mechanical engineering, and then a PhD in microtechnology, from the University of Birmingham in 2003. He then joined GE Sensing, where he worked on the design and development of the company's high-performance TERPS range of pressure sensors. He left GE Sensing to take a lectureship in manufacturing and metrology at The University of Nottingham in 2007, where he worked on metrology for high-precision manufacturing. Then, in 2010, he joined Loughborough University as a Senior Lecturer in Metrology. At Loughborough university, he is an associate director of the Intelligent Automation Centre, where he leads the Robust Intelligent Metrology Laboratory.

Editorial

Special Issue on Manufacturing Metrology

Kuang-Chao Fan [1,2,*] and Peter Kinnell [3]

1 School of Mechanical Engineering, Dalian University of Technology, 2 Linggong Rd., Dalian 116023, China
2 Department of Mechanical Engineering, National Taiwan University, 1, Sec. 4, Roosevelt Rd., Taipei 10617, Taiwan
3 Wolfson School of Mechanical, Electrical and Manufacturing Engineering, Loughborough University, Loughborough LE11 3TU, UK; P.Kinnell@lboro.ac.uk
* Correspondence: fan@ntu.edu.tw

Citation: Fan, K.-C.; Kinnell, P. Special Issue on Manufacturing Metrology. *Appl. Sci.* **2021**, *11*, 10660. https://doi.org/10.3390/app112210660

Received: 26 October 2021
Accepted: 27 October 2021
Published: 12 November 2021

1. Introduction

Metrology is the science of measurement and can be divided into three overlapping activities: (1) the definition of units of measurement, (2) the realization of units of measurement, and (3) the traceability of measurement units. Manufacturing metrology originally referred to the measurement of components and inputs in a manufacturing process to ensure that they are within the required specifications. It also referred to measuring the performance of manufacturing equipment. This Special Issue presents a wide selection of papers on novel measurement methodologies and instrumentations for manufacturing metrology, from conventional industries to frontiers in advanced, hi-tech industries. Twenty-five papers are included in this Special Issue. These published papers can be categorized into four main groups, including length measurement, surface profile and form measurements, angle measurement, and laboratory systems. Detailed descriptions of these groups are introduced below.

2. Length Measurement

For the distance between two parallel surfaces, such as long gauge blocks, three methods are proposed. Using a combination of a laser interferometer and white light interference, the practical positioning method in end-plate surface distance measurement can achieve nanometer-scale precision [1]. Using a combination of laser triangulation sensors and a contact probe, not only the distance of two parallel surfaces but also the difference between surface shape contours can be measured to micrometer-scale accuracy [2]. With a multi-path laser interferometer, a new computational model for step gauge calibration was proposed based on synthesis technology [3], and a differential quadrature Fabry–Pérot interferometer was proposed separately [4]. In this measurement system, the nonlinearity error can be improved effectively and the DC offset during the measurement procedure can be eliminated. The turbine blade vibration was obtained by the blade tip timing (BTT) technique using the time of arrival (ToA) of the blade tip passing the casing mounted probes [5]. A fast laser adjustment-based laser triangulation displacement sensor was designed for dynamic measurement of a dispensing robot [6]. An invited paper contributed by Prof. S.W Kim's group from KAIST reported an absolute interferometer configured with a 1 GHz microwave source photonically synthesized from a fiber mode-locked laser with a 100 MHz pulse repetition rate [7]. This photonic microwave interferometer is expected to replace conventional incremental-type interferometers in diverse long-distance measurement applications, particularly for large machine axis control, precision geodetic surveying, and inter-satellite ranging in space.

3. Surface and Profile Measurement

On this topic, 12 papers were collected. A novel design for a surface topography measurement system was proposed in a relatively large area of 100 mm × 100 mm to achieve nanometer-scale accuracy [8]. The motion error of the stage was separated by a differential measurement configuration for a confocal sensor and a film interferometry module. The same group further designed an internal scanning mechanism for a confocal sensor [9]. By synchronizing the local scan, enabled by the internal actuator in the confocal sensor, and the global scans, enabled by external positioners, the developed system was able to perform noncontact line scans and area scans. Thus, this system was able to measure both surface roughness and surface uniformity. A non-scanning 3D imaging system with a single-pixel detector was reported to achieve 3D imaging of a target via compressed sensing to overcome the shortcomings of existing laser 3D imaging technology [10]. In order to solve the problems in the accuracy and adaptability of existing methods for blade twist measurement, a high-precision and form-free metrological method of blade twist based on the parameter evaluation of twist angular position and twist angle was proposed [11]. A rapid optical gear measurement system was developed for measuring the irregular tooth contours of large ring parts: the tooth root, tooth height, and tooth thickness of the workpiece [12]. The measured diameter was approximately 200 mm, and the radial inspection accuracy was within ±20 μm. Compared with the conventional stylus contacting method, which takes a long time, this image processing method can be performed in one minute. An off-axis differential method for improvement of a femtosecond laser differential chromatic confocal probe was set up to obtain the normalized chromatic confocal output with a better signal-to-noise ratio. It achieved a Z-directional measurement range of approximately 46 μm as well as a measurement resolution of 20 nm [13]. For the purpose of in situ measurement, a design framework for optimizing spectral-domain low-coherence interferometric sensors was proposed for profilometry measurements to optimize system performance [14]. Another in situ measurement system dealing with the on-machine precision form truing of resin-bonded spherical diamond wheels was also proposed [15]. A novel nanotechnology presented a method for measuring the high-precision cutting edge radius of single point diamond tools using an atomic force microscope (AFM) and a reverse cutting edge artifact based on the edge reversal method [16]. For precision measurement of miniature internal structures with high aspect ratios, a spherical scattering electrical field probe (SSEP) was proposed based on charge signal detection. The developed SSEP has great potential to be the ideal solution for precision measurement of miniature internal structures with high aspect ratios [17]. For the evaluation of roundness of small cylinders, the linear-scan surface form stylus profilometer was employed. The technique used to compensate for the influences of measuring angular misalignments was also proposed, and the measurement uncertainty was analyzed [18].

Finally, in this category, the quality of wafer fabrication was studied in two papers. The first one was on the development of a dynamic pad monitoring system (DPMS) for measuring the surface topography of wafer polishing pads using a chromatic confocal sensor. It is applicable to monitoring the pad dressing process and CMP parameter evaluation to produce IC devices [19]. The second one used deep learning methods to develop a set of algorithms to detect wafer die particle defects from the captured images [20]. Both are practical in industrial applications.

4. Angle Measurements

Three papers on angle measurement technology were included. An innovative dual-axis precision level based on the light passing through a liquid container was designed and commercialized [21]. It works better than the current commercial levels as it is light weight, has a low cost, and has two axes for pitch and roll error measurement of machine tool stages. For the squareness measurement of five-axis machine tools, a technique using circular trajectories to identify the eccentricity and squareness error of the B and C axes was proposed [22]. Prof. Wei Gao's group from Tohoku University provided a review article on the optical angle sensor techniques based on the mode-locked femtosecond laser [23], which includes (1) the angle scale comb, which can be generated by combining the dispersive characteristic of a scale grating and the discretized modes in a mode-locked femtosecond laser, and (2) mode-locked femtosecond laser autocollimators. This is so far the most cited paper in this Special Issue.

5. Laboratory Systems

Two papers were related to laboratory metrology systems. A GD&T-based benchmark for evaluating the performance of different CMM operators in computer-aided inspection (CAI) was proposed [24]. This, in turn, emphasized the importance of GD&T training and certification in order to ensure a uniform understanding among different operators, combined with a fully automated inspection code generator for GD&T purposes. Another paper dealt with the development of an enhanced circulating cooling water (CCW) machine [25]. It can simultaneously achieve high temperature stability and dynamic performance in CCW temperature control. The developed machine can satisfy the challenging requirements in precision manufacturing.

Author Contributions: For the writing—original draft preparation, K.-C.F.; writing—review and editing, P.K. All authors have read and agreed to the published version of the manuscript.

Funding: This research received no external funding.

Acknowledgments: This issue was mainly executed during 2020, which was a difficult time in terms of the COVID-19 pandemic. The contributions of earnest authors, professional reviewers, and the dedicated editorial team of *Applied Sciences* led to the success of this Special Issue. Many thanks to J.B Tan and Wei Gao, who supported us by contributing several papers. Special thanks to all reviewers, who always provided fast and responsible services to our requests. Finally, we take this opportunity to express our gratitude to the editorial team of *Applied Sciences* and special thanks to Daria Shi from the Beijing Branch Office of MDPI.

Conflicts of Interest: The authors declare no conflict of interest.

References

1. Gao, H.; Wang, Z.; Cheng, Y.; Li, Y.; Sun, S.; Shang, Z. A practical positioning method in end-plate surface distance measurement with Nano-meter precision. *Appl. Sci.* **2019**, *9*, 4970. [CrossRef]
2. Lu, Z.; Zhang, B.; Li, Z.; Zhao, C. A method of on-site describing the positional relation between two horizontal parallel surfaces and two vertical parallel surfaces. *Appl. Sci.* **2020**, *10*, 2152. [CrossRef]
3. Ren, G.; Qu, X.; Chen, X. A new computational model of step gauge calibration based on the synthesis technology of multi-path laser interferometers. *Appl. Sci.* **2020**, *10*, 2089. [CrossRef]
4. Shih, Y.; Tung, P.; Jywe, W.; Chang, C.; Shyu, L.; Hsieh, T. Investigation on the differential quadrature Fabry–Pérot interferometer with variable measurement mirrors. *Appl. Sci.* **2020**, *10*, 6191. [CrossRef]
5. Liu, Z.; Duan, F.; Niu, G.; Ma, L.; Jiang, J.; Fu, X. An improved Circumferential Fourier Fit (CFF) method for blade tip timing measurements. *Appl. Sci.* **2020**, *10*, 3675. [CrossRef]
6. Nan, Z.; Tao, W.; Zhao, H.; Lv, N. A fast laser adjustment-based laser triangulation displacement sensor for dynamic measurement of a dispensing robot. *Appl. Sci.* **2020**, *10*, 7412. [CrossRef]
7. Kim, W.; Fu, H.; Lee, K.; Han, S.; Jang, Y.; Kim, S. Photonic microwave distance interferometry using a mode-locked laser with systematic error correction. *Appl. Sci.* **2020**, *10*, 7649. [CrossRef]
8. Cheng, F.; Zou, J.; Su, H.; Wang, Y.; Yu, Q. A differential measurement system for surface topography based on a modular design. *Appl. Sci.* **2020**, *10*, 1536. [CrossRef]
9. Cheng, F.; Fu, S.; Chen, Z. Surface texture measurement on complex geometry using dual-scan positioning strategy. *Appl. Sci.* **2020**, *10*, 8418. [CrossRef]
10. Shi, G.; Zheng, L.; Wang, W.; Lu, K. Non-scanning three-dimensional imaging system with a single-pixel detector: Simulation and experimental study. *Appl. Sci.* **2020**, *10*, 3100. [CrossRef]
11. Li, X.; Shi, Z. A form-free and high-precision metrological method for the twist of aeroengine blade. *Appl. Sci.* **2020**, *10*, 4130. [CrossRef]
12. Chen, Y.; Liang, X.; Ye, Z.; Hsu, Q. Development of a rapid optical measurement system for circular workpieces with irregular tooth contours after broaching process. *Appl. Sci.* **2020**, *10*, 4418. [CrossRef]
13. Chen, C.; Shimizu, Y.; Sato, R.; Matsukuma, H.; Gao, W. An off-axis differential method for improvement of a femtosecond laser differential chromatic confocal probe. *Appl. Sci.* **2020**, *10*, 7235. [CrossRef]
14. Hovell, T.; Petzing, J.; Justham, L.; Kinnell, P. From light to displacement: A design framework for Optimising spectral-domain low-coherence interferometric sensors for in situ measurement. *Appl. Sci.* **2020**, *10*, 8590. [CrossRef]

15. Wang, J.; Zhao, Q.; Zhang, C.; Guo, B.; Yuan, J. On-machine precision form truing and in-situ measurement of resin-bonded spherical diamond wheel. *Appl. Sci.* **2020**, *10*, 1483. [CrossRef]
16. Zhang, K.; Cai, Y.; Shimizu, Y.; Matsukuma, H.; Gao, W. High-precision cutting edge radius measurement of single point diamond tools using an atomic force microscope and a reverse cutting-edge artifact. *Appl. Sci.* **2020**, *10*, 4799. [CrossRef]
17. Bian, X.; Cui, J.; Lu, Y.; Zhao, Y.; Cheng, Z.; Tan, J. Quantitative investigation of surface charge distribution and point probing characteristics of spherical scattering electrical field probe for precision measurement of miniature internal structures with high aspect ratios. *Appl. Sci.* **2020**, *10*, 5268. [CrossRef]
18. Li, Q.; Shimizu, Y.; Saito, T.; Matsukuma, H.; Gao, W. Measurement uncertainty analysis of a stitching linear-scan method for the evaluation of roundness of small cylinders. *Appl. Sci.* **2020**, *10*, 4750. [CrossRef]
19. Chen, C.; Li, J.; Liao, W.; Ciou, Y.; Chen, C. Dynamic pad surface metrology monitoring by swing-arm chromatic confocal system. *Appl. Sci.* **2021**, *11*, 179. [CrossRef]
20. Chen, S.; Kang, C.; Perng, D. Detecting and measuring defects in wafer die using GAN and YOLOv3. *Appl. Sci.* **2020**, *10*, 8725. [CrossRef]
21. Huang, Y.; Fan, Y.; Lou, Z.; Fan, K.; Sun, W. An innovative dual-axis precision level based on light transmission and refraction for angle measurement. *Appl. Sci.* **2020**, *10*, 6019. [CrossRef]
22. Yao, Y.; Nishizawa, K.; Kato, N.; Tsutsumi, M.; Nakamoto, K. Identification method of geometric deviations for multi-tasking machine tools considering the squareness of translational axes. *Appl. Sci.* **2020**, *10*, 1811. [CrossRef]
23. Shimizu, Y.; Matsukuma, H.; Gao, W. Optical angle sensor technology based on the optical frequency comb laser. *Appl. Sci.* **2020**, *10*, 4047. [CrossRef]
24. Aidibe, A.; Tahan, S.; Kamali Nejad, M. Interlaboratory empirical reproducibility study based on a GD&T benchmark. *Appl. Sci.* **2020**, *10*, 4704.
25. Lu, Y.; Cui, J.; Tan, J.; Bian, X. Quick response circulating water cooling of ± 3 mK using dynamic thermal filtering. *Appl. Sci.* **2020**, *10*, 5483. [CrossRef]

Article

A Practical Positioning Method in End-Plate Surface Distance Measurement with Nano-Meter Precision

Hongtang Gao [1,2], Zhongyu Wang [1], Yinbao Cheng [1,*], Yaru Li [1], Shuanghua Sun [2] and Zhendong Shang [3]

1 School of Instrumentation and Optoelectronic Engineering, Beihang University, Beijing 100191, China; htgao@nim.ac.cn (H.G.); mewan@buaa.edu.cn (Z.W.); Liyr022518@163.com (Y.L.)
2 Division of Metrology in Length and Precision Engineering, National Institute of Metrology, Beijing 100013, China; sunshh@nim.ac.cn
3 School of Mechatronics Engineering, Henan University of Science and Technology, Luoyang 471003, China; cnlyszd@163.com
* Correspondence: chengyinbao@buaa.edu.cn; Tel.: +86-10-6452-6097

Received: 15 October 2019; Accepted: 15 November 2019; Published: 19 November 2019

Abstract: End-plate surface distance is important for length value dissemination in the field of metrology. For the measurement of distance of two surfaces, the positioning method is the key for realizing high precision. A practical method with nanometer positioning precision is introduced in consideration of the complexity of positioning laser sources of the traditional methods and new methods. The surface positioning is realized by the combination of laser interference and white light interference. In order to verify the method, a 0.1 mm height step is made, and an experiment system based on the method is established. The principle and the basic theory of the method are analyzed, and the measures to enhance the repeatability from optical and mechanical factors and signal processing methods are presented. The experimental result shows that the surface positioning repeatability is in the order of 10 nm. The measurement uncertainty evaluation shows that the standard uncertainty is 21 nm for a 0.1 mm step. It is concluded that the method is suitable to be applied to the length measurement standard of the lab.

Keywords: white light interference; laser interference; surface positioning; end-plate surface distance measurement

1. Introduction

End-plate surface distance is one of the important geometric parameters that are widely used in industry and science. The end-plate surface distance objective standards such as gauge block, step master or step height gauge, and step gauge are extensively applied to the calibration of length instruments. Besides, the end-plate surface distance such as the length of glass cavity is one of important parameters for the frequency stabilization of the laser and the performance of the F–P interferometer, so accurately measuring the end-plate surface distance measurement is needed. The end-plate surface distance of objective length standards is in the range from a few tens of nanometers to one meter. Different sizes of objective length standards have respective calibration applications. For example, the step gauge with nanometer size is generally used for calibration of relevant measurement instruments such as AFM(Atomic Force Microscope) and SPM(Scanning Probe Microscope) [1–3]. The step master with micrometer size is generally used for the z-axis (vertical direction) calibration of optical instruments. No matter what size of gauge block it is, the positioning of surface is the key to realizing high precision. Generally, interference methods are used for the measurement of end-plate surface distance with high precision. The traditional interference method is the excess fraction method [4], known from the CMCs(Calibration and Measurement Capabilities) published at the website of BIPM

(International Bureau of Weights and Measures Metre Convention signatories); this method has about 20 nm positioning expansion uncertainty for end surface distance measurement of the gauge block. Although the positioning precision is very high for the excess fraction method, it needs at least two kinds of frequency stabilized lasers with different wavelength. Nowadays new methods using a femtosecond mode-locked pulse laser have appeared with the development of laser technology; researchers in [5] gave a description of this method and showed a measurement repeatability of 19 nm. Due to use of the femtosecond mode-locked pulse laser, the system is very complex. The common shortfall of both methods is that they are not easy to popularize because of their high cost and the complexity. For these reasons, a practical method realized by the combination of white light interference and laser interference is presented in this paper. The feature of this method is that it takes advantage of the positioning function of white light interference and good coherence of the laser. An experimental system was created to verify this method, by optimizing the different parts of the system such as the signal processing part, and by using dynamic measurement, 10 nm measurement repeatability was obtained in ordinary laboratory conditions. The measurement repeatability should be better if using this method to the standard device.

2. System Description

The system block diagram is shown in Figure 1. The system is mainly composed of two parts that realize positioning and measuring function. The positioning part is the key of the system. The measured 0.1 mm step is made first, as shown in Figure 1, where A, B, and C are the standard gauge blocks. The optical path of positioning interference is the simple Michelson type. The laser beam and white light beam are polarized first by polarizer and then separated by PBS and NPBS (PBS is polarizing beam splitter, NPBS is non-polarizing beam splitter) to transmit to each receiver. Three photoelectric receivers are used for receiving three interference signals, i.e., the positioning signals of white light interference and laser interference as well as laser interference signal. The positioning laser and displacement measurement laser use the same laser source of semiconductor laser. The positioning laser and displacement measurement laser are separated by a splitter and transmitted into each optical path. The light beam collimating and expanding unit is used for reducing the error caused by the diffraction of the Gaussian laser beam. The signal processing system can simultaneously sample and record the position interference signal and laser interference displacement signal, and then process these signals to get the value of the height of the step. There are two design features. The first feature is the position optical path, which is designed to be suitable for the cooperation of laser and white light interference, and the other feature is the signal processing system designed for dynamic measurement.

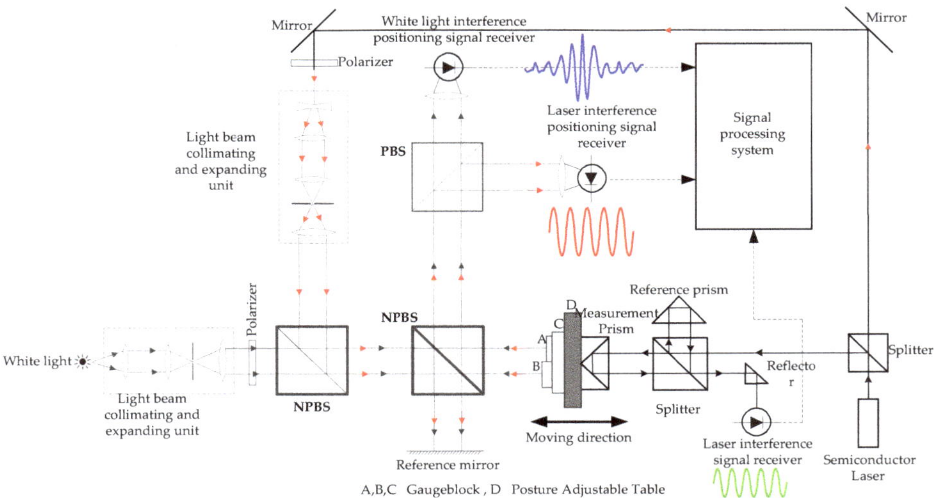

Figure 1. Block diagram of the system.

3. Positioning Principle

White light interferometry is used to position the surface. In order to describe the process clearly, it is necessary to give a theoretical introduction about white interference. For single wavelength interference, the interference pattern can be expressed as Equation (1), where γ is the contrast ratio of the interference pattern, I_0 is the background intensity of light, and d is the difference between the measurement surface and the reference surface in the interferometer.

$$I_\lambda(d) = I_0(1 + \gamma \cos \frac{2\pi d}{\lambda}) \tag{1}$$

The interference pattern of white light is considered as the incoherent superposition of three colors of lights. Assuming that γ and I_0 are the same, the RGB light interference pattern can then be expressed as Equation (2).

$$\begin{cases} I_{\lambda_r}(d) = I_0(1 + \gamma \cos \frac{2\pi d}{\lambda_r}) \\ I_{\lambda_g}(d) = I_0(1 + \gamma \cos \frac{2\pi d}{\lambda_g}) \\ I_{\lambda_b}(d) = I_0(1 + \gamma \cos \frac{2\pi d}{\lambda_b}) \end{cases} \tag{2}$$

$$I(d) = I_m\left\{1 + \sin c[\frac{2(d-d_0)}{\lambda^2/\Delta\lambda})] \cdot \cos[2\pi(d-d_0)/\lambda_0 + \phi_0]\right\} \tag{3}$$

where $I(d)$ is the intensity of white light interference; I_m is the maximum intensity of the interference signal; d_0 is the position of maximum intensity of interference signal; d is the position of the surface of measured; λ_0 is the average wavelength of white light; ϕ_0 is the initial phase; and $\Delta\lambda$ is the wavelength range above half intensity of light.

The intensity of white light interference is expressed with Equation (3), which can be derived from Equation (2). From Equations (2) and (3), the interference patterns of three colors of lights and white light are shown in Figure 2, and the white light interference intensity signal is shown in Figure 3. As seen from the curve of Figure 3, a maximum point represents the surface position, i.e., the zero optical path difference position. Then by scanning to find the zero order interference fringe of white light, the position of the surface can be assured rapidly. Figure 3 shows two positioning signals of the 0.1 mm step surface obtained by the white light interference.

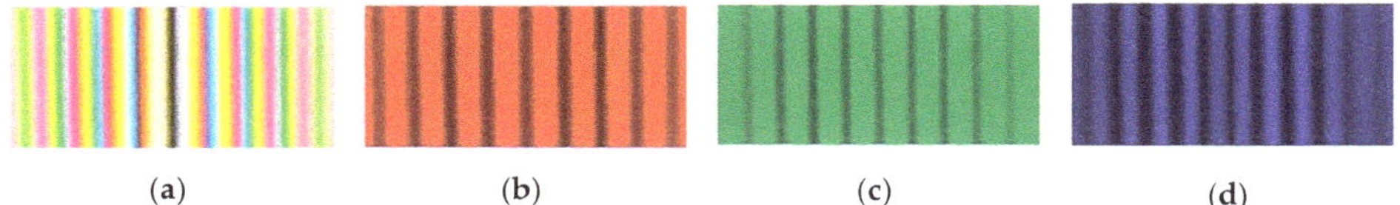

Figure 2. The interference pattern of white light and RGB light interference components. (**a**) White light interference pattern; (**b**) red light interference pattern; (**c**) green light interference pattern; (**d**) blue light interference pattern.

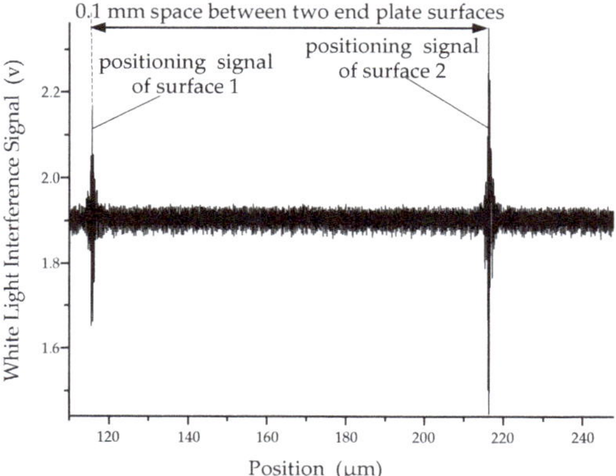

Figure 3. Positioning signal of 0.1 mm step surface.

In fact, the noise and error are inevitable, and actual white interference signal is shown in Figure 4; the noises and shape defection is obvious by comparison with the theoretical signal shown in Figure 5. The noise and shape defection are the main factors that reduce the positioning repeatability of measurement. How to reduce the error caused by the noise and the defection is to be studied. The basic measure is to enhance the signal to noise ratio of the interference signal by optimizing the system and filtering the signal. For this purpose, the surface is positioned by the combination of laser interference and white light interference, the rough position is obtained by the white light interference signal, and the accurate position is obtained from the laser interference positioning signal.

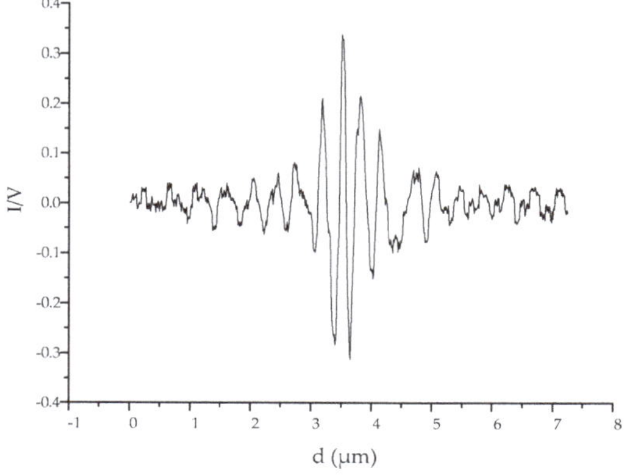

Figure 4. Actual white light positioning signal for surface.

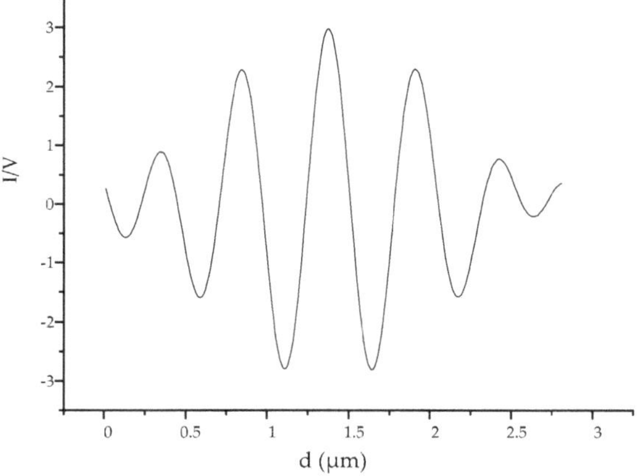

Figure 5. Theoretical white light interference signal.

4. Accuracy Enhancing Methods

Known from experience of the length measurement, as for a small space of 0.1 mm of length, the measurement repeatability component is the main part of measurement uncertainty. Thus, the measurement to improve the measurement repeatability is important. The two aspects of optical and mechanical factors and electronic factors to improve measurement repeatability are mainly studied in this paper.

4.1. Optical and Mechanical Aspects

The relevant errors generated from optical and mechanical parts of the experiment system include the mechanical vibration and the stability of moving table, the quality of optics, and the laser wavelength stability.

4.1.1. Vibration

Among the mechanical error sources, the vibration is a leading factor that affects the repeatability. The vibration isolation is a necessary step that is relatively easy to do. Generally, it is effective to eliminate the vibration coming from outside the system when the measurement system is mounted in a whole on the isolation platform. For the vibration generated from inside the system, it is not easy to eliminate vibration by isolating only. The vibration elimination in this case is studied. Figure 6 is the layout of experiment system.

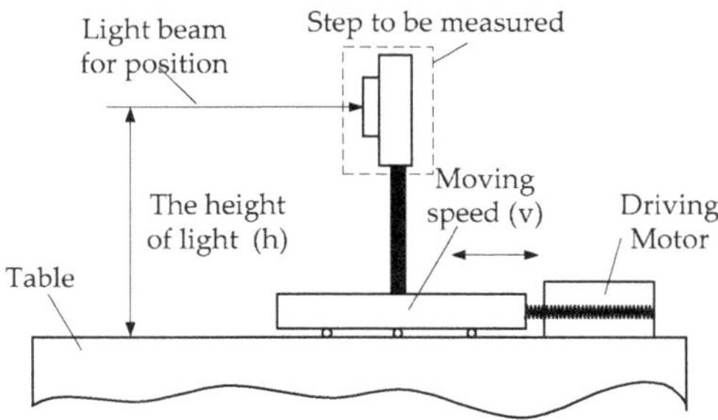

Figure 6. The layout of positioning unit.

A kinetic characteristic of the system is expressed by Equation (4).

$$M\ddot{\varphi} + C\dot{\varphi} + K\varphi = Q \tag{4}$$

where C is the system damping matrix; K is the system stiffness matrix; Q is the systematic general force matrix; M is the mass matrix; ϕ is the modal coordinates describing the flexible deformation of the system; and $\dot{\phi}$ is the first derivative and second derivative of ϕ.

Based on Equation (4), the equation for solving the natural frequency of the system can be obtained as Equation (5).

$$\det(p^2 I - M^{-1}K) = 0 \tag{5}$$

where p is the natural frequency of the system and I is the unit matrix.

As seen from Equation (5), the natural frequency of the system is determined by the mass matrix M and the stiffness matrix K. M and K are related to the structural and motion parameters of the system, respectively. For structural parameters, if the height of the light beam (the height of step mounting) is decreased, the stiffness of the system will increase, and the same goes for natural frequency of the corresponding system. The repeatability experimental results confirmed that when the height of step mounting was decreased from 50 mm to 30 mm, the repeatability was better than 20 nm. For motion parameters, the moving speed V is one of the adjustable parameters to obtain the better signal. The experiment result showed that when the moving speed V was 50 μm/s, the best signal was obtained. The positioning signal at different moving speeds is presented in Figures 7–9. In the case of a relative lower speed at 30 μm/s, it was easy to be affected by different frequencies of vibration. In case of a relative higher speed at 80 μm/s, although the signal SNR was good, the symmetrical characteristics of the signal were poor, which also affected the signal to reduce the positioning repeatability.

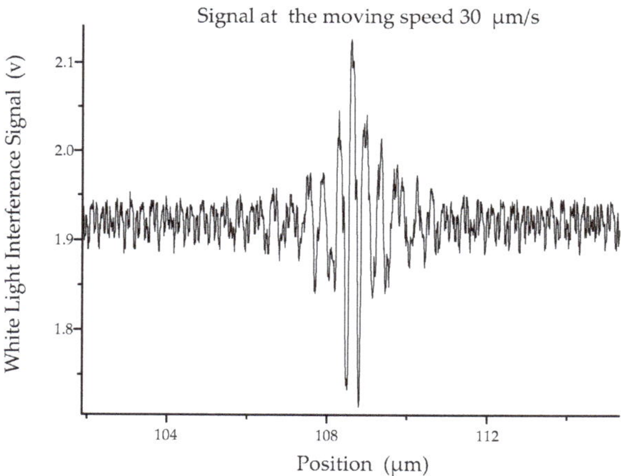

Figure 7. Positioning signal at moving speed of 30 μm/s.

The moving stability of table is another factor to affect repeatability. Figure 10 shows the length interference signals of different motion tables. Automatic interference comparator (AIC) is a specialized length measurement with a high precision. The uniformity and smoothness of the interference signal are very good because the motion slide is driven by hydraulics [6,7]. The vibration from the driven unit is negligible. The motion driven part used in this paper is a step motor, and the vibration from the motor is inevitable. Then, if the driven mode changes, the uniformity and smoothness of the motion will be improved. Estimating from the measurement experience of AIC, at least 5 nm of repeatability will be reduced.

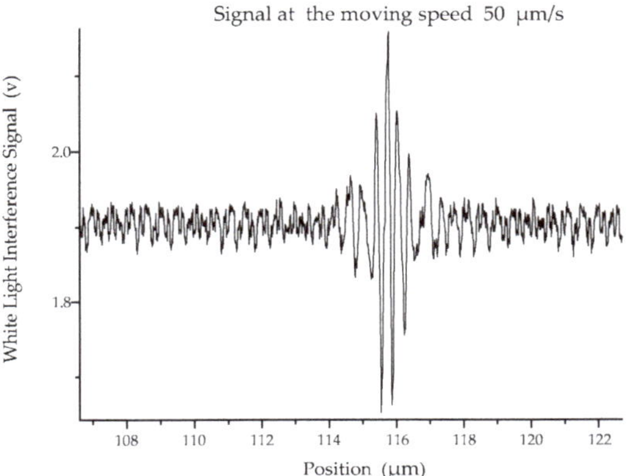

Figure 8. Positioning signal at moving speed of 50 μm/s.

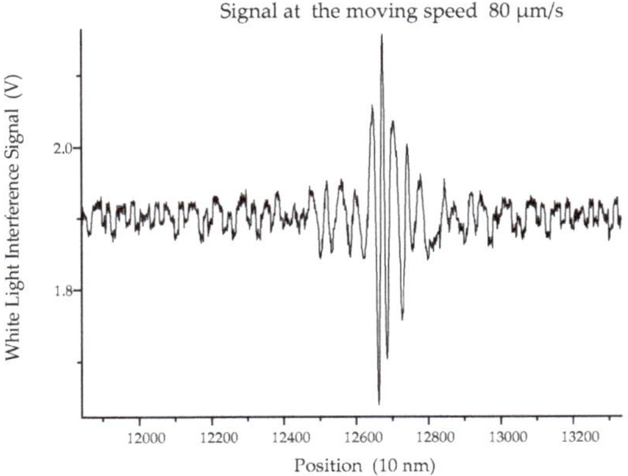

Figure 9. Positioning signal at moving speed of 80 μm/s.

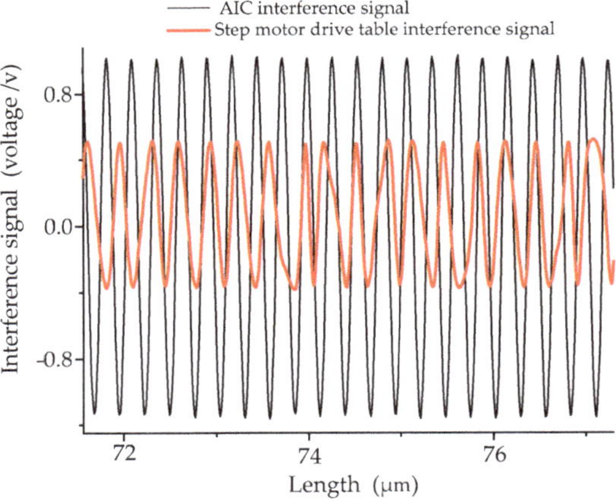

Figure 10. Signal difference in different driving modes. AIC is automatic interference comparator.

4.1.2. Laser Wavelength

The laser wavelength is an important parameter for length measurement with a high precision. In order to ensure the length measurement accuracy, the length measurement system should use the frequency stabilized laser. Furthermore, the interference system should be placed in vacuum environment for length measurement with nanometer precision [8–11]. The experiment measurement repeatability was affected by laser wavelength stability. Both an ordinary semiconductor laser and HP5517 frequency stabilized laser as input light sources were experienced, respectively. The results show that the measurement repeatability was better than 5 nm when using the HP5517 frequency stabilized laser. The wavelength of the semiconductor laser was monitored by the laser wavelength meter with a precision of 2×10^{-8}. The result is shown in Figure 11 that there was about a maximum of 0.06 nm variation for about 40 min of continuously monitoring. It was indicated that for the average wavelength of 634.286 nm, the correspondent relative variation was 9×10^{-5}, and for about 15 min of repeatability measurement, the relative variation of the wavelength was 4×10^{-5}. For the length of 0.1 mm, the correspondent variation was about 4 nm. Analytical results are consistent with the experiment results.

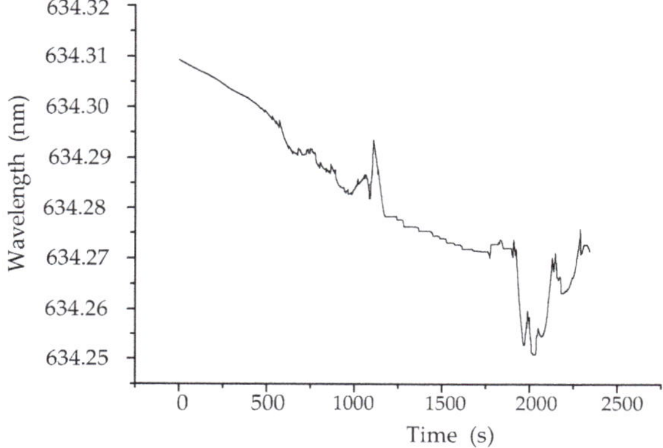

Figure 11. The semiconductor laser wavelength stability monitoring.

4.1.3. Combination of Laser Interference and White Light Interference

As described in Section 3, the laser interference and white light interference have respective advantages. The white light interference has surface positioning advantages for its incoherence and the laser interference has a good interference fringe because of its good coherence. The combination of two ways of interference has advantages for improving measurement results. The signal obtained by the combination of two ways of interference is shown in Figure 12.

The signal center of zero-order interference fringes of white light represents the position of surface. It is easy to ensure the rough position by processing the white light interference signal first. Then according to the rough position obtained from white light interference signal, accurate position is further calculated from the laser interference signal. The processing method for signal is critical to obtaining good repeatability. The methods include center strategy, the signal filtering, etc., which will be described in the following section.

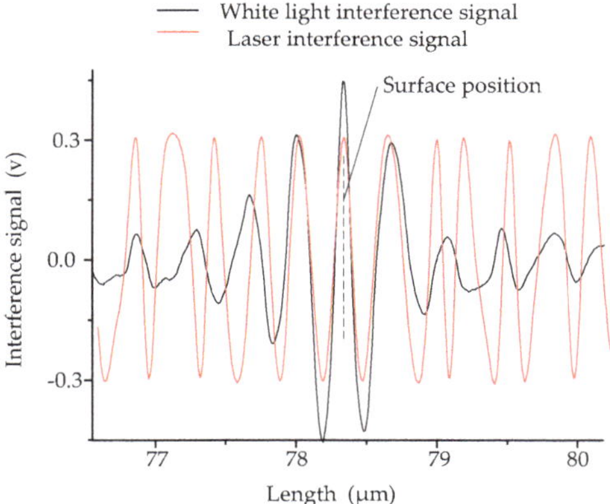

Figure 12. The signal of the combination of laser interference and white light interference.

4.2. Signal Processing System

Signal processing system is also import for measurement repeatability. As seem in Figure 13, the feature of the system has both manual and automatic signal processing functions in the soft layer of the system. Every unit realizes its corresponding function. Among them, signal shaping is the basic function causing the trigging signal for synchronous sampling. Its working principle is shown in Figure 14. The white light interference fringes generate a group of pulse outputs. The rising edge of first pulse is used to start signal sampling, and the falling edge of the last pulse is used to stop signal sampling. The signal processing methods are also critical to obtain good measurement repeatability. The signal processing methods include signal smooth filtering, the positioning strategy, bidirectional measurement, and averaging the multiple measurement results, etc.

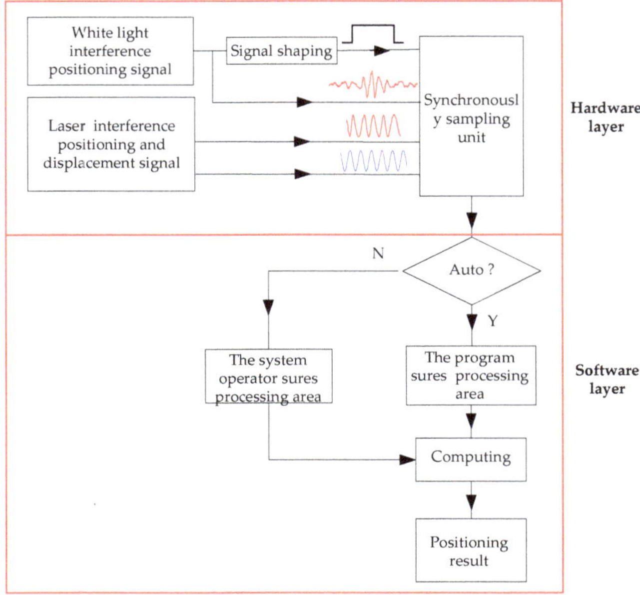

Figure 13. Block diagram of signal processing system.

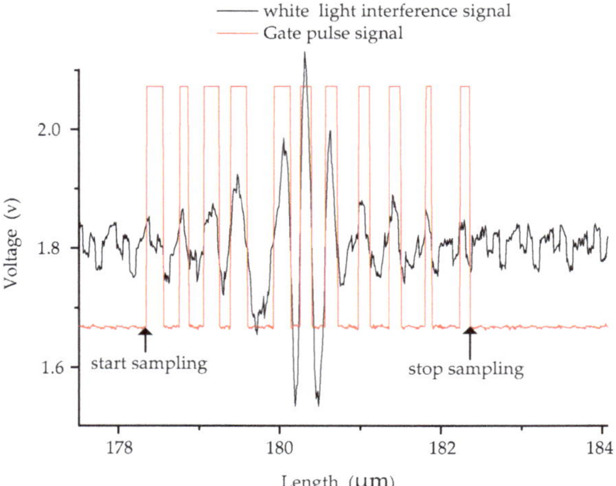

Figure 14. The shaping of white light interference signal.

4.2.1. Signal Processing Methods

The center of the zero-order fringe in white light interference means the position of the surface. The methods to obtain the center from the positioning signal are studied. There are many kinds of methods for line scale signal processing. Generally, they are classified as two classes. One is determined by a few points to the left and right edges of the signal [12], and the other is determined by multiple points to the left and right edges of the line scale signal [13]. Since the positioning signal of the surface is familiar to the positioning signal of the line scale, the signal processing method used for line scale is suitable for surface positioning. Due to the advantage of reliability, the multiple point method was adopted in this paper. For the multiple point method, one important step is to ensure the section of signal for processing in an appropriate section, beneficial to enhance the positioning accuracy Figures 15 and 16 show the positioning signal of the first and second surface of the step tested, respectively. The section of signal to be processed was the rectangular area below the peak of signal, determined by the value V_1 and V_2. As showed in Figure 16, when V_1 and V_2 were appropriate, the area A1 was a good area, and when V_1 and V_2 were not appropriate, the area was the bad area A_2. In the case of A_1, the computed result was accurate to the center, while in the case of A_2, the computed result was left to the center, causing the positioning error. In fact, there are many cases that were worse than the case of A_2, so the parameters V_1 and V_2 should be automatically adjusted to fit different poor positioning signals. Furthermore, for more complex cases, when it is difficult to select automatically, the processing section should be decided by the system operator. After confirming the processing section of signal, the result is automatically computed as Equation (6).

$$C(l) = \frac{\int_{l_1}^{l_2} s(l) l dl}{\int_{l_1}^{l_2} s(l) dl} \tag{6}$$

where l is the length; $C(l)$ is the center computed; l_1 is the length l when $s(l)$ is V_1; and l_{12} is the length l when $s(l)$ is V_2.

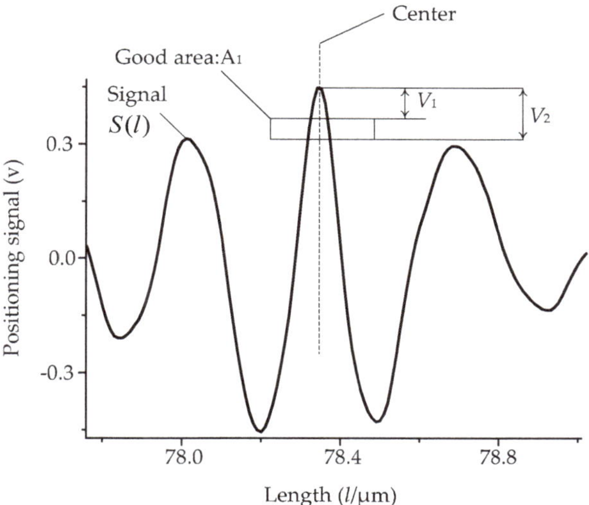

Figure 15. Positioning signal of first surface.

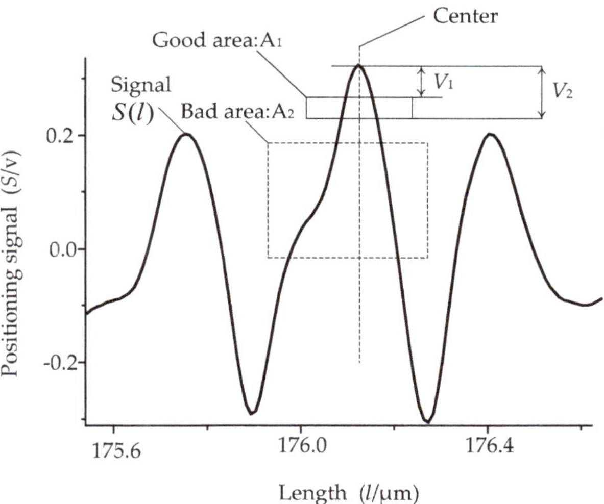

Figure 16. Positioning signal of second surface.

Considering the importance of the values V_1 and V_2, the measurement repeatability affected with value V_1 and V_2 were also done. The experiment result was that if the signal was good in uniformity, the optimal values of V_1 and V_2 were 20% and 40% of the peak-to-peak value, respectively. However, if the signal was bad in uniformity, the suitable value of V_1 and V_2 were those as close as possible to the peak value of the signals.

4.2.2. Measurements

Besides the signal processing methods, the bidirectional measurement and averaging multiple measurement were adopted. The measurement experiment for the same step in two measurement directions was done in two days. In Figure 17, it is obvious that when the measurement direction was different, the result was relatively changed. On the first day, the measurement results were unsymmetrical to the direction in an approximate stair shape. On the second day, the measurement result was symmetrical to the direction in an approximate triangle shape. A system error was coupled into the measurement result, and it was changed with the time and the measurement direction. If there was no averaging, the measurement result difference was divergent and the maximum different was more than 0.7 μm. If averaging, the final result was convergent, and the difference was less than 0.03 μm.

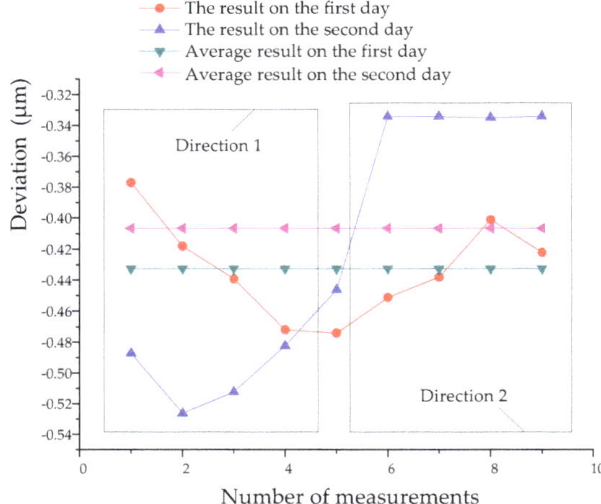

Figure 17. Bidirectional measurement results and the convergence of average result.

5. Measurement Uncertainty Evaluation

Acquiring optimal measurement uncertainty is the target of a measurement system. In this paper, the methods to decrease every possible error were the emphasis. Limited by the experiment condition and hardware of the system, the measurement uncertainty was a few tens of nanometers, obtained by estimation. The analysis on measurement uncertainty of the system is given in this section.

Firstly, the mathematical model is given in Equation (7):

$$l = \frac{N\lambda_0}{2n(p, t_{air}, f)} + \alpha L(20 - t_s) + \delta l_{dif} + \delta l_{Abbe} + \delta h + \delta l_{\cos} \tag{7}$$

where λ_0 is vacuum wavelength of laser, $n(p, t_{air}, f)$ is air refractive index obtained by Edlen's formula (1998 Version), p is air pressure, f is air humidity, t_{air} is air temperature, a is the linear thermal expansion coefficient, t_s is the material temperature, δl_{dif} is the correction for the laser beam diffraction, δl_{Abbe} is the correction for the Abbe error, δh is the correction for the flatness derivative of the step, and $\delta l_{\cos}$ is the correction for the cosine error.

In Equation (7), δl_{dif} is expressed as follows:

$$\delta l_{dif} = \frac{\lambda^2}{4\pi^2\omega_0^2}L \tag{8}$$

where, L is the measured length, λ is laser wavelength, and ω_0 is the radius of waist of laser beam.

Since the laser source used for length measurement was a semiconductor laser and the laser beam directly output without any collimation, the beam waist radius of the laser is estimated to be 0.1 mm, δl_{dif} 0.1 nm for L 0.1 mm was obtained by Equation (8).

Table 1 gives the information used for the measurement uncertainty evaluation. As a whole, error sources can be classified into two types: random errors and system errors. For random errors, the Type A evaluation method based on statistics is used. For example, the measurement uncertainty component caused by the repeatability is obtained by Type A method. For system errors, the Type B evaluation method is used [14–16]. The detailed evaluation information and combined standard uncertainty are also given in Table 1. Since the length value was 0.1 mm, the length-dependent components were so small to be negligible, so the length-independent components were the main contributions to total measurement uncertainty. These components included repeatability, resolution, Abbe error, and the surface flatness of the step measured.

Table 1. Measurement uncertainty evaluation.

Error Sources	x_i	$u(x_i)$	Prob.	$c_i{=}\delta l/\delta x_i$	Unit	$u_i(l)$
Repeatability	S	10.0 nm	N	1.0		10.0 nm
Resolution	N	0.58 nm	R	10		5.8 nm
Laser wavelength	λ_0	$2.3 \times 10^{-5} \times \lambda_0$	R	L/λ_0		$(2.3 \times 10^{-5}) \times L$
Diffraction of laser	δl_{dif}	0.1 nm	R	1.0×10^{-6}	L/nm	$(1.0 \times 10^{-7}) \times L$
Edlen formula	n	1.0×10^{-8}	R	1.0	L	$(1.0 \times 10^{-8}) \times L$
Air pressure	p	10 Pa		2.70×10^{-9}	L/Pa	$(2.7 \times 10^{-9}) \times L$
Air temperature	t_{air}	0.5 °C	R	0.923×10^{-6}	L/°C	$(0.5 \times 10^{-6}) \times L$
Air humidity	f	30 Pa	R	0.367×10^{-9}	L/Pa	$(11 \times 10^{-9}) \times L$
Thermal linear expansion coefficient	α	2.0×10^{-6} °C	R	0.5	L/°C	$(1.0 \times 10^{-6}) \times L$
Material temperature	t_s	0.5 °C	R	11.5×10^{-6}	L/°C	$(5.8 \times 10^{-6}) \times L$
Abbe error	δl_{Abbe}	17.3 nm	R	1.0		17.3
Flatness of surface	δh	5 nm	R	1.0		5
Cosine error	$\delta l_{\cos}$	29×10^{-6} rad	R	50×10^{-6}	L/rad	$(1.45 \times 10^{-9}) \times L$
Standard uncertainty (when L is 0.1 mm)						21

6. Conclusions

The length value disseminated in the way of end-plate surface distance is widely used in industry and science. Surface positioning is an important step for end-plate surface distance measurement. In this paper, a practical method is introduced for realizing the surface positioning with nano-meter precision. The method was used experimentally by making 0.1 mm step sizes in the measurement system. The measurement uncertainty analysis shows that the measurement uncertainty is 21 nm. Since the 0.1 mm length is relatively small, the main contribution of measurement uncertainty is the repeatability, and the measures to enhance repeatability are studied as the emphasis of the paper. The measures include the optimization of optical and mechanical parts of the system, using appropriate signal processing methods, and the appropriate measurement methods, etc. In conclusion, the combination of laser interference and white light interference for positioning of the surface is the effective measure to obtain good measurement repeatability. The method is verified to be effective. It will be applied to the length measurement primary standard (AIC) of NIM(National Institute of Metrology of China) in future studies.

Author Contributions: Conceived the Methods and Wrote the Paper, H.G. and Y.C.; Data curation, Y.L.; Performed some Confirmatory Experiments, S.S. and Z.S.; Resources, Z.W.; Edited the Manuscript, Y.C.

Funding: This research was supported by the Plan of Enhancing the Ability of National Standard Devices of General Administration of Quality Supervision, Inspection and Quarantine of China, Grant Number 24-ANL1804; Special Scientific Research on Civil Aircraft of Ministry of Industry and Information Technology of China, Grant Number MJ-2018-J-70; Sichuan Province Key Research and Development Plan of China, Grant Number 2019YFSY0039; National Key Research and Development Plan of China, Grant Number 2016YFF0203801.

Conflicts of Interest: The authors declare no conflict of interest.

References

1. Li, C.S.; Yang, S.M.; Wang, Y.M.; Wang, C.Y.; Ren, W.; Jiang, Z.D. Measurement and characterization of a nano-scale multiple-step sample using a stylus profiler. *Appl. Surf. Sci.* **2016**, *387*, 732–735. [CrossRef]

2. Orji, N.G.; Dixson, R.G.; Fu, J.; Vorburger, T.V. Traceable pico-meter level step height metrology. *Wear* **2004**, *257*, 1264–1269. [CrossRef]
3. Vorburger, T.V.; Hilton, A.M.; Dixson, R.G.; Orji, N.G.; Powell, J.A.; Trunek, A.J. Calibration of 1 nm sic step height standards. In Proceedings of the SPIE Advanced Lithography Conference, San Jose, CA, USA, 22–26 February 2010; Volume 7638.
4. Agurok, I.P. Rigorous decision of the excess fraction method in absolute distance interferometry. In Proceedings of the SPIE 3134, Optical Manufacturing and Testing II, San Diego, CA, USA, 1 November 1997; pp. 502–511.
5. Chanthawong, N.; Takahashi, S.; Takamasu, K.; Matsumoto, H. A new method for high-accuracy gauge block measurement using 2 GHz repetition mode of a mode-locked fiber laser. *Meas. Sci. Technol.* **2012**, *23*, 054003. [CrossRef]
6. Ye, X.Y.; Gao, H.T.; Sun, S.H.; Zou, L.D.; Shen, X.P.; Gan, X.C.; Chang, H.T.; Zhang, X. The Establishment of 2m length Measurement Primary Standard. *Acta Metrol. Sin.* **2012**, *33*, 193–197.
7. Sun, S.H.; Ye, X.Y.; Gao, H.T. Slide Motion Control System of 2 m Automatic Interference Length Comparator. *Acta Metrol. Sin.* **2008**, *29*, 123–126.
8. Bosse, H.; Bodermann, B.; Dai, G.; Flügge, J.; Frase, C.G.; Grob, H. Challenges in nanometrology: High precision measurement of position and size. *Tm-Tech. Mess.* **2015**, *82*, 346–358. [CrossRef]
9. Tiemann, I.; Spaeth, C.; Wallner, G.; Metz, G.; Israel, W.; Yamaryo, Y. An international length comparison using vacuum comparators and a photoelectric incremental encoder as transfer standard. *Precis. Eng.* **2008**, *32*, 1–6. [CrossRef]
10. Takahashi, A. Long-term dimensional stability of a line scale made of low thermal expansion ceramic nexcera. *Meas. Sci. Technol.* **2012**, *23*, 035001. [CrossRef]
11. *Key Comparison: Calibration of Line Scales, Euromet.L-K7*; Metrology Institute of the Republic of Slovenia: Maribor, Slovenia, 2006; pp. 123–125.
12. *Wgdm-7: Preliminary Comparison on Nanometrology, Nano 3: Line Scale Standards, Final Report, Annex C: Description of the Measurement Methods and Instruments of the Participants*; Physikalisch-Technische Bundesanstalt: Braunschweig, Germany, 2003; pp. 2–10.
13. Gao, H.T.; Ye, X.Y.; Zou, L.D. Study of automatic measurement system for line space measurement with nanometer accuracy in 2 m length comparator. *Acta Metrol. Sin.* **2012**, *33*, 98–103.
14. Gao, H.T.; Wang, Z.Y.; Wang, H. A scanning measurement method of the pitch of grating based on photoelectric microscope. In Proceedings of the SPIE AOPC 2015: Advances in Laser Technology and Applications, Beijing, China, 15 October 2015; Volume 9671.
15. Cheng, Y.B.; Wang, Z.Y.; Chen, X.H.; Li, Y.R.; Li, H.Y.; Li, H.L.; Wang, H.B. Evaluation and Optimization of Task-oriented Measurement Uncertainty for Coordinate Measuring Machines Based on Geometrical Product Specifications. *Appl. Sci.* **2019**, *9*, 6. [CrossRef]
16. Bich, W. Revision of the 'Guide to the Expression of Uncertainty in Measurement'. Why and how. *Metrologia* **2014**, *51*, S155–S158. [CrossRef]

Article

On-Machine Precision Form Truing and In-Situ Measurement of Resin-Bonded Spherical Diamond Wheel

Jinhu Wang [1,*], Qingliang Zhao [2,*], Chunyu Zhang [3], Bing Guo [2] and Julong Yuan [1]

1 Ultra-Precision Machining Center, Zhejiang University of Technology, Hangzhou 310024, China; jlyuan@zjut.edu.cn
2 Center for Precision Engineering, School of Mechatronics Engineering, Harbin Institute of Technology, Harbin 150001, China; guobing@hit.edu.cn
3 Research Center of Laser Fusion, China Academy of Engineering Physics, Mianyang 621900, China; zhangchunyu880420@163.com
* Correspondence: wangjinhu@zjut.edu.cn (J.W.); zhaoqingliang@hit.edu.cn (Q.Z.); Tel.: +86-571-85290933 (J.W.); +86-451-86402683 (Q.Z.)

Received: 17 January 2020; Accepted: 18 February 2020; Published: 21 February 2020

Abstract: The resin-bonded spherical diamond wheel is widely used in arc envelope grinding, where the demands for form accuracy are high and the form truing process is challenging. In this paper, on-machine precision form truing of the resin-bonded spherical diamond wheel is accomplished by using a coarse-grained diamond roller, and in-situ measurement of the form-truing error is conducted through a laser scan micrometer. Firstly, a novel biarc curve-fitting method is proposed based on the in-situ measurement results to calculate the alignment error between the diamond roller and the spherical diamond wheel. Then, on-machine precision form truing of a D46 resin-bonded spherical diamond wheel is completed after alignment error compensation. The in-situ measurement results show that the low-frequency form-truing error is approximately 5 μm. In addition, the actual form-trued diamond wheel has been employed in grinding a test specimen, and the resulting form accuracy is approximately 1.6 μm without any compensation. The ground surface profile shared similar characteristics with the roller-trued diamond wheel profile, confirming that the diamond roller truing and in-situ measurements methods are accurate and feasible.

Keywords: spherical diamond wheel; diamond roller; form truing; in-situ measurements

---

1. Introduction

The spherical diamond wheels can be used for the grinding of complex surfaces, such as aspheric surfaces, off-axis surfaces, free-form surfaces, and also play a significant role in the optical manufacturing industry [1,2]. However, ultra-precision grinding is only possible provided that a periodic form truing of the diamond wheels is conducted over the whole grinding cycle [3,4]. The resin-bonded spherical diamond wheels are especially beneficial for obtaining better ground surface quality but are more susceptible to wear [5]. A high-efficiency on-machine precision form truing process and a form-truing error in-situ measurement method are therefore necessary to reduce the influence of grinding wheel wear on the resulting accuracy of the grinding surface profile, thus allowing for efficient and precision grinding of complex surfaces of hard and brittle materials [6].

On-machine precision form truing is primarily aimed to eliminate the errors caused by installation, manufacturing, wear, and other factors of the grinding wheel [7]. The error elimination could improve the rotation and profile accuracy of the grinding wheels, which is crucial for obtaining good surface accuracy and quality [8,9]. The common form-truing methods that can be used for resin-bonded diamond wheels include diamond pen truing, pulsed laser truing, abrasive block truing, diamond roller truing, etc. [10]. The single point diamond pen can achieve the purpose of truing by extruding it on the grinding wheel surface and shedding the bond materials [11]. This method is simple, but the diamond pen wears out quickly when truing large diamond wheels. In addition, the single point diamond pen is not suitable for truing grinding wheels with complex profile [12]. The pulsed laser truing method can sharpen and profile the grinding wheel in radial and tangential directions, respectively [13,14]. Although it can be used for truing resin-bonded grinding wheels, the resulting form accuracy is usually not impressive. The main difference between abrasive block truing and diamond roller truing lies in the truing tools. For abrasive block truing, in order to make the grinding wheel bond materials wear and fall off to achieve the truing purpose, sintered corundum and green silicon carbide truing tools on the diamond wheel are often used [15]. Due to the rapid wear of the abrasive block truing tools, it is difficult to control the form-truing accuracy of the grinding wheels. However, abrasive block truing has the advantages of low cost, small truing force, and good sharpening effect [16]. The diamond roller can achieve from truing of the grinding wheels by both grinding and rolling effects under high rotation speed [17]. The coarse-grained electroplated diamond wheels are generally used for diamond roller truing, which results in a large truing force and has high requirement on the rigidity of the truing spindle, tool spindle, and machine tool. As the diamond roller is resistant to wear, it is easy to control the grinding wheel shape, and the resulting truing efficiency is high [18].

Although the truing and dressing techniques of resin-bonded diamond wheels have been widely reported, relatively few reports focusing on the precision form truing of spherical diamond wheels. Chen et al. [19] applied the generating method to precision form truing of a micro-spherical diamond wheel, where the cup truer and the grinding wheel rotate around their respective axes and make axial feed to generate the spherical surface of the grinding wheel. By this method, the profile PV (peak to valley) value of the grinding wheel with a diameter of 1 mm is approximately 1–2 μm. Based on the generating method, the cup truer can be replaced with an electrode tool for truing the metal bonded spherical diamond wheels. Wang et al. [20] applied the electrical discharge truing method to profile a metal-bonded spherical diamond wheel with a diameter of 3.8 mm. The profile graph showed that the achieved form-truing error can be less than 2 μm under the optimal process parameters. More research on precision form truing of spherical diamond wheel is of great significance to further improve the grinding surface form accuracy and reduce the machining allowance in the subsequent polishing process.

The in-situ measurements of the dimension and form accuracies of the grinding wheels are the critical process of on-machine form truing. The common measurement methods of grinding wheel profile include contact measurement, image acquisition method, graphite copy method, and so on [21,22]. The detection probe in contact measurement is highly liable to be damaged since sharp diamond abrasive particles are randomly distributed on the grinding wheel surface, which could substantially affect the detection result [23]. The image acquisition and graphite copy methods are not suitable for profile measurement of grinding wheels with large-size and complex surface. Chromatic confocal displacement sensor and laser displacement sensor are commonly used in non-contact measurement systems. Chromatic confocal displacement sensor is based on wavelength displacement modulation technique, and possesses a sub-micron meter accuracy over a millimeter order range [24]. Chromatic confocal profilometer has been widely used in science investigation and industry fields for its high precision and great measurement range [25]. Zou et al. [26] integrated a chromatic confocal measurement probe with an ultra-precision turning machine, and on-machine measurement with nanometer-level accuracy was achieved. Laser scan micrometer is based on the triangulation measurement principle, which is characterized by good stability, large measurement range, and high efficiency [27]. It has become widely available in measurements of dimension, surface profile, and even 3D shape [28–30]. However, the measurement accuracy is affected by the geometrical and optical conditions [31]. For non-contact measurement of grinding wheel profile accuracy, the sensor needs to meet the requirements of precision, range, and working distance at the same time.

In this paper, the coarse-grained diamond roller is applied to perform precision form truing of the resin-bonded spherical diamond wheel. The non-contact in-situ measurement method of the form-truing error and the alignment error are proposed and verified by using a laser scan micrometer.

2. System and Methods

2.1. Inclined Axis Grinding System

Figure 1 shows the inclined axis grinding system. The machine tool has a self-leveling system for air vibration isolation dampers. The run-out errors of the two ultra-precision aerostatic spindles are less than 10 nm, and the minimum infeed resolution of slides is 1 nm. The ground surface can be measured by an on-machine workpiece measurement system with an air bearing linear variable differential transformer (LVDT) measurement probe. The surface form error that can be measured and corrected is less than 0.1 μm. Precision grinding of various axisymmetric optical elements can be realized through a spherical diamond wheel and the 3-axis linkage platform. The truing spindle and laser scan micrometer are located on both sides of the workpiece spindle and move along the x- and y-axis. The tool spindle is fixed on the z-axis guide rail, and adjusted perpendicular to the truing spindle. The inclined angle between the tool spindle and the z-axis is 15° to avoid the low speed zone of the spherical diamond wheel from participating in grinding. The tool spindle is extended to accommodate concave surface grinding, and the extended range (ER) collets are adopted to connect the spherical diamond wheels of different sizes and models. The laser scan micrometer is installed on the positioning surface, and can be removed in the grinding and truing processes. The on-machine precision form truing and form error measurement is carried out by using a coarse-grained electroplated diamond roller and a laser scan micrometer fixed on both sides of the workpiece, respectively.

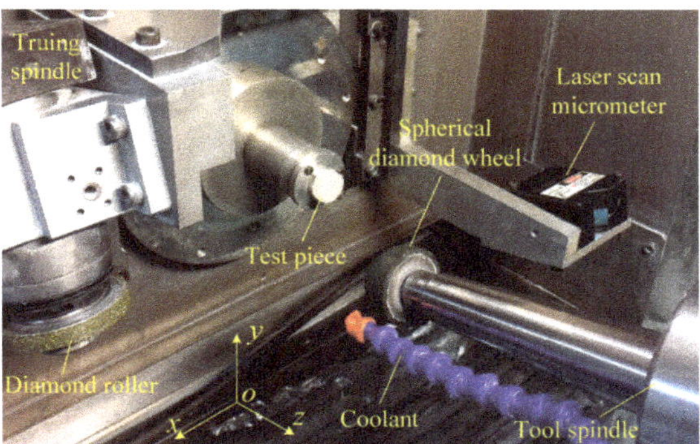

Figure 1. Inclined axis grinding system with a spherical diamond wheel.

2.2. In-situ Measurement System

The system composition and working principle of the in-situ measurement of the grinding wheel form-truing error is illustrated in Figure 2. The operation parameters of the laser scan micrometer, such as the sampling frequency and data storage capacity, are set in the host computer and sent to the controller. The numerical control (NC) programs are coded by the host computer. The initial relative position between the laser scan micrometer and the diamond wheel is adjusted manually. Then, the computer numerical control (CNC) system controls the machine tool to move according to the NC programs to realize the laser scan micrometer to scan the grinding wheel surface and collect the profile information. Finally, the laser scan micrometer collected profile dates are transferred back to the host computer by the controller and analyzed to determine the form accuracy of the diamond wheel.

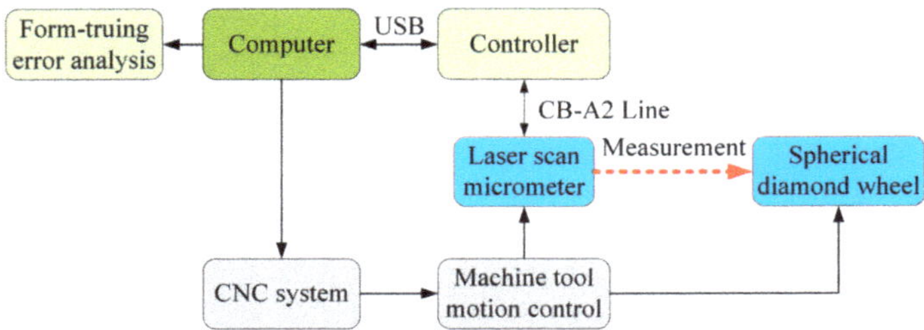

Figure 2. Principle of in-situ measurement system.

Due to the random distribution of sharp diamond grains on the surface of the grinding wheel, the contact measurement probe is easily damaged, which affects the accuracy of the measurement results. The laser scan micrometer used in this research needs to meet the requirements of precision, range, and working distance. The model and working parameters are listed in Table 1.

Table 1. Working parameters of the laser scan micrometer.

Details	Values
Type	Keyence LK-G5000
Sampling frequency/kHz	10
Resolution/μm	0.1
Measuring range/mm	6
Working distance/mm	20
Grinding wheel RPM/rpm	0 and 600
Scanning speed/mm/min	30

2.3. Precision Form Truing of Spherical Diamond Wheel

Figure 3 shows the diamond roller truing process of the resin-bonded spherical diamond wheel. In the *yoz*-plane of the machine tool coordinate system, the rotary axes of the tool spindle and the truing spindle are perpendicular to each other. During the truing process, the hemispherical diamond wheel and the diamond roller rotate at speed ω_t and ω_d, respectively. The diamond roller cuts in along the x-axis direction from the top of the spherical diamond wheel, then feeds along the generatrix of the spherical surface through the x-, y-, and z-axes interpolation motion to achieve the truing process. Finally, the diamond roller cuts out parallel to the tool spindle axis. The truing depth should be set in both x and z directions, and the detailed on-machine form truing parameters are shown in Table 2. The diamond roller is a disc-shaped coarse-grained electroplated grinding wheel (type 1A1). As the diamond roller has good wear resistance, the truing process can be approximated as a deterministic material removal process.

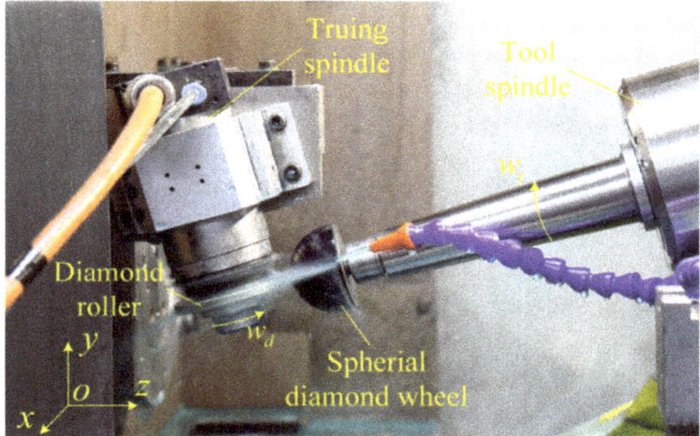

Figure 3. Precision form truing of the spherical diamond wheel.

Table 2. Form truing parameters.

Details	Values
Truing Spindle RPM ω_d/rpm	7200
Tool spindle RPM ω_t/rpm	900
Depth of truing/μm/pass	5–20
Feed rate/mm/min	10
Diameter of grinding wheel/mm	≈50
Grain size and concentration	D46 C100
Coolant, Challenge 300-HT	3 vol% in water

According to the diamond roller truing principle shown in Figure 3, the form-truing error is not introduced from tool setting in the y- and z-axes directions as the 1A1 type diamond roller is used. However, the alignment error in the x direction has a great influence on the form-truing accuracy of the spherical diamond wheel. The tool setting in the x direction involves finding the x-axis coordinates of the machine tool that make the diamond roller rotation axis and the diamond wheel rotation axis coplanar. The difference between the practical tool position and the ideal tool position Δx is the alignment error in the x direction, as shown in Figure 4.

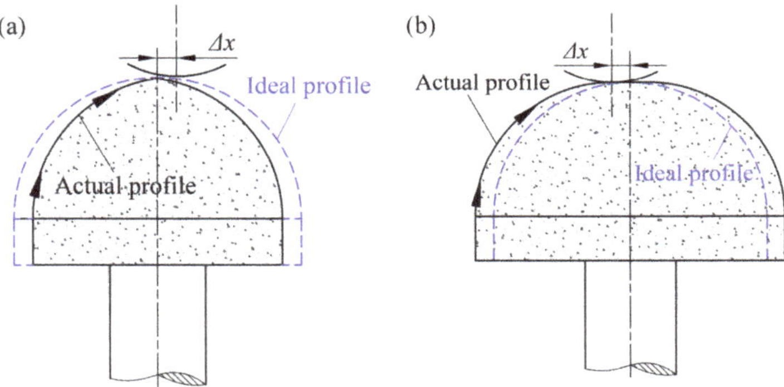

Figure 4. Influence of the alignment error on the profile of the spherical diamond wheel: (**a**) diamond roller not to center, (**b**) diamond roller past center.

3. Measurement and Compensation of the Alignment Error

The position of the spherical diamond wheel and the laser scan micrometer in the measuring process is shown in Figure 5a. The laser beam by the laser scan micrometer is adjusted to parallel to the *yoz*-plane of the machine tool coordinate system, and coplanar with the axis of the diamond wheel. Then, the laser scan micrometer is controlled by the CNC system to scan the diamond wheel surface bottom-up linearly. The scanning direction is perpendicular to the axis of the diamond wheel. The sampling points of the laser scan micrometer are the height information of the diamond wheel surface, and the continuous sampling data constitute the profile of the spherical diamond wheel. As shown in Figure 5b, the arc AC in the measurement results corresponds to the AC section profile of the spherical diamond wheel. After truing the spherical diamond wheel, the AC section is tested, and the least square fitting is performed. The fitting radius and residual value are the measuring radius and profile error of the spherical diamond wheel, respectively.

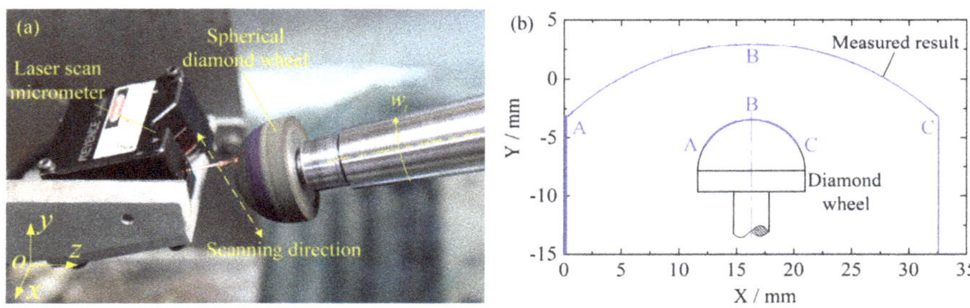

Figure 5. In-situ measurement of spherical diamond wheel profile: (**a**) measurement, (**b**) results.

Based on the measurement results shown in Figure 5b, a biarc curve-fitting method is proposed to calculate the alignment error between the truing spindle and tool spindle. As displayed in Figure 6, the arc A′B′C′ represents the ideal spherical diamond wheel profile, and the arc ABC is the actual diamond roller trued profile. The DE arc is intercepted and fitted from the AB segment of the detected profile, and the coordinates (x_l, y_l) of the circle center O_l as well as the arc radius r_l, are then calculated. Similarly, the GF arc is intercepted and fitted from the BC segment of the detected profile, and the position coordinates (x_r, y_r) of the circle center O_r as well as the arc radius r_r are then obtained. The distance O_lO_r between the center of the two circles is the alignment error in the *x* direction. When $x_r > x_l$, the roller is not trued to the center of the spherical diamond wheel, as shown in Figure 6a. When $x_r < x_l$, the roller has passed by the center of the spherical diamond wheel, as shown in Figure 6b.

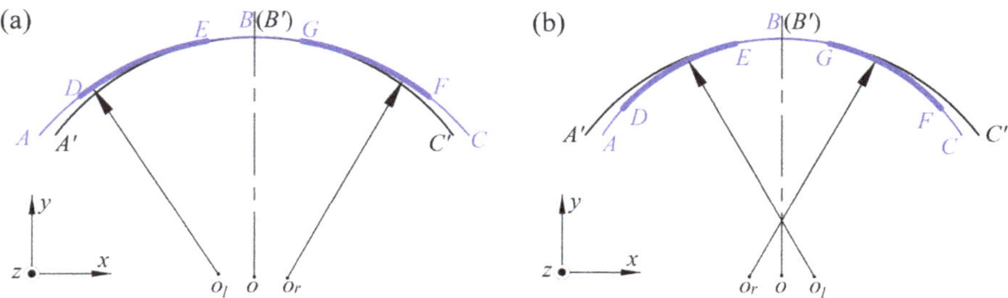

Figure 6. Schematic of the biarc curve-fitting method: (**a**) diamond roller is before the center, (**b**) diamond roller is past the center.

Figure 7 shows a set of center coordinates and radius fitting values obtained by the biarc curve-fitting method. It is seen that the alignment error between the roller and the spherical diamond wheel is approximately 0.28 mm, and the diamond roller is not trued to the center of the spherical diamond wheel. The alignment error can be eliminated by introducing a "truing-detection-compensation" process. Then, the precision form truing of the resin-bonded spherical diamond wheel can be achieved. The diameter of diamond roller can be approximately considered as fixed in the actual truing process of the resin-bonded diamond wheel.

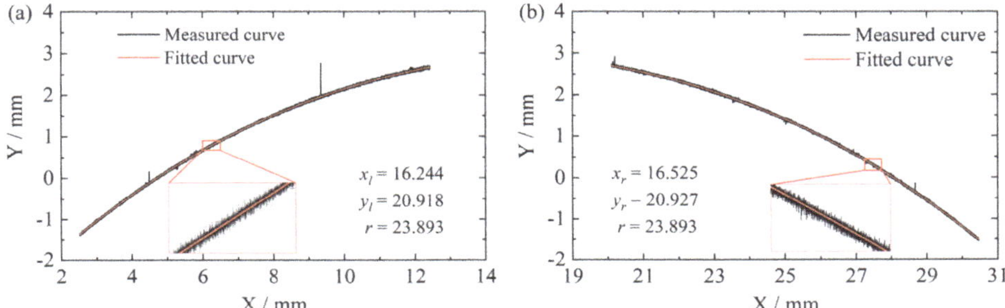

Figure 7. Principle of biarc curve-fitting error analysis for the spherical diamond wheel: (**a**) profile fitting of the left side, (**b**) profile fitting of the right side.

Figure 8 displays scanning electron microscope (SEM) micrographs showing the surface topography of the resin-bonded spherical diamond wheel after the diamond roller truing. The flat and regular surface of the spherical diamond wheel proves the good truing effect of the diamond roller. Part of the diamond grains were broken during the truing process, and the micro-breakage of the abrasive grains generates new sharp cutting edges. The diamond roller truing method can achieve the purposes of both shaping and sharpening. It is conducive to enhancing the grinding ability of the spherical diamond wheel, and improving both the ground surface form accuracy and the surface quality.

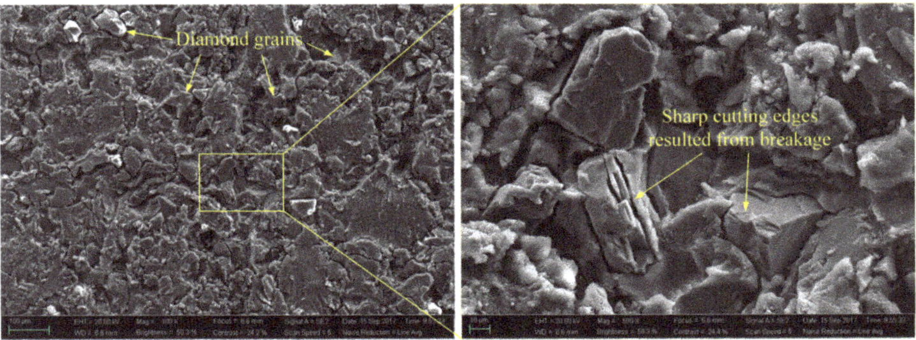

Figure 8. Surface morphology of the form-trued spherical diamond wheel.

4. Results and Discussion

Figure 9a shows the measured line-profile of the spherical diamond wheel after diamond roller truing. The spherical diamond wheel is stationary during the measuring process. The measuring radius and the two-dimensional (2D, along the generatrix direction) form-truing error of the spherical diamond wheel can be obtained by least square fitting. As affected by the abrasive grain size and the surface micro-roughness, the measuring radius possesses a certain deviation from the actual radius, which needs to be compensated according to the ground surface form error of the test specimen. The 2D form-truing error of the spherical diamond wheel is the deviation between the measured surface profile and the fitted curve. As shown in Figure 9b, the obtained 2D form-truing error is approximately 40 μm, which is slightly less than the average grain size of the spherical diamond wheel.

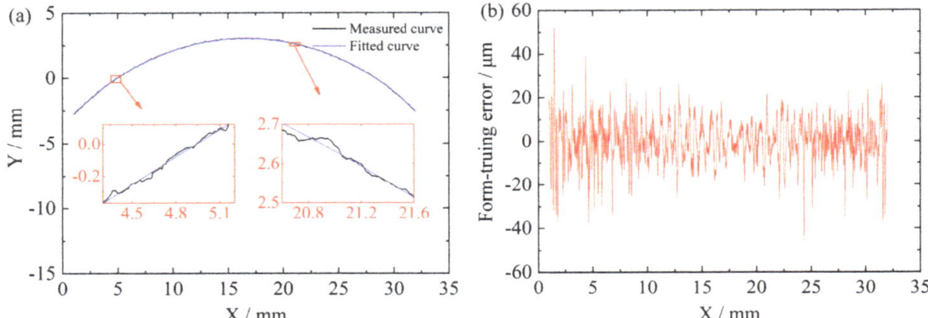

Figure 9. Line-profile of the form-trued spherical diamond wheel: (**a**) measurement results and data fitting, (**b**) 2D from-truing error.

In order to validate the accuracy of the in-situ measurement results, the 2D form-truing error of the spherical diamond wheel are also measured by stylus profilometer (Form Talysurf PGI 1240, Taylor Hobson Ltd., Leicester, UK), as shown in Figure 10. By comparing the error curves in Figures 9b and 10, it is observed that the two measurement results have a high degree of consistency, especially the high-frequency form error of the spherical diamond wheel. The high-frequency error signals in the measurement results mainly originate from the protruding diamond grains and bond pits on the surface of the spherical diamond wheel. These grains and pits are extremely random and can be expressed by the signal amplitude in the measurement results. In addition, the amplitude detected by the stylus profilometer is less than that of the laser scan micrometer. Due to the limitation of the profiler probe size and shape, it cannot detect the bottom of the pits in the contact measurement process, this may generate an error in the peak to valley value. The 2D form-truing errors obtained by both the laser scan micrometer and the stylus profilometer cannot accurately characterize the form accuracy of the grinding wheel, as only the protruding diamond grains on the whole grinding wheel surface is functional for the ground surface form accuracy.

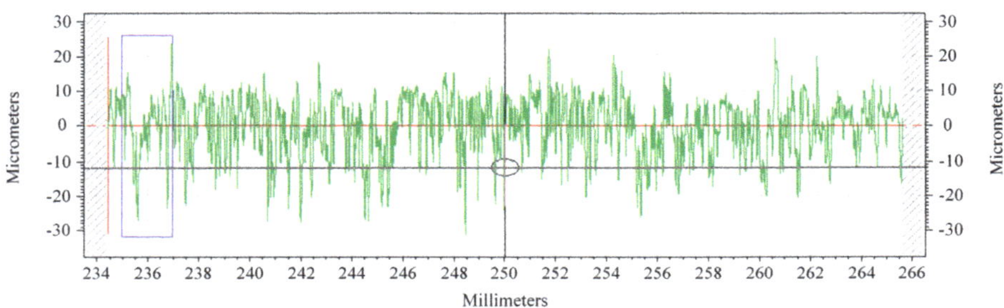

Figure 10. The form-truing error detected by a stylus profilometer.

The highest point of the protruding diamond grains on the rotation circumference of the spherical diamond wheel is the actual effective profile in the grinding process. Therefore, the surface profile can better reflect the form accuracy of the spherical diamond wheel rather than the line-profile. One of the advantages of non-contact measurement is that the spherical diamond wheel can rotate in the detection process to obtain the surface profile. Figure 11 shows the surface profile information of the form-trued spherical diamond wheel detected by the laser scan micrometer under a rotation speed of 600 rpm. It can be seen from Figure 11a that the measured profile of the spherical diamond wheel is in good consistency with the fitted curve. However, the rotation of the spherical diamond wheel causes continuous change and rapid fluctuation of the measured surface, the high-frequency signals are introduced into the measured results. These high frequency signals contain not only the form error in the rotation direction but also noise signals. According to the original form error signals shown in Figure 11b, the maximum amplitude exceeds 150 μm. In order to obtain the effective information about the enveloping profile of the protruding diamond grains, the noise signals are removed by low-pass filtering. Along with cut-off frequency increasing, more information of form-truing error in the rotation direction as well as more noise signals will be retained. However, the cut-off frequency needs to be greater than the rotational frequency of the spherical diamond wheel to avoid the radial run-out error signals being filtered out. The rotational frequency f_r of the diamond wheel is expressed as:

$$f_r = \frac{n_t}{v_f}, \tag{1}$$

where, n_t is the rotation speed of the spherical diamond wheel (rpm), and v_f is the scanning movement speed in the measurement process (mm/min). According to Equation (1), under the experimental conditions of a grinding wheel rotation speed of 600 rpm and a scanning feed rate of 30 mm/min, the rotational frequency is 20 Hz. As shown in Figure 11c, the 3D (along the generatrix and the rotation direction) form-truing error of the spherical diamond wheel is approximately 5 μm by setting the cut-off frequency of the low-pass filter to 32 Hz. The form-truing error is distributed symmetrically on both sides of the grinding wheel center.

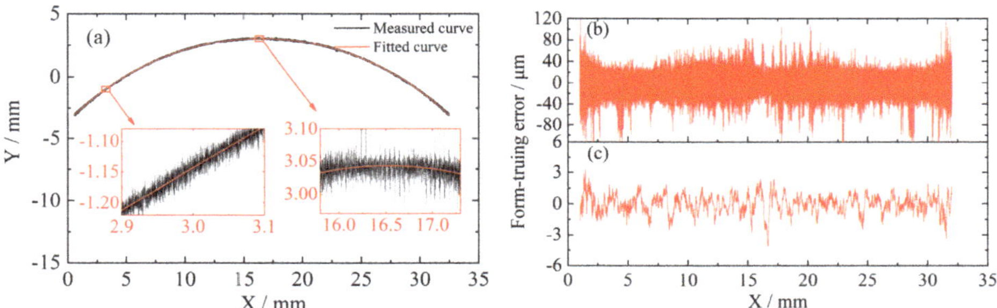

Figure 11. Surface profile of the form-trued spherical diamond wheel: (**a**) measurement results and data fitting, (**b**) original form error signals, (**c**) 3D form-truing error after low-pass filtering.

The form accuracy of the spherical diamond wheel can be verified by grinding a test specimen, and the actual radius of the spherical diamond wheel can also be calculated based on dimension error of the ground surface. In this paper, the hot-pressed ZnS with a diameter of 20 mm is selected as the test specimen, which incurs little wear to the spherical diamond wheel. The grinding wheel rotation speed is 12,276 rpm, the feeding speed is 5 mm/min, and the grinding depth is 2 μm. The grinding surface of the test specimen is a sphere with a target radius of 25 mm in the NC program. As shown in Figure 12a, the ground surface form error is approximately 1.6 μm without any compensation, which is determined mainly by the form-truing accuracy of the spherical diamond wheel. It can be found from Figure 12b that the surface form error of the test specimen is mainly resulted from low-frequency fluctuations with a space period of about 2 mm, which is equivalent to the low-frequency fluctuation period of the spherical diamond wheel surface profile shown in Figure 11c. The test specimen grinding results verified the accuracy of the 3D form-truing error that was obtained by laser scan micrometer non-contact measurement. By comparing the obtained 3D form-truing error and ground surface form error values, it is confirmable that the uncertainty of the measured form-truing error is about 3 μm for aspherical diamond wheel with a diameter of about 50 mm. In addition, the sphere radius of the test specimen is 24.9896 mm, which is 10.4 μm less than the target radius. Therefore, the measuring radius of the spherical diamond wheel is 10.4 μm smaller than the actual value. The uncertainty of the measuring radius is high related to the size and surface state of the grinding wheel, but the measuring error can be compensated by test specimen grinding. In conclusion, the accurate form-truing error as well as the actual size of the spherical diamond wheel can be obtained by combining the laser scan micrometer non-contact measurement and test specimen grinding.

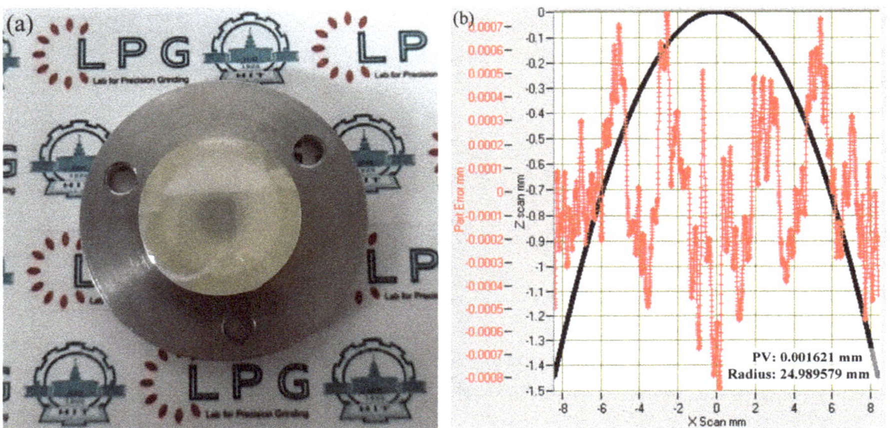

Figure 12. Grinding results of the ZnS test workpiece: (**a**) grinding surface, (**b**) surface form error.

5. Conclusions

In this paper, the coarse-grained diamond roller is applied to perform on-machine precision form truing of the resin-bonded spherical diamond wheel. The non-contact in-situ measurement based on a laser scan micrometer is proposed to achieve the measurement of form and alignment errors. The main conclusions include:

(1) The diamond roller truing is an efficient and feasible method for conditioning the resin-bonded spherical diamond wheel. The grinding wheel surface after truing is flat and regular, and micro-breakage of the abrasive grains generated new sharp cutting edges. The form truing of a resin-bonded spherical diamond wheel with a radius of about 24 mm and particle size of D46 was conducted after alignment error compensation, and the obtained form-truing error is approximately 5 μm.

(2) The alignment error between the diamond roller and the spherical diamond wheel can be accurately calculated by applying the proposed biarc curve-fitting method. The spherical diamond wheel with good form accuracy can be obtained by introducing a "truing-detection-compensation" process to eliminate the alignment error.

(3) The laser scan micrometer is used for non-contact in-situ measurement of the form-truing error of the spherical diamond wheel. The accuracy of the measurement results is validated by comparing with the results of the stylus profilometer. Besides, the actual form-trued diamond wheel has been employed in grinding a ZnS spherical surface, and the form accuracy is approximately 1.6 μm without any compensation. The ground surface profile shared similar characteristics with the roller-trued diamond wheel profile, confirming that the diamond roller truing and in-situ measurements methods are accurate and feasible.

Author Contributions: Conceptualization, J.W. and C.Z.; investigation, J.W.; methodology, Q.Z. and B.G.; resources, Q.Z.; software, C.Z.; validation, B.G.; writing—original draft, J.W.; writing—review and editing, J.Y. All authors have read and agreed to the published version of the manuscript.

Funding: This research was funded by the National Natural Science Foundation of China [No. 51805484, 51875135, and 51575492], and the China Postdoctoral Science Foundation [No. 2019M652137].

Conflicts of Interest: The authors declared no potential conflicts of interest with respect to the research, authorship and/or publication of this article.

References

1. Wang, T.; Cheng, J.; Liu, H.; Chen, M.; Wu, C.; Su, D. Ultra-precision grinding machine design and application in grinding the thin-walled complex component with small ball-end diamond wheel. *Int. J. Adv. Manuf. Technol.* **2019**, *101*, 2097–2110. [CrossRef]
2. Yan, G.; You, K.; Fang, F. Three-Linear-Axis Grinding of Small Aperture Aspheric Surfaces. *Int. J. Precis. Eng. Manuf. Green Technol.* **2019**, 1–12. [CrossRef]
3. Kannan, K.; Arunachalam, N. Grinding wheel redress life estimation using force and surface texture analysis. *Procedia CIRP* **2018**, *72*, 1439–1444. [CrossRef]
4. Tran, T.H.; Le, X.H.; Nguyen, Q.T.; Le, H.K.; Hoang, T.D.; Luu, A.T.; Banh, T.L.; Vu, N.P. Optimization of Replaced Grinding Wheel Diameter for Minimum Grinding Cost in Internal Grinding. *Appl. Sci.* **2019**, *9*, 1363. [CrossRef]
5. Zhang, Z.; Yang, X.; Zheng, L.; Xue, D. High-performance grinding of a 2-m scale silicon carbide mirror blank for the space-based telescope. *Int. J. Adv. Manuf. Technol.* **2017**, *89*, 463–473. [CrossRef]
6. Chen, B.; Guo, B.; Zhao, Q. On-machine precision form truing of arc-shaped diamond wheels. *J. Mater. Process. Technol.* **2015**, *223*, 65–74. [CrossRef]
7. Liu, Z.; Tang, Q.; Liu, N.; Liang, P.; Liu, W. A Novel Optimization Design Method of Form Grinding Wheel for Screw Rotor. *Appl. Sci.* **2019**, *9*, 5079. [CrossRef]
8. Hwang, Y.; Kuriyagawa, T.; Lee, S.K. Wheel curve generation error of aspheric microgrinding in parallel grinding method. *Int. J. Mach. Tools Manuf.* **2006**, *46*, 1929–1933. [CrossRef]
9. Wegener, K.; Hoffmeister, H.W.; Karpuschewski, B.; Kuster, F.; Hahmann, W.C.; Rabiey, M. Conditioning and monitoring of grinding wheels. *CIRP Ann.* **2011**, *60*, 757–777. [CrossRef]
10. Deng, H.; Xu, Z. Dressing methods of superabrasive grinding wheels: A review. *J. Manuf. Process.* **2019**, *45*, 46–69. [CrossRef]
11. Qiu, Z.; Zou, D.; Yan, G. Mechanical dressing of resin bold diamond grinding wheel based on dressing force. *Opt. Precis. Eng.* **2015**, *23*, 996–1003.
12. Pombo, I.; Cearsolo, X.; Sánchez, J.A.; Cabanes, I. Experimental and numerical analysis of thermal phenomena in the wear of single point diamond dressing tools. *J. Manuf. Process.* **2017**, *27*, 145–157. [CrossRef]
13. Ding, W.; Li, H.; Zhang, L.; Xu, J.; Fu, Y.; Su, H. Diamond wheel dressing: A comprehensive review. *J. Manuf. Sci. Eng.* **2017**, *139*, 121006. [CrossRef]
14. Deng, H.; Chen, G.Y.; Zhou, C.; Li, S.C.; Zhang, M.J. Processing parameter optimization for the laser dressing of bronze-bonded diamond wheels. *Appl. Surf. Sci.* **2014**, *290*, 475–481. [CrossRef]
15. Lin, X.H.; Wang, Z.Z.; Guo, Y.B.; Peng, Y.F.; Hu, C.L. Research on the error analysis and compensation for the precision grinding of large aspheric mirror surface. *Int. J. Adv. Manuf. Technol.* **2014**, *71*, 233–239. [CrossRef]
16. Xie, J.; Zhou, R.M.; Xu, J.; Zhong, Y.G. Form-truing error compensation of diamond grinding wheel in CNC envelope grinding of free-form surface. *Int. J. Adv. Manuf. Technol.* **2010**, *48*, 905–912. [CrossRef]
17. Xu, L.M.; Fan, F.; Zhang, Z.; Chao, X.J.; Niu, M. Fast on-machine profile characterization for grinding wheels and error compensation of wheel dressing. *Precis. Eng.* **2019**, *55*, 417–425. [CrossRef]
18. Palmer, J.; Ghadbeigi, H.; Novovic, D.; Curtis, D. An experimental study of the effects of dressing parameters on the topography of grinding wheels during roller dressing. *J. Manuf. Process.* **2018**, *31*, 348–355. [CrossRef]
19. Chen, W.K.; Kuriyagawa, T.; Huang, H.; Yosihara, N. Machining of micro aspherical mould inserts. *Precis. Eng.* **2005**, *29*, 315–323. [CrossRef]
20. Wang, T.; Wu, C.; Liu, H.; Chen, M.; Cheng, J.; Dingning, S. On-machine electric discharge truing of small ball-end fine diamond grinding wheels. *J. Mater. Process. Technol.* **2020**, *277*, 116472. [CrossRef]

21. Wan, D.; Hu, D.; Wu, Q.; Zhang, Y. Online grinding wheel wear compensation by image based measuring techniques. *Chin. J. Mech. Eng.* **2006**, *19*, 509–513. [CrossRef]
22. Yang, Z.; Yu, Z. Grinding wheel wear monitoring based on wavelet analysis and support vector machine. *Int. J. Adv. Manuf. Technol.* **2012**, *62*, 107–121. [CrossRef]
23. Darafon, A.; Warkentin, A.; Bauer, R. Characterization of grinding wheel topography using a white chromatic sensor. *Int. J. Mach. Tools Manuf.* **2013**, *70*, 22–31. [CrossRef]
24. Bai, J.; Li, X.; Wang, X.; Zhou, Q.; Ni, K. Chromatic Confocal Displacement Sensor with Optimized Dispersion Probe and Modified Centroid Peak Extraction Algorithm. *Sensors* **2019**, *19*, 3592. [CrossRef]
25. Bai, J.; Li, X.; Zhou, Q.; Ni, K.; Wang, X. Improved chromatic confocal displacement-sensor based on a spatial-bandpass-filter and an X-shaped fiber-coupler. *Opt. Express* **2019**, *27*, 10961–10973. [CrossRef]
26. Zou, X.; Zhao, X.; Li, G.; Li, Z.; Sun, T. Non-contact on-machine measurement using a chromatic confocal probe for an ultra-precision turning machine. *Int. J. Adv. Manuf. Technol.* **2017**, *90*, 2163–2172. [CrossRef]
27. Sun, B.; Li, B. Laser displacement sensor in the application of aero-engine blade measurement. *IEEE Sens. J.* **2015**, *16*, 1377–1384. [CrossRef]
28. Miyasaka, T.; Okamura, H. Dimensional change measurements of conventional and flowable composite resins using a laser displacement sensor. *Dent. Mater. J.* **2009**, *28*, 544–551. [CrossRef]
29. Nishikawa, S.; Ohno, K.; Mori, M.; Fujishima, M. Non-contact type on-machine measurement system for turbine blade. *Procedia Cirp* **2014**, *24*, 1–6. [CrossRef]
30. Oya, S. Measurement of the vibrating shape of a bimorph deformable mirror using a laser displacement sensor. *Opt. Eng.* **2009**, *48*, 033601. [CrossRef]
31. Vukašinović, N.; Bračun, D.; Možina, J.; Duhovnik, J. The influence of incident angle, object colour and distance on CNC laser scanning. *Int. J. Adv. Manuf. Technol.* **2010**, *50*, 265–274.

Article

A Differential Measurement System for Surface Topography Based on a Modular Design

Fang Cheng *, Jingwu Zou, Hang Su, Yin Wang and Qing Yu

College of Mechanical Engineering and Automation, Huaqiao University, Xiamen 361021, China; zoujingwu96@163.com (J.Z.); 18351965378@163.com (H.S.); yin.wangyin@hqu.edu.cn (Y.W.); yuqing@hqu.edu.cn (Q.Y.)
* Correspondence: chf19chf19@hotmail.com or chengfang@hqu.edu.cn

Received: 15 January 2020; Accepted: 21 February 2020; Published: 24 February 2020

Abstract: In this paper, a novel design of a surface topography measurement system is proposed, to address the challenge of accurate measurement in a relatively large area. This system was able to achieve nanometer-scale accuracy in a measurement range of 100 mm × 100 mm. The high accuracy in a relatively large area was achieved by implementing two concepts: (1) A static coordinate system was configured to minimize the Abbe errors. (2) A differential measurement configuration was developed by setting up a confocal sensor and a film interferometry module to separate the motion error. In order to accommodate the differential measurement probes from both sides of the central stage and ensure the system rigidity with balanced supports, separate linear guides were introduced in this system. Therefore, the motion Degree of Freedom (DoF) was analyzed in order to address the challenge of an over-constrained mechanism due to multiple kinematic pairs. An optimal configuration and a quick assembly process were proposed accordingly. The experimental results presented in this paper showed that the proposed modular measurement system was able to achieve 10 nm accuracy in measuring the surface roughness and 100 nm accuracy in measuring the step height in the range of 100 mm × 100 mm. In summary, the novel concept of this study is the build of a high-accuracy system with conventional mechanical components.

Keywords: topography measurement; differential measurement system; modular design; confocal sensor; film interferometry; over-constrained mechanism

1. Introduction

Surface topography measurement [1–3] is an important metrological tool for quality assessment in industries that include semiconductors [4,5], micro-electromechanical system (MEMS) [6], additive manufacturing [7], and aerospace [8]. A common challenge with developing surface metrology technologies is the achievement of multi-scale measurement capability, which means high accuracy over a large measurement area.

Optical approaches that are playing important roles in this area [9,10], such as confocal microscopy [11,12], coherence scanning interferometry [13], and focus variation [14], are able to meet nanometer accuracy, but the measurement range is within a few millimeters or even smaller. The data stitching technique is able to expand the measurement range, but it relies heavily on common features of the overlapping area from neighboring data sets [15]. If the overlapping area does not show distinct features, the stitching accuracy will rely heavily on the performance aspects of the motion system, such as the straightness and flatness [16].

Therefore, high-precision 2D/3D positioning [17] is a key technical enabler for multi-scale measurement systems. In reference [18], an optical probe was mounted on a Coordinate Measuring Machine (CMM). In references [19,20], Scanning Probe Microscopy (SPM) probes were equipped with high-precision positioning systems, to achieve a sub-nanometer resolution in a millimeter measurement area.

When integrating a high-accuracy probe with a motion system, there are two types of critical measurement errors that researchers have to consider. They are the motion error, introduced by the imperfect straightness of the guideway [21,22], and the Abbe error, introduced by the system design [23,24]. Motion errors can be minimized by using high-precision components as geometric references [25]. However, the accuracy loss of mechanical references is inevitable after heavy usage. The motion error can also be compensated by implementing differential measurement [26,27] or by introducing additional optical measurement systems [28,29]. The Abbe error can be minimized by employing a static coordinate system in which the position sensors track the central stage along the coordinate axis [30,31]. In reference [30], three laser interferometers that formed a static coordinate system were used to track the position of the central stage. In reference [31], the central stages were mechanically tracked by optical encoders. However, both concepts have their drawbacks. Optical tracking is very sensitive to environmental disturbances such as temperature and airflow. Mechanical tracking systems require a very complicated assembly and adjustment process due to the over-constrained mechanism. Furthermore, the geometric accuracy of the mechanical components will significantly affect the measurement accuracy.

In order to address the challenges discussed in the above paragraph, a modular system for surface topography measurement was developed using conventional components. The central stages were mechanically tracked by optical encoders with consideration of the cost and robustness [31]. An optimal system design was proposed in order to address the challenge of the over-constrained mechanism. A film interferometer was developed to compensate for the straightness error from the motion system.

The motivation and scope of the presented work can be illustrated by the graphic abstract shown in Figure 1.

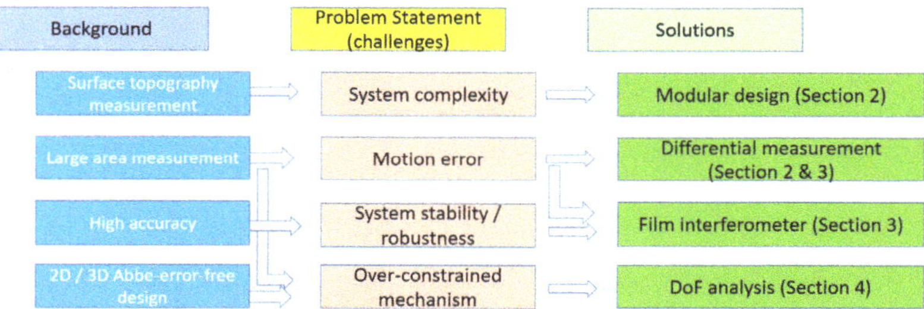

Figure 1. Graphic abstract.

2. Overall System Design

The system configuration of the proposed design is shown in Figure 2. The overall dimension of the system is L 890 mm × W 820 mm × H 675 mm. The measurement range is 100 mm × 100 mm × 50 mm. Like the first prototype, this system was built of aluminum alloy. The central stage was supported by two sets of 2D sliders and driven by a pair of actuators (Zolix uKSA 100, Zolix Instruments CO.,LTD, Beijing, China). Each uKSA 100 is able to achieve a travel range of 100 mm, which determined the measurement range of the presented system.

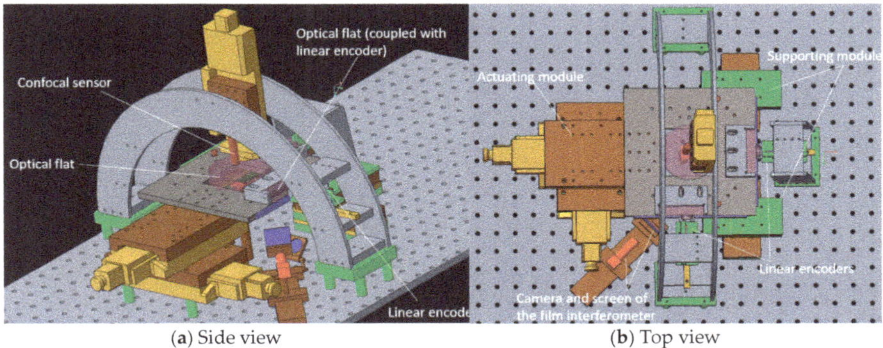

(a) Side view (b) Top view

Figure 2. System design: (**a**) Side view, (**b**) Top view.

A confocal sensor, Precitec CHRocodile SE (Precitec Group, Germany), was mounted on the main spindle to measure the height at each location of the test piece. This confocal sensor was able to achieve 3 nm resolution in a measurement range of 600 μm. An optical flat [32] (φ 300 mm × T 30 mm, K9 glass) was placed in the center of the stage as the geometric datum. A film interferometer was developed in order to configure a differential measurement system with the confocal sensor. During the lateral movement, the film interferometer was used to compensate for the straightness error by measuring the fluctuation of the optical plat from the bottom.

The XY positions were measured by linear encoders, which were coupled with the central stage through contact kinematic pairs. Each contact kinematic pair consisted of two optical flats, which helped to minimize the coupling errors. The linear encoders only moved along the measurement axes. In the XY plane, therefore, a static coordinate system was formed by the linear encoders, and the system complied with the Abbe principle [23,24].

In this system, the position measurement accuracy relied on the datum features formed by the optical flats, instead of the mechanical translation components. All of the linear guides used in this system were within the ordinary accuracy range.

The geometric error factors for the height measurement are shown in Figure 3. The actual height of the measurement point could be expressed as:

$$h = h_1 - h_2 - h_3 - \left(\frac{T}{cos\theta} - T\right). \tag{1}$$

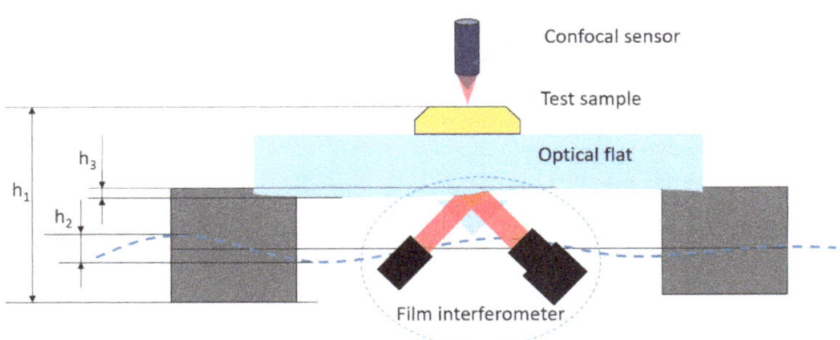

(a) Straightness and flatness

Figure 3. *Cont.*

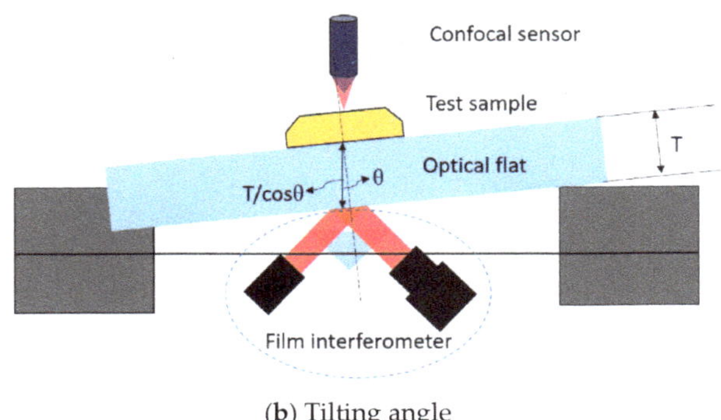

(**b**) Tilting angle

Figure 3. Geometric error factors: (**a**) Straightness and flatness, (**b**) Tilting angle.

In Equation (1) h_1 is the height measured by the confocal sensor, h_2 represents the motion error caused by the linear guides, and h_3 is the height fluctuation due to the flatness error of the optical flat. T is the thickness of the optical flat and θ is its tilting angle during the movement.

In this system, the flatness of the optical flat was ±25 nm. The thickness of the optical flat was 30 mm. By assuming that the angular motion error was no greater than 20″, the maximum height measurement error $T(1 - cos\theta)$ could be calculated to be 0.14 nm. Therefore, the values of h_3 and $T(1 - cos\theta)$ were very insignificant and Equation (1) could be simplified as:

$$h = h_1 - h_2, \tag{2}$$

where h_2 was mainly contributed by the straightness error of the linear guide and it could be measured by the film interferometer.

Instead of an all-in-one structure, the concept of modular design was employed in the proposed system. As shown in Figure 2, the system was assembled with a few functional modules:

(1) An actuating module, which consisted of two motorized stages.
(2) Supporting modules, each of which consisted of pre-assembled 2D sliders
(3) A differential height measurement module, which consisted of a confocal sensor and a film interferometer
(4) A displacement measurement module, which consisted of an optical gratings couple with the central stage through optical flats.
(5) A datum module, which consisted of three optical flats.

Since these functional modules were all commercially available components, no special design or assembly processes were needed in the proposed system.

3. Motion Error Compensation Based on a Film Interferometer

As described in the section above, the motion error caused by the linear guide represented by h_2 was measured by an in-house developed film interferometer. The principle is shown in Figure 4.

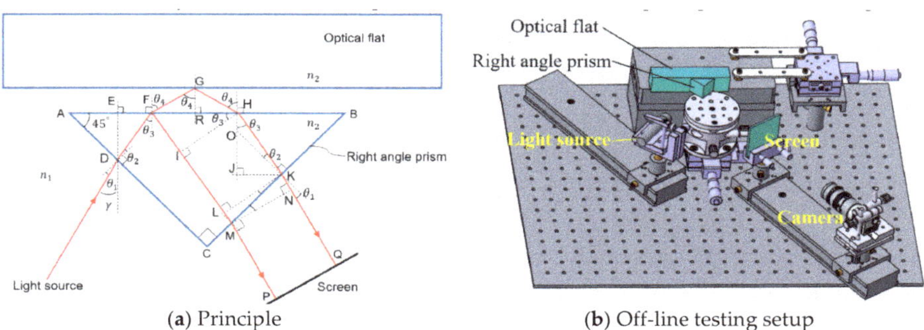

(a) Principle (b) Off-line testing setup

Figure 4. Principle of the film interferometer: (a) Principle, (b) Off-line testing setup.

The key specifications of the components used in this system are listed in Table 1.

Table 1. Component specifications in the film interferometer.

Components	Specifications
Camera	Basler acA2000-165um, frame rate 165 fps, resolution 2048 × 1088 (2 MP)
Lens	Moritex ML - MC50HR, 0.8×, focal length 50 mm
Right-angle prism	25.4 mm × 25.4 mm × 25.4 mm, K9 glass, refractive index 1.5163

As shown in Figure 4a, the film interferometer included two optical components, the optical flat and the right-angle prism. The optical flat moved with the stage and the right-angle prism was stationary. The two optical surfaces formed an air gap between themselves with a gradually changing thickness. The incident laser beam was split and reflected by the two optical surfaces. When the angle of the air gap was very small, the reflected laser beams propagated along the same path and generated an interferogram that was detected by the camera. By analyzing the phase shift of the interferogram, the distance variation between the optical flat and the right-angle prism could be measured. This distance variation represented the motion error caused by the imperfect straightness of the linear guide.

The root cause of the interferogram was the optical path difference of the reflected beams from the two optical surfaces. As shown in Figure 4a, the optical path difference D could be expressed as:

$$\mathrm{D} = \mathrm{n}_1(\mathrm{FG} + \mathrm{GH} + \mathrm{KN}) + n_2 \cdot \mathrm{HK} - \mathrm{n}_2(\mathrm{FI} + \mathrm{IL} + \mathrm{LM}), \quad (3)$$

where n_1 and n_2 are the relative refractive indexes of the air and glass, respectively, and $\lambda/2$ represents the half-wave rectification. Based on geometrically analyzing the optical path, it could be concluded that:

(1) Since the angle of the air gap was very small, FG was approximately equal to GH:

$$FG = GH. \quad (4)$$

(2) Since HK and IL could be seen as the same light beam in the same medium, HK was equal to IL:

$$HK = IL. \quad (5)$$

(3) Since LM and KN could be seen as the same light beam in different media, they had the same length of the optical path:

$$n_2 \cdot LM = n_1 \cdot KN \quad (6)$$

It should be highlighted that Equations (4)–(6) were approximately true because the optical flat was not ideally parallel with the bottom of the right-angle prism. When the angle between the two reflective surfaces was very small, the optical path differences $FG - GH$, $HK - IL$, and $n_2{\cdot}LM - n_1{\cdot}KN$ could be considered as a higher-order infinitesimal of the angle.

With Equations (5)–(7) substituted into Equation (4), the optical path difference could be expressed as:

$$\mathrm{D} = 2\mathrm{n}_1{\cdot}FG - n_2{\cdot}FI. \tag{7}$$

By assuming that the current distance between the two optical surfaces was GR = d, D could be modified as:

$$\mathrm{D} = \frac{2\mathrm{n}_1 d}{\cos\theta_4} - \frac{2n_2 d\sin\theta_3}{\tan\theta_4}. \tag{8}$$

Based on Shell's law,

$$\begin{cases} n_2 sin\theta_3 = n_1 sin\theta_4 \\ n_1 sin\theta_1 = n_2 sin\theta_2 \end{cases}. \tag{9}$$

Substituting into Equation (8):

$$\begin{aligned} \mathrm{D} &= 2d\mathrm{n}_1\left(\tfrac{1}{\cos\theta_4} - \tfrac{sin^2\theta_4}{cos\theta_4}\right), \\ &= 2d\mathrm{n}_1 cos\theta_4, \\ &= 2d\sqrt{n_1{}^2 - n_2{}^2 sin^2\theta_3}, \\ &= 2d\sqrt{n_1{}^2 - n_2{}^2 sin^2\left(\tfrac{\pi}{4} - \theta_2\right)}, \\ &= d\sqrt{4n_1{}^2 - 2n_2{}^2 + 4n_1 n_2 sin\theta_1\sqrt{1 - \tfrac{n_1{}^2}{n_2{}^2}sin^2\theta_1}} \end{aligned} \tag{10}$$

If half-wave rectification was considered, then

$$\mathrm{D} = \frac{\lambda}{2} + d\sqrt{4n_1{}^2 - 2n_2{}^2 + 4n_1 n_2 sin\theta_1\sqrt{1 - \frac{n_1{}^2}{n_2{}^2}sin^2\theta_1}}. \tag{11}$$

Therefore, the interferogram, represented by D, was dependent on the distance d, the refractive indexes n_1 and n_2, and the incidence angle θ_1. The accurate measurement of n_1, n_1, and θ_1 was difficult. In this project, therefore, a calibration process was conducted to model the relationship between D and d. When a constant C was defined as:

$$\mathrm{C} = \sqrt{4n_1{}^2 - 2n_2{}^2 + 4n_1 n_2 sin\theta_1\sqrt{1 - \frac{n_1{}^2}{n_2{}^2}sin^2\theta_1}} \tag{12}$$

Equation (11) could be modified as:

$$\mathrm{D} = \frac{\lambda}{2} + C{\cdot}d. \tag{13}$$

During the movement, the phase shift of the interferogram was correlated with the changing of D, instead of the absolute value. Thus, Equation (13) could be modified as:

$$\Delta D = C{\cdot}\Delta d. \tag{14}$$

This linear relationship could be determined by observing the phase shift of the interferogram and measuring the distance h with a reference sensor:

As shown in Figure 5, a small angle was intentionally set between the optical flat and the moving axis of the stage. During the axial movement of the stage, the distance between the two optical surfaces gradually changed. The distance was measured by both the confocal sensor and the film interferometer.

The images of the interferogram were captured continuously and the phase value of the central point was computed based on image processing, as shown in Figure 6.

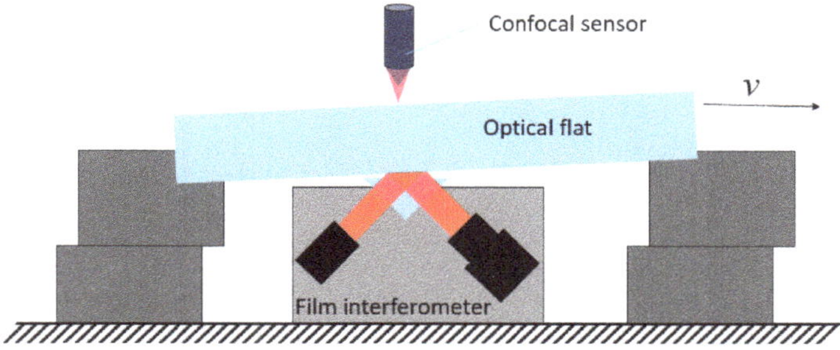

Figure 5. Calibration setup of the film interferometer.

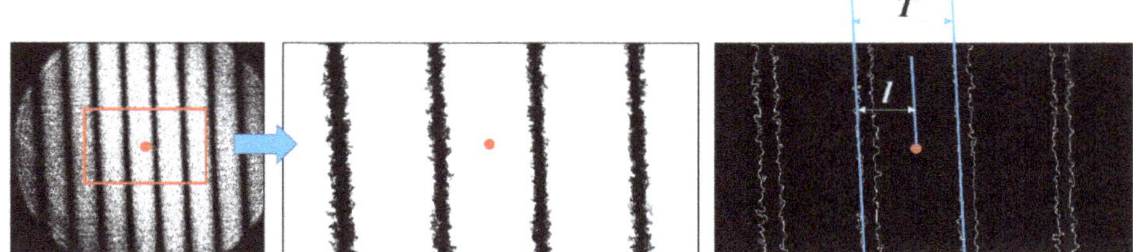

Figure 6. Image processing for phase calculation.

The interfergram did not show an ideal sinusoidal wave cycle due to high-order harmonics. Therefore, the edge detection technique was applied to identify the signal cycle. The position of the central point indicated its phase shift ω:

$$\omega = \left(\mathrm{m} + \frac{l}{T}\right)\cdot 2\pi, \tag{15}$$

where m is the number of wave cycle shifts.

Although the angular motion errors may have changed the density and orientation of the interference stripes, the calculation of the phase using Equation (16) was not affected. The phase shifts calculated with Equation (16) and the distance change measured by the confocal sensor were plotted in the characteristic curve, as shown in Figure 7.

In order to generate the fringe-type interferogram, the two reflective surfaces could not be exactly parallel with each other. A small angle was set to form an air wedge. During the movement, this angle might change due to the imperfect straightness of the guideway. In order to verify the model consistency when the angle changed, a simulation-based on Zemax was conducted. Different angles, 20″, 30″ and 40″ were set. As shown in Figure 7b, despite the different fringe densities, the relationship between distance and phase shift remained consistent. Therefore, the angle between the two reflective surfaces won't significantly affect the distance measurement at the central point.

It worth highlighting that the proposed film interferometer showed better stability compared with conventional Michelson interferometry designs. As shown in Figure 4, the interference occurred in the thin gap between the two optical surfaces, which was considered a relatively enclosed space. This optical structure, therefore, showed better robustness against temperature change and airflow.

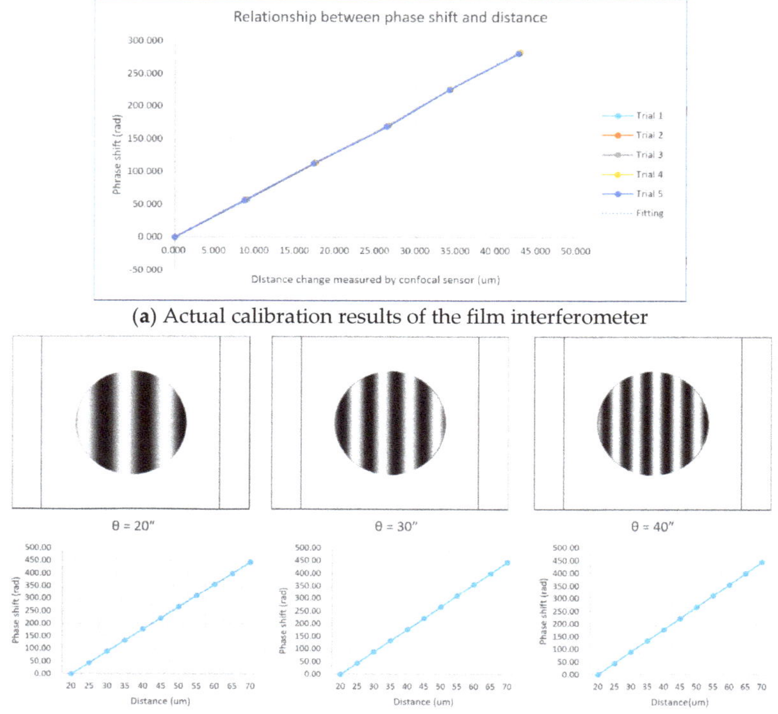

(**a**) Actual calibration results of the film interferometer

(**b**) Simulation results with different angles set between the reflective surfaces

Figure 7. Relationship between phase shift and distance.

4. Analysis of the Degree of Freedom

In order to allow a long travel range and to accommodate a differential measurement setup, the proposed system needed separate supporting and guiding mechanisms, as discussed in Section 2. Furthermore, the Abbe error-free design may have also introduced a separate guiding mechanism [31,33,34]. To avoid an over-constrained mechanism, a strict alignment and adjustment process was needed. The straightness and parallelism of the linear guides are very critical to such systems. In order to address this challenge, the proposed modular system was developed based on DoF analysis:

As shown in Figure 8, the key points of designing and assembling the system were those listed below:

(1) There were one actuating module and two supporting modules in this system. Each supporting module consisted of two perpendicularly placed sliders, which were pre-assembled. This 2D sliding pair could be considered a plane contacting pair that allowed free movement in a horizontal plane despite possible in-plane placement errors.

(2) The sliders were well leveled to the horizontal reference, assisted by an LVDT (Linear Variable Differential Transformer)

(3) The stage top was rigidly connected with the actuating module, and it was coupled with the two supporting modules through ball hinges.

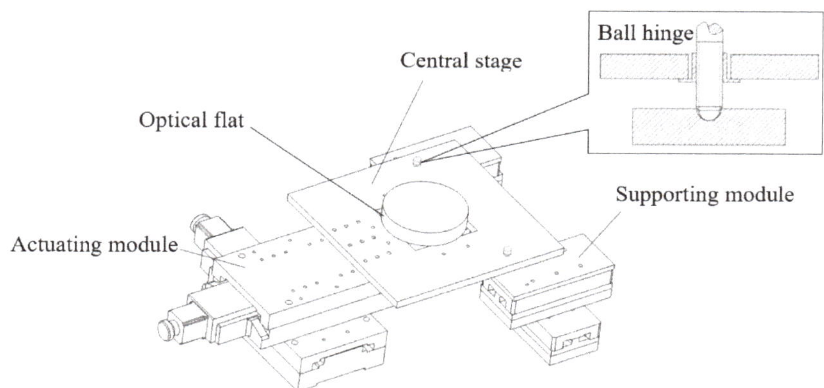

Figure 8. XY motion system.

The following paragraphs provide the DoF analysis of the mechanism.

Traditional DoF theory does not cover the scenarios of imperfect straightness or parallelism, and parallel linear guides are often considered redundant constraints [35]. This methodology, however, does not apply to precision positioning systems. Based on the traditional theory, the structure shown in Figure 8 could be simplified to the mechanism shown in Figure 9a. When imperfect straightness and parallelism were considered, the linear guides had to be represented by two-point contacting pairs instead of sliding pairs, as shown in Figure 9b. The DoF of the mechanism shown in Figure 9b could be expressed by:

$$\mathrm{DoF} = 3 \cdot \mathrm{n} - 2 \cdot \mathrm{p} - 1 \cdot \mathrm{q}. \quad (16)$$

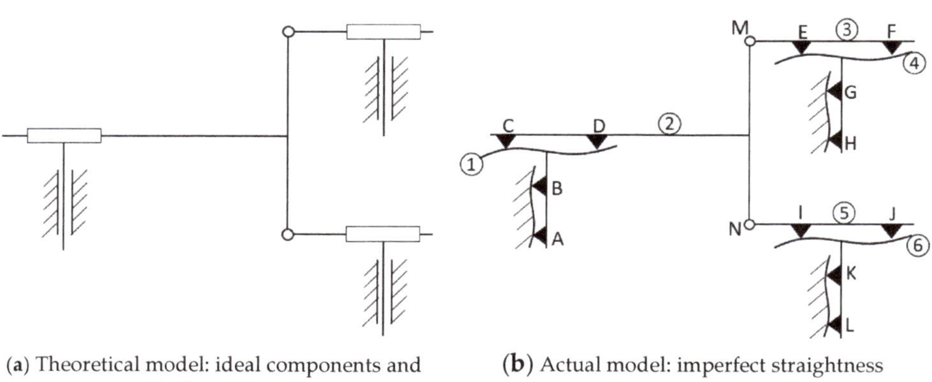

(**a**) Theoretical model: ideal components and ideal assembly

(**b**) Actual model: imperfect straightness parallelism

Figure 9. Mechanical model for DoF analysis: (**a**) Theoretical model: ideal components and ideal assembly, (**b**) Actual model: imperfect straightness parallelism.

In Equation (16), n is the number of components: $n = 6$. p is the number of revolving pairs (ball hinges M and N), each of which was constrained to 2 DoF: $p = 2$. q is the number of contacting pairs A to L, each of which was constrained to 1 DoF: $q = 12$. The DoF of this mechanism could then be calculated to be two, which met the requirement of 2D motion.

Therefore, it was concluded that when the sliders were all well leveled and the 3D model could be projected to a 2D plane, an over-constrained mechanism would not be a concern for the proposed system. When a bigger measurement range is demanded in the future, this system can be easily reconfigured by replacing the actuating and supporting module with longer travel ranges.

With the assistance of an LVDT, the error of leveling could be controlled within a micrometer level, which would not affect the motion smoothness. The motion error due to the imperfect leveling can be detected by the film interferometer and compensated accordingly.

When a bigger measurement range is demanded, the system may need more supporting modules. As shown in Figure 9, one more supporting module would introduce two more components, one more revolving pair, and four more contacting pairs. By using Equation (16), it could be calculated that the additional DoF introduced by one more supporting module was zero, which meant more additional supporting modules would not affect the system DoF. With this modular design, therefore, the system had good scalability.

5. Experimental Verification

In order to verify the system performance, experimental tests were conducted. This section provides the details for the experimental methodology and data analysis.

5.1. D Topography Measurement

A coin was scanned using the proposed system, with a scan range of 2 mm × 2 mm and a spacing of 10 μm. The experimental setup and the scanned image are shown in Figure 10.

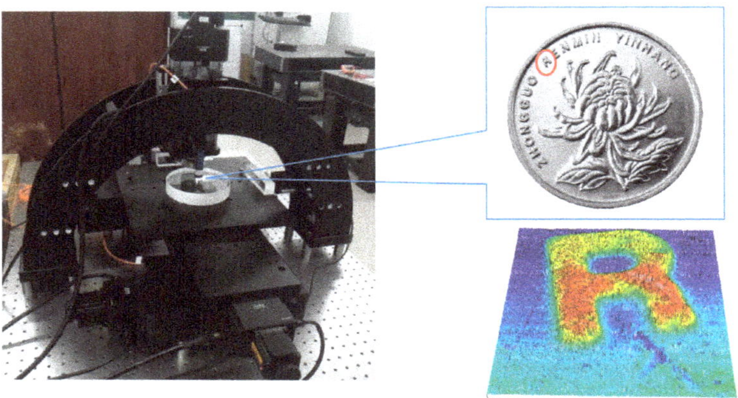

Figure 10. Experimental setup and coin scanning results.

In order to quantitatively evaluate the topography measurement performance, a set of roughness comparators was measured. Table 2 shows the comparison of the roughness parameters measured by the proposed system, namely, the HQU and a reference system, Mahr XR20. Each measurement was repeated five times. The average values are listed in Table 2. The standard deviation of each measurement was within 0.005 μm.

It was shown that the proposed system was able to achieve nanometer accuracy when measuring surface roughness.

Table 2. Roughness measurement results.

Sample	Ra (µm)		Deviation (µm)	Deviation (%)	Rq (µm)		Deviation (µm)	Deviation (%)
	HQU	Mahr XR20			HQU	Mahr XR20		
1	0.589	0.602	0.602	2.2%	0.735	0.714	0.021	2.9%
2	1.493	1.471	1.471	1.5%	1.746	1.737	0.009	0.5%
3	2.736	2.728	2.728	0.2%	3.238	3.234	0.004	0.1%
4	5.842	5.810	5.810	0.6%	6.753	6.728	0.025	0.4%

5.2. Step Height Measurement in a Large Area

The step height made by gauge blocks was measured at four different locations in the system measurement area of 100 mm × 100 mm. The scan length of each measurement was 50 mm. At each location, the measurements were repeated five times. Figure 11a shows the experimental setup and Figure 11b shows the profile measured by the differential sensors. The differential setup is discussed in Section 2. Sensor 1 (confocal sensor) was used to measure the heights from the top, while sensor 2 (film interferometer) was used to measure the straightness error from the bottom.

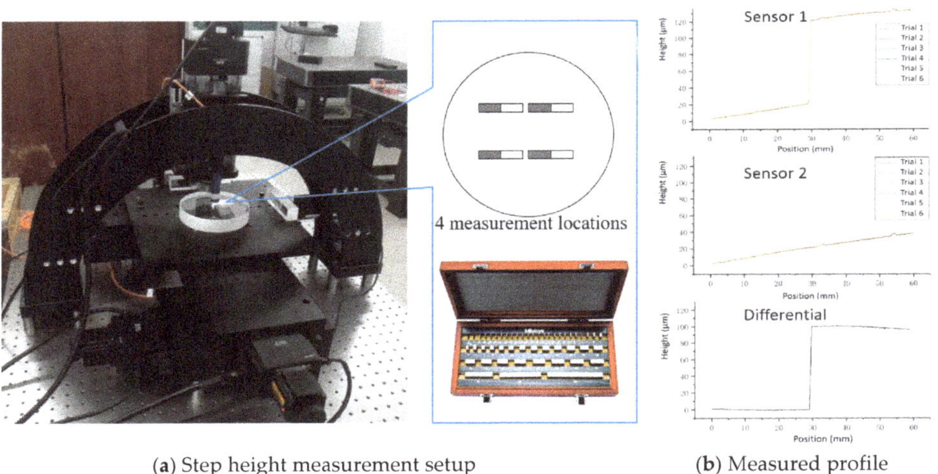

(**a**) Step height measurement setup (**b**) Measured profile

Figure 11. Step height measurement: (**a**) Step height measurement setup, (**b**) Measured profile.

It was shown that the straightness error of the linear guide was effectively compensated for. Table 3 shows the measurement results.

Table 3. Step height measurement results.

	Location 1	Location 2	Location 3	Location 4
Step height measured (µm) (Nominal: 100.080 µm)	100.1910	100.1073	99.9984	99.9834
	100.1284	100.2008	99.9857	99.9637
	100.1179	100.1414	99.9284	99.9806
	100.2042	100.1783	100.0852	100.0523
	100.2156	100.1290	99.985	100.0361
Average (µm)	100.1715	100.1514	99.9965	100.0032
Repeatability σ (µm)	0.0451	0.0378	0.05648	0.03859
Reproducibility σ′ (µm)	0.09293			

The standard deviations of the repeated measurements at the same location and different locations were calculated, in order to represent the system's repeatability and reproducibility. The data in Table 3 shows that the repeatability and the reproducibility were within 100 nm.

6. Conclusions and Discussion

In this paper, a concept of modular design for a high-accuracy measurement system is proposed. In the presented system, all of the components were standard components and easily available in the market. Instead of customizing high-accuracy built-in references, optical flats were applied in the proposed system to separate the geometric error caused by the mechanical components. The DoF analysis showed that this system could be easily assembled without special adjustment process. This system could also be easily scaled up for a bigger measurement range by introducing sliders with longer travel lengths.

The experimental results showed that the developed system was able to achieve nanometer repeatability and reproducibility for a surface topography measurement in the range of 100 mm × 100 mm.

The motivation for developing this system was to explore a generic methodology of setting up systems for high-accuracy topography measurements in a large area. Table 4 is the comparison between the presented system and other topography measurement technologies.

Table 4. Technology comparison.

Technologies	Measurement Resolution	Measurement Area	Probe	Motion Error Compensation
Confocal microscopy	~10 nm	~1 mm	Optical	No
coherence scanning interferometry	~1 nm	~1 mm	Optical	No
Focus variation	~20 nm	~1 mm	Optical	No
Stylus profilometry	~1 nm	~100 mm	Contact	No
HQU developed system	~10 nm	100 mm	Optical	Yes
STM with high-precision stage [19]	Sub-nanometer	~1 mm	STM	Not stated
AFM with high-precision stage [20]	Sub-nanometer	~1 mm	AFM	Yes

As shown in Table 4, the optical topography measurement technologies, such as confocal microscopy, coherence scanning interferometry and focus variation are able to achieve nanometer vertical resolution in a millimeter measurement area. The measurement resolution and area are determined by the optical DOV (depth of view) and FOV (field of view) respectively. Stylus profilometry is able to perform the high-resolution measurements in a larger area but the contact probing mechanism limited its applications if the target surfaces are soft or have cleanliness requirements.

Compared with the measurement systems based on scanning tunneling microscopy (STM) or atomic force microscopy (AFM) [19,20], The proposed system is not an ideal solution for ultra-precision applications. The objective of this study was to develop a measurement system with sub-micron level accuracy in a measurement area up to 100 mm × 100 mm. Targeted applications of the proposed system include the metrological tasks in advanced manufacturing industries, such as semiconductor, consumer electronics, aerospace, etc.

This system is now working as a research platform for the authors' team. The authors' team has published some relevant work, including a confocal measurement system with expanded vertical measurement range [36], fast chromatic confocal sensor development [37], and image grating development for displacement measurement as an alternative of linear encoders [38]. These in-house developed technologies will be integrated into the prototype presented in this paper.

Author Contributions: Conceptualization, F.C.; Methodology, F.C.; Software, J.Z. and H.S.; Validation, F.C., Y.W. and Q.Y.; Formal Analysis, F.C., J.Z. and H.S.; Investigation, J.Z. and H.S.; Resources, Q.Y. and Y.W.; Data Curation, J.Z..; Writing-Original Draft Preparation, F.C.; Writing-Review & Editing, F.C..; Visualization, J.Z.; Supervision,

F.C.; Project Administration, Y.W. and Q.Y.; Funding Acquisition, Q.Y. All authors have read and agreed to the published version of the manuscript.

Funding: This research was funded by the National Natural Science Foundation of China (Grant No. 51505162) and the Science and Technology Program of Fujian, China (Grant No. 2019I0013).

Conflicts of Interest: The authors declare no conflict of interest.

References

1. Davim, J.P. *Surface Integrity in Machining*; Davim, J.P., Ed.; Springer: London, UK, 2010; ISBN 9781848828735.
2. Whitehouse, D. *Surfaces and Their Measurement*; Butterworth-Heinemann: London, UK, 2004; ISBN 978-1-903996-01-0.
3. Tay, C.; Wang, S.; Quan, C.; Shang, H. In situ surface roughness measurement using a laser scattering method. *Opt. Commun.* **2003**, *218*, 1–10. [CrossRef]
4. Bain, L.E.; Collazo, R.; Hsu, S.; Latham, N.P.; Manfra, M.J.; Ivanisevic, A. Surface topography and chemistry shape cellular behavior on wide band-gap semiconductors. *Acta Biomater.* **2014**, *10*, 2455–2462. [CrossRef]
5. Müller, T.; Kumpe, R.; Gerber, H.A.; Schmolke, R.; Passek, F.; Wagner, P. Techniques for analysing nanotopography on polished silicon wafers. *Microelectron. Eng.* **2001**, *56*, 123–127. [CrossRef]
6. Lior, K. The influence of surface topography on the electromechanical characteristics of parallel-plate MEMS capacitors. *J. Micromech. Microeng.* **2005**, *15*, 1068.
7. Townsend, A.; Senin, N.; Blunt, L.; Leach, R.K.; Taylor, J.S. Surface texture metrology for metal additive manufacturing: A review. *Precis. Eng.* **2016**, *46*, 34–47. [CrossRef]
8. Mediratta, R.; Ahluwalia, K.; Yeo, S.H. State-of-the-art on vibratory finishing in the aviation industry: An industrial and academic perspective. *Int. J. Adv. Manuf. Technol.* **2016**, *85*, 415–429. [CrossRef]
9. Yoshizawa, T. (Ed.) *Handbook of Optical Metrology: Principles and Applications*; CRC Press: Boca Raton, FL, USA, 2009; ISBN 978-0-8493-3760-4.
10. Vorburger, T.V.; Rhee, H.G.; Renegar, T.B.; Song, J.F.; Zheng, A. Comparison of optical and stylus methods for measurement of surface texture. *Int. J. Adv. Manuf. Technol.* **2007**, *33*, 110–118. [CrossRef]
11. Sandoz, P.; Tribillon, G.; Gharbi, T.; Devillers, R. Roughness measurement by confocal microscopy for brightness characterization and surface waviness visibility evaluation. *Wear* **1996**, *201*, 186–192. [CrossRef]
12. Paddock, S.W. *Confocal Microscopy*; Humana Press: Totowa, NJ, USA, 1998; Volume 122, ISBN 1-59259-722-X.
13. Viotti, M.R.; Albertazzi, A.; Fantin, A.V.; Pont, A.D. Comparison between a white-light interferometer and a tactile formtester for the measurement of long inner cylindrical surfaces. *Opt. Lasers Eng.* **2008**, *46*, 396–403. [CrossRef]
14. Danzl, R.; Helmli, F.; Scherer, S. Focus variation—A robust technology for high resolution optical 3D surface metrology. *J. Mech. Eng.* **2011**, *57*, 245–256. [CrossRef]
15. Ebtsam, A.; Mohammed, E.; Hazem, E. Image Stitching based on Feature Extraction Techniques: A Survey. *Int. J. Comput. Appl.* **2014**, *99*, 1–8.
16. Henzold, G. *Geometrical Dimensioning and Tolerancing for Design, Manufacturing and Inspection*, 2nd ed.; Elsevier: Oxford, UK, 2006; ISBN 978-0750667388.
17. Xue, Y.; Cheng, T.; Xu, X.; Gao, Z. High-accuracy and real-time 3D positioning, tracking system for medical imaging applications based on 3D digital image correlation. *Opt. Lasers Eng.* **2017**, *88*, 82–90. [CrossRef]
18. Bradley, C. Automated Surface Roughness Measurement. *Int. J. Adv. Manuf. Technol.* **2000**, *16*, 668–674. [CrossRef]
19. Sawano, H.; Gokan, T.; Yoshioka, H.; Shinno, H. A newly developed STM-based coordinate measuring machine. *Precis. Eng.* **2012**, *36*, 538–545. [CrossRef]
20. Wang, S.H.; Tan, S.L.; Xu, G.; Koyama, K. Measurement of deep groove structures using a self-fabricated long tip in a large range metrological atomic force microscope. *Meas. Sci. Technol.* **2011**, *22*, 094013. [CrossRef]
21. Yang, P.; Takamura, T.; Takahashi, S.; Takamasu, K.; Sato, O.; Osawa, S.; Takatsuji, T. Development of high-precision micro-coordinate measuring machine: Multi-probe measurement system for measuring yaw and straightness motion error of *XY* linear stage. *Precis. Eng.* **2011**, *35*, 424–430. [CrossRef]
22. Hsieh, T.; Chen, P.; Jywe, W.; Chen, G.; Wang, M. A Geometric Error Measurement System for Linear Guideway Assembly and Calibration. *Appl. Sci.* **2019**, *9*, 574. [CrossRef]

23. Huang, Q.; Wu, K.; Wang, C.; Li, R.; Fan, K.C.; Fei, Y. Development of an Abbe Error Free Micro Coordinate Measuring Machine. *Appl. Sci.* **2016**, *6*, 97. [CrossRef]
24. Bryan, J.B. The Abbe principle revisit: An updated interpretation. *Precis. Eng.* **1979**, *1*, 129–132. [CrossRef]
25. Okuyama, E.; Akata, H.; Ishikawa, H. Multi-probe method for straightness profile measurement based on least uncertainty propagation (1st report): Two-point method considering cross-axis translational motion and sensor's random error. *Precis. Eng.* **2010**, *34*, 49–54. [CrossRef]
26. Chen, X.; Sun, C.; Fu, L.; Liu, C. A novel reconstruction method for on-machine measurement of parallel profiles with a four-probe scanning system. *Precis. Eng.* **2019**, *59*, 224–233. [CrossRef]
27. Jin, T.; Ji, H.; Hou, W.; Le, Y.; Shen, L. Measurement of straightness without Abbe error using an enhanced differential plane mirror interferometer. *Appl. Opt.* **2017**, *56*, 607–610. [CrossRef] [PubMed]
28. Feng, W.L.; Yao, X.D.; Azamat, A.; Yang, J.G. Straightness error compensation for large CNC gantry type milling centers based on B-spline curves modeling. *Int. J. Mach. Tools Manuf.* **2015**, *88*, 165–174. [CrossRef]
29. Chen, B.; Xu, B.; Yan, L.; Zhang, E.; Liu, Y. Laser straightness interferometer system with rotational error compensation and simultaneous measurement of six degrees of freedom error parameters. *Opt. Express* **2015**, *23*, 9052–9073. [CrossRef] [PubMed]
30. Jäger, G.; Hausotte, T.; Manske, E.; Büchner, H.J.; Mastylo, R.; Dorozhovets, N.; Hofmann, N. Nanomeasuring and nanopositioning engineering. *Measurement* **2010**, *43*, 1099–1105. [CrossRef]
31. Fan, K.-C.; Cheng, F. "The System and Mechatronics of a Pagoda Type Micro-CMM", Special Issue on Precision Micro- and Nano-Metrology for Nanomanufacturing. *Int. J. Nanomanufacturing* **2011**, *8*, 107–112.
32. Wirotrattanaphaphisan, K.; Buajarern, J.; Butdee, S. Uncertainty evaluation for absolute flatness measurement on horizontally aligned fizeau interferometer. *J. Phys. Conf. Series* **2019**, *1183*, 012009. [CrossRef]
33. Vermeulen, M.; Rosielle, P.C.J.N.; Schellekens, P.H.J. Design of a high-precision 3D-coordinate measuring achine. *CIRP Ann. Manuf. Technol.* **1998**, *47*, 447–450. [CrossRef]
34. Fan, K.C.; Fei, Y.T.; Yu, X.F.; Chen, Y.J.; Wang, W.L.; Chen, F.; Liu, Y.S. Development of a low-cost micro-CMM for 3D micro/nano measurements. *Meas. Sci. Tech.* **2006**, *17*, 524–532. [CrossRef]
35. Uicker, J.J.; Pennock, G.R.; Shigley, J.E. *Theory of Machines and Mechanisms*; Oxford University Press: New York, NY, USA, 2003.
36. Fu, S.; Cheng, F.; Tegoeh, T. A Non-Contact Measuring System for In-Situ Surface Characterization Based on Laser Confocal Microscopy. *Sensors* **2018**, *18*, 2657. [CrossRef]
37. Yu, Q.; Zhang, K.; Cui, C.; Zhou, R.; Cheng, F. Calibration of a Chromatic Confocal Microscope for Measuring a Colored Specimen. *IEEE Photonics J.* **2018**, *10*, 6901109. [CrossRef]
38. Fu, S.; Cheng, F.; Tegoeh, T.; Liu, M. Development of an Image Grating Sensor for Position Measurement. *Sensors* **2019**, *19*, 4986. [CrossRef] [PubMed]

Article

Identification Method of Geometric Deviations for Multi-Tasking Machine Tools Considering the Squareness of Translational Axes

Yan Yao [1,*], Keisuke Nishizawa [1], Noriyuki Kato [2], Masaomi Tsutsumi [3] and Keiichi Nakamoto [1]

1 Department of Mechanical Systems Engineering, Tokyo University of Agriculture and Technology, Tokyo 184-8588, Japan; asasbag1@gmail.com (K.N.); nakamoto@cc.tuat.ac.jp (K.N.)
2 Technical Staff, Monodzukuri Center, Osaka Institute of Technology, Osaka 535-8585, Japan; noriyuki.kato@oit.ac.jp
3 Tokyo University of Agriculture and Technology, 2-24-16 Nakacho Koganei, Tokyo 184-8588, Japan; tsutsumi@cc.tuat.ac.jp
* Correspondence: s334967u@st.go.tuat.ac.jp; Tel.: +81-42-388-7372

Received: 18 February 2020; Accepted: 3 March 2020; Published: 6 March 2020

Abstract: Some methods to identify geometric deviations of five-axis machining centers have been proposed until now. However, they are not suitable for multi-tasking machine tools because of the different configuration and the mutual motion of the axes. Therefore, in this paper, an identification method for multi-tasking machine tools with a swivel tool spindle head in a horizontal position is described. Firstly, geometric deviations are illustrated and the mathematical model considering the squareness of translational axes is established according to the simultaneous three-axis control movements. The influences of mounting errors of the measuring instrument on circular trajectories are investigated and the measurements for the B axis in the Cartesian coordinate system and the measurements for the C axis in a cylindrical coordinate system are proposed. Then, based on the simulation results, formulae are derived from the eccentricities of the circular trajectories. It is found that six measurements are required to identify geometric deviations, which should be performed separately in the B axis X-direction, in B axis Y-direction, in C axis axial direction, and three times in C axis radial direction. Finally, a numerical experiment is conducted and identified results successfully match the geometric deviations. Therefore, the proposed method is proved to identify geometric deviations effectively for multi-tasking machine tools.

Keywords: geometric deviations; multi-tasking machine tools; identification method; squareness of translational axes

1. Introduction

In recent years, multi-tasking machine tools have become widely popular in industry because of their growing capabilities in performing complex motions and in reducing machining time and cost. Therefore, many researchers research their machining capabilities and processing technology [1–3]. Based on the basic configuration of a lathe or turning machine, multi-tasking machine tools are developed by equipping with a swivel tool spindle head, which can perform not only a turning operation but also a drilling or milling operation [4]. With the increase of the functionality and the number of simultaneously controlled axes, multi-tasking machine tools are difficult to achieve high machining accuracy and efficiency. Therefore, it is essential to investigate the factors affecting the accuracy of finished products which are machined by multi-tasking machine tools. There are two kinds of factors which could affect the machining accuracy, which are geometric errors of machines operating under no-load or quasi-static conditions and kinematic errors during processing. The geometric errors of machines include the geometric errors of the components and the accuracy of assembly of machine tool executive units. The kinematic errors during processing may be caused by thermal distortion [5], chatter vibrations [6], cutting force or component stiffness [7]. The research in this paper is to establish an effective measurement method for identifying geometric errors of multi-tasking machine tools to improve the motion accuracy under a no-load condition.

Up to now, identification methods of the geometric deviations for five-axis machining centers by using the ball bar, the R-test, a touch-trigger probe or other measuring instruments have been proposed by many researchers. For example, J.R.R. Mayer et al. proposed five tests by the ball bar with a single setup to assess the axis motion errors of a trunnion-type A-axis [8]. W.T. Lei et al. used the ball bar to inspect motion errors of the rotary axes on five-axis machining centers. As a result, the servo mismatch of the rotary axes was successfully detected and the gain mismatch errors could be eliminated by tuning the velocity gains of the position control loops of all servo-controlled linear and rotary axes [9,10]. Tsutsumi et al. proposed an algorithm for identifying particular deviations relating to rotary axes in five-axis machining centers [11]. Tsutsumi et al. applied the ball bar to diagnose the motion accuracy of simultaneous four-axis control movements for identifying the eight deviations inherent to five-axis machining centers [12]. Tsutsumi et al. also investigated the kinematic accuracy of five-axis machining centers with a tilting rotary table by two different settings of the ball bar in simultaneous three axis motion [13]. They corrected the squareness deviations of three translational axes for identifying the geometric deviations inherent to five-axis machining centers with an inclined A-axis [14]. S.H. Yang et al. measured and verified the position-independent geometric errors of a five-axis machining center using the ball bar [15]. In addition, other researchers also used the ball bar to explore the measurement and identification methods for geometric errors of five-axis machining centers with a tilting rotary table [16–18] or in universal spindle head type five-axis machining centers [19]. On the other hand, R-test has been applied recently to investigate the geometric deviations identification method for the five-axis machining centers with a swiveling head [20,21]. Ibaraki et al. identified the kinematic errors of five-axis machining centers by developing a simulator and a set of machining tests [22,23]. The simulator graphically presented the influence of rotary axis geometric errors on the geometry of a finished workpiece measured by R-test [24–28]. Furthermore, other measuring instruments and methods are also developed to investigate the geometric deviations for five-axis machining centers. Ibaraki et al. applied a touch-trigger probe to calibrate the error map of the rotary axes for five-axis machining centers by means of on-the-machine measuring of test pieces [29]. J.R.R. Mayer et al. estimated all axis to axis location errors and some axis component errors of a five-axis horizontal machining center by probing a scale enriched reconfigurable uncalibrated master balls artefact [30]. Beñat Iñigo et al. proposed a new strategy to simulate the calibration and compensation of volumetric error in milling machines of medium and large size and laser trackers are used to optimize volumetric error calibration processes [31]. E.Diaz-Tena et al. studied a radical new 'multitasking' machine model to give a useful outcome regarding the sensitivity of the machine with respect to the feasible assembly errors or errors produced by light misalignments caused by the machine tool continuous use [32].

However, identification methods for multi-tasking machine tools are seldom reported in this field. In fact, due to the special topological structure of the multi-tasking machine tools, the identification method of geometric deviations is different from that of five-axis machining centers, which are introduced in the aforementioned works. Therefore, the accuracy measurement method for multi-tasking machine tools has not clarified and standardized. This is still a critical issue to be solved in field of precision machining.

In this paper, geometric deviations of multi-tasking machine tools are investigated in two different coordinate systems. Simulation results clarify that the measurements for the B axis in Cartesian coordinate system and the measurements for the C axis in cylindrical coordinate system are proposed to eliminate the influence of mounting errors of the ball bar on circular trajectories. Moreover, the formulae and the identification procedures for geometric deviations are concluded in consideration with the squareness of translational axes. The numerical experiment is conducted to verify that the proposed method is effective to identify the deviations accurately for multi-tasking machine tools.

2. Coordinate System and Geometric Deviations of Multi-Tasking Machine Tools

Figure 1 shows the schematic view of a multi-tasking machine tool with a swivel tool spindle head in horizontal position. The structural configuration of this multi-tasking machine tool can be expressed as w-C′bZYXB (C1)-t if it is displayed from the work spindle side (w) to the tool spindle side (t), through the bed (b).

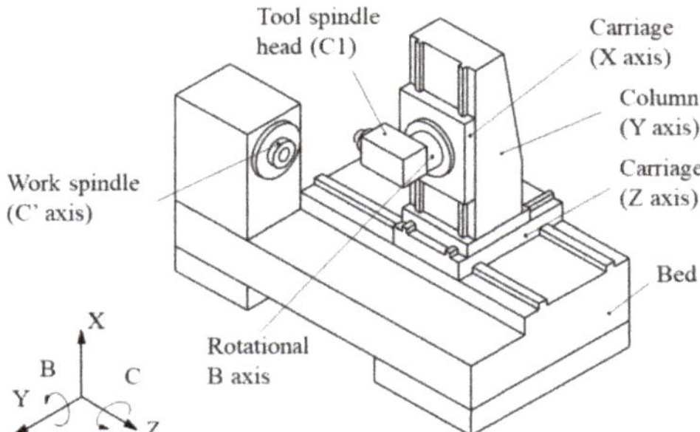

Figure 1. Schematic view of a multi-tasking machine tool with a swivel tool spindle head in horizontal position.

Although there are five controlled axes in the multi-tasking machine tools, the number of geometric deviations is different from that of five-axis machining centers according to the axis configuration and the mutual motion of the axes. Figure 2 shows the geometric deviations and the relation of each axis of the considered machine. Based on the theory of form-shaping system for machine tools [33], the negligible deviations of each axis are deleted from the possible deviations in order from ① to ⑤ shown in Figure 2. Therefore, there are 10 geometric deviations related with two axes of rotation—B and C axes, and three geometric deviations between three translational axes—X, Y, and Z axes, which should be identified to improve the motion accuracy of multi-tasking machine tools. In Figure 2, δx, δy, and δz represent the positional deviations in X-, Y-, and Z-direction, respectively. Similarly, α, β, and γ represent the angular deviations around X-, Y- and Z-axis, respectively. The large suffixes indicate two neighboring axes. For example, δx_{BT} presents the positional deviation in X-direction of the tool spindle axis of rotation with respect to B axis origin. The variable α_{BT} presents the squareness error of B axis with respect to the tool spindle axis of rotation about X axis.

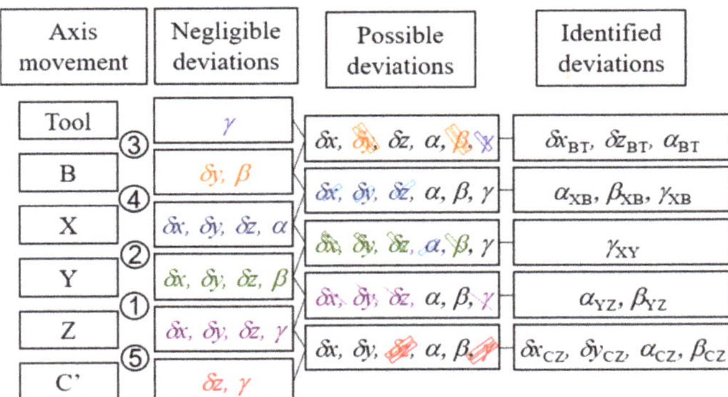

Figure 2. Definition of geometric deviations according to the considered multi-tasking machine tool.

The definitions of thirteen geometric deviations are summarized in Table 1 and illustrated in Figure 3. There are four coordinate systems, which are machine coordinate system (O_M-XYZ), B axis coordinate system (O_B-$X_BY_BZ_B$), C axis coordinate system (O_C-$X_CY_CZ_C$) and tool spindle axis coordinate system (O_T-$X_TY_TZ_T$). The machine coordinate system is defined as the reference system, whose origin is the intersection of the rotational center of B axis and the rotational center of C axis when the geometric deviations are all zero and the command values for each axis are set to its initial values. Since B axis and C axis are not directly connected, the upper surface of the C axis table is set as the origin of the machine coordinate system for the convenience.

Table 1. Symbols of geometric deviations and descriptions.

Symbol	Description
δx_{BT}	X-direction offset of tool spindle axis of rotation with respect to B axis origin
δz_{BT}	Z-direction offset of tool spindle axis of rotation with respect to B axis origin
α_{BT}	Squareness error of B axis with respect to tool spindle axis of rotation about X axis
α_{XB}	Squareness error of B axis of rotation with respect to Z axis motion
β_{XB}	Initial angular position error of B axis of rotation with respect to X (Z) axis motion
γ_{XB}	Squareness error of B axis of rotation with respect to X axis motion
γ_{XY}	Squareness error between X axis motion and Y axis motion
α_{YZ}	Squareness error between Y axis motion and Z axis motion
β_{YZ}	Squareness error between Z axis motion and X axis motion
δx_{CZ}	X-direction offset of C axis origin with respect to machine coordinate origin
δy_{CZ}	Y-direction offset of C axis origin with respect to machine coordinate origin
α_{CZ}	Parallelism error of C axis of rotation with respect to Z axis about X axis
β_{CZ}	Parallelism error of C axis of rotation with respect to Z axis about Y axis

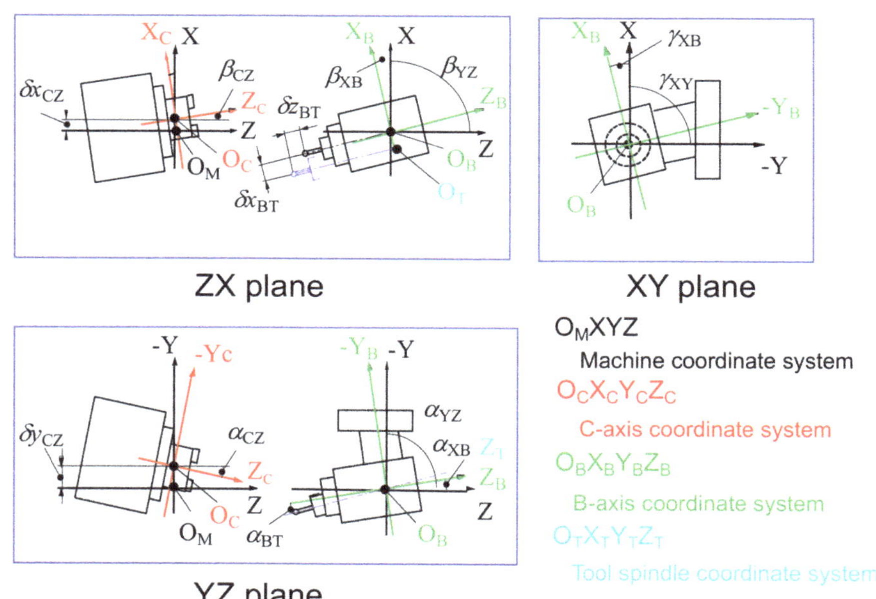

Figure 3. Illustration of coordinate systems and geometric deviations.

3. Simultaneous Three-Axis Control Movements and Mathematical Model

3.1. Simultaneous Three-Axis Control Movements

Simultaneous three-axis control movements which include two linear axes and one rotary axis can be conducted by means of the ball bar both in cylindrical coordinate system and in Cartesian coordinate system. The motions are named according to the sensitive direction of the ball bar as radial direction, tangential direction, axial direction in cylindrical coordinate system, illustrated in Figure 4, and X-, Y- and Z-direction in Cartesian coordinate system, illustrated in Figure 5.

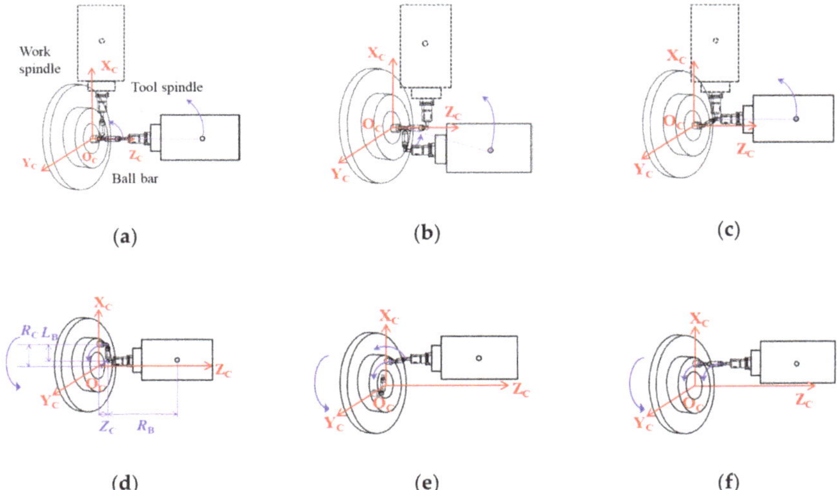

Figure 4. (**a**) B axis radial measurement; (**b**) B axis tangential measurement; (**c**) B axis axial measurement; (**d**) C axis radial measurement; (**e**) C axis tangential measurement; (**f**) C axis axial measurement.

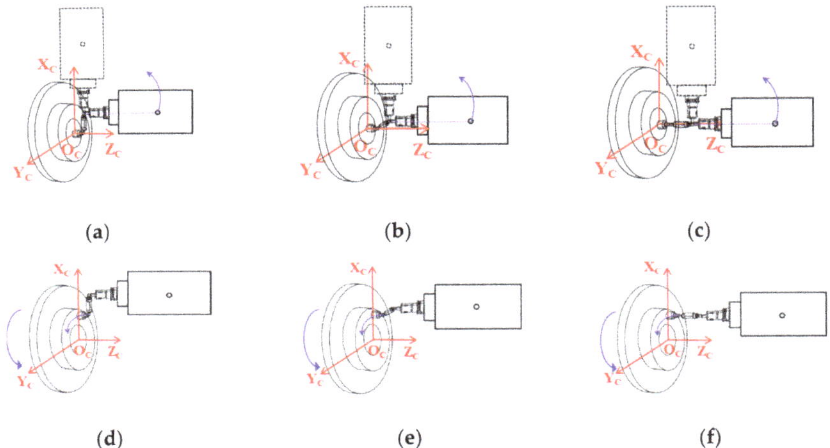

Figure 5. (**a**) B axis X-direction measurement; (**b**) B axis Y-direction measurement; (**c**) B axis Z-direction measurement; (**d**) C axis X-direction measurement; (**e**) C axis Y-direction measurement; (**f**) C axis Z-direction measurement.

3.2. Mathematical Model

Since the angular deviations are generally less than 1°, the small angle approximation ($\sin\theta \approx \theta$ $\cos\theta \approx 1$, if the angle $\theta < 0.244$ radians (14°), the relative error does not exceed 1%) is assumed and second order errors are neglected. Moreover, for small rotation matrices, the order of rotation matrix could be interchanged and it is possible to add and subtract matrices. Therefore, according to this minute rotation approximation theory, homogeneous transformation matrix (HTM) can be simplified in the mathematical model to express the center coordinates of both the tool spindle side ball (T-side ball) and the work spindle side ball (W-side ball) viewed from the machine coordinate system.

3.2.1. Determination of Center Coordinate T of the T-Side Ball Viewed from the Machine Coordinate System

When the distance from the rotational center of B axis to the center of the T-side ball is R_B, the center coordinates T_T of the T-side ball in the tool spindle axis coordinate system are expressed by Equation (1).

$$T_T = \begin{bmatrix} 0 & 0 & -R_B & 1 \end{bmatrix}^T \quad (1)$$

As there are angular deviation α_{BT} and positional deviations δx_{BT}, δz_{BT} between B axis and the tool spindle axis, the homogeneous transformation matrix M_{BT} from the tool spindle axis coordinate system to the B axis coordinate system is expressed by Equation (2) if the above simplified homogeneous coordinate transformation is used.

$$M_{BT} = \begin{bmatrix} 1 & 0 & 0 & \delta x_{BT} \\ 0 & 1 & -\alpha_{BT} & 0 \\ 0 & \alpha_{BT} & 1 & \delta z_{BT} \\ 0 & 0 & 0 & 1 \end{bmatrix} \quad (2)$$

In the same way, the homogeneous transformation matrix M_{XB} between the X axis and B axis is defined as following.

$$M_{XB} = \begin{bmatrix} 1 & -\gamma_{XB} & \beta_{XB} & 0 \\ \gamma_{XB} & 1 & -\alpha_{XB} & 0 \\ -\beta_{XB} & \alpha_{XB} & 1 & 0 \\ 0 & 0 & 0 & 1 \end{bmatrix} \quad (3)$$

The T-side ball circularly moves around the B axis with radius R_B. The circular motion of the T-side ball is expressed by the transformation matrix E_B shown in Equation (4), using the rotation angle φ of the B axis.

$$E_B = \begin{bmatrix} \cos\varnothing & 0 & \sin\varnothing & 0 \\ 0 & 1 & 0 & 0 \\ -\sin\varnothing & 0 & \cos\varnothing & 0 \\ 0 & 0 & 0 & 1 \end{bmatrix} \quad (4)$$

The squareness of the X, Y and Z translational axes can be expressed by the homogeneous transformation matrix M_{YZ} and M_{XY} based on the theory of form-shaping system, as followings.

$$M_{YZ} = \begin{bmatrix} 1 & 0 & \beta_{YZ} & 0 \\ 0 & 1 & -\alpha_{YZ} & 0 \\ -\beta_{YZ} & \alpha_{YZ} & 1 & 0 \\ 0 & 0 & 0 & 1 \end{bmatrix} \quad (5)$$

$$M_{XY} = \begin{bmatrix} 1 & -\gamma_{XY} & 0 & 0 \\ \gamma_{XY} & 1 & 0 & 0 \\ 0 & 0 & 1 & 0 \\ 0 & 0 & 0 & 1 \end{bmatrix} \quad (6)$$

If the translational motion controlled by the command values of X, Y and Z axis is respectively expressed by E_X, E_Y and E_Z, the center coordinate of the T-side ball viewed from the machine coordinate system is denoted as following.

$$T = E_Z M_{YZ} E_Y M_{XY} E_X M_{XB} E_B M_{BT} T_T \quad (7)$$

3.2.2. Determination of Center Coordinate W of the W-Side Ball Viewed from the Machine Coordinate System

If the initial position of the center coordinate of the W-side ball in the C axis coordinate system is W_C (x_{wC}, y_{wC}, z_{wC}), the center coordinates W of the W-side ball viewed from the machine coordinate system are calculated as following.

The homogeneous transformation matrix M_{CZ} between the C axis and Z axis is expressed by Equation (8).

$$M_{CZ} = \begin{bmatrix} 1 & 0 & \beta_{CZ} & \delta x_{CZ} \\ 0 & 1 & -\alpha_{CZ} & \delta y_{CZ} \\ -\beta_{CZ} & \alpha_{CZ} & 1 & 0 \\ 0 & 0 & 0 & 1 \end{bmatrix} \tag{8}$$

The rotation of C axis around Z axis is defined by the transformation matrix E_C when the rotation angle θ of the C axis is used.

$$E_C = \begin{bmatrix} \cos\theta & -\sin\theta & 0 & 0 \\ \sin\theta & \cos\theta & 0 & 0 \\ 0 & 0 & 1 & 0 \\ 0 & 0 & 0 & 1 \end{bmatrix} \tag{9}$$

Therefore, the center coordinate W of the W-side ball viewed from the machine coordinate system is denoted as following.

$$W = M_{CZ} E_C W_C \tag{10}$$

3.2.3. Determination of the Initial Position W_C of Center Coordinate of the W-Side Ball

The W-side ball is positioned based on the center coordinate of the T-side ball, so it is necessary to add the influence of geometric deviations to the initial position of the W-side ball.

The mounting position W'_C of the W-side ball in the machine coordinate system is same with the center coordinate T of the T-side ball in the machine coordinate system. Further, the initial position of the W-side ball is decided only by the command values of the translational axes without the B axis rotation. Therefore, it can be calculated by removing E_B from the Equation (7), expressed as the following Equations (11) and (12) for setup of the B and C axes measurements, respectively.

When W-side ball is set for the B axis measurement,

$$W'_C = \begin{bmatrix} 1 & 0 & 0 & 0 \\ 0 & 1 & 0 & 0 \\ 0 & 0 & 1 & Z_C + R_B \\ 0 & 0 & 0 & 1 \end{bmatrix} \begin{bmatrix} 1 & -\gamma_{XY} & \beta_{YZ} & 0 \\ \gamma_{XY} & 1 & -\alpha_{YZ} & 0 \\ -\beta_{YZ} & \alpha_{YZ} & 1 & 0 \\ 0 & 0 & 0 & 1 \end{bmatrix} \begin{bmatrix} 1 & 0 & 0 & 0 \\ 0 & 1 & 0 & 0 \\ 0 & 0 & 1 & 0 \\ 0 & 0 & 0 & 1 \end{bmatrix} \begin{bmatrix} 1 & -\gamma_{XB} & \beta_{XB} & \delta x_{BT} \\ \gamma_{XB} & 1 & -\alpha_{XB} - \alpha_{BT} & 0 \\ -\beta_{XB} & \alpha_{XB} + \alpha_{BT} & 1 & \delta z_{BT} \\ 0 & 0 & 0 & 1 \end{bmatrix} \begin{bmatrix} 0 \\ 0 \\ -R_B \\ 1 \end{bmatrix} \tag{11}$$

When W-side ball is set for the C axis measurement,

$$W'_C = \begin{bmatrix} 1 & 0 & 0 & 0 \\ 0 & 1 & 0 & 0 \\ 0 & 0 & 1 & Z_C + R_B \\ 0 & 0 & 0 & 1 \end{bmatrix} \begin{bmatrix} 1 & -\gamma_{XY} & \beta_{YZ} & 0 \\ \gamma_{XY} & 1 & -\alpha_{YZ} & 0 \\ -\beta_{YZ} & \alpha_{YZ} & 1 & 0 \\ 0 & 0 & 0 & 1 \end{bmatrix} \begin{bmatrix} 1 & 0 & 0 & R_C \\ 0 & 1 & 0 & 0 \\ 0 & 0 & 1 & 0 \\ 0 & 0 & 0 & 1 \end{bmatrix} \begin{bmatrix} 1 & -\gamma_{XB} & \beta_{XB} & \delta x_{BT} \\ \gamma_{XB} & 1 & -\alpha_{XB} - \alpha_{BT} & 0 \\ -\beta_{XB} & \alpha_{XB} + \alpha_{BT} & 1 & \delta z_{BT} \\ 0 & 0 & 0 & 1 \end{bmatrix} \begin{bmatrix} 0 \\ 0 \\ -R_B \\ 1 \end{bmatrix} \tag{12}$$

The transformation matrix from the machine coordinate system to the C axis coordinate system is performed by the inverse transformation of Equation (8). Moreover, when the W-side ball is mounted, the command value of the C axis is zero, so there is no need to consider the C axis rotation in the Equation (10). Therefore, the conversion from W'_C to W_C is expressed by Equation (13).

$$W_C = \begin{bmatrix} 1 & 0 & -\beta_{CZ} & -\delta x_{CZ} \\ 0 & 1 & \alpha_{CZ} & -\delta y_{CZ} \\ \beta_{CZ} & -\alpha_{CZ} & 1 & 0 \\ 0 & 0 & 0 & 1 \end{bmatrix} W'_C \tag{13}$$

3.2.4. Calculation of the Difference ΔL between Reference Length and Measured Length of Ball Bar

From the above equations, the relative distance L between the T-side ball and the W-side ball can be calculated by the center coordinate T (x_t, y_t, z_t) of the T-side ball and the center coordinate W (x_w, y_w, z_w) of the W-side ball when a command value is given to each axis. The relative distance L is different from the reference length L_B of the ball bar because of the existence of geometric deviations in the multi-tasking machine tools. Therefore, the ball bar length change amount ΔL can be obtained from the relative distance L by subtracting the ball bar reference length L_B as written in Equation (14).

$$\Delta L = \sqrt{(x_t - x_w)^2 + (y_t - y_w)^2 + (z_t - z_w)^2} - L_B \tag{14}$$

4. Simulation

Influence of each deviation on the eccentricity is investigated by using the above mathematical model. The commands separately given to each axis during simulation are shown in Table 2.

Table 2. Commands given to each axis during simulation.

Cylindrical	B Axis			C Axis		
	Radial	Tangential	Axial	Radial	Tangential	Axial
Command X	$(L_B + R_B) \sin \varphi$	$R_{BT} \sin (\varphi + \tan^{-1} L_B/R_B)$ [a]	$R_B \sin \varphi$	$(R_C - L_B) \cos \theta$	$R_{CT} \cos (\theta + \tan^{-1} L_B/R_C)$ [b]	$R_C \cos \theta$
Command Y	0	0	L_B	$(R_C - L_B) \sin \theta$	$R_{CT} \sin (\theta + \tan^{-1} L_B/R_C)$ [b]	$R_C \sin \theta$
Command Z	$Z_C + (L_B + R_B) \cos \varphi$	$Z_C + R_{BT} \cos (\varphi + \tan^{-1} L_B/R_B)$ [a]	$Z_C + R_B \cos \varphi$	$Z_C + R_B$	$Z_C + R_B$	$Z_C + L_B + R_B$
Command B	φ	φ	φ	0	0	0
Command C	0	0	0	θ	θ	θ
Cartesian	X-direction	Y-direction	Z-direction	X-direction	Y-direction	Z-direction
Command X	$L_B + R_B \sin \varphi$	$R_B \sin \varphi$	$R_B \sin \varphi$	$L_B + R_C \cos \theta$	$R_C \cos \theta$	$R_C \cos \theta$
Command Y	0	L_B	0	$R_C \sin \theta$	$L_B + R_C \sin \theta$	$R_C \sin \theta$
Command Z	$Z_C + R_B \cos \varphi$	$Z_C + R_B \cos \varphi$	$L_B + Z_C + R_B \cos \varphi$	$Z_C + R_B$	$Z_C + R_B$	$L_B + Z_C + R_B$
Command B	φ	φ	φ	0	0	0
Command C	0	0	0	θ	θ	θ

[a] $R_{BT} = \sqrt{R_B^2 + L_B^2}$; [b] $R_{CT} = \sqrt{R_C^2 + L_B^2}$.

The simulation of the simultaneous three-axis control movements is conducted at the condition of L_B = 100 mm, Z_C = 100 mm, R_B = 200 mm, R_C = 50 mm. ±0.005 degrees and ±20 μm are given as angular deviations and positional deviations, respectively. Then, simulation results are obtained as shown in Table 3. The dotted circle represents theoretical trajectory when there is no geometric deviation and the red or blue one represents changed trajectory affecting by geometric deviations. The figures show that if only one of the thirteen geometric deviations exist, the red or blue circular trajectory will appear and reflect the effect of the given deviation on the eccentricity. The blank part shows that there is no influence of the geometric deviation on trajectory. For example, for B axis radial direction measurement, when a value of +20 μm is given to δx_{BT} while other twelve geometric deviations are all zero, eccentricity of trajectory occurs in -X axis direction. On the contrary, when a value of −20 μm is given to δx_{BT}, eccentricity of trajectory occurs in +X axis direction.

Table 3. Effect of geometric deviations on the eccentricities of circular trajectories in cylindrical coordinate system and in Cartesian coordinate system.

	δx_{BT}	δz_{BT}	α_{XB}	β_{XB}	γ_{XB}	α_{BT}	δx_{CZ}	δy_{CZ}	α_{CZ}	β_{CZ}	α_{YZ}	β_{YZ}	γ_{XY}
B axis Radial	X Z	X Z		X Z								X Z	
B axis Tangential	X Z	X Z		X Z								X Z	
B axis Axial			X Z		X Z						X Z		
B axis X-direction	X Z	X Z		X Z								X Z	
B axis Y-direction			X Z		X Z						X Z		
B axis Z-direction	X Z	X Z		X Z									
C axis Radial	Y X		Y X	Y X		Y X	Y X	Y X	Y X	Y X	Y X	Y X	Y X
C axis Tangential	Y X		Y X	Y X		Y X	Y X	Y X	Y X	Y X	Y X	Y X	Y X
C axis Axial									Y X	Y X	Y X	Y X	
C axis X-direction	Y X		Y X	Y X		Y X	Y X	Y X	Y X	Y X	Y X	Y X	
C axis Y-direction	Y X		Y X	Y X		Y X	Y X	Y X	Y X	Y X	Y X	Y X	Y X
C axis Z-direction									Y X	Y X	Y X	Y X	

———: +20 μm, +0.005°, ———: −20 μm, −0.005°, Blank: No influence.

The eccentricities occur by the following three reasons.

- The position of trajectory center is changed; for example, the effect of δx_{BT} on eccentricity in case of the B axis radial measurement.
- The size of trajectory radius is changed; for example, the effect of δz_{BT} on eccentricity in case of the B axis radial measurement.
- The shape of trajectory is changed; for example, the effect of β_{YZ} on eccentricity in case of the B axis radial measurement.

4.1. Influence of Mounting Errors of Ball Bar on Circular Trajectories

Considering the influence of mounting errors of the T-side ball, the center offset of the T-side ball (x_T, y_T, z_T) with respect to the tool spindle axis is added to the initial center coordinate T_{T} in the Equation (1), expressed as the following Equation (15). In the same way, the center offset of the W-side ball (x_W, y_W, z_W) with respect to the C axis origin is added to the initial center coordinate W_{C} in the

Equation (13), expressed as the following Equation (16). The simulation is performed in two coordinate systems at the condition that the offset is 20 μm and the obtained results are shown in Table 4.

$$T_{\mathrm{T}} = \begin{bmatrix} x_T \\ y_T \\ z_T \\ 0 \end{bmatrix} + \begin{bmatrix} 0 \\ 0 \\ -R_{\mathrm{B}} \\ 1 \end{bmatrix} \tag{15}$$

$$W_{\mathrm{C}} = \begin{bmatrix} x_W \\ y_W \\ z_W \\ 0 \end{bmatrix} + \begin{bmatrix} 1 & 0 & -\beta_{CZ} & -\delta x_{CZ} \\ 0 & 1 & \alpha_{CZ} & -\delta y_{CZ} \\ \beta_{CZ} & -\alpha_{CZ} & 1 & 0 \\ 0 & 0 & 0 & 1 \end{bmatrix} W'_{\mathrm{C}} \tag{16}$$

Table 4. Influence of mounting errors of ball bar on the eccentricity of circular trajectories.

		x_W	y_W	z_W	x_T	y_T	z_T
Cylindrical coordinate system	B axis Radial	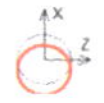 X Z		 X Z	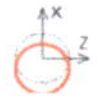 X Z		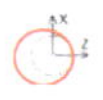 X Z
	B axis Tangential	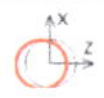 X Z		 X Z	 X Z		 X Z
	B axis Axial		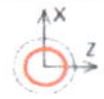 X Z				
Cartesian coordinate system	B axis X-direction	 X Z			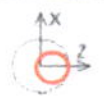 X Z		 X Z
	B axis Y-direction		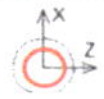 X Z				
	B axis Z-direction			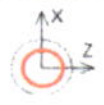 X Z	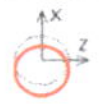 X Z		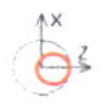 X Z
Cylindrical coordinate system	C axis Radial	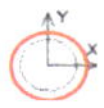 Y X			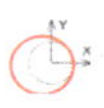 Y X	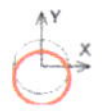 Y X	
	C axis Tangential		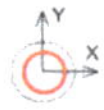 Y X		 Y X	 Y X	
	C axis Axial			 Y X			
Cartesian coordinate system	C axis X-direction	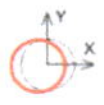 Y X	 Y X		 Y X	 Y X	
	C axis Y-direction	 Y X	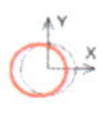 Y X		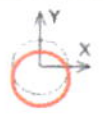 Y X	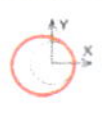 Y X	
	C axis Z-direction			 Y X			

In Table 4, a dotted circle represents theoretical trajectory when there is no mounting errors of ball bar and the red circle represents changed trajectory affecting by one mounting error for each measurement. The blank indicates the trajectory has not changed. It is found that the eccentricity of circular trajectories in these two coordinate systems are strongly affected by the mounting errors of the T-side ball. Therefore, it is crucial to coincide the center of the T-side ball to the tool spindle before conducting measurements. However, the mounting errors of the W-side ball do not affect the eccentricity in Cartesian coordinate system for the B axis measurements, and in cylindrical coordinate system for the C axis measurements. Therefore, to eliminate the influence of mounting errors of the W-side ball on the eccentricities of circular trajectories, the B axis measurements in Cartesian coordinate system and C axis measurements in cylindrical coordinate system are proposed to identify the geometric deviations of multi-tasking machine tools.

4.2. Influence of Squareness of Translational Axes

The influence of squareness deviations of translational axes, α_{YZ}, β_{YZ}, and γ_{XY}, on the eccentricity of circular trajectories are shown at the last three columns in Table 3. It is observed that the squareness deviation γ_{XY} only affect the eccentricity of circular trajectory in case of the C axis Y-direction measurement in Cartesian coordinate system. Therefore, it is indispensable to conduct C axis Y-direction measurement for identifying the squareness deviations γ_{XY}. However, measurement accuracy of the eccentricity for the C axis Y-direction is strongly affected by the mounting errors of W-side ball and T-side ball, shown in Table 4. Besides, γ_{XY} is very small theoretically and it is difficult to identify γ_{XY} correctly by my proposed identification method. Therefore, the identification for γ_{XY} will not be researched in this study.

5. Identification Procedures and Validity

5.1. Mathematical Expressions between Eccentricities and Geometric Deviations

Mathematical expressions between eccentricities of circular trajectories and geometric deviations are summarized in Table 5 according to the simulation results. *ex*, *ey* and *ez* represent the components of eccentricities in X-, Y- and Z-direction, respectively. The subscripts indicate the type of measurements, for example, CA and BY represent measurements of the C axis axial direction and B axis Y- direction, respectively. Among the eccentricity, the positional deviation appears as eccentricity is, while the angular deviation, multiplied by coefficient Z_C, R_B or R_C, appears in eccentricity. The deviation is positive if the direction of the deviation is identical with the direction of the eccentricity. On the contrary, the deviation is negative.

Table 5. Mathematical expressions between eccentricities and geometric deviations.

	ex	ey	ez
B: X-direction	δz_{BT}	—	$\delta x_{BT} - R_B\beta_{XB} - R_B\beta_{YZ}$
B: Y-direction	$-R_B\gamma_{XB}$	—	$R_B\alpha_{XB} + R_B\alpha_{YZ}$
B: Z-direction	$R_B\beta_{XB} - \delta x_{BT}$		δz_{BT}
C: Radial	$-\delta x_{BT} + R_B\beta_{XB} + \delta x_{CZ} + Z_C\beta_{CZ} + R_B\beta_{YZ}$	$-R_B\alpha_{XB} - R_B\alpha_{BT} + \delta y_{CZ} - Z_C\alpha_{CZ} - R_B\alpha_{YZ}$	—
C: Tangential	$R_B\alpha_{XB} + R_B\alpha_{BT} - \delta y_{CZ} + Z_C\alpha_{CZ} + R_B\alpha_{YZ}$	$-\delta x_{BT} + R_B\beta_{XB} + \delta x_{CZ} + Z_C\beta_{CZ} + R_B\beta_{YZ}$	
C: Axial	$R_C\beta_{CZ} - R_C\beta_{YZ}$	$R_C\alpha_{YZ} - R_C\alpha_{CZ}$	—

It is found that the expressions of the C axis tangential measurement have only opposite arithmetic signs comparing to those of the C axis radial measurement, and the expressions of the B axis X-direction measurement is similar with those of the B axis Z-direction measurement. Furthermore, it has been known that a pitch error of the worm gear affects the eccentricity of circular trajectory measured in the tangential direction. Thus, the C axis tangential measurement and B axis Z-direction measurement are not discussed to identify geometric deviations in this study.

In summary, considering the squareness of translational axes, twelve geometric deviations for multi-tasking machine tools can be identified by measuring the eccentricities of circular trajectories of the B axis X-direction, B axis Y-direction, C axis radial direction and C axis axial direction.

5.2. Formulae Considering Squareness of Translational Axes to Calculate Geometric Deviations

To determine 12 variables from eight expressions obtained from above four sensitive direction measurements, measurement conditions should be devised as following.

Firstly, as only one geometric deviation affects the X axis component of eccentricity in the B axis X-direction and Y-direction measurements, two deviations γ_{XB} and δz_{BT} are directly identified based on the corresponding eccentricities measured from the B axis X-direction and Y-direction, expressed as Equations (17) and (18).

$$\gamma_{XB} = -\frac{ex_{BY}}{R_B} \quad (17)$$

$$\delta z_{BT} = ex_{BX} \quad (18)$$

Secondly, when the distance between the center of the W-side ball and the workbench surface of the C axis is changed from Z_C to Z'_C, new eccentricities ex'_{CR} and ey'_{CR} are obtained by measuring the radial direction of the C axis again. Then, another six geometric deviations δx_{CZ}, β_{CZ}, β_{YZ}, α_{CZ}, α_{YZ} and α_{XB} are calculated by the following Equations (19)–(24).

$$\delta x_{CZ} = ex_{CR} + ez_{BX} - \frac{Z_C\left(ex'_{CR} - ex_{CR}\right)}{Z'_C - Z_C} \quad (19)$$

$$\beta_{CZ} = \frac{ex'_{CR} - ex_{CR}}{Z'_C - Z_C} \quad (20)$$

$$\beta_{YZ} = \frac{ex'_{CR} - ex_{CR}}{Z'_C - Z_C} - \frac{ex_{CA}}{R_C} \quad (21)$$

$$\alpha_{CZ} = \frac{ey_{CR} - ey'_{CR}}{Z'_C - Z_C} \quad (22)$$

$$\alpha_{YZ} = \frac{ey_{CR} - ey'_{CR}}{Z'_C - Z_C} + \frac{ey_{CA}}{R_C} \quad (23)$$

$$\alpha_{XB} = \frac{ez_{BY}}{R_B} - \frac{ey_{CA}}{R_C} - \frac{ey_{CR} - ey'_{CR}}{Z'_C - Z_C} \quad (24)$$

Thirdly, when the radius value is changed from R_B to R''_B, new eccentricities ex''_{CR} and ey''_{CR} are obtained from the C axis radial measurement. The remaining deviations δx_{BT}, β_{XB}, α_{BT} and δy_{CZ} are calculated by the following Equations (25)–(28).

$$\delta x_{BT} = ez_{BX} + \frac{R_B\left(ex''_{CR} - ex'_{CR}\right)}{R''_B - R_B}. \quad (25)$$

$$\beta_{XB} = \frac{ex'_{CR} - ex''_{CR}}{R_B - R''_B} - \frac{ex'_{CR} - ex_{CR}}{Z'_C - Z_C} + \frac{ex_{CA}}{R_C}. \quad (26)$$

$$\alpha_{BT} = \frac{ey''_{CR} - ey'_{CR}}{R_B - R''_B} - \frac{ez_{BY}}{R_B} \quad (27)$$

$$\delta y_{CZ} = \frac{R_B\left(ey'_{CR} - ey''_{CR}\right)}{R''_B - R_B} + \frac{Z'_C ey_{CR} - Z_C ey'_{CR}}{Z'_C - Z_C} \quad (28)$$

5.3. Measurement to Identify Geometric Deviations

To identify 12 geometric deviations based on the eccentricity of circular trajectory controlled by simultaneous three-axis motions, the following six measurements by means of the ball bar are indispensable, as shown in Figure 6.

Step 1: Four measurements, which are the B axis X-direction, B axis Y-direction, C axis axial direction and C axis radial direction, are conducted at the condition of L_B, Z_C and R_B.

Step 2: The C axis radial measurement is conducted for the second time at the condition of L_B, Z'_C and R_B.

Step 3: The C axis radial measurement is conducted for the third time at the condition of L_B, Z'_C and R''_B.

As a result, a ball bar measurement should be conducted once in the B axis X-direction, B axis Y-direction, C axis axial direction, and three times in C axis radial direction. After that, including the squareness of translational axes, 12 geometric deviations can be calculated by Equations (17)–(28).

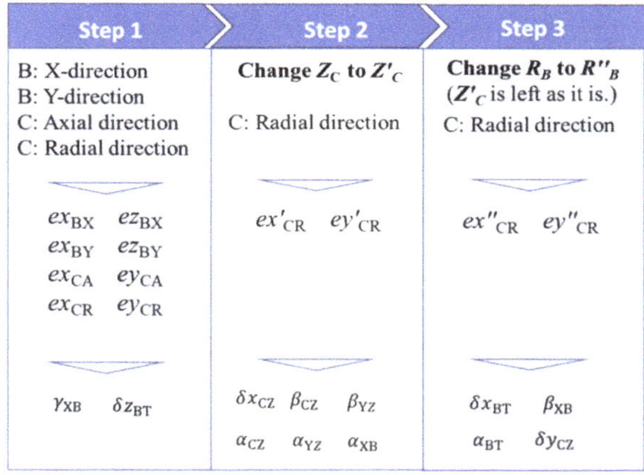

Figure 6. Measurement procedures to identify twelve geometric deviations.

5.4. Validity of the Proposed Identification Method

Numerical experiments are performed to confirm the validity of the formulae for geometric deviations. At first, by using a random number table, the initial values of geometric deviations are selected and substituted into the mathematical model. Six numerical experiments are conducted at $R_C = 50$ mm, $R_B = 360$ mm, $R''_B = 410$ mm, $Z_C = 300$ mm and $Z'_C = 340$ mm. Table 6 summarizes the eccentricities obtained at each measurement. Finally, the eccentricities are substituted into the above formulae to calculate the geometric deviations. The differences between the initial values and the identified values are shown in Table 7.

From the results, it is found that the difference of angular deviations is less than 0.12 arcsecond and the difference of positional deviations is smaller than 0.37 μm. Therefore, 12 geometric deviations of multi-tasking machine tools, including the squareness deviations, α_{YZ} and β_{YZ}, can be identified correctly by using the above formulae.

Table 6. Eccentricities of circular trajectories in numerical experiments.

Direction	B Axis (μm)		C Axis (μm)			
	X [a]	Y [a]	Axial [a]	Radial [a]	Radial [b]	Radial [c]
ex	16.63	−11.24	0.29	25.73	26.89	28.97
ey	-	-	0.05	−19.59	−20	−23.26
ez	−4.02	10.05	-	-	-	-

[a] R_B = 360 mm; Z_C = 300 mm; [b] Z'_C = 340 mm; [c] R''_B = 410 mm.

Table 7. Difference between initial values and identified values.

	Initial Values	Identified Values	Difference
δx_{BT} (μm)	11.00	10.96	−0.04
δz_{BT} (μm)	17.00	16.63	−0.37
α_{BT} (″)	7.63	7.69	0.06
α_{XB} (″)	3.51	3.44	−0.07
β_{XB} (″)	3.92	3.80	−0.12
γ_{XB} (″)	6.39	6.44	0.05
α_{YZ} (″)	2.27	2.32	0.05
β_{YZ} (″)	4.74	4.79	0.05
δx_{CZ} (μm)	13.00	13.01	0.01
δy_{CZ} (μm)	7.00	6.96	−0.04
α_{CZ} (″)	2.06	2.11	0.05
β_{CZ} (″)	5.98	5.98	0.00

6. Conclusions

In this paper, a method to identify geometric deviations which exist in multi-tasking machine tools on the basis of the trajectories of simultaneous three-axis control motions is investigated. The proposed method is applied to a multi-tasking machine tool with a swivel tool spindle head in horizontal position and the identification of the deviations is carried out. From the numerical experiments, the validity of the proposed identification method is clarified. Conclusions are summarized as following.

(1) The identification method for 12 geometric deviations, in which two squareness deviations of translational axes α_{YZ} and β_{YZ} are included, is proposed.
(2) From the simulation results, it is confirmed that in order to eliminate the influence of the mounting errors of the W-side ball on the eccentricities of the circular trajectories, measurements for the B axis should be performed in Cartesian coordinate system and those for the C axis should be performed in cylindrical coordinate system.
(3) Considering the squareness of translational axes, six measurements by means of the ball bar are necessary to identify twelve geometric deviations.
(4) The results of numerical experiments agree well with the given intentional deviations. Therefore, the influence of the analysis accuracy of the formulae on the identification could be considered to be negligible.

It can be concluded that the proposed identification method and the measurement procedure can be sufficiently utilized to identify geometric deviations for multi-tasking machine tools, in which the squareness of translational axes is taken into consideration.

At the next stage, the validity of the proposed method will be verified by an actual measurement of a multi-tasking machine tool. According to the proposed identification method, six measurements by using a ball bar will be conducted firstly and the eccentricities in each translational axis are obtained. Then, 12 geometric deviations of the considered machine tool will be calculated based on the formulae proposed in this paper. Finally, the measurements by a ball bar will be conducted again with compensation of deviations. If the eccentricities would disappear after the compensation, it can be concluded that the proposed method is effective to identify the geometric deviations accurately for multi-tasking machine tools.

Author Contributions: Conceptualization, M.T. and K.N. (Keiichi Nakamoto); methodology, N.K. and M.T.; software, N.K., K.N. (Keisuke Nishizawa) and Y.Y.; validation, M.T. and K.N. (Keisuke Nishizawa); formal analysis, Y.Y. and K.N. (Keisuke Nishizawa); investigation, K.N. (Keisuke Nishizawa) and Y.Y.; resources, K.N. (Keiichi Nakamoto); data curation, N.K., K.N. (Keisuke Nishizawa) and Y.Y.; writing—original draft preparation, Y.Y.; writing—review and editing, K.N. (Keiichi Nakamoto) and M.T.; visualization, Y.Y. and M.T.; supervision, K.N. (Keiichi Nakamoto) and M.T.; project administration, K.N. (Keiichi Nakamoto); funding acquisition, K.N. (Keiichi Nakamoto). All authors have read and agreed to the published version of the manuscript.

Funding: This research received no external funding.

Conflicts of Interest: The authors declare no conflict of interest. The funders had no role in the design of the study; in the collection, analyses, or interpretation of data; in the writing of the manuscript, or in the decision to publish the results.

References

1. Álvarez, Á.; Calleja, A.; Ortega, N.; De Lacalle, L. Five-Axis Milling of Large Spiral Bevel Gears: Toolpath Definition, Finishing, and Shape Errors. *Metals* **2018**, *8*, 353. [CrossRef]
2. Calleja, A.; Bo, P.; González, H.; Bartoň, M.; De Lacalle, L.N.L. Highly accurate 5-axis flank CNC machining with conical tools. *Int. J. Adv. Manuf. Technol.* **2018**, *97*, 1605–1615. [CrossRef]
3. Bo, P.; Bartoň, M.; Plakhotnik, D.; Pottmann, H. Towards efficient 5-axis flank CNC machining of free-form surfaces via fitting envelopes of surfaces of revolution. *Comput. Aided Des.* **2016**, *79*, 1–11. [CrossRef]
4. Calleja, A.; Gonzalez Barrio, H.; Polvorosa Teijeiro, R.; De Lacalle Marcaide, L.N.L. MÁQUINAS MULTITAREA: EVOLUCIÓN, RECURSOS, PROCESOS Y PROGRAMACIÓN. *DYNAII* **2017**, *92*, 637–642. [CrossRef]
5. Gomez-Acedo, E.; Olarra, A.; Orive, J.; De Lacalle, L.N.L. Methodology for the design of a thermal distortion compensation for large machine tools based in state-space representation with Kalman filter. *Int. J. Mach. Tools Manuf.* **2013**, *75*, 100–108. [CrossRef]
6. Pelayo, G.U.; Olvera, D.; De Lacalle, L.N.L.; Beranoagirre, A.; Elías-Zúñiga, A.; De Lacalle, L.N.L. Prediction Methods and Experimental Techniques for Chatter Avoidance in Turning Systems: A Review. *Appl. Sci.* **2019**, *9*, 4718.
7. Olvera, D.; De Lacalle, L.N.L.; Compeán, F.I.; Fz-Valdivielso, A.; Lamikiz, A.; Campa, F.J. Analysis of the tool tip radial stiffness of turn-milling centers. *Int. J. Adv. Manuf. Technol.* **2012**, *60*, 883–891. [CrossRef]
8. Zargarbashi, S.H.H.; Mayer, J.R.R. Assessment of machine tool trunnion axis motion error, using magnetic double ball bar. *Int. J. Mach. Tools Manuf.* **2006**, *46*, 1823–1834. [CrossRef]
9. Lei, W.T.; Paung, I.M.; Yu, C.-C. Total ballbar dynamic tests for five-axis CNC machine tools. *Int. J. Mach. Tools Manuf.* **2009**, *49*, 488–499. [CrossRef]
10. Lei, W.T.; Sung, M.P.; Liu, W.L.; Chuang, Y.C. Double ballbar test for the rotary axes of five-axis CNC machine tools. *Int. J. Mach. Tools Manuf.* **2007**, *47*, 273–285. [CrossRef]
11. Tsutsumi, M.; Saito, A. Identification and compensation of systematic deviations particular to 5-axis machining centers. *Int. J. Mach. Tools Manuf.* **2003**, *43*, 771–780. [CrossRef]
12. Tsutsumi, M.; Saito, A. Identification of angular and positional deviations inherent to 5-axis machining centers with a tilting-rotary table by simultaneous four-axis control movements. *Int. J. Mach. Tools Manuf.* **2004**, *44*, 1333–1342. [CrossRef]
13. Tsutsumi, M.; Tone, S.; Kato, N.; Sato, R. Enhancement of geometric accuracy of five-axis machining centers based on identification and compensation of geometric deviations. *Int. J. Mach. Tools Manuf.* **2013**, *68*, 11–20. [CrossRef]
14. Tsutsumi, M.; Miyama, N.; Tone, S.; Saito, A.; Cui, C.; Dasssanayake, K.M. Correction of Squareness of Translational Axes for Identification of Geometric Deviations Inherent to Five-Axis Machining Centres with a Tilting-Rotary Table. *Trans. Jpn. Soc. Mech. Eng. Ser. C* **2013**, *79*, 759–774. [CrossRef]
15. Lee, K.-I.; Yang, S.-H. Measurement and verification of position-independent geometric errors of a five-axis machine tool using a double ball-bar. *Int. J. Mach. Tools Manuf.* **2013**, *70*, 45–52. [CrossRef]

16. Xiang, S.; Deng, M.; Li, H.; Du, Z.; Yang, J. Volumetric error compensation model for five-axis machine tools considering effects of rotation tool center point. *Int. J. Adv. Manuf. Technol.* **2019**, *102*, 4371–4382. [CrossRef]
17. Yang, J.; Ding, H. A new position independent geometric errors identification model of five-axis serial machine tools based on differential motion matrices. *Int. J. Mach. Tools Manuf.* **2016**, *104*, 68–77. [CrossRef]
18. Chen, J.; Lin, S.; He, B. Geometric error measurement and identification for rotary table of multi-axis machine tool using double ballbar. *Int. J. Mach. Tools Manuf.* **2014**, *77*, 47–55. [CrossRef]
19. Dassanayake, K.M.M.; Tsutsumi, M.; Saito, A. A strategy for identifying static deviations in universal spindle head type multi-axis machining centers. *Int. J. Mach. Tools Manuf.* **2006**, *46*, 1097–1106. [CrossRef]
20. Li, J.; Mei, B.; Shuai, C.; Liu, X.; Liu, D. A volumetric positioning error compensation method for five-axis machine tools. *Int. J. Adv. Manuf. Technol.* **2019**, *103*, 3979–3989. [CrossRef]
21. Li, J.; Xie, F.; Liu, X.-J.; Dong, Z.; Song, Z.; Li, W. A geometric error identification method for the swiveling axes of five-axis machine tools by static R-test. *Int. J. Adv. Manuf. Technol.* **2017**, *89*, 3393–3405. [CrossRef]
22. Ibaraki, S.; Sawada, M.; Matsubara, A.; Matsushita, T. Machining tests to identify kinematic errors on five-axis machine tools. *Precis. Eng.* **2010**, *34*, 387–398. [CrossRef]
23. Uddin, M.S.; Ibaraki, S.; Matsubara, A.; Matsushita, T. Prediction and compensation of machining geometric errors of five-axis machining centers with kinematic errors. *Precis. Eng.* **2009**, *33*, 194–201. [CrossRef]
24. Ibaraki, S.; Yoshida, I. A Five-Axis Machining Error Simulator for Rotary-Axis Geometric Errors Using Commercial Machining Simulation Software. *Int. J. Autom. Technol.* **2017**, *11*, 179–187. [CrossRef]
25. Ibaraki, S.; Nagai, Y.; Otsubo, H.; Sakai, Y.; Morimoto, S.; Miyazaki, Y. R-Test Analysis Software for Error Calibration of Five-Axis Machine Tools—Application to a Five-Axis Machine Tool with Two Rotary Axes on the Tool Side. *Int. J. Autom. Technol.* **2015**, *9*, 387–395. [CrossRef]
26. Hong, C.; Ibaraki, S. Observation of Thermal Influence on Error Motions of Rotary Axes on a Five-Axis Machine Tool by Static R-Test. *Int. J. Autom. Technol.* **2012**, *6*, 196–204. [CrossRef]
27. Hong, C.; Ibaraki, S.; Oyama, C. Graphical presentation of error motions of rotary axes on a five-axis machine tool by static R-test with separating the influence of squareness errors of linear axes. *Int. J. Mach. Tools Manuf.* **2012**, *59*, 24–33. [CrossRef]
28. Ibaraki, S.; Oyama, C.; Otsubo, H. Construction of an error map of rotary axes on a five-axis machining center by static R-test. *Int. J. Mach. Tools Manuf.* **2011**, *51*, 190–200. [CrossRef]
29. Ibaraki, S.; Ota, Y. Error Calibration for Five-Axis Machine Tools by On-the-Machine Measurement Using a Touch-Trigger Probe. *Int. J. Autom. Technol.* **2014**, *8*, 20–27. [CrossRef]
30. Mayer, J.R.R. Five-axis machine tool calibration by probing a scale enriched reconfigurable uncalibrated master balls artefact. *CIRP Ann.* **2012**, *61*, 515–518. [CrossRef]
31. Iñigo, B.; Ibabe, A.; Aguirre, G.; Urreta, H.; De Lacalle, L.N.L. Analysis of Laser Tracker-Based Volumetric Error Mapping Strategies for Large Machine Tools. *Metals* **2019**, *9*, 757. [CrossRef]
32. Díaz-Tena, E.; Ugalde, U.; López de Lacalle, L.N.; De La Iglesia, A.; Calleja, A.; Campa, F.J. Propagation of assembly errors in multitasking machines by the homogenous matrix method. *Int. J. Adv. Manuf. Technol.* **2013**, *68*, 149–164. [CrossRef]
33. Portman, V.; Inasaki, I.; Sakakura, M.; Iwatate, M. Form-Shaping Systems of Machine Tools: Theory and Applications. *CIRP Ann.* **1998**, *47*, 329–332. [CrossRef]

Article

A New Computational Model of Step Gauge Calibration Based on the Synthesis Technology of Multi-Path Laser Interferometers

Guoying Ren [1,2,*], Xinghua Qu [1] and Xiangjun Chen [1]

1 State Key Laboratory of Precision Measurement Technology and Instruments, Tianjin University, Tianjin 300072, China; quxinghua@tju.edu.cn (X.Q.); jun689074@163.com (X.C.)
2 Length Division, National Institute of Metrology, Beijing 100029, China
* Correspondence: rengy@nim.ac.cn

Received: 20 February 2020; Accepted: 17 March 2020; Published: 19 March 2020

Featured Application: Inspired by the principle of closed-loop vector model, this paper proposes a new computational model to realize the high precision calibration of the step gauge based on the commercial CMM platform and laser interference technology. According to this method, only real-time measurements of the attitude and displacement changes in the movement of the step gauge are needed for the accurate length value of the step gauge to be obtained.

Abstract: A step gauge is a commonly used length standard for international comparison, and its calibration accuracy is often used as a sign to measure a country's length Calibration and Measurement Capability (CMC). Based on this, some developed countries and developing countries all over the world have been carrying out the research of precision calibration technology for step gauge. On the basis of summarizing the current situation of step gauge calibration technology in other countries, this paper presents a new computational model of step gauge calibration based on the Synthesis Technology of Multi-Path Laser Interferometers (SMLI) and an auto-collimator, which can synthesize the three laser light paths into the measured centerline of step gauge. It is very important to obtain a good measurement accuracy for the step gauge, conformed to the Abbe principle, no matter where it is installed on the CMM measurement platform. In this paper, the development of the mathematical model, the data collection algorithms, data analysis techniques, and measurement uncertainty budgets are discussed. Finally, the experimental measurement is carried out and the measurement accuracy is verified to be effective. The results show that this method can effectively avoid the influence of Abbe error in length measurement, and significantly enhance the calibration accuracy of the step gauge.

Keywords: metrology; step gauge; length calibration; multi-path laser synthesis technology

---

1. Introduction

As a universal length standard for calibrating instruments, step gauges are widely used in the precision calibration and error compensation of high-precision instruments, such as the Coordinate Measuring Machine (CMM), numerical control machine tool, optical instruments and so on.

Moreover, the International Bureau of Metrology (BIPM) has also listed the step gauge as a key length comparison, making it an important component in measuring the metrological capacity of a national geometric measurement calibration laboratory [1–4]. Many ambitious countries have established their own step gauge calibration devices [5–7].

On the basis of Leitz CMM, the National Metrology Institute of Japan (NMIJ) realize their step gauge calibration by use of the large displacement measurement of a laser interference measuring system, and the micro displacement of the probe of the CMM, thereby obtaining the dimension values of the measured step gauge. The measurement uncertainty of this device is U= (0.06 + 0.22L) μm [8].

Based on the Moore M48 CMM, the National Institute of Standards and Technology (NIST) has developed a step gauge calibration device using a specially designed induction type probe that provides a short-term repeatability of 7 nm [9].

The above two calibration systems are based on the measurement platform of CMM, but the Abbe error, pitch and yaw errors are considered only as the error sources of measurement uncertainty, rather than being directly measured and compensated.

The National Institute of Metrology of Finland (VTT MIKES) has constructed a step gauge calibration system on a vibration isolated stone table with a 1D movable guide rail, rather than CMM, to eliminate mechanical disturbances [10]. The measurement uncertainty of the system is U = Q[0.090; 0.14L] μm. This system also considers the pitch and yaw errors as the error sources of measurement uncertainty, and so does not directly measure them.

At Physikalisch-Technische Bundesanstalt (PTB), the Nanometer Comparator is upgraded with a tactile sensor system to add the capability to measure step gauges. The 1σ repeatability of the device is better than 7 nm [11]. The system adopts the precise displacement measurement method based on the principle of vacuum interference, which has high precision, but is relatively complex.

Based on the analysis of the advantages and disadvantages of the above-mentioned instruments, this paper proposes a new computational model of step gauge calibration based on the Synthesis Technology of Multi-Path Laser Interferometers (SMLI), in which three of four interferometers are used for distance measurement, and the fourth channel is used for the wavelength compensation of the laser [12–14], thereby establishing our own step gauge calibration device on the platform of Leitz CMM. By adding four laser interferometers in the measurement platform, the spatial position relationship model of three out of four lasers is constructed, and the synthetic optical path of these lasers is translated to the measurement line of the measured step gauge, which can effectively eliminate the influence of the Abbe error of the CMM on the measurement result, and compensate the errors of pitch and yaw. Based on the presented calculation model of the closed-loop vector principle, the error caused by pitch and yaw can be corrected to the measured length in real time by measuring attitude change with an auto-collimator to obtain a better calibration accuracy.

This paper will firstly introduce the composition and structure of the developed step gauge measurement system by in detail. Then, the principle and implementation of the measurement method are be addressed systematically. Finally, the experimental results are given and discussed to provide a conclusion.

2. The Mathematical Modeling and Simulation

In order to eliminate the Abbe measurement error, we designed a mathematical model of SMLI which can synthesize the three laser light paths into the one measured center line of the step gauge, as shown in Figure 1. The P_1, P_2 and P_3 are the installation points of three reflector mirrors, and the O is the measurement point of the step gauge. The three reflector mirrors and the step gauge are installed in the same moveable platform. The $\overline{P_1P_1'}$ is the first laser measurement path during the measurement that is marked as l_1u_1 (l_1 is the measurement value of the first interferometer, and u_1 is its unit directional vector), the $\overline{P_2P_2'}$ is the second laser measurement path that is marked as l_2u_2 (l_2 is the measurement value of the second interferometer, and u_2 is its unit directional vector), the $\overline{P_3P_3'}$ is the third laser measurement path that is marked as l_3u_3 (l_3 is the measurement value of the third interferometer, and u_3 is its unit directional vector), and the $\overline{OO'}$ (the O point is the first measurement position and the O′ is the nth measurement position) is the synthesized measurement path of the measured step gauge that is marked as r (r is the only unknown quantity). The β_i is the angle between the direction of OPi and X axis of coordinate system, which is known and definite. The θ_i is the angle between the direction of PiPi′ (i = 1,2,3)and the Z axis of the coordinate system, which represents the change of pitch angle and yaw angle during the measurement and is measured by the auto-collimator.

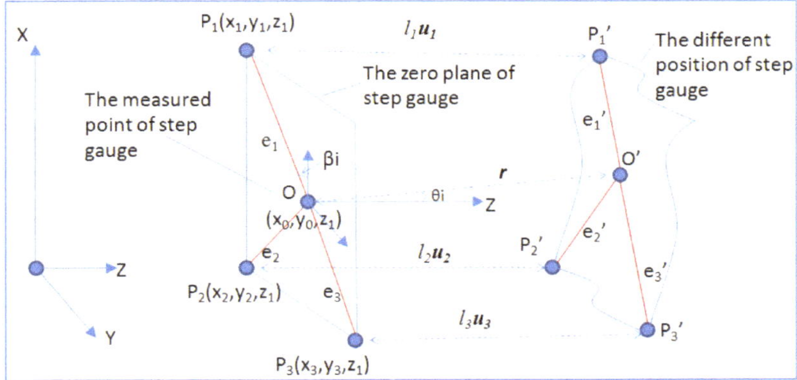

Figure 1. Mathematical model of Synthesis Technology of Multi-Path Laser Interferometers (SMLI).

According to Figure 1, the mathematical model based on the closed-loop vector principle [15–17] is as follows.

$$\vec{r} = \vec{e}_i + l_i\vec{u}_i - \vec{e'}_i,\ i = 1,2,3, \tag{1}$$

where $\vec{e_i} = e_i\begin{pmatrix} \cos\beta_i & \sin\beta_i & 0 \end{pmatrix}^T (i = 1,2,3),\|\overrightarrow{OP_i}\| = \|\vec{e_i}\|$,it is known.

where $\vec{u_i} = (\sin\theta_i\cos\beta_i\ \sin\theta_i\sin\beta_i\ \cos\theta_i)^T$,it is known. Then, l_i is also known.

Changing Equation (1) to Equation (2):

$$(\boldsymbol{r} - \boldsymbol{e}_i - l_i\boldsymbol{u}_i)^T(\boldsymbol{r} - \boldsymbol{e}_i - l_i\boldsymbol{u}_i) = e_i^2. \tag{2}$$

Expanding Equation (2) to Equation (3) is done as follows.

$$\left[(\boldsymbol{r} - \boldsymbol{e}_i)^T - (l_i\boldsymbol{u}_i)^T\right][(\boldsymbol{r} - \boldsymbol{e}_i) - (l_i\boldsymbol{u}_i)] - e_i^2 = 0. \tag{3}$$

Then, Equation (3) is expanded as follows.

$$\Rightarrow (\boldsymbol{r} - \boldsymbol{e}_i)^T(\boldsymbol{r} - \boldsymbol{e}_i) - (\boldsymbol{r} - \boldsymbol{e}_i)^T(l_i\boldsymbol{u}_i) - (l_i\boldsymbol{u}_i)^T(\boldsymbol{r} - \boldsymbol{e}_i) + l_i^2 - e_i^2 = 0,$$
$$\Rightarrow (\boldsymbol{r} - \boldsymbol{e}_i)^T(\boldsymbol{r} - \boldsymbol{e}_i) + l_i^2 - e_i^2 - (\boldsymbol{r} - \boldsymbol{e}_i)^T(l_i\boldsymbol{u}_i) - (l_i\boldsymbol{u}_i)^T(\boldsymbol{r} - \boldsymbol{e}_i) = 0.$$

Suppose the coordinate of point O is (0,0,0), and the coordinate of point O′ is (x,y,z), then:

$$G_i - l_i\boldsymbol{r}^T\boldsymbol{u}_i + l_i\boldsymbol{e}_i^T\boldsymbol{u}_i - l_i\boldsymbol{u}_i^T\boldsymbol{r} + l_i\boldsymbol{u}_i^T\boldsymbol{e}_i = 0, \tag{4}$$

where:

$$G_i = (\boldsymbol{r}-\boldsymbol{e}_i)^T(\boldsymbol{r}-\boldsymbol{e}_i) + l_i^2 - e_i^2 = \boldsymbol{r}^T\boldsymbol{r} - \boldsymbol{r}^T\boldsymbol{e}_i - \boldsymbol{e}_i^T\boldsymbol{r} + \boldsymbol{e}_i^T\boldsymbol{e}_i + l_i^2 - e_i^2$$
$$= x^2+y^2+z^2 - \begin{pmatrix} x & y & z \end{pmatrix}\begin{pmatrix} e_i\cos\beta_i \\ e_i\sin\beta_i \\ 0 \end{pmatrix} - \begin{pmatrix} e_i\cos\beta_i & e_i\sin\beta_i & 0 \end{pmatrix}\begin{pmatrix} x \\ y \\ z \end{pmatrix} + e_i^2 + l_i^2 - e_i^2$$
$$= x^2+y^2+z^2 - 2xe_i\cos\beta_i - 2ye_i\sin\beta_i + l_i^2$$

$$\begin{aligned} l_i\boldsymbol{r}^T\boldsymbol{u}_i &= l_i\begin{pmatrix} x & y & z \end{pmatrix}\begin{pmatrix} \sin\theta_i\cos\beta & \sin\theta_i\sin\beta_i & \cos\theta\ i \end{pmatrix}^T \\ &= l_i(x\sin\theta_i\cos\beta_i + y\sin\theta_i\sin\beta_i + z\cos\theta_i) \end{aligned}$$

$$\begin{aligned} l_i\boldsymbol{e}_i^T\boldsymbol{u}_i &= 1_i\begin{pmatrix} e_i\cos\beta_i & e_i\sin\beta_i & 0 \end{pmatrix}\begin{pmatrix} \sin\theta_i\cos\beta_i & \sin\theta_i\sin\beta_i & \cos\theta_i \end{pmatrix}^T \\ &= l_i(e_i\sin\theta_i\cos^2\beta_i + e_i\sin\theta_i\sin^2\beta_i) \\ &= e_il_i\sin\theta_i \end{aligned}$$

$$\begin{aligned} l_i\boldsymbol{u}_i^T\boldsymbol{r} &= l_i\begin{pmatrix} \sin\theta_i\cos\beta_i & \sin\theta_i\sin\beta_i & \cos\theta_i \end{pmatrix}\begin{pmatrix} x & y & z \end{pmatrix}^T \\ &= l_i(x\sin\theta_i\cos\beta_i + y\sin\theta_i\sin\beta_i + z\cos\theta_i) \end{aligned}$$

$$\begin{aligned} l_i\boldsymbol{u}_i^T\boldsymbol{e}_i &= l_i\begin{pmatrix} \sin\theta_i\cos\beta_i & \sin\theta_i\sin\beta_i & \cos\theta_i \end{pmatrix}\begin{pmatrix} e_i\cos\beta_i & e_i\sin\beta_i & 0 \end{pmatrix}^T \\ &= l_i(e_i\sin\theta_i\cos^2\beta_i + e_i\sin\theta_i\sin^2\beta_i) \\ &= e_il_i\sin\theta_i \end{aligned}$$

Then, Equation (4) can be extended to Equation (5).

$$x^2+y^2+z^2 - 2xe_i\cos\beta_i - 2ye_i\sin\beta_i + l_i^2 - 2l_i(x\cos\beta_i + y\sin\beta_i - e_i)\sin\theta_i - 2l_iz\cos\theta_i = 0 \quad (5)$$

Considering that Equation (5) is the function of (x,y,z), we can get Equation (6):

$$f_i(x,y,z) = x^2+y^2+z^2+l_i^2 - 2(l_i\sin\theta_i + e_i)\cos\beta_i x - 2(l_i\sin\theta_i + e_i)\sin\beta_i y + 2l_ie_i\sin\theta_i - 2l_iz\cos\theta_i = 0. \quad (6)$$

Let t=(x y z), t (k) is the k-th approximation of the solution t, f_i(t) is expanded in vector form by Taylor series to Equation (7).

$$\boldsymbol{f}_i(t) \approx \boldsymbol{f}_i(t^{(k)}) + \nabla\boldsymbol{f}_i(t^{(k)})^T(t - t^{(k)}), \quad (7)$$

that is:

$$\boldsymbol{f}'(t) = \begin{pmatrix} \nabla f_1(t)^T \\ \nabla f_2(t)^T \\ \vdots \\ \nabla f_n(t)^T \end{pmatrix} = \begin{pmatrix} \frac{\partial f_1(t)}{\partial t_1} & \frac{\partial f_1(t)}{\partial t_2} & \cdots & \frac{\partial f_1(t)}{\partial t_n} \\ \frac{\partial f_2(t)}{\partial t_1} & \frac{\partial f_2(t)}{\partial t_2} & \cdots & \frac{\partial f_2(t)}{\partial t_n} \\ \vdots & \vdots & \cdots & \vdots \\ \frac{\partial f_m(t)}{\partial t_1} & \frac{\partial f_m(t)}{\partial t_2} & \cdots & \frac{\partial f_m(t)}{\partial t_n} \end{pmatrix}.$$

By using the Newton iterative method to solve the formula, the solution t can easily be calculated:$t^{(k+1)} = t^{(k)} + \Delta t^{(k)}$.

The procedure of the MATLAB simulation is as follows.

(1) The initial values of coordinate point $t_0 = \begin{pmatrix} x_0 & y_0 & z_0 \end{pmatrix}^T$ are assigned for the Newton iterative method;

(2) Calculate the initial value of the function according to the initial value t_0;

(3) Calculating the initial Hesse matrix $\boldsymbol{f}'_0(t)$;

(4) According to the formula $\boldsymbol{f}'(t^{(k)})\Delta t^{(k)} = -\boldsymbol{f}(t^{(k)})$, the value $\Delta t^{(0)}$ can be calculated;

(5) According to the formula $t^{(k+1)} = t^{(k)} + \Delta t^{(k)}$, the value $t^{(1)}$ can be obtained;

(6) Judge whether the end condition of the iteration is met. If not, return to step (2) to continue iteration.

In order to verify the accuracy of the mathematical model, according to the actual operation state and the actual situation of the measuring device, we design two sets of the angles θ, $\theta \in [-2°, 0°]$

and $\theta \in [0°, 2°]$ respectively, and generate the deflection angle according to the interval of 0.5°. The simulation results are shown in Figure 2.

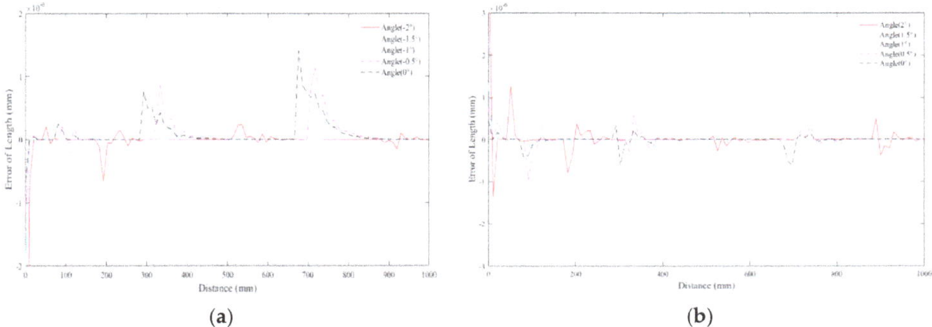

Figure 2. Simulation distribution of measurement length deviation of different distances of step gauge with different attitude angle θ; (a). θ is a negative angle, which means the normal direction of the measured surface and the starting zero plane of the step gauge is opposite; (b) when θ is a positive angle, the normal direction of the measured surface and the starting zero plane of the step gauge are the same.

It can be seen from Figure 2 that the maximum length error of the simulation results is less than 3 nm.

3. Device and Experiment

According to the above-mentioned mathematical model, the step gauge calibration system is developed, as shown in Figure 3.

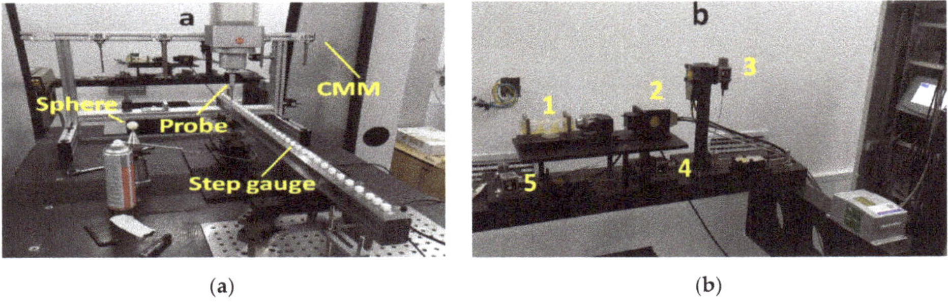

(a) (b)

Figure 3. The step gauge measurement system; (a) actual calibration scenario of step gauge measurement system; (b) the critical optical system. Among this, 1 is the refractive index tracker to compensate the laser wavelength change, 2 is the plane interferometer, 3, 4 and 5 are the three-way length measuring interferometers from a single-frequency laser.

The trigger signal of the probe of CMM is used to determine the position of the measuring surface of the step gauge.

Before the experiment, it is necessary to calibrate the trigger accuracy of CMM probe. In this system, we adopt the trigger probe with a parallel leaf spring structure, and the displacement change of the probe can be measured by a Linear Variable Differential Transformer (LVDT).

The measuring principle of the probe is shown in Figure 4. Given the U_{prim} signal at the input end of LVDT, when the probe moves to the left side, it drives the core to move to the left, which leads to the greater voltage U_{left} in the left than the voltage U_{right} in the right, so the output voltage U_{sec} is displayed as a sinusoidal positive signal; otherwise, it is a sinusoidal negative signal. After shaping, subdividing and amplifying the output voltage U_{sec}, it is converted into digital voltage signal via an 18-bit analog-to-digital conversion(AD) card first, and then converted into the moved displacement value.

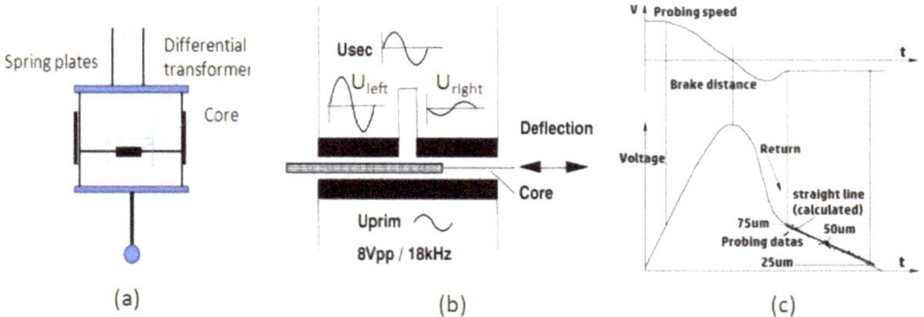

Figure 4. Measuring principle diagram of precision displacement probe for Coordinate Measuring Machine (CMM). (**a**) Mechanical schematic diagram of the CMM probe; (**b**) circuit schematic diagram of the CMM probe; (**c**) schematic diagram of the voltage of the probe signal changing with time during the contact measurement of the probe.

In this system, the trigger sampling interval of the probe is in the linear region of the falling edge of the probe voltage signal shown in Figure 4c.

The key technology of the step gauge measurement system in realizing the high measurement accuracy is to realize the synchronous sampling and signal synthesis of the probe voltage signal of the CMM and the displacement measurement data of the laser interferometer. The probe of CMM produces the voltage signal with linear change in the measurement process, as shown in Figure 5a. Through synchronous dynamic sampling, the corresponding laser interferometer displacement measurement signal is obtained, as shown in Figure 5b. After the two signals are synthesized, the constant coordinates of the measuring points at the time of triggering are obtained, as shown in Figure 5c.

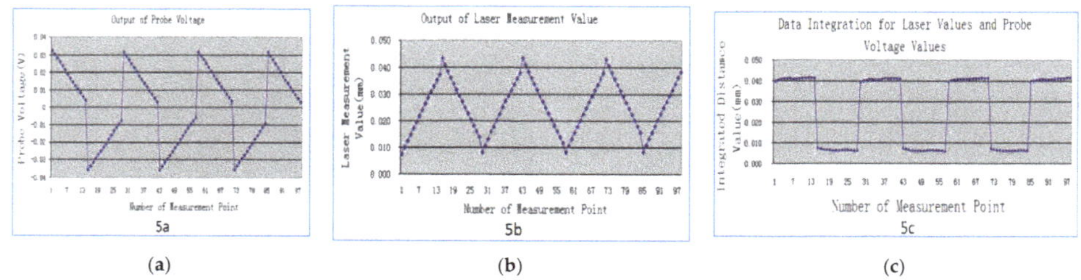

Figure 5. Signal synthetic diagram of probe voltage and displacement of laser interferometer: (**a**) the sampled voltage value via AD card after triggering during measurement; (**b**) the sampled laser value after triggering; (**c**) the combined value of laser value and voltage value obtained by synchronous sampling after triggering.

Before calibration, we first use the known high-level standard ball (its sphericity is 0.03 μm, its diameter is 29.98604 mm) to calibrate the diameter of the CMM probe. The two-side measurement points of the standard ball are measured 10 times, and the measurement results are shown in Table 1 below.

Table 1. The measurement results for calibrate the diameter of the CMM probe.

Measurement Number	The Measured Laser Value(mm)	
	left side	right side
1	713.81023	678.82778
2	713.81020	678.82771
3	713.81024	678.82777
4	713.81029	678.82778
5	713.81028	678.82775
6	713.81026	678.82778
7	713.81031	678.82779
8	713.81024	678.82773
9	713.81023	678.82773
10	713.81030	678.82777
Std. Deviation	0.000034	0.000026
Average Value	713.81026	678.82776

Then, the calibrated diameter of the probe can be calculated by Equation (8).

$$D_{\text{probe}} = L_{\text{leftLaser}} - R_{\text{rightLaser}} - D_{\text{sphereNom.}} = 713.81026 - 678.82776 - 29.98604 = 4.99646 \text{ mm}. \quad (8)$$

In order to verify the correctness of the mathematic model and this device, a step gauge made by the KOBA company of Germany is calibrated. The maximum deviation of measurement results between this device and the Deutscher Kalibrierdienst (DKD) device is less than 0.5 μm, as shown in Figure 6.

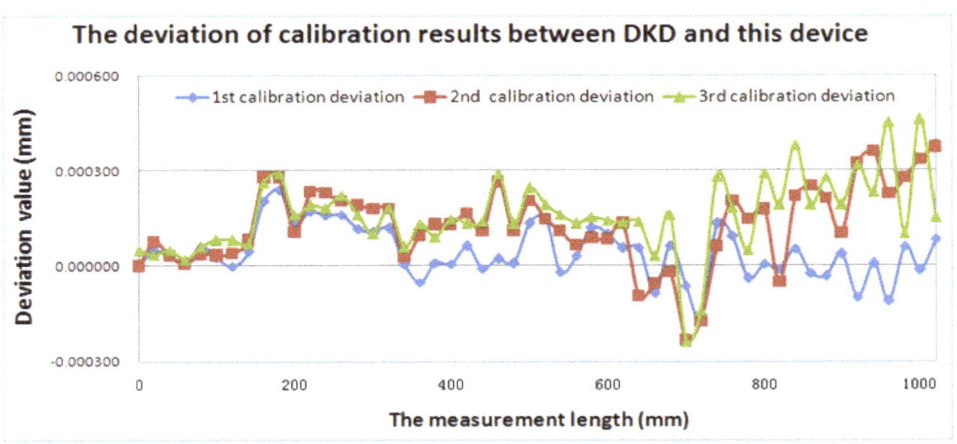

Figure 6. The variation trend of differences between Deutscher Kalibrierdienst (DKD) data and our device with measurement length.

In order to verify the measurement accuracy of this device, a second-grade standard gauge block of 500 mm was repeatedly calibrated 10 times. The deviation between the measurement results of this device and the nominal value of the gauge block is shown in Figure 7.

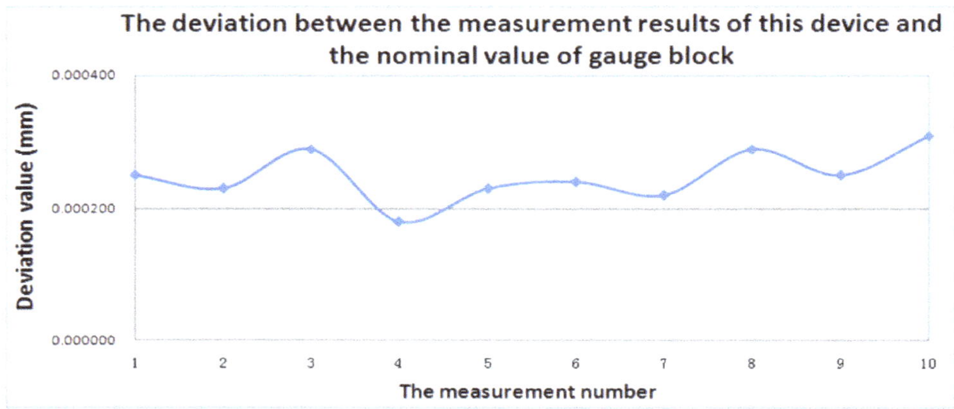

Figure 7. The trend chart of deviation of measurement result of this device from nominal value.

4. Measurement Uncertainty

From this device, we can evaluate the uncertainty in measurement in accordance with the GUM (ISO, "Guide to the Expression of Uncertainty in Measurement," Switzerland, 1993, corrected and reprinted 1995). The calibration value is obtained by Equation (9) [8,18].

$$L_{stepgage} = L_{measured} - L_{measured}\alpha_{stepgage}\Delta t_{stepgage} + \delta_{Astepgage} + \delta_{Alaser} + \delta_{probing} + \delta_{sphere}, \quad (9)$$

where:

$L_{stepgage}$: the measured distance of the step gauge at the reference temperature of 20 °C.

$L_{measured}$: the distance of the step gauge from the calculated measurement value of SMLI.

$\alpha_{stepgage}$: the thermal expansion coefficient of the step gauge.

$\Delta t_{stepgage}$: the difference between the step gauge temperature and the reference temperature of 20 °C.

$\delta_{Astepgage}$: the uncertainty of the step gauge alignment.

δ_{Alaser}: the uncertainty of the laser alignment.

$\delta_{\mathrm{Probing}}$: the reproducibility of the probing.

δ_{sphere}: the uncertainty of the standard sphere calibration.

The uncertainty of the step gauge calibration is calculated by Equation (10).

$$\begin{aligned} u(L_{stepgage})^2 = \; & u(L_{measured})^2 + L^2{}_{Nom.} \times u(\alpha_{stepgage})^2 \times \Delta t^2{}_{stepgage} \\ & + L^2{}_{Nom.} \times \alpha^2{}_{stepgage} \times u(\Delta t_{stepgage})^2 + u(\delta_{Astepgage})^2 + u(\delta_{Alaser})^2 \\ & + u(\delta_{\mathrm{Probing}})^2 + u(\delta_{sphere})^2. \end{aligned} \quad (10)$$

Table 2 shows the uncertainty components in the step gauge calibration.

Table 2. Uncertainty budget in step gauge calibration.

No.	Sources of Nncertainty	Magnitude	Type	Uncertainty
1	Probing repeatability	0.034 μm	A	0.034 μm
2	Calibration of probe	0.030 μm	B	0.030 μm
3	Uncertainty of standard sphere	0.030 μm	B	0.030 μm
4	Temperature measurement	5 mK	B	0.143 Lμm
5	Temperature distribution	10 mK	A	0.286 Lμm
6	Wavelength (frequency)	1.5 MHz	B	0.002 Lμm
7	Wavelength (temperature)	95 mK	A	0.053 Lμm
8	Wavelength (air pressure)	53 Pa	A	0.081 Lμm
9	Wavelength (humidity)	0.64 %	A	0.008 Lμm
10	Wavelength (CO2)	50 ppm	B	0.004 Lμm
11	Thermal expansion coefficient	1.0×10^{-6} /K	B	0.365 Lμm
12	Error of cosine (step gauge)	0.1 mm/ 1020 mm	B	0.005 Lμm
13	Error of cosine (laser)	0.1 mm/1020 mm	B	0.005 Lμm

In Table 2, 1–3 are the independent sources of uncertainty, and 4–13 are dependent on thermal expansion uncertainties and have been converted to the length L.

Then, the uncertainty in measurement of the step gauge calibration using our instrument is as follows.

$$U = (k \times \sqrt{0.054^2 + (0.50L)^2})\mu\text{m};\ L\text{:m};\ k\text{:coverage factor.}$$

5. Conclusions

In order to achieve the high calibration accuracy of the step gauge, we proposed acomputational model of step gauge calibration based on SMLI technology, completed the corresponding mathematical modeling and simulation, and established a set of calibration device. We also carried out experimental verification and uncertainty analysis.

From the measurement results, we find that, although the measurement repeatability of this calibration device is stable over a short distance, it has a divergence trend over a long distance. Although the deviation of the measurement results is in accordance with the uncertainty analysis conclusion claimed by the device, the specific causes of this phenomenon will be further analyzed and studied.

In the future, we will further reduce the measurement uncertainty and improve the calibration accuracy according to the error source. It is hoped that this paper will provide a new measurement method reference for related researchers.

Author Contributions: Conceptualization, G.R. and X.Q.; data curation, X.C.; methodology, G.R.; software, G.R.; writing – original draft, G.R. All authors have read and agreed to the published version of the manuscript.

Funding: This research was funded by the National Key Research and Development Program of China, grant number: 2018YFF0212702.

Acknowledgments: This study was supported by the National Key Research and Development Program of China (Grant No.: 2018YFF0212702). The authors would like to thank the other members of the research team for their contributions to this study.

Conflicts of Interest: The authors declare no conflict of interest.

References

1. Jusko, O.; Brown, N.; Thalmann, R.; Reyes, E.A.; Phillips, S.; Eom, C.I.; Shaoxi, S.; Takatsuji, T.; Prieto, E. CCL-K5: CMM 1D: Step gauge and ball bars: Final report. *Metrologia* **2006**, *43*, 04006. [CrossRef]
2. Eom, T.; Sitian, G.; Wong, S.Y.; Chaudhary, K.P.; Drijarkara, A.P.; Howick, E.; Tonmeanwai, A.; Pan, S.-P.; Takatsuji, T.; Cox, P.; et al. Final report on APMP Regional Comparison APMP.L-K5: Calibration of step gauge. *Metrologia* **2012**, *49*, 04007. [CrossRef]
3. Prieto, E.; Brown, N.; Lassila, A.; Lewis, A.; Matus, M.; Vailleau, G.; De Chiffre, L.; Kotte, G.W.J.L.; Frennberg, M.; McQuoid, H.; et al. Final report on inter-RMO Key Comparison EUROMET.L-K5.2004: Calibration of a step gauge. *Metrologia* **2012**, *49*, 04008. [CrossRef]
4. Abe, M.; Drijarkara, A.P.; Babu, V. Bi-lateral comparison APMP.L-K5.2006.1—calibration of step gauge. *Metrologia* **2017**, *54*, 4006. [CrossRef]
5. Hemming, B.; Esala, V.-P.; Laukkanen, P.; Rantanen, A.; Viitala, R.; Widmaier, T.; Kuosmanen, P.; Lassila, A. Interferometric step gauge for CMM verification. *Meas. Sci. Technol.* **2018**, *29*, 074012. [CrossRef]
6. A Kruger, O. High-accuracy interferometric measurements of flatness and parallelism of a step gauge. *Metrologia* **2001**, *38*, 237–240. [CrossRef]
7. Osawa, S.; Takatsuji, T.; Noguchi, H.; Kurosawa, T. Development of a ball step-gauge and an interferometric stepper used for ball-plate calibration. *Precis. Eng.* **2002**, *26*, 214–221. [CrossRef]
8. Osawa, S.; Takatsuji, T.; Kurosawa, T. STEP-gauge calibration using an interferometric coordinate measuring machine and the uncertainty. In Proceedings of the XVII imeko World Congress, Dubrovnik, Croatia, 22–27 June 2003.
9. Stoup, J.; Faust, B. Measuring Step Gauges Using the NIST M48 CMM. *NCSLI Meas.* **2011**, *6*, 66–73. [CrossRef]
10. Byman, V.; Jaakkola, T.; Palosuo, I.; Lassila, A. High accuracy step gauge interferometer. *Meas. Sci. Technol.* **2018**, *29*, 054003. [CrossRef]
11. Weichert, C.; Bütefisch, S.; Köning, R.; Flügge, J. Integration of a step gauge measurement capability at the PTB Nanometer Comparator—Concept and preliminary tests. In Proceedings of the Conference "MacroScale 2017—Recent Developments in Traceable Dimensional Measurements", VTT MIKES Espoo, Finnland, 17–19 October 2017. [CrossRef]
12. Ren, G.; Qu, X.; Ding, S. A Real-Time Measurement Method of Air Refractive Index Based on Special Material Etalon. *Appl. Sci.* **2018**, *8*, 2325. [CrossRef]
13. Guo-Ying, R.; Wei-Nong, W.; Zi-Jun, W. Uncertainty Analysis for Measurement of Step Gauges Based on Coordinate Measuring Machine. *Acta Metrol. Sin.* **2008**, *29* (Suppl. S1), 151–153.
14. Bönsch, G.; Potulski, E. Measurement of the refractive index of air and comparison with modified Edlen's formula. *Metrologia* **1998**, *35*, 133–139. [CrossRef]
15. Garcia-Murillo, M.A.; Sanchez-Alonso, R.E.; Gallardo-Alvarado, J. Kinematics and Dynamics of a Translational Parallel Robot Based on Planar Mechanisms. *Machines* **2016**, *4*, 22. [CrossRef]
16. Zhang, X.; Dong, Y.; Zhang, L. kinematics modeling and simulation of a five-bar Cabot by using closed loop vector method. *Control Theory Appl.* **2006**, *25*. [CrossRef]
17. Buschhaus, A.; Blank, A.; Franke, J. Vector based closed-loop control methodology for industrial robots. In Proceedings of the 2015 International Conference on Advanced Robotics (ICAR), Istanbul, Turkey, 27–31 July 2015; pp. 452–458.
18. BIPM; IEC; IFCC; ISO; IUPAC; IUPAP; OIML. Evaluation of measurement data — Guide to the expression of uncertainty in measurement. *Int. Organ. Stand. Geneva ISBN* **2008**, *50*, 134.

Article

A Method of On-Site Describing the Positional Relation between Two Horizontal Parallel Surfaces and Two Vertical Parallel Surfaces

Zechen Lu [1], Bao Zhang [2], Zhenjun Li [1] and Chunyu Zhao [1,*]

1 School of Mechanical Engineering and Automation, Northeastern University, Shenyang 110819, China; 1610091@stu.neu.edu.cn (Z.L.); 1510103@stu.neu.edu.cn (Z.L.)

2 School of Mechanical and Electrical Engineering, Xinyu University, Xinyu 338004, China; 1710101@stu.neu.edu.cn

* Correspondence: chyzhao@mail.neu.edu.cn

Received: 19 February 2020; Accepted: 13 March 2020; Published: 21 March 2020

Abstract: The position error of two parallel surfaces is generally constrained by parallelism. However, as a range of change, it cannot represent the positional relation between two parallel surfaces. Large-scale equipment such as machine tools are complex and difficult to move. It is also an engineering problem to perform field measurements on it. To this end, a method of describing the positional relation between two horizontal parallel surfaces and two vertical parallel surfaces on-site is proposed in this paper, which is a novel kind of position error, enriching the form of parallel surface position error, and solves the inconvenience problem of large equipment position error measurement. The measurement mechanisms are designed, and the measurement principle is given. Firstly, the combined projection waveform of the measured surface can be obtained by the geometric relationship between the measurement mechanism and the measured surface. Secondly, an algorithm is studied to process the obtained waveform, and the combined projection curve of the measured surface is acquired. Then, under the condition of considering the shape contours of the two surfaces, an algorithm is developed to acquire the calculated shape contour of the measured surface. According to the difference between the calculated surface shape contour and the known shape contour of the measured surface, the positional relation of the two surfaces can be determined. Meanwhile, the mathematical models of algorithms are established, and the measurement experiments are carried out, and the algorithms are verified by the mutual measurement method of the two surfaces. The results show that this method can accurately obtain the positional relation of two horizontal parallel surfaces and two vertical parallel surfaces.

Keywords: measurement mechanism; machine tool; surface shape contour; on-site measurement; positional relation

1. Introduction

The CNC machine tool is an important piece of equipment in the traditional manufacturing industry. Two positioning surfaces of the machine tool serve as the datum surface for the mounting rails, and their position error plays a vital role in the accuracy of the machine tool. In the modern manufacturing industry, the position error of two parallel surfaces is generally constrained by parallelism, and parallelism is orientation error, which is the total allowable variation of the measured element in the direction relative to the datum. Therefore, the orientation error has the function of controlling the direction, that is, controlling the direction of the measured elements relative to the datum elements. However, the calculation method of parallelism is very mature. Firstly, we need to establish a datum direction on the datum plane. The method is to fit a straight line (as shown in

Figure 1a) or a surface (as shown in Figure 1b) as a datum direction. The obtained parallelism is a number. The key to parallelism evaluation is to determine the datum direction.

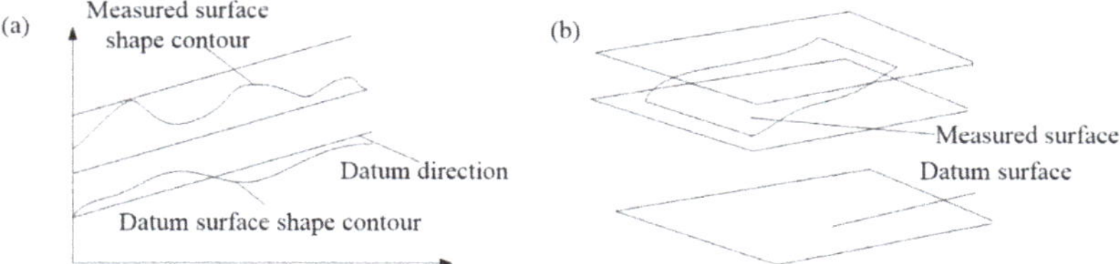

Figure 1. Evaluation method of parallelism: (**a**). Line-to-line (**b**). Surface to surface.

With the development of the manufacturing industry, the requirements for machine tools are increasing. In order to improve the performance of the machine tool, we study the impact of the positioning accuracy of the machine tool on the accuracy of the machined parts and maintain the accuracy of the machine tool itself. The more we know about position errors, the better for us. Therefore, this paper proposes a method of describing the positional relation between two parallel surfaces on-site, which is a novel kind of position error. It has no datum direction, enriching the form of parallel surface position error. The significance of obtaining the positional relation curve of the two positioning surfaces on the machine tool is that, on one hand, the positional relation curve can guide the subsequent modification, grinding, bed fitting and quantitative analysis of the machined parts. On the other hand, the positional relation curve can also be used to evaluate the torsional state of the surface.

In the development of shape error detection methods, there are several novel methods are used to obtain straightness [1–4], flatness [5–10], roundness [11,12]. However, because the method of evaluating the parallelism in position error is relatively simple, there is relatively little research on this aspect. Hsieh et al. [13] proposed a method for measuring the parallelism of static and dynamic linear guideways. Hwang et al. [14] proposed a three-probe system, which can be used to acquire the parallelism and straightness of a pair of guideways simultaneously. The parallelism measurements are based on an expanded reversal method and the straightness measurements are based on a sequential two-point method. Bhattacharya et al. [15] described an interferometric method for measuring the parallelism of an end face of a transparent material. Lee et al. [16] developed a double ball-bar method to measure parallelism errors of the spindle axis of machine tools. Maurizio et al. [17] discussed two methods for measuring the parallelism error of gauge blocks. It can be known from searching literatures that in the study of shape and position errors, there is very little description of the positional relation between two parallel surfaces. Moreover, the above methods need to be carried out using specific equipment and in a specific environment, and the layout of the measurement mechanism is very complex and needs enough space, but the operating space of machine tools is often limited. Especially for two vertical surfaces, it is more difficult to perform field measurement. However, Jywe et al. [18] performed an online measurement of the machine tool's two-axis straightness error. Hsien et al. [19] designed an online measurement system for measuring the linearity and parallelism of linear guides. Unfortunately, these two measurement systems cannot measure the positional relation of two parallel surfaces.

The parallelism measurement in industry commonly uses specific tools, such as micrometer, deflection instrument, coordinate measuring machines etc. In fact, micrometer has great randomness, which is greatly influenced by the operator's proficiency and experience. In the detection process, the operator can only judge the quality by observing the change range of the micrometer, and the detection data cannot be directly saved to the computer. The coordinate measuring machine can record the test data with high precision, but it can't measure the machine tool and large equipment on site. Therefore, from the above introduction, for machine tools with limited operating space, developing a simple and easy-to-operate measuring instrument will be an effective method to on-site obtain the positional relation of two positioning surfaces.

In this paper, a method of describing the positional relation of the two horizontal parallel surfaces and two vertical parallel surfaces on-site is proposed. This method can separate the shape contours of the two surfaces and obtain the real positional relation curve of the two surfaces. At the same time, the laser displacement sensor, which can record the detection data to the computer, is selected as the detection element, a continuous measuring mechanism suitable for manual operation is developed, and the corresponding algorithms are studied. Finally, the calculation results are drawn into a curve, which makes the description of the positional relation more intuitive. This method can also be used to measure two working surfaces and two guideways of large equipment.

The rest of this paper is organized as follows. In Section 2 the principle of the measurement mechanism is depicted. In Section 3, the method for describing the positional relation between two surfaces is introduced. In Section 4, mathematical models of the algorithms are established. In Section 5, the measurement experiment is carried out and the results are analyzed. Finally, a conclusion is given in Section 6.

2. Principle of the Measurement Mechanism

2.1. Measuring Mechanism for Measuring Two Horizontal Parallel Surfaces

Figure 2 shows the measurement mechanism for measuring the positional relation between two horizontal parallel surfaces. The laser displacement sensor (triangulation) ILD-2300 is selected as the detection element and the technical parameters of the sensor are shown in Table 1.

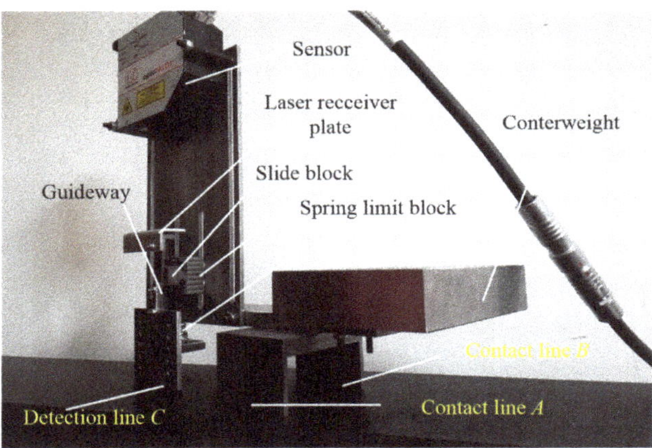

Figure 2. Measurement mechanism for measuring two horizontal parallel surfaces.

Table 1. Laser displacement sensor parameters.

Model	Parameter
Frequency	20 kHz
Resolution	0.3 μm
Absolute error	4 μm (≤±0.02%) full scale
Starting measurement range	50 mm
Light source	semiconductor laser (1 mW, 670 nm red)
Impact resistance	15 g/6 ms
Operating temperature	0–50 °C
Power requirements	24 VDC (11 . . . 30 VDC), Max 150 mA
Security Level	Class 2 according to DIN EN 60825-1: 2001-11
Spot diameter	SMR140 × 200 μm, MMR46 × 45 μm, EMR140 × 200 μm

The measurement mechanism consists of slider, guideway, spring limit block, laser receiving plate, counterweight block, spring, contact line *A* and *B*, and detection line *C*. The length of contact line *A*, *B* and detection line *C* are 50 mm, and the distance between contact line *A* and detection line *C* is 60 mm. The contact line *A*, *B* and detection line *C* are ground after processing, and the processing accuracy grade is IT3.The measurement mechanism is in equilibrium with the counterweight, and contact lines *A* and *B* are fixed contact lines. Detection line *C* is connected with the measurement mechanism through micro-guideway and springs, the slider block is fixed on the measurement mechanism, the laser receiving plate is integrated with the micro-guideway and detection line *C*. In the process of measurement, the measurement mechanism is manually driven to move in a fixed direction, under the function of the spring, the detection line *C* fluctuates up and down with the change of shape contour the measured surface, the fluctuation is transmitted to the laser receiving plate through the micro-guideway. At this time, the laser displacement sensor detects the fluctuation and the computer records the measurements.

It is known that the surface contour consists of roughness contour, waviness contour and surface shape contour. The surface shape contour is macroscopic. In this paper, the three-dimensional shape of the measured surface is projected on a plane perpendicular to it, and the macroscopic contour curve obtained is the surface shape contour, as shown in Figure 3. Therefore, the motion trajectories of the contact lines *A*, *B*, and detection lines *C* are surface shape contour.

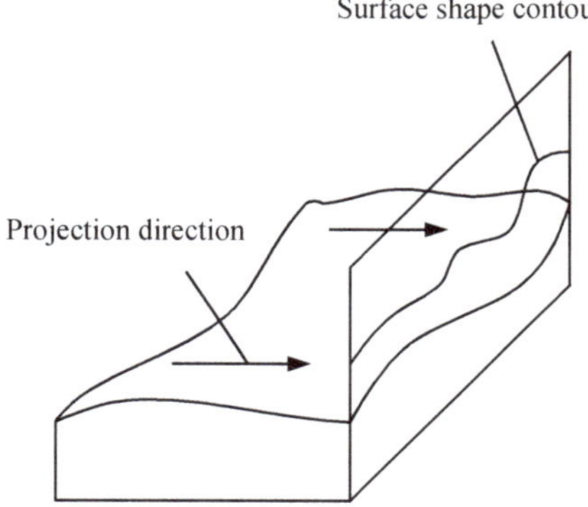

Figure 3. Detection path.

As shown in Figure 4a, according to Table 1, the initial measuring range is 50 mm. When the detection line *C* and the contact line *A* are collinear, the detection data should be a fixed value, defining it as the initial value X_0. When considering the shape contours of the two surfaces, the position of the detection line C may change, as shown in Figure 4b. The difference between the measurements and the initial value is the displacement of the detection line C relative to the contact line *A*.

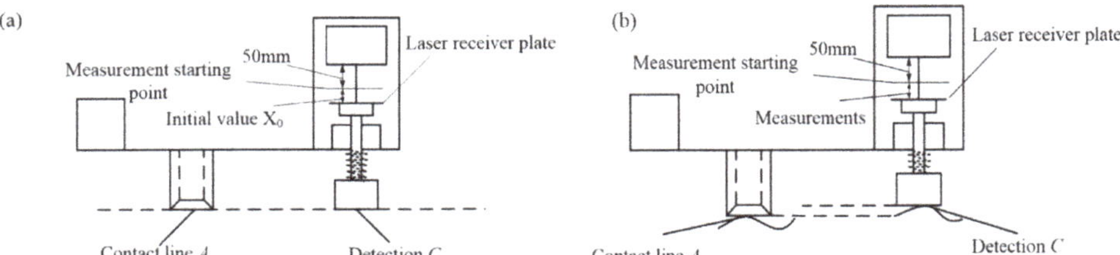

Figure 4. The positional relation between contact line *A* and detection line C: (**a**). Contact line *A* and detection line C are collinear; (**b**). Contact line *A* and detection line C are not collinear.

The advantage of this mechanism is that it easy to operate, and suitable for field measurement. It can continuously obtain the stored data which change with the shape contour of the measured surface, instead of some discrete points on the same straight line along the measuring direction. Meanwhile, it can separate the shape contours of the two surfaces, which cannot be achieved by ordinary testing equipment.

2.2. Measuring Mechanism for Measuring Two Vertical Parallel Surfaces

Figure 5 shows the measuring mechanism for measuring the positional relation of two vertical parallel surfaces. Contact lines *A* and *B* are fixed, and the detection line *C* is movable. The length of contact lines *A* and *B* are 70 mm, and detection line *C* is 50 mm. The contact line *A*, *B*, and detection line *C* are ground after processing, and the processing accuracy grade is IT3.

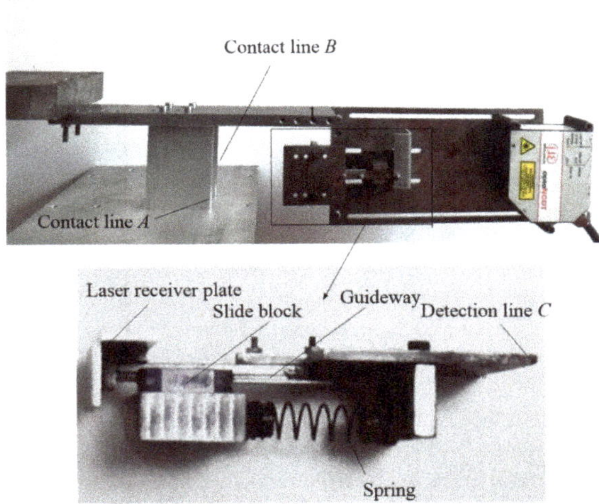

Figure 5. Measuring mechanism for measuring two vertical parallel surfaces.

During the detection process, the fixed contact lines *A* and *B* move in one surface, the variable contact line *C* moves in another detection surface, and the contact lines *A*, *B* and the detection line *C* form a caliper. Under the clamping action of the caliper, the spring generates tension and compression with the shape contour of the measured surface. At the same time, the fluctuate is transmitted to the laser receiving plate along the guide rail, and the laser displacement sensor detects the weak change of the laser receiving plate and records the change through the computer.

As shown in Figure 6a, according to Table 1, start measuring range is 50 mm. When the surface where the contact line A and the detection line C are located are absolutely parallel and the surfaces are absolutely smooth, the detection data is a fixed value and is defined as the initial value X_1. When the shape contours of the two surfaces are considered, the position of the detection line C may change, as shown in Figure 6b. At this time, fluctuating measurements will be generated, and the difference between the measurements and the initial value is the displacement of the detection line C relative to the contact line A.

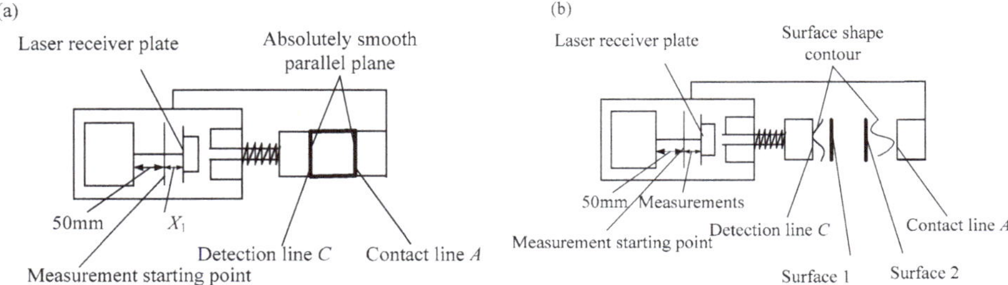

Figure 6. The positional relation between contact line A and detection line C: (**a**) The distance between the contact line A and the detection line C is unchanged (**b**) Contact line A and detection line C are located in two surfaces with shape contours.

3. The Method for Describing the Positional Relation of Two Parallel Surfaces

3.1. The Description Method of Positional Relation between Two Horizontal Parallel Surfaces

On the basis of the measurement mechanism in Section 2.1, when A, B, and C are in an absolute smooth surface, the spring is in the initial compression state, this moment, the data recorded by laser displacement sensor is X_0, and X_0 is defined as the initial value.

Figure 7 shows the measurement model of two horizontal parallel surfaces. With PA as the datum surface, PB as the measured surface. L_s is defined as the starting position, L_e is defined as the terminal position. By coinciding contact line A with L_s, the measurement mechanism is driven to move in the direction indicated in Figure 7, and the measurement mechanism is clinging to the auxiliary surface to ensure that there will be no sliding during the measurement process. When contact line A reaches to L_e, the measurement is completed, and the data obtained in this process represent effective measurements.

Figure 8a shows the geometric model of PA and PB cross section direction. The shape contour of PB is determined by the shape contour of PA, the relative height difference of PA surface shape contour relative to PB surface shape contour and the positional relation of the two surfaces. According to the mathematical relationship between the effective measurements and X_0, the extraction algorithm for combined projection waveform is developed and the combined projection waveform can be obtained. Next, since the measurements of the laser displacement sensor change with time, a time-space conversion algorithm is developed. This algorithm is used to map the time series waveform to the coordinate position of PB surface, which is called the spatial sequence waveform. Then the trend curve of this waveform is found by the method of neural network fitting (a fitting method in the toolbox of MATLAB), which is named as the combined projection curve. This curve is the height difference of PB surface shape contour relative to PA surface shape contour. Due to the shape contour of each surface have been measured before, so they are known quantities. According to the combined projection curve of the PB and the surface shape contour of PA, the separation algorithm for the combined projection curve is studied. The shape contour of PB surface can be obtained by this algorithm, which is called the calculated surface shape contour. When the two surfaces PA and PB are coplanar, as shown in

Figure 8b, the positional relation between the two surfaces is zero, and the calculated surface shape contour is the same as the known shape contour of *PB* surface. If the two results are different, it proves there exists positional relation between two surfaces. Therefore, a description algorithm for positional relation curve between two surfaces is studied. According to the difference between the calculated surface shape contour and the known shape contour of the measured surface, the positional relation between two surfaces can be determined.

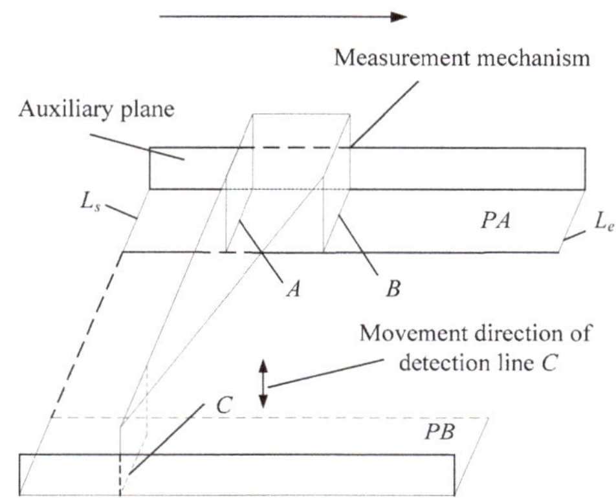

Figure 7. Detection model of two horizontal parallel surfaces.

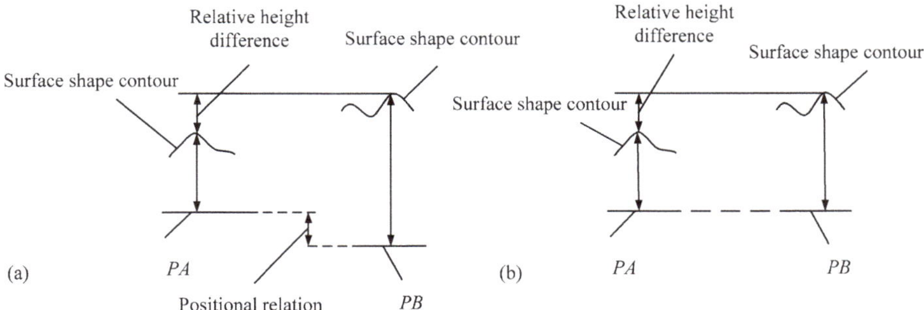

Figure 8. Geometric model of *PA* and *PB* cross section direction: (**a**) Positional relation exists between *PA* and *PB* surfaces. (**b**)Positional relation does not exists between *PA* and *PB* surfaces.

3.2. The Description Method of Positional Relation between Two Vertical Parallel Surfaces

Figure 9 shows the detection model of two vertical parallel surfaces. The detection process and algorithm flow are the same as the detection of two horizontal parallel surfaces

Measurement mechanism

l_s l_e PC A B PD C

Movement direction of detection line C

Figure 9. Detection model of two vertical parallel surfaces.

Figure 10 shows the geometric model of *PC* and *PD* cross section direction. When *PD* is the measured surface with *PC* as the datum surface. If there is no positional relation between the *PC* and *PD* surfaces, as shown in Figure 10a. The existed geometric relationship is the surface shape contour and relative displacement of the two surfaces. Since both are known. Therefore, the calculated surface shape contour is the same as the known surface shape contour. If there is an unknown positional relation between the *PC* and *PD* surfaces, the relative distance between the shape contours of the two surfaces will change, as shown in Figure 10b. At this time, the shape contour of *PD* surface cannot be obtained only from the known shape contour of *PC* surface and relative displacements. Therefore, the calculated shape contour of *PD* surface is different from the known shape contour of *PD* surface. Their shape difference is the positional relation curve.

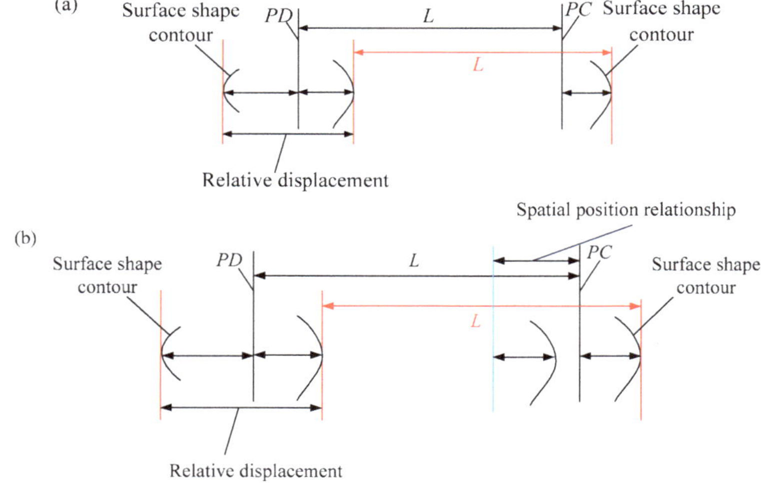

Figure 10. Geometric model of *PC* and *PD* cross section direction: (**a**) Positional relation does not exists between *PC* and *PD* surfaces. (**b**)Positional relation exists between *PC* and *PD* surfaces.

Due to the existing methods can only perform the parallelism evaluation, but the spatial position curve of the two parallel surfaces cannot be depicted. Therefore, it is difficult to carry out a comparative test to verify the test results in this paper.

By definition, parallelism is different to the positional relation of this article. Parallelism requires a datum direction as a datum, which is an orientation error. The positional relation has no datum direction. In this paper, the method of mutual detection by the two surfaces is used to verify the correctness of this detection method. The significance of mutual detection is that, ideally, the positional relation curve obtained by *PA*, *PB* mutual detection must be the same shape and opposite direction. The positional relation curve obtained by *PC*, *PD* mutual detection must be the same shape. When different measuring mechanisms are used to measure two horizontal parallel surfaces and two vertical parallel surfaces, the same positional relation description method is used to obtain the positional relation curves. However, the datum surface and the measured surface have changed, that is, the movement trajectories of the detection lines *A*, *B*, and the contact line *C* have changed. Under this condition, if the results of the mutual detection are all consistent with the objective law, the correctness of the description method of the positional relation can also be proved.

4. Mathematical Model of the Algorithm

4.1. Mathematical Model for Describing Positional Relation between Two Horizontal Parallel Surfaces

4.1.1. Extraction Algorithm for Combined Projection Waveform

When measuring *PB* with *PA* as the datum, the measurements is X_{s1}. When the shape contour of *PB* surface is higher than *PA*, the guideway moves upwards, $X_{s1} < X_0$. When the shape contour of *PB* surface is lower than *PA*, the guideway moves downward, $X_{s1} > X_0$, then the combined projection waveform can be defined as

$$C_1 = X_{s1} - X_0 \tag{1}$$

When measuring *PA* with *PB* as the datum, the measurements is X_{s2}, and the moving direction of guideway is opposite to that of measuring *PB* with *PA* as datum. When the shape contour of *PB* surface is higher than *PA*, the guideway moves downward, $X_{s2} > X_0$. When the shape contour of *PB* surface is lower than *PA*, the guideway moves upwards, $X_{s2} < X_0$. Then, the combined projection waveform can be defined as

$$C_2 = X_{s2} - X_0 \tag{2}$$

4.1.2. Time-Space Conversion Algorithm

Because the combined projection waveform is time series, it is necessary to converse the measurement waveform of time series into the measurement waveform of spatial sequence. It is known that the sampling frequency of the sensor is as high as 20 kHz, which means that only 0.00005 s is needed to obtain measurement data, indicating that the sampling density is extremely high. At such a high sampling density, the measurements at different times are the same, which means that duplicate measurements are obtained at some detection position, as shown in Figure 11.

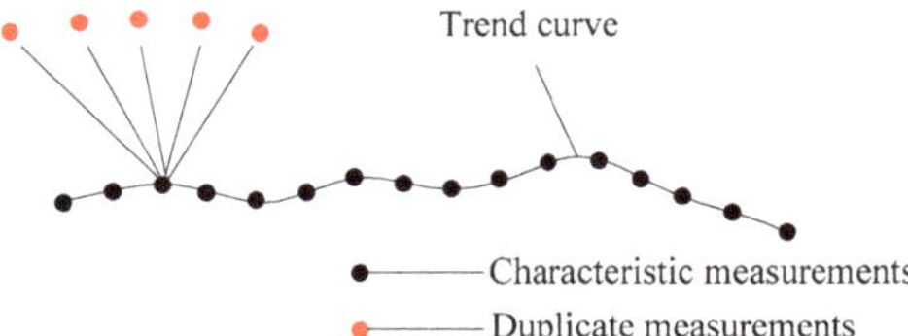

Figure 11. Structure of measurements.

Further, due to the high resolution of the sensor, the two adjacent measurements may be at the same position even if the data are not identical. In fact, each detection position only corresponds to one real measurements. Duplicate measurements will interfere with the shape recognition of waveforms. Therefore, the purpose of the time-space conversion algorithm is to extract the measurements at different positions that can recognize the shape of the measurements. These extracted measurements are defined as characteristic measurements.

It is known that sampling frequency and the total number of measurement data, the measurement time t can be expressed as

$$t = \frac{N_{L_p}}{f} \tag{3}$$

where f is the sampling frequency, N_{L_p} is the total number of sampling data corresponding to the moving distance L_p.

Since the speed of the detection mechanism is randomly changed, it is assumed that the average speed of the mechanism travel is 100 mm/s. Taking the sampling interval of the characteristic measurements is 1 mm. At this moment, the time between the two characteristic measurements is 0.01 s. When the velocity between the two characteristic measurements is 50 mm/s and the passing time is 0.01 s, the sampling distance between the two characteristic measurements is 0.5 mm, which is 0.5 mm different from the sampling interval under the average speed, which is insignificant for the overall curve shape of long surface. By extracting the trend curve through the neural network, the effect of the above errors can be eliminated, and only the shape of the whole curve of the waveform can be retained. Therefore, the influence of the driving speed on the detection position can be neglected. Therefore, the average velocity can be selected as the calculation parameter. The average speed of the measurement mechanism is

$$v = \frac{L_p}{t} \tag{4}$$

where L_p represents the moving distance of the measurement mechanism.

Let M represent the data set of the combined projection waveform. Then, the variables a, j, i, s are created, the initial value of a is 1, $j = 1, 2, 3, \ldots, N_{L_p}$, $i = a, \ldots, N_{L_p}$.

Then, Equation (5) is defined as the objective function.

$$|M(i) - s| < d \tag{5}$$

where d is a given value; d is selected according to the actual time series waveform.

If d is too large, the sampling interval of the spatial sequence waveform will be too large, resulting in distortion of the curve shape. If the d is too small, the redundant data cannot be filtered out. $M(i)$ represents the i-th data in the set M. The initial value of s is $M(1)$.

When $j = 1$, Equation (5) is executed to perform data search, when the searched data meets the objective function condition, $a = a + 1$, when the condition of the objective function is not met, the data is stored, at this time, the redundant data is filtered out, and the equation can be expressed as follows

$$\begin{cases} F(j) = a \\ K(j) = M(a) \end{cases} \tag{6}$$

where $F(j)$ represents the position of $K(j)$ in M; $K(j)$ represents the set of characteristic measurements.

Then, perform new calculations on s, and the equation can be expressed as

$$s = M(a) \tag{7}$$

After Equation (7) is completed, $j = j + 1$, next, repeat Equations (5)–(7). When $a = N_{L_p}$, the loop is completed. The flow chart of the algorithm is shown in Figure 12.

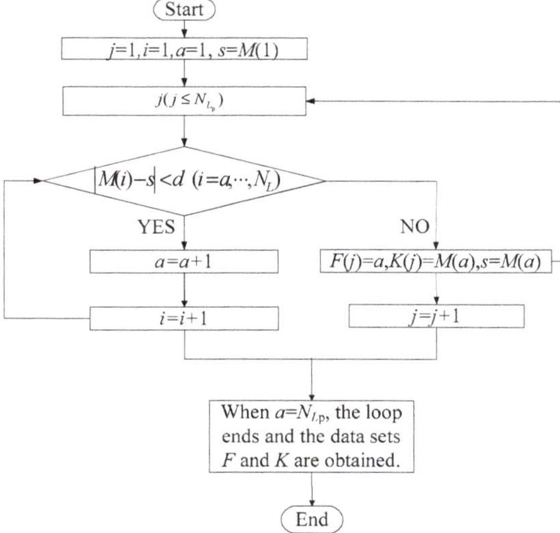

Figure 12. Flow chart of the algorithm.

At this time, the number of data in the sets F and K is N_c. It is known that the sampling interval of the measured surface is two characteristic measurements with a certain value difference, so the measurement time between the two characteristic measurements can be expressed as

$$t_c(u) = \frac{F(u)t}{N_{L_P}} \tag{8}$$

where $F(u)$ represents the u-th data in set F, $u = 1, 2, 3, \ldots, N_c$;

The measurement waveform of spatial sequence corresponding to the coordinate position on the measured surface can be obtained according to Equation (9), so the time-space conversion of the data can be realized and Equation (9) can be expressed as follows

$$\begin{array}{ll} vt_c(1) & \to K(1) \\ vt_c(2) & \to K(2) \\ \cdot & \cdot \\ \cdot & \cdot \\ \cdot & \cdot \\ vt_c(N_c) & \to K(N_c) \end{array} \tag{9}$$

where $t_c(1, \ldots, N_c)$ is derived from Equation (8); $K(1, \ldots, N_c)$ is derived from Equation (6).

At this time, the spatial sequence waveform corresponding to the combined projection waveform C_1 can be represented by R_1, and the spatial sequence waveform corresponding to the combined projection waveform C_2 can be represented by R_2. After that, the trend curves Δ_1 and Δ_2 of R_1 and R_2 are extracted respectively, and Δ_1 is defined as the combined projection curve of R_1, Δ_2 is defined as the combined projection curve of R_2.

4.1.3. Separation Algorithm for Combined Projection Curve

The separation algorithm model of the combined projection curve is shown in Figure 13.

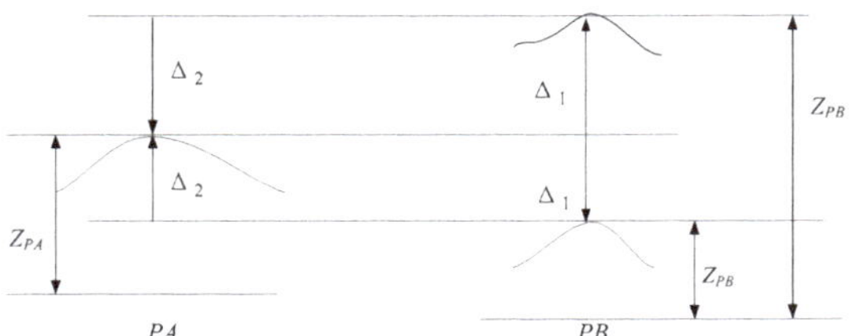

Figure 13. Separation algorithm model of the combined projection curve.

With *PA* as the datum surface and *PB* as the measured surface, the calculated shape contour can be defined as

$$S_{PB} = Z_{PA} \pm |\Delta_1| \quad (10)$$

where Z_{PA} is the shape contour of *PA* surface, when $\Delta_1 < 0$, uses '+', when $\Delta_1 > 0$, uses '−'.

With *PB* as the datum surface and *PA* as the measured surface, the calculated shape contour can be defined as

$$S_{PA} = Z_{PB} \pm |\Delta_2| \quad (11)$$

where Z_{PB} is the shape contour of *PB* surface, when $\Delta_2 < 0$, uses '+', when $\Delta_2 > 0$, uses '−'.

4.1.4. Description Algorithm for Positional Relation

When *PB* is the measured surface, *PA* as the datum, the positional relation curve of *PB* relative to *PA* can be expressed as

$$G_B = S_{PB} - Z_{PB} \quad (12)$$

where Z_{PB} is the shape contour of *PB* surface.

When *PA* is the measured surface, *PB* as the datum, the positional relation curve of *PA* relative to *PB* can be expressed as

$$G_A = S_{PA} - Z_{PA} \quad (13)$$

where Z_{PA} is the shape contour of *PA* surface.

4.2. Mathematical Model for Describing Positional Relation between Two Vertical Parallel Surfaces

4.2.1. Extraction Algorithm for Combined Projection Waveform

Three kinds of situations may occur during the measurement of two vertical parallel surfaces. As the detection model shown in Figure 14, when *PD* is measured surface with the *PC* as the datum surface, the measurements is X_{s3}. When the contact line *A* is at P_0 and the detection line *C* is at P_1, the distance between the surface where the contact line *A* is located and the surface where the detection line *C* is located is the same as the distance between the two surfaces which are absolutely parallel, and both are *L*. At this time, the measurement is fixed, defined as X_1. Ideally, in the case where there is no positional relation between the two surfaces, the detection line *C* should be located at P_2, and the contact line *A* should be located at P_0. Since two absolutely parallel surfaces do not exist, the actual detection line *C* should be located at P_3.

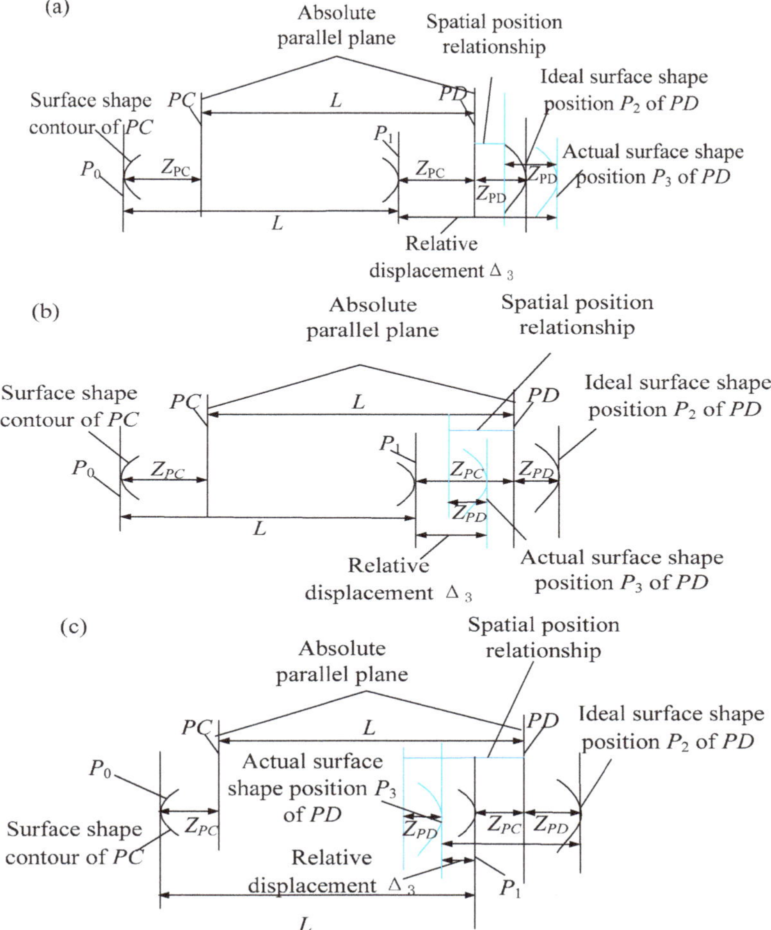

Figure 14. Separation algorithm model of the combined projection curve: (**a**) Model 1; (**b**) Model 2; (**c**) Model 3.

In the detection model shown in Figure 14a,b, the spring is in compression state, at this time, $X_{s3} < X_1$. In the detection model shown in Figure 14c, the spring is in the stretching state, at this time, $X_{s3} > X_1$. Then, the combined projection waveform can be defined as

$$C_3 = X_{s3} - X_1 \tag{14}$$

When measuring PC with PD as datum, the measurement is X_{s4}, when the spring is in compression state, $X_{s4} < X_1$, and when the spring is in stretching state, $X_{s4} > X_1$. Then, the combined projection waveform can be defined as

$$C_4 = X_{s4} - X_1 \tag{15}$$

At this time, the spatial sequence waveform corresponding to the combined projection waveform C_3 can be represented by R_3, and the spatial sequence waveform corresponding to the combined projection waveform C_4 can be represented by R_4. After that, the trend curves Δ_3 and Δ_4 of R_3 and

R_4 are extracted respectively, and Δ_3 is defined as the combined projection curve of R_3, while Δ_4 is defined as the combined projection curve of R_4.

4.2.2. Separation Algorithm for Combined Projection Curve

The separation algorithm model of the combined projection curve is shown in Figure 14. With *PC* as the datum surface, *PD* as the measured surface, the calculated shape contour can be defined as

$$S_{PD} = -\Delta_3 - Z_{PC} \tag{16}$$

where Z_{PC} is the shape contour of *PC* surface.

With *PD* as the datum surface, *PC* as the measured surface, the calculated shape contour can be defined as

$$S_{PC} = -\Delta_4 - Z_{PD} \tag{17}$$

where Z_{PD} is the shape contour of *PD* surface.

4.2.3. Description Algorithm for Positional Relation

When *PC* is the measured surface, *PD* as the datum, the positional relation curve of *PD* relative to *PC* can be expressed as

$$G_D = S_{PD} - Z_{PD} \tag{18}$$

When *PC* is the measured surface, *PD* as the datum, the positional relation curve of *PD* relative to *PC* can be expressed as

$$G_C = S_{PC} - Z_{PC} \tag{19}$$

5. Experimental Validation

5.1. Measuremen Texperiment of Two Horizontal Parallel Surfaces

5.1.1. Experimental Step

In order to verify the correctness of this measurement method, the measurement experiment was carried out. Before measuring, a marble flat ruler with class 000 precision as shown in Figure 2 was selected as the datum surface, the size of the marble flat ruler selected in this paper is 500 × 100 × 50 mm, and its flatness is 1.5μm. Therefore, it can be regarded as an approximately absolute horizontal plane. Then, the measurement mechanism was placed on the surface of marble ruler to calibrate the initial value X_0. It is assumed that contact line *A* and detection line *C* are on the same horizontal surface. Place the detection mechanism at different positions on the marble flat ruler and find the average of the five measurements, then the initial value X_0 can be obtained, and X_0 is 32.92014 mm.

Taking *PA* and *PB* surfaces with a length of 500 mm and a width of 50 mm, the distance between the two surfaces is 60 mm. Calibrating the starting and ending positions, ensure that the spring is in compression. Firstly, the *PB* surface was detected by using *PA* surface as the datum surface, as shown in Figure 15a, and then the *PA* surface was detected by using *PB* surface as the datum surface, as shown in Figure 15b. The measurement method is described in Section 3.1. Each surface was measured four times. During the measurement process, the measurement mechanism moves along the auxiliary surface toward the arrow direction. Under the constraint of an auxiliary surface, the detection mechanism will not slide.

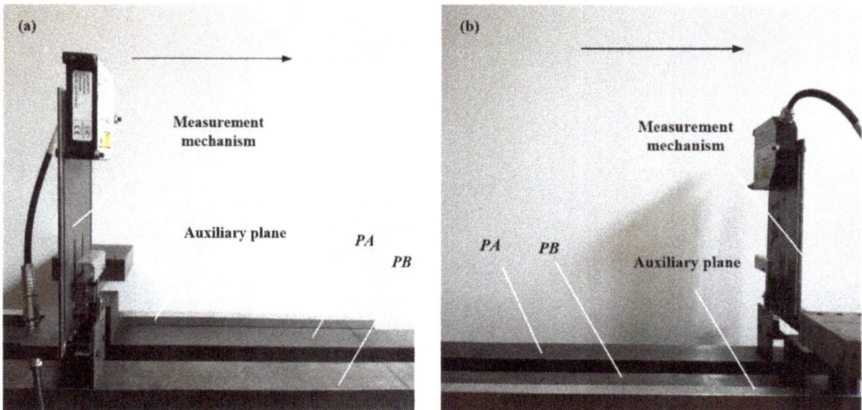

Figure 15. Measurement system of two horizontal parallel surfaces. (**a**) *PB* is the measured surface. (**b**) *PA* is the measured surface.

5.1.2. Results and Analysis

The measurements of laser displacement sensor are shown in Figures 16 and 17. When the contact line *A* is located at the starting position, computer begins to record the data. After the contact line *A* reached the end position, where it necessary to save the data, the measurement mechanism is in a static state, and the measurements of this process represent a straight line approaching the level. It is known that the measurements between starting position and end position is defined as valid measurements. Therefore, the suspended time series waveforms need to be separated and the effective measurements were obtained.

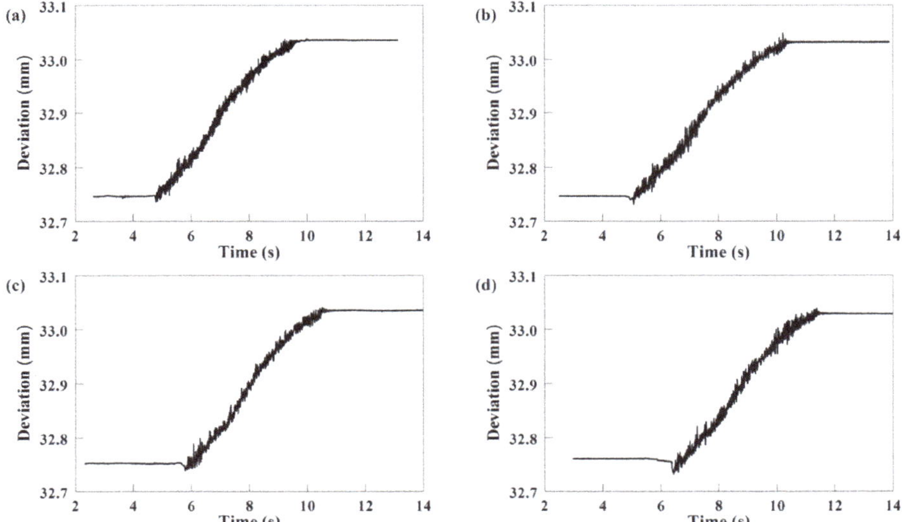

Figure 16. The measurements when *PB* is the measured surface (**a**) measurements 1 (**b**) measurements 2 (**c**) measurements 3 (**d**) measurements 4.

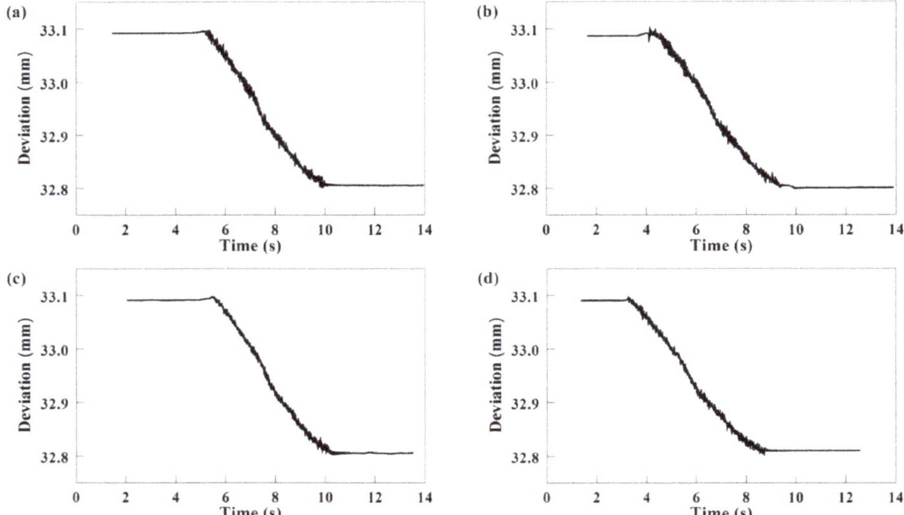

Figure 17. The measurements when *PA* is the measured surface (**a**) measurements 1 (**b**) measurements 2 (**c**) measurements 3 (**d**) measurements 4.

When the effective measurements and X_0 were obtained, the combined projection waveforms were calculated according to Equations (1) and (2). Figure 18 shows the combined projection waveforms. It can be seen that the waveforms change with sampling time, and the measurement time is different, which has greater randomness.

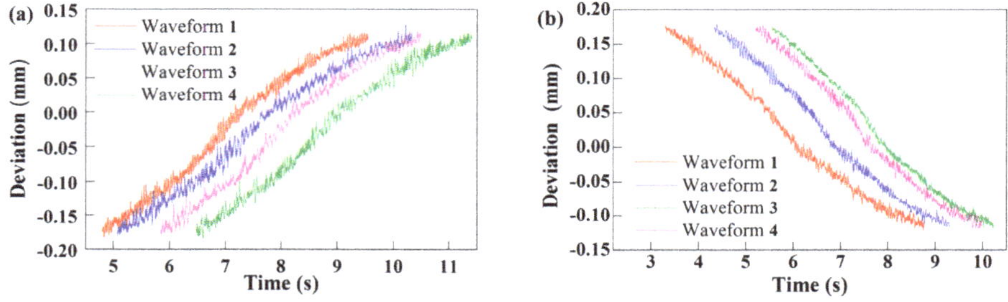

Figure 18. Combined projection waveforms. (**a**) *PB* is the measured surface (**b**) *PA* is the measured surface.

The measurements in Figure 18 are sequenced according to the measurement time from short to long. Then, FFT is carried out to obtain the amplitude frequency characteristics, as shown in Figure 19. It can be seen from the analysis of the amplitude of the low-frequency band that the amplitudes of the spectrum graphs corresponding to different waveforms are basically identical. Further, the peak mainly occurs near 0 Hz and there is no frequency doubling relationship in the amplitude frequency characteristics. This is because the measurements changes with the shape of the measured plane and is not affected by other factors. When 2–28 Hz is amplified, it is found that the amplitude does not increase or decrease monotonously with the increase of measurement time. It shows that the measurement time has no effect on the amplitude frequency characteristics of the measurements. It can also be understood that when four waveforms are converted into spatial sequences, the overall shape of the waveforms does not change significantly.

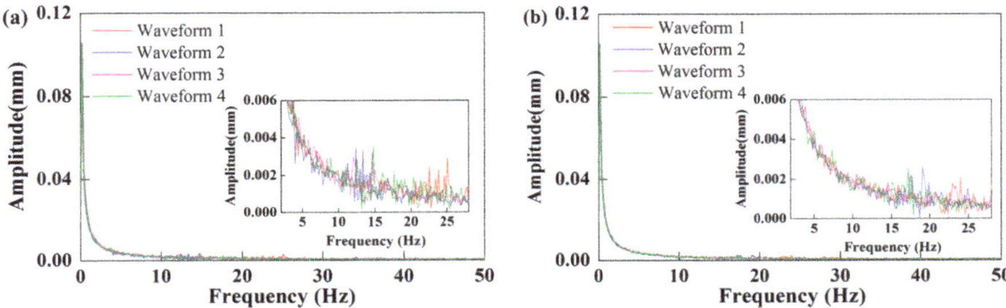

Figure 19. Amplitude frequency characteristics of measurements. (**a**) *PB* is the measured plane (**b**) *PA* is the measured plane.

Extracting the amplitude corresponding to the 2 Hz, 4 Hz, 6 Hz, 8 Hz, as shown in Figure 20, it can be seen that the amplitude corresponding to the same frequency has little difference with varying measurement time, indicating that different driving speed has little impact on the attribute of measurements. Therefore, the average velocity can be used as the calculation parameter.

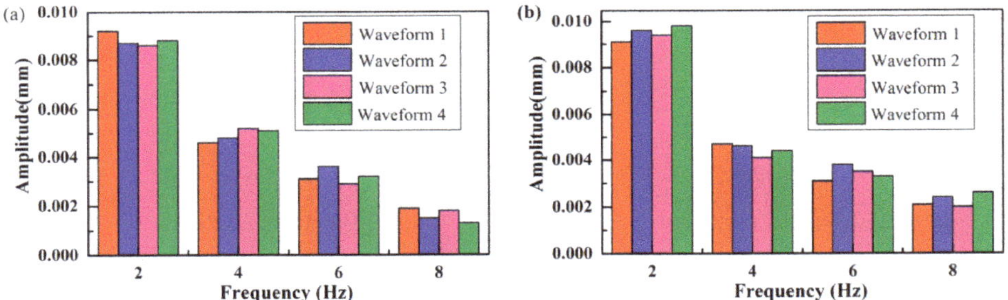

Figure 20. Amplitude corresponding to frequency 2 Hz, 4 Hz, 6 Hz, 8 Hz. (**a**) *PB* is the measured plane (**b**) *PA* is the measured plane.

Then the waveforms in Figure 18 are processed by the time-space conversion algorithm. The converted spatial sequence waveforms are shown in Figure 21. Because the positional relation curve is macro geometric, therefore, the trend curves of the waveforms in Figure 21 are extracted, which are called the combined projection curves, as shown in Figure 22. It can be seen from Figures 21 and 22, when the measurement mechanism detects the identical surface, *A*, *B*, and *C* always pass the shape contours of the identical surfaces. Although the characteristic measurements at the same position are not exactly the same, but the overall shape of the waveform trend curve is constant, the combined projection curves on the identical measured surfaces are close to coincidence. This shows that the detection mechanism has good repeatability and stability, and the time–space conversion algorithm successfully converts the waveforms of time–deviation into the position–deviation.

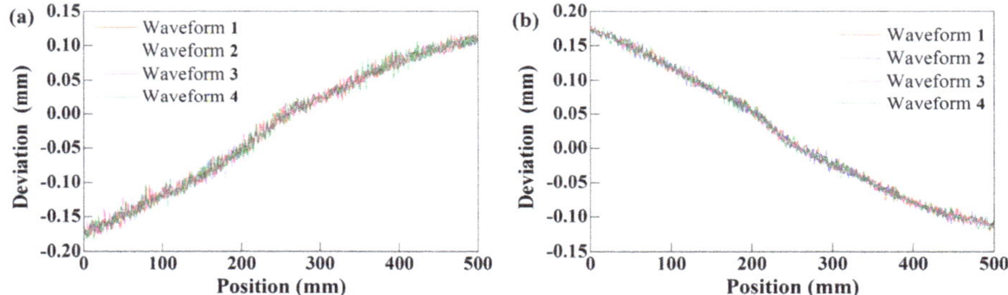

Figure 21. Combined projection waveforms of spatial sequence (**a**) *PB* is the measured surface. (**b**) *PA* is the measured surface.

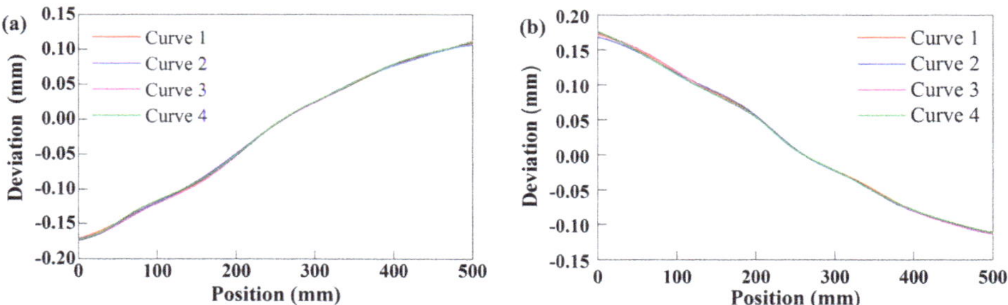

Figure 22. Combined projection curves. (**a**) *PB* is the measured surface. (**b**) *PA* is the measured surface.

Figure 23 shows the average of combined projection curve of the *PA* and *PB* surfaces. Here, the data structure is the same as that of the known surface shape contours in Figure 24, and each datum corresponds to the identical detection position. As can be seen from Figure 23 the combined projection curves obtained by mutual measurement of the two surfaces are symmetrically distributed with 0 as the center, which accords with the mathematical relationship of the data corresponding to the same position in the two surfaces in Figure 13.

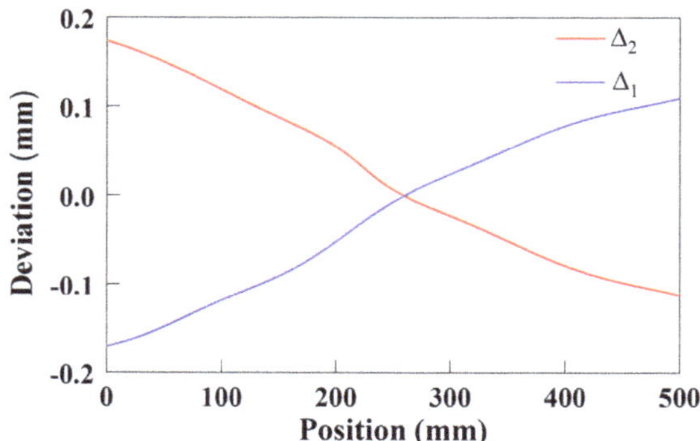

Figure 23. The average of the combined projection curves of the *PA* and *PB* surfaces.

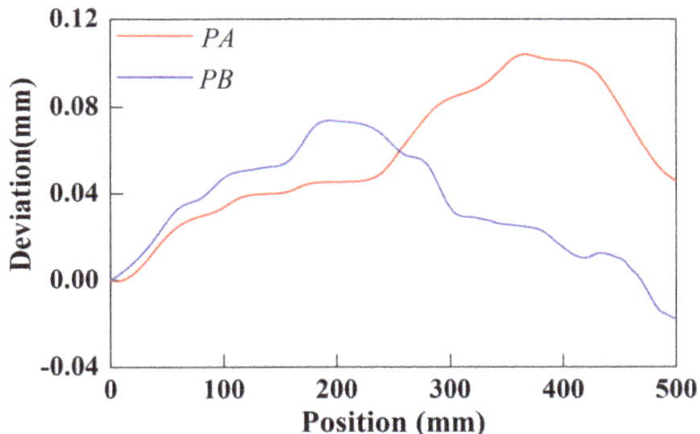

Figure 24. Known surface shape contours.

Based on the above analysis, substituting the data in Figures 23 and 24 into Equations (10) and (11), the calculated surface shape contours are obtained, as shown in Figure 25. It can be seen from Figure 25 that the calculated surface shape contours are different from the known surface shape contours, which indicates that there exists a positional relation between two surfaces *PA* and *PB*.

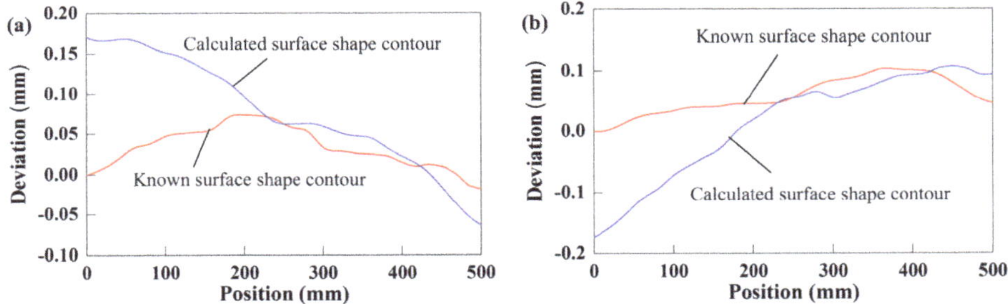

Figure 25. Comparison between the calculated surface shape contour and known surface shape contour: (**a**) *PB* is the measured surface. (**b**) *PA* is the measured surface.

Finally, the data in Figure 25 were substituted into Equations (12) and (13) to obtain the difference curve between the calculated surface shape contour and the known surface shape contour, as shown in Figure 26. The curve describes the positional relation of the two surfaces. It can be seen from Figure 26. The positional relation curve obtained by mutual check has the same shape and symmetric distribution, which is consistent with the principle of a positional relation between two horizontal parallel planes. The maximum error is 4.07 μm, which has little effect on the overall shape of the curve. The correctness of this detection method can be proved.

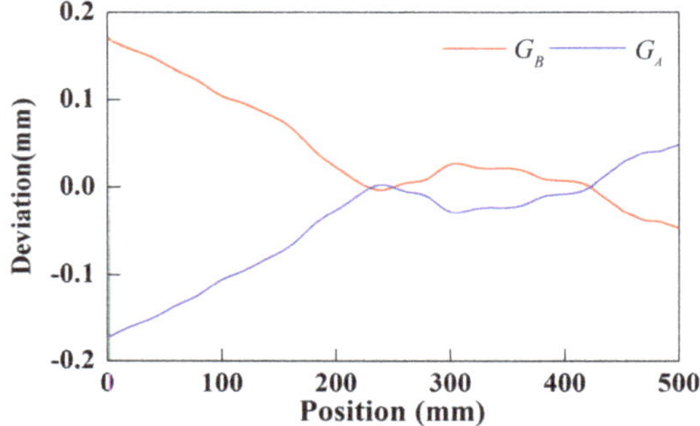

Figure 26. Two positional relation curves.

5.2. Measuremen Texperiment of Two Vertical Parallel Surfaces

5.2.1. Experimental Step

In order to verify the correctness of the algorithm for describing the positional relation between two vertical parallel surfaces, a measurement experiment was performed using a homemade measuring instrument, and the measurement site is shown in Figure 27. Before measurement, choosing a class 000 marble ruler with length of 100 mm, width of 60 mm, and height of 50 mm. As its parallelism is 1 μm, the surfaces on both sides of the marble ruler can be approximately regarded as absolutely smooth parallel surfaces. Firstly, place the contact lines *A*, *B*, and the detection line *C* on both sides of the marble ruler, and tightly clamp the marble ruler. Then, the initial value X_1 can be obtained and X_1 is 24.1379 mm.

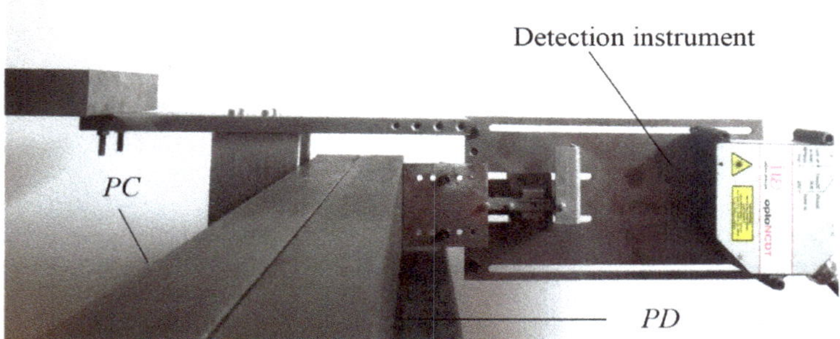

Figure 27. Measurement system of two vertical parallel surfaces.

The method for detecting the positional relation between two vertical parallel surfaces is the same as the method for detecting the positional relationship between two horizontal surfaces. Firstly, use the *PC* as the datum surface to detect the *PD* surface, and then use the *PD* surface as the datum surface to detect the *PC* surface. Repeat the measurement four times for each surface.

5.2.2. Results and Analysis

The original measurement data recorded by the computer is shown in Figures 28 and 29. When the contact line *A* is located at the starting position, computer begins to record the data. After the contact line *A* reached the end position, where it is also necessary to save the data, the measurement mechanism is in a static state, and the measurement of this process is a straight line approaching the level. It is known that the measurement between starting position and end position is defined as a valid measurement. Therefore, the suspended time series waveforms need to be separated and the effective measurements were obtained.

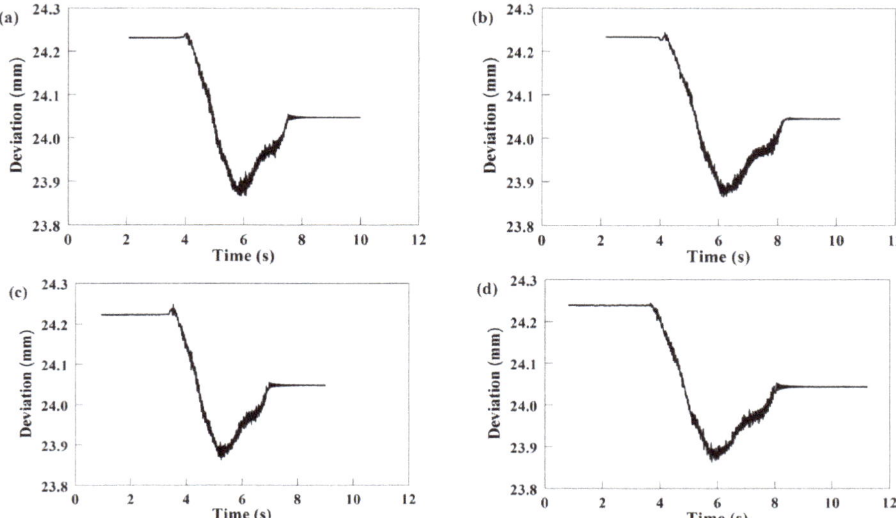

Figure 28. The measurements when *PD* is the measured surface (**a**) measurements 1 (**b**) measurements 2 (**c**) measurements 3 (**d**) measurements 4.

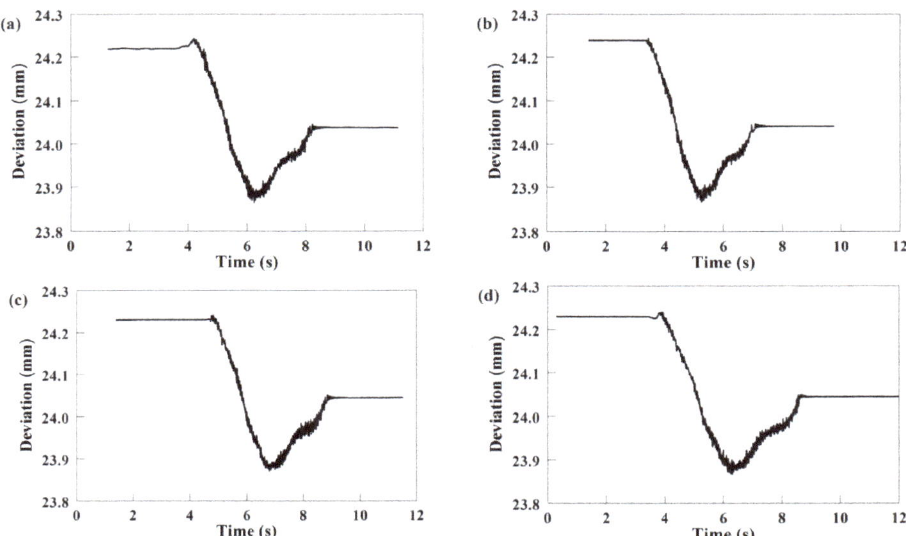

Figure 29. The measurements when *PC* is the measured surface (**a**) measurements 1 (**b**) measurements 2 (**c**) measurements 3 (**d**) measurements 4.

Extract the effective measurements, according to Equations (14) and (15), and the combined projection waveform will be obtained, as shown in Figure 30.

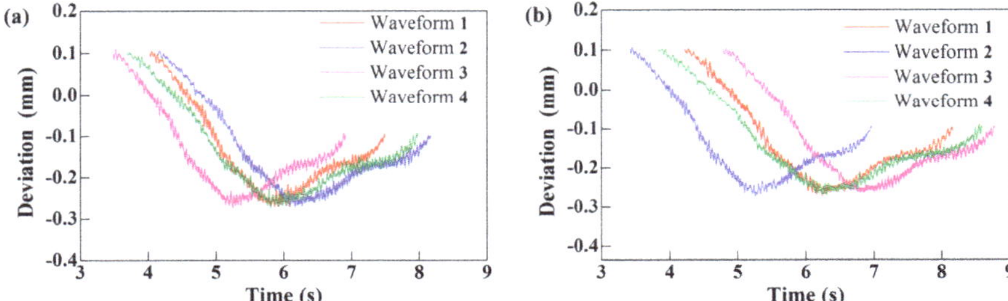

Figure 30. Combined projection waveforms: (**a**) *PD* is the measured surface (**b**) *PC* is the measured surface.

It can be seen from the Figures 19 and 20 that the measurement time has no effect on the overall shape of the waveforms. Then, use the space-time conversion algorithm to obtain the spatial sequence waveform, as shown in Figure 31. The trend curve of the waveform in Figure 31 is extracted to obtain a combined projection curve, as shown in Figure 32. The projection curves obtained from multiple measurement are close to coincidence, which shows that the detection mechanism also has high repeatability and stability when applied to the detection of vertical surface, which proves the rationality of the design of the detection mechanism in the vertical direction.

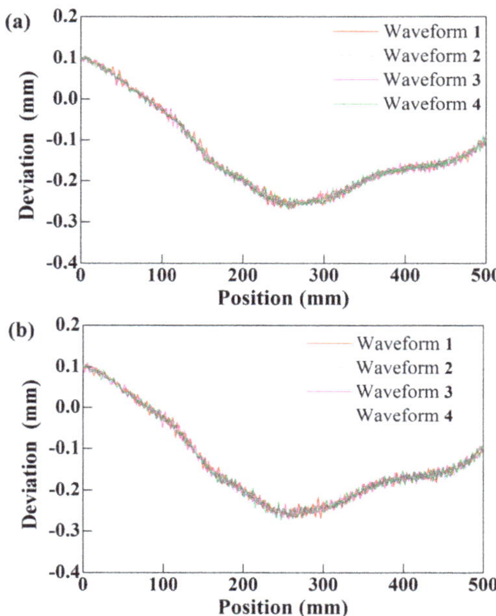

Figure 31. Combined projection waveforms of spatial sequence: (**a**) *PD* is the measured surface (**b**) *PC* is the measured surface.

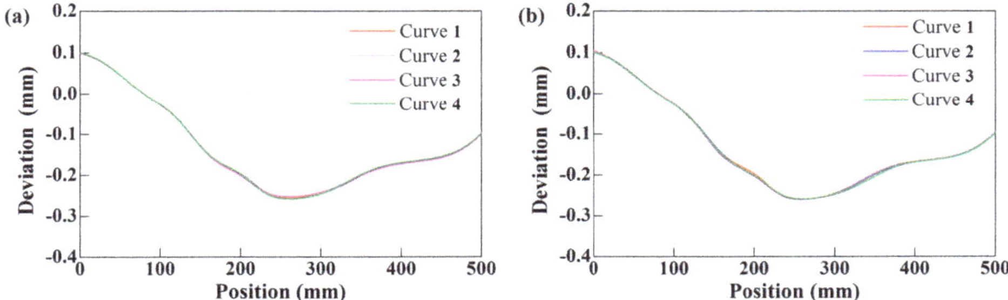

Figure 32. Combined projection curves (**a**) *PD* is the measured surface (**b**) *PC* is the measured surface.

Figure 33 shows the average of the combined projection curve of *PC* and *PD* surfaces. It can be seen from Figure 33 that the combined projection curves are close to coincide. It is proved that the relative displacement of the two vertical surfaces is only related to the distance of the detection line *C* relative to the datum surface, independent of the direction, and in line with the geometric relationship in the detection model in Figure 14.

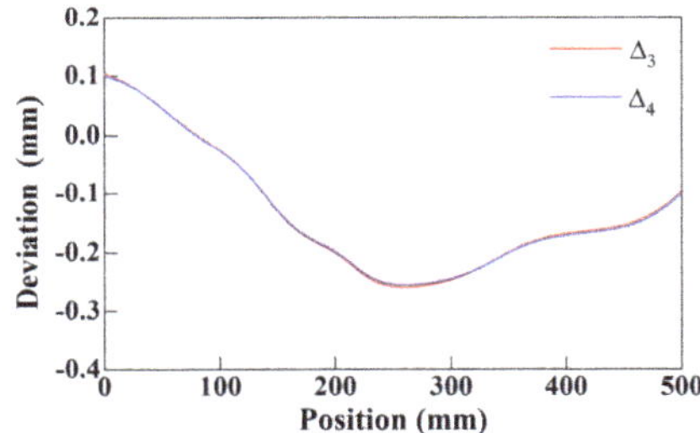

Figure 33. The average of the combined projection curves of the *PC* and *PD* surfaces.

Based on the above analysis, the data in Figures 24 and 33 are substituted into Formulas (16) and (17) to obtain the calculated surface shape contours, as shown in Figure 34. It can be seen from Figure 34 that the calculated surface shape contour is different from the known surface shape contour, indicating that there exists a positional relation between the two surfaces.

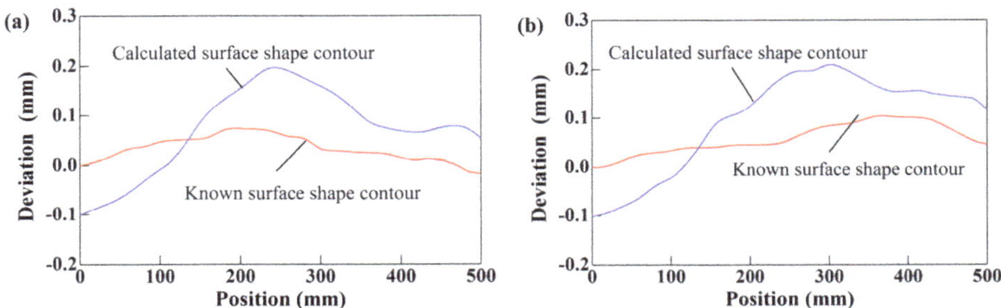

Figure 34. Comparison between calculated surface shape contour and known surface shape contour: (**a**) *PD* is the measured surface (**b**) *PC* is the measured surface.

Finally, the data in Figure 34 are substituted into Equations (18) and (19) to obtain the difference curve between the calculated surface shape contour and the known surface shape contour, as shown in Figure 35. The curve describes the spatial relationship between *PC* and *PD* surfaces. As can be seen from Figure 35, the shape of positional relation curves obtained by mutual detection are close to coincidence, which id consistent with the principle of positional relation between two vertical parallel surfaces. The maximum error of the two positional relation curves is 4.56 μm, which has little effect on the overall shape of the curve. The correctness of this detection method can be proved.

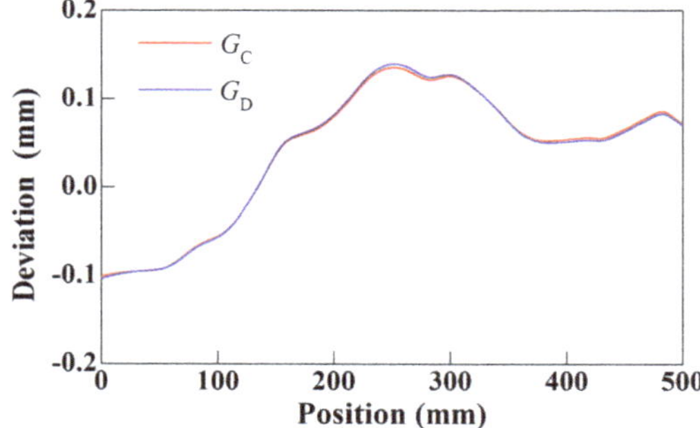

Figure 35. Two positional relation curves.

5.3. Discussion

In order to verify whether this method is correct and whether the detection instruments are designed reasonably, only short span tests are carried out in this paper. When measuring the two surfaces with large span, as long as the detection mechanisms are in a balanced state, there will be no sliding. Therefore, this method can still be used for measurement of large span surfaces.

Because the contact lines *A*, *B*, and detection line *C* have shape errors, and the combined projection waveform is equal to the measurements minus the initial value, therefore the influence of the shape errors of the contact lines *A*, *B*, and detection line *C* on the measurement results can be eliminated.

During the detection progress, the measurements will be affected by motion error, surface roughness, spring damping and other factors. However, the influence of these factors accounts for a small proportion of the measurements. Since the position relation curve is macro, the trend curve reflects the overall shape of the measurements. The extracted trend curve has a certain inclusiveness to the peak and valley of the waveform. When the trend curve of the waveform is extracted, the error components in the measurements will be largely separated. Therefore, the error has little effect on the shape of trend curve and the error of the final positional relation curve mainly comes from the method of extracting the trend curve.

The article focuses on verifying the correctness of the detection method and whether the detection mechanism can be applied to actual measurement. The next work will focus on the analysis of the uncertainty of detection methods and the detection process.

6. Conclusions

This paper proposed a method for describing the positional relation of two parallel surfaces on-site. The extraction algorithm for combined projection waveform, time-space conversion algorithm, the separation algorithm for combined projection curve, and the description algorithm for positional relation are successfully developed. It can be seen from the experimental results that the combined projection curves obtained by detecting the same surface are nearly coincident, indicating that the detection mechanism has high repeatability. The extracted trend curve has a certain inclusiveness to the peak and valley of the waveform. Although the measurements affected by motion error, surface roughness, spring damping, and others factor, the error has little effect on the shape of trend curve. The results of the mutual detection of the two examples are in accordance with the objective law.

The maximum error of the positional relation curve obtained by the mutual detection of the two horizontal parallel surfaces is 4.07 μm, and the maximum error of the position relation curve obtained by the mutual detection of the two vertical parallel surfaces is 4.56 μm, which has little effect on the curve shape, indicating that this method can accurately obtain the positional relation between two horizontal parallel planes and two vertical parallel planes. The measurement mechanisms are designed reasonably, and the measurement method is correct. It is operable in practical application.

Moreover, it can be completed in two minutes from the acquisition of the detection data to the drawing of the positional relation curve of two surfaces, which means high efficiency. Compared with the existing parallelism evaluation method, the positional relation curve obtained by this method makes the description of the position error of two parallel planes more specific. Compared with the existing measurement instruments, the mechanisms are more suitable for field measurement of large equipment. This research demonstrates innovation, and it is also of great significance to the requirements of modern industries such as field measurement and the digitalization of assembly processes.

Author Contributions: Conceptualization, Z.L. (Zechen Lu); Data curation, Z.L. (Zechen Lu); Formal analysis, Z.L. (Zechen Lu); Funding acquisition, C.Z.; Investigation, Z.L. (Zhenjun Li); Methodology, C.Z.; Software, B.Z.; Visualization, B.Z.; Writing—original draft, Z.L. (Zechen Lu); Writing—review & editing, C.Z. All authors have read and agreed to the published version of the manuscript.

Funding: This work was funded by the National Natural Science Foundation of China (No. 51775094).

Conflicts of Interest: The authors declare no conflicts of interest.

Data Availability: The data used to support the findings of this study are included within the article.

References

1. Zha, J.; Xue, F.; Chen, Y. Straightness error modeling and compensation for gantry type open hydrostatic guideways in grinding machine. *Int. J. Mach. Tools Manuf.* **2017**, *112*, 1–6. [CrossRef]
2. Kang, C.S.; Kim, J.A.; Jin, J. An optical straightness measurement sensor for the KRISS watt balance. *Measurement* **2015**, *61*, 257–262. [CrossRef]
3. Borisov, O.; Fletcher, S.; Longstaff, A. New low cost sensing head and taut wire method for automated straightness measurement of machine tool axes. *Opt. Lasers Eng.* **2013**, *51*, 978–985. [CrossRef]
4. Wei, X.; Su, Z.; Yang, X.; Lv, Z.; Yang, Z.; Zhang, H.; Li, X.; Fang, F. A Novel Method for the Measurement of Geometric Errors in the Linear Motion of CNC Machine Tools. *Appl. Sci.* **2019**, *9*, 3357. [CrossRef]
5. Li, P.; Ding, X.M.; Tan, J.B. A hybrid method based on reduced constraint region and convex-hull edge for surface evaluation. *Precis. Eng.* **2016**, *45*, 168–175. [CrossRef]
6. Radlovački, V.; Hadžistević, M.; Štrbac, B. Evaluating minimum zone surface using new method—Bundle of plains through one point. *Precis. Eng.* **2016**, *43*, 554–562. [CrossRef]
7. Giusca, C.L.; Claverley, J.D.; Sun, W. Practical estimation of measurement noise and flatness deviation on focus variation microscopes. *CIRP Ann. Manuf. Technol.* **2014**, *63*, 545–548. [CrossRef]
8. Han, Z.; Chen, L.; Wulan, T. The absolute flatness measurements of two aluminum coated mirrors based on the skip flat test. *Opt. Int. J. Light Electron Opt.* **2013**, *124*, 3781–3785. [CrossRef]
9. Haijun, L.; Zhigang, D.; Han, H. A new method for measuring the flatness of large and thin silicon substrates using a liquid immersion technique. *Meas. Sci. Technol.* **2015**, *26*, 115008.
10. Zhou, P.; Xu, K.; Wang, D. Rail Profile Measurement Based on Line-structured Light Vision. *IEEE Access* **2018**, *6*, 16423–16431. [CrossRef]
11. Srinivasu, D.S.; Venkaiah, N. Minimum zone evaluation of roundness using hybrid global search approach. *Int. J. Adv. Manuf. Technol.* **2017**, *92*, 2743–2754. [CrossRef]
12. Cao, Z.; Wu, Y.; Han, J. Roundness deviation evaluation method based on statistical analysis of local least square circles. *Meas. Sci. Technol.* **2017**, *28*, 105017.
13. Hsieh, T.H.; Huang, H.L.; Jywe, W.Y. Development of a machine for automatically measuring static/dynamic running parallelism in linear guideways. *Rev. Sci. Instrum.* **2014**, *85*, 035115. [CrossRef] [PubMed]
14. Hwang, J.; Park, C.H.; Gao, W. A three-probe system for measuring the parallelism and straightness of a pair of rails for ultra-precision guideways. *Int. J. Mach. Tools Manuf.* **2017**, *47*, 1053–1058. [CrossRef]

15. Bhattacharya, J.C. Measurement of parallelism of the surfaces of a transparent sample. *Opt. Lasers Eng.* **2001**, *35*, 27–31. [CrossRef]
16. Lee, K.I.; Shin, D.H.; Yang, S.H. Parallelism error measurement for the spindle axis of machine tools by two circular tests with different tool lengths. *Int. J. Adv. Manuf. Technol.* **2017**, *88*, 2883–2887. [CrossRef]
17. Vannoni, M.; Bertozzi, R. Parallelism error characterization with mechanical and interferometric methods. *Opt. Lasers Eng.* **2007**, *45*, 719–722. [CrossRef]
18. Jywe, W.Y.; Hsieh, T.H.; Chen, P.Y.; Wang, M.S. An Online Simultaneous Measurement of the Dual-Axis Straightness Error for Machine Tools. *Appl. Sci.* **2018**, *8*, 2130. [CrossRef]
19. Hsieh, T.H.; Chen, P.Y.; Jywe, W.Y.; Chen, G.W.; Wang, M.S. A Geometric Error Measurement System for Linear Guideway Assembly and Calibration. *Appl. Sci.* **2019**, *9*, 574. [CrossRef]

Article

Non-Scanning Three-Dimensional Imaging System with a Single-Pixel Detector: Simulation and Experimental Study

Guang Shi *, Leijue Zheng, Wen Wang and Keqing Lu

School of Mechanical Engineering, Hangzhou Dianzi University, Hangzhou 310018, China; zhengleijue@163.com (L.Z.); wangwn@hdu.edu.cn (W.W.); lkq@hdu.edu.cn (K.L.)

* Correspondence: shiguang@hdu.edu.cn

Received: 3 April 2020; Accepted: 27 April 2020; Published: 29 April 2020

Abstract: Existing scanning laser three-dimensional (3D) imaging technology has slow measurement speed. In addition, the measurement accuracy of non-scanning laser 3D imaging technology based on area array detectors is limited by the resolution and response frequency of area array detectors. As a result, applications of laser 3D imaging technology are limited. This paper completed simulations and experiments of a non-scanning 3D imaging system with a single-pixel detector. The single-pixel detector can be used to achieve 3D imaging of a target by compressed sensing to overcome the shortcomings of the existing laser 3D imaging technology. First, the effects of different sampling rates, sparse transform bases, measurement matrices, and reconstruction algorithms on the measurement results were compared through simulation experiments. Second, a non-scanning 3D imaging experimental platform was designed and constructed. Finally, an experiment was performed to compare the effects of different sampling rates and reconstruction algorithms on the reconstruction effect of 3D imaging to obtain a 3D image with a resolution of 8 × 8. The simulation results show that the reconstruction effect of the Hadamard measurement matrix and the minimum total variation reconstruction algorithm performed well.

Keywords: scanless 3D imaging; compressed sensing; depth detection; single-pixel detector

1. Introduction

At present, using three-dimensional (3D) imaging technology to obtain 3D information about the surrounding environment is important in many application fields, particularly in the fields of autonomous driving, 3D printing, machine vision, and virtual reality [1,2]. Traditional laser 3D imaging is divided into two main types: scanning and non-scanning. Scanning technology performs a point-by-point measurement of the target through a mechanical scanning device. The entire system occupies a large volume and has low measurement efficiency [3,4]. Non-scanning technology usually uses an area array detector to measure the reflected light at every point in the target area simultaneously. However, imaging using an area array detector has problems, such as a low signal-to-noise ratio, limited imaging resolution, and limited response frequency; these lead to low imaging resolution and accuracy [5–7].

Nyquist sampling theory requires that the sampling rate must be more than twice the signal bandwidth to reconstruct the original signal without distortion from the discrete sampling signal, but a large amount of redundant information occupies the resources of the signal sampling system. In 2006, Donoho, Candes, and Tao proposed the theory of compressed sensing [8–10]. It was proven that for sparse or sparsely expressed signals, the original signal can be reconstructed accurately with a high probability using a small number of measurement times. This resolves the contradiction between measurement resolution and measurement efficiency in traditional 3D imaging signal sampling and makes 3D imaging possible with a single-pixel detector. In 2008, Takhar of Rice University developed a single-pixel camera using the theory of compressed sensing; Takhar first applied compressed sensing to imaging systems [11]. In 2011, Howland of the University of Rochester used a photon-counting method to achieve 3D compressed sensing imaging [12] using a pulsed laser with a wavelength of 780 nm and achieved a distance resolution of 30 cm. In 2014, Guo of Beijing Institute of Technology proposed a 3D compressed sensing imaging method based on the phase method ranging principle [13] and conducted preliminary simulation verification of the entire system but did not study 3D imaging experiments further. In 2015, Sun of Beihang University used a slicing method and a pulsed laser to realize 3D compressed sensing imaging [14] and accurately reconstructed a target scene with a resolution of 128 × 128, achieving an accuracy of 3 mm within a distance of 5 m.

In this paper, a single-pixel non-scanning 3D imaging system was designed, and the system was investigated in terms of theories, simulations, and experiments. First, the effects of different sampling rates, sparse transform bases, measurement matrices, and reconstruction algorithms on the measurement results were compared through simulation experiments. Second, a non-scanning 3D imaging experimental platform, based on a single-pixel detector, was designed and constructed. Finally, an experiment was performed to compare the effects of different sampling rates and reconstruction algorithms on the reconstruction effect of 3D imaging to obtain a 3D image with a resolution of 8 × 8. The results of the experiment showed that the 3D imaging system works, and its reconstruction of the Hadamard measurement matrix and the minimum total variation reconstruction algorithm performed well.

2. Three-Dimensional Imaging Theory

2.1. System Structure

In this study, the non-scanning 3D imaging system with a single-pixel detector is based on the traditional compressed sensing two-dimensional imaging. Meanwhile, modulation information is added to the laser light intensity. We can obtain the distance information about the target area by phase detection to complete the 3D imaging. The system structure is shown in Figure 1.

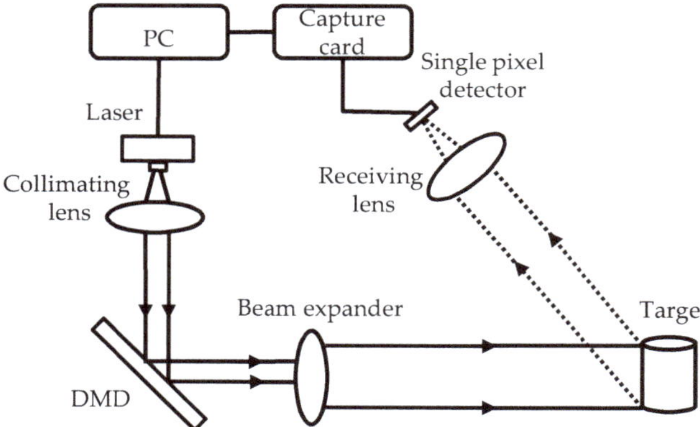

Figure 1. System structure.

The 3D imaging process of the system works as follows. The laser emits a continuous modulated beam with a sine wave modulation. The continuous laser enters the mirror area of a digital micro-mirror device (DMD) after collimation. The DMD's micro-mirror array flips according to the elements of the first row in the measurement matrix, which are input in advance. The laser beam reflected by the DMD is irradiated onto the target through an expanded beam. The reflected laser from the target surface is focused by the condensing lens on the photo-sensitive element of the single-pixel detector, which then receives an electrical signal after photo-electric conversion. The signal is collected by the capture card, and a measured value is obtained after the phase detection process.

The remaining elements of the measurement matrix are entered in sequence, and the above operation is repeated to obtain a number of measured values that correspond to the required sampling rate. After performing the compressed sensing reconstruction processing on the obtained measurement values, the delay phases of the modulated laser at various points in the target area are obtained. Finally, according to the phase method ranging principle, the distance of each point in the target area is calculated, and the 3D information about the target can be obtained.

2.2. Measurement Principle

The two-dimensional signal of the target image, with a resolution of $n \times n$, can be connected in columns or rows, and converted to a one-dimensional discrete signal x. There are $N = n^2$ elements in x, of which there are K non-zero values. After obtaining the linear measurement values $y \in R^M$ under the measurement matrix $\Phi \in R^{M \times N}(M < N)$, the compressed sensing sampling process can be described as follows [15]:

$$y = \Phi x. \tag{1}$$

Equation (1) is solved by the reconstruction algorithm; a one-dimensional signal x can be reconstructed from y, and then restored to a two-dimensional signal by x.

According to the above-mentioned compressed sensing imaging theory, M measurements must be performed by the system used in this study. That is, the DMD needs to be flipped M times. Each row element (1 or 0) in the pre-selected measurement matrix Φ ($M \times N$) corresponds to the switching state of the micro-mirror at each pixel of the DMD in a single measurement. The resolution of the DMD micro-mirror array is $n \times n$. Let the horizontal and vertical coordinates of the DMD pixel be i and j, respectively. In the kth measurement of these M measurements, the state of a single micro-mirror in (i, j) in DMD is

$$\Phi_{ij}^{k} = \begin{cases} 1 \\ 0 \end{cases}. \tag{2}$$

$\Phi_{ij}^{k} = 1$ means that the micro-mirror is on (+ 12° flip), and $\Phi_{ij}^{k} = 0$ means that the micro-mirror is off (−12° flip). The transmitted laser signal after sine wave modulation is

$$X_T^k(t) = M_T^k \cos(wt + \varphi_0), \tag{3}$$

where M_T^k is the amplitude of the transmitted signal, w is the modulation angular frequency of the transmitted signal, and φ_0 is the initial phase of the transmitted signal. The optical signal after DMD modulation can be expressed as

$$X_T^k(t) = \Phi_{ij}^{k} \cdot M_T^k \cos(wt + \varphi_0). \tag{4}$$

Because the distance from each point of the target surface to the transmitting end is different, the transmitted signal will have different phase delays at each point. After the emitted light is reflected by the target, the reflected signal at the corresponding pixel point (i, j) is

$$X_{R_{ij}}^k(t) = \Phi_{ij}^{k} \cdot M_{R_{ij}}^k \cos(wt + \varphi_0 + \Delta\varphi_{ij}), \tag{5}$$

where $M_{R_{ij}}$ is the amplitude of the reflected signal at (i, j) due to the attenuation of light intensity, and $\Delta\varphi_{ij}$ is the delay phase of the reflected signal at (i, j) relative to the transmitted signal.

The total reflected light received at the kth time step can be expressed as the sum of the reflected light at each point:

$$
\begin{aligned}
X_R^k(t) &= \sum_{i=1}^{n}\sum_{j=1}^{n} \Phi_{ij}^k \cdot X_{R_{ij}}(t) \\
&= \cos(wt+\varphi_0)\cdot \sum_{i=1}^{n}\sum_{j=1}^{n} (\Phi_{ij}^k \cdot M_{R_{ij}} \cos\Delta\varphi_{ij}) \\
&- \sin(wt+\varphi_0)\cdot \sum_{i=1}^{n}\sum_{j=1}^{n} (\Phi_{ij}^k \cdot M_{R_{ij}} \sin\Delta\varphi_{ij})
\end{aligned} \tag{6}
$$

The total reflected light can be received by a single-pixel photo-detector and measured. The total reflected signal amplitude is M_R^k, and the delay phase relative to the transmitted signal is $\Delta\Phi^k$. An alternative expression of the total reflected light is

$$
\begin{aligned}
X_R^k(t) &= M_R^k \cos(wt+\varphi_0+\Delta\Phi^k) \\
&= M_R^k\left[\cos(wt+\varphi_0)\cos\Delta\Phi^k - \sin(wt+\varphi_0)\sin\Delta\Phi^k\right]
\end{aligned} \tag{7}
$$

Comparing Equations (6) and (7), the following can be obtained:

$$
\begin{cases}
\sum_{i=1}^{n}\sum_{j=1}^{n} \Phi_{ij}^k \cdot M_{R_{ij}} \cos\Delta\varphi_{ij} = M_R^k \cos\Delta\Phi^k = a^k \\
\sum_{i=1}^{n}\sum_{j=1}^{n} \Phi_{ij}^k \cdot M_{R_{ij}} \sin\Delta\varphi_{ij} = M_R^k \sin\Delta\Phi^k = b^k
\end{cases}. \tag{8}
$$

After M measurements, two sets of measurement values, A and B, are obtained from Equation (8):

$$
\begin{cases}
A = \left[a^1, a^2, a^3, \cdots\cdots, a^k, \cdots\cdots, a^M\right]^T \\
B = \left[b^1, b^2, b^3, \cdots\cdots, b^k, \cdots\cdots, b^M\right]^T
\end{cases}. \tag{9}
$$

Equation (9) expresses the product of the sine and cosine of the phase difference of each pixel and the signal amplitude as two column vectors:

$$
\begin{cases}
x_1 = \begin{bmatrix} M_{R_{11}}\sin\Delta\varphi_{11}, M_{R_{12}}\sin\Delta\varphi_{12}, \cdots\cdots, M_{R_{1n}}\sin\Delta\varphi_{1n}, \cdots \\ \cdots, M_{R_{n1}}\sin\Delta\varphi_{n1}, \cdots\cdots, M_{R_{nn}}\sin\Delta\varphi_{nn} \end{bmatrix}^T \\
x_2 = \begin{bmatrix} M_{R_{11}}\cos\Delta\varphi_{11}, M_{R_{12}}\cos\Delta\varphi_{12}, \cdots\cdots, M_{R_{1n}}\cos\Delta\varphi_{1n}, \cdots \\ \cdots, M_{R_{n1}}\cos\Delta\varphi_{n1}, \cdots\cdots, M_{R_{nn}}\cos\Delta\varphi_{nn} \end{bmatrix}^T
\end{cases}. \tag{10}
$$

According to Equations (8)–(10), and the measurement matrix Φ corresponding to M measurements, we can obtain

$$
\begin{cases}
A = \Phi x_1 \\
B = \Phi x_2
\end{cases}. \tag{11}
$$

The two equations in Equation (11) are solved by the compressed sensing reconstruction algorithm, and the vectors x_1 and x_2 are reconstructed. We can eliminate the unknown parameter M_{Rij} in x_1 and x_2, and we then get the delay phase $\Delta\varphi_{nn}$ of the target. The distance of each point is calculated according to the phase method ranging principle to complete the compressed sensing 3D imaging of the target.

3. Simulation Experiments

3.1. Simulation Process

To study the effects of different sampling rates, sparse transform bases, measurement matrices, and reconstruction algorithms on the effect of 3D imaging reconstruction in compressed sensing, simulation experiments were performed in the MATLAB software environment. Figure 2 is the simulation target used in the simulation experiment. In the figure, the distance d_1 of the grayscale value 255 (white) of the pixel is 1 m, and the distance d_2 of the grayscale value 0 (black) is 6 m.

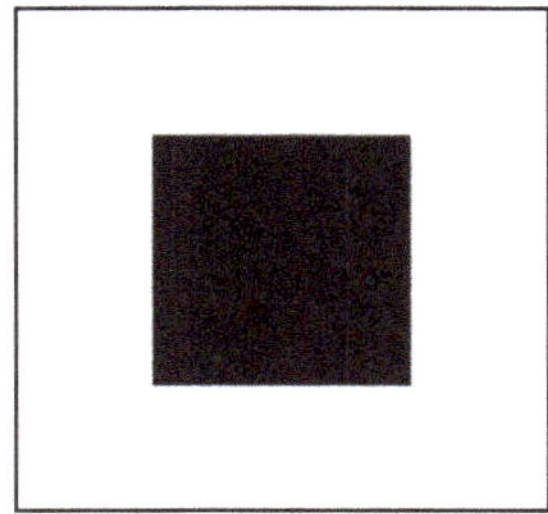

Figure 2. An 8 × 8 target map.

The laser modulation frequency was f = 5 × 10^6 Hz, and the sampling frequency was fs = 1 × 10^9 Hz. Comparative experiments were conducted using the following compressed sensing parameters:

1. Sampling rate: 0.2, 0.4, 0.6, and 0.8.
2. Sparse transformation basis: Discrete cosine transform (DCT) and fast Fourier transform (FFT).
3. Measurement matrix: Partial Hadamard matrix (HA) [16] and Bernoulli matrix (BE) [17].
4. Reconstruction algorithm: Basis pursuit (BP) [18], orthogonal matching pursuit (OMP) [19], and minimum total variation (TV) [10,20].

Because the DMD micro-mirror can only represent 1 or 0, we select as the measurement matrix the partial Hadamard matrix and Bernoulli matrix, and we replace -1 in the matrix by 0 [21]. TV is an image restoration method based on the discrete gradient of an object. The discrete gradient of most natural images is sparse, so this algorithm does not need to perform sparse transformation. The simulation uses the root mean square error (RMSE) of the delay phase obtained after reconstruction, at each point, as a metric to evaluate the reconstruction effect.

3.2. Simulation Experiment Conclusion

The simulation results are shown in Figure 3.

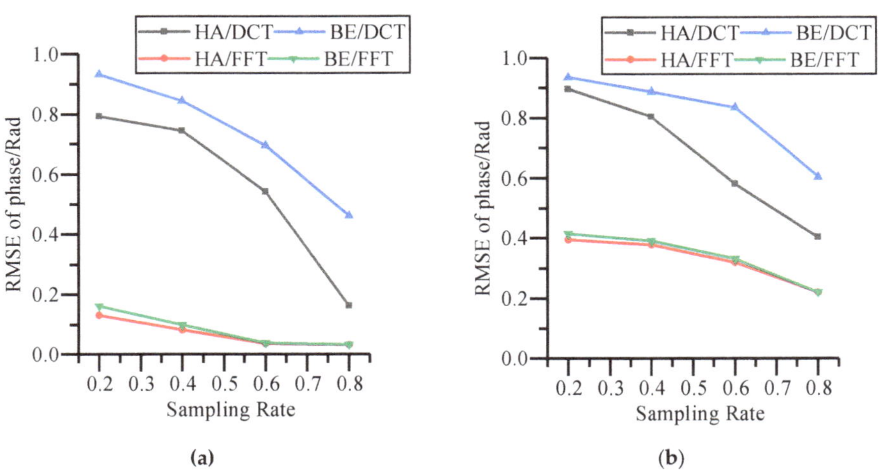

Figure 3. *Cont.*

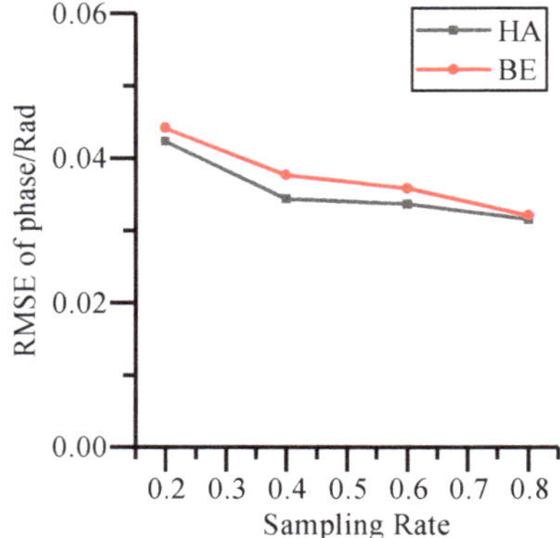

Figure 3. Reconstruction results of three reconstruction algorithms, based on different measurement matrices and sparse transforms. (**a**) Basis pursuit (BP) algorithm; (**b**) orthogonal matching pursuit (OMP) algorithm; (**c**) total variation (TV) algorithm.

The results show that under the same sparse transform basis, measurement matrix, and reconstruction algorithm, increasing the sampling rate reduces the phase RMSE and improves the reconstruction effect. When the TV algorithm has the same sampling rate and a different measurement matrix, the phase RMSE after reconstruction is not much different, indicating that the two measurement matrices have a considerable impact on the reconstruction effect. The reconstruction effect of the TV algorithm is better than the effect of the BP or OMP algorithm. The reconstruction effect of the BP algorithm is better than that of the OMP algorithm. The reconstruction effect using FFT sparse transform is clearly better than when using DCT sparse transform. When the sampling rate is the same and the measurement matrix is different, the Hadamard matrix is slightly better than the Bernoulli matrix.

A comparison of the three reconstruction algorithms is shown in Figure 4 using the Hadamard matrix and FFT sparse transform.

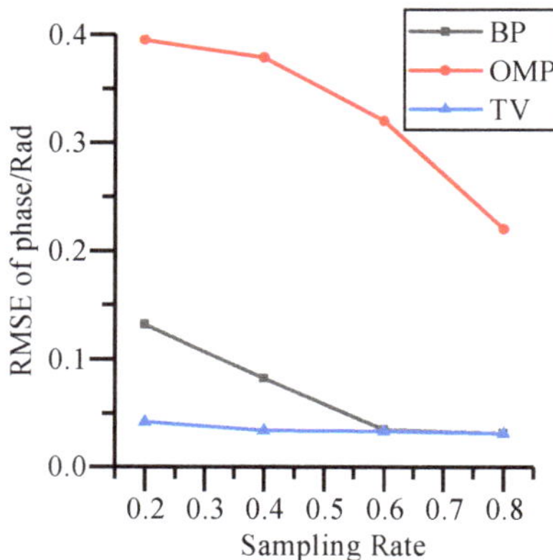

Figure 4. Comparison of the results of three reconstruction algorithms.

Figure 4 shows that the BP algorithm can reconstruct the target map more accurately at a higher sampling rate (0.8 or 0.6), but its phase RMSE is larger at a low sampling rate. The OMP algorithm cannot reconstruct the target image accurately. The phase RMSE of the TV algorithm is maintained at approximately 0.03, and the reconstruction effect is good.

The simulation results show that the reconstruction effect of the FFT sparse transform is obviously better than that of the DCT sparse transform; the reconstruction effect of the Hadamard measurement matrix is slightly better than that of the Bernoulli measurement matrix; and the TV reconstruction algorithm is slightly better than BP, with OMP producing the worst results.

Figure 5 shows the 3D image recovered by the TV reconstruction algorithms and Hadamard matrix. When the sampling rate is 0.8 or 0.6, the 3D imaging effect is good.

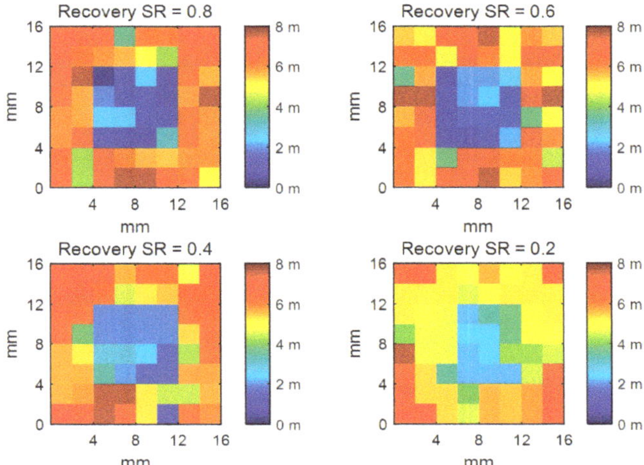

Figure 5. Three-dimensional image reconstructed by TV reconstruction algorithms when the sampling rate is 0.8, 0.6, 0.4, and 0.2.

In order to further verify that this scheme is also feasible for a complex target, we increased the resolution of imaging and the complexity of the imaging target in the simulation experiments. Figure 6 shows the target with a resolution of 16 × 16. Figure 7 shows the reconstructed 3D image using TV reconstruction algorithms. This illustrates that the complex target can also be successfully reconstructed at a sampling rate of 0.8.

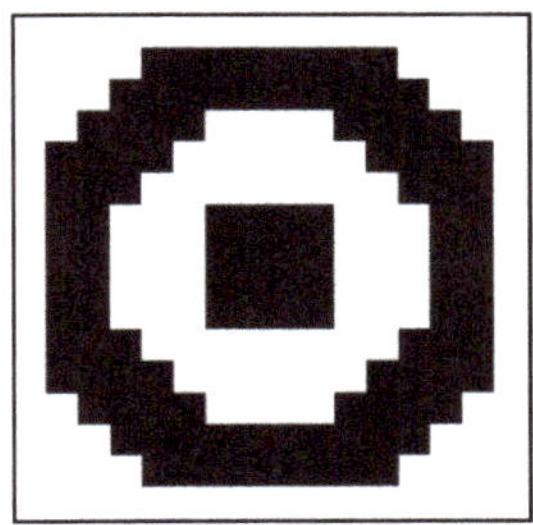

Figure 6. A 16 × 16 target map.

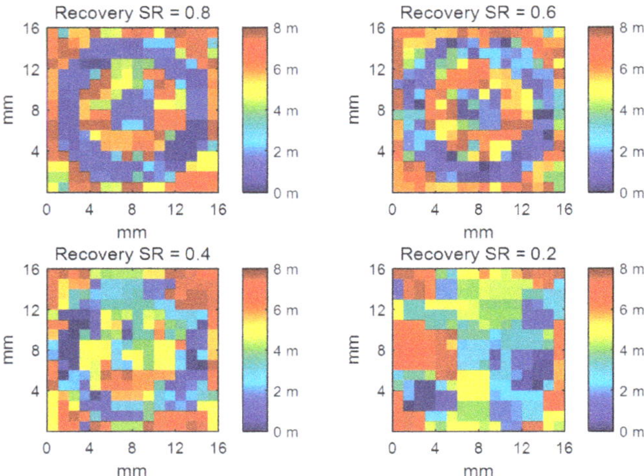

Figure 7. Reconstructed three-dimensional image of a complex target when the sampling rates are 0.8, 0.6, 0.4, and 0.2.

4. Imaging Experiments

4.1. Experimental Process

According to the system structure described in Section 2.1, we constructed an experimental system platform, as shown in Figure 8. The experimental device was installed on a vibration isolation platform, and the position and relative distance of the lens and other experimental devices were manually adjusted.

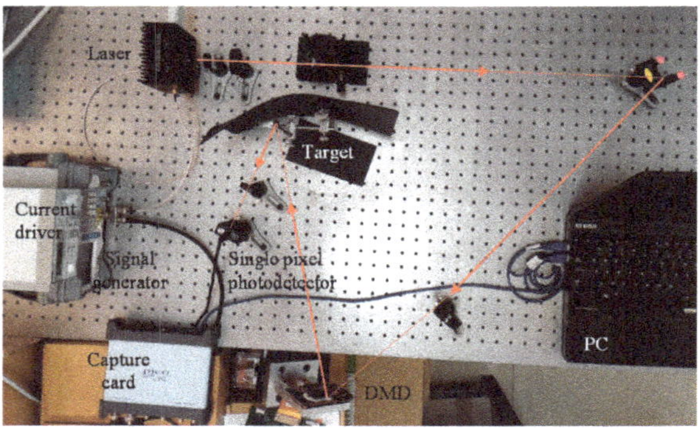

Figure 8. Experimental system platform.

In the experiment, four different sampling rates (0.2, 0.4, 0.6, and 0.8) were selected, and compressed sensing 3D imaging was performed at these sampling rates. The test target is shown in Figure 9.

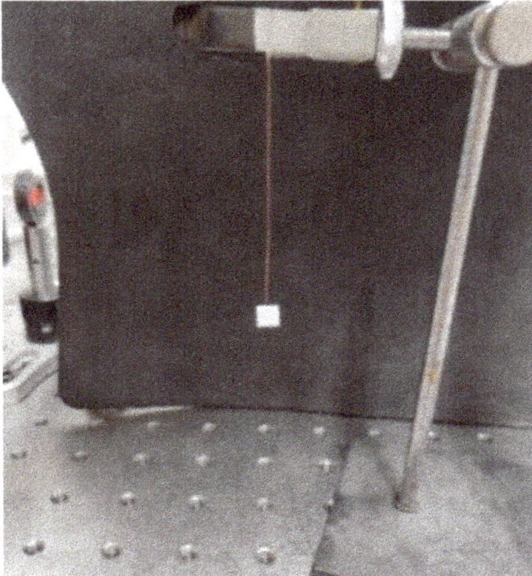

Figure 9. Experimental test target.

The target to be measured was a square target with a set distance of 2 m and a modulation frequency of 5×10^6 Hz. The measurement matrix was a Hadamard matrix, the sparse transform was FFT, and the reconstruction algorithms were BP, OMP, and TV.

4.2. Experimental Results and Discussion

After multiple experiments and post-processing, 3D imaging phase RMSE comparison maps of the three reconstruction algorithms were obtained, as shown in Figure 10. Figure 11 shows the 3D image recovered by the three reconstruction algorithms.

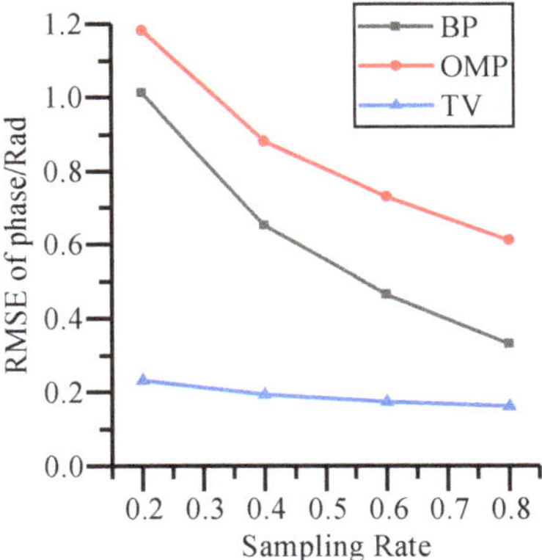

Figure 10. Comparison of 3D imaging phase root mean square error (RMSE) with three reconstruction algorithms.

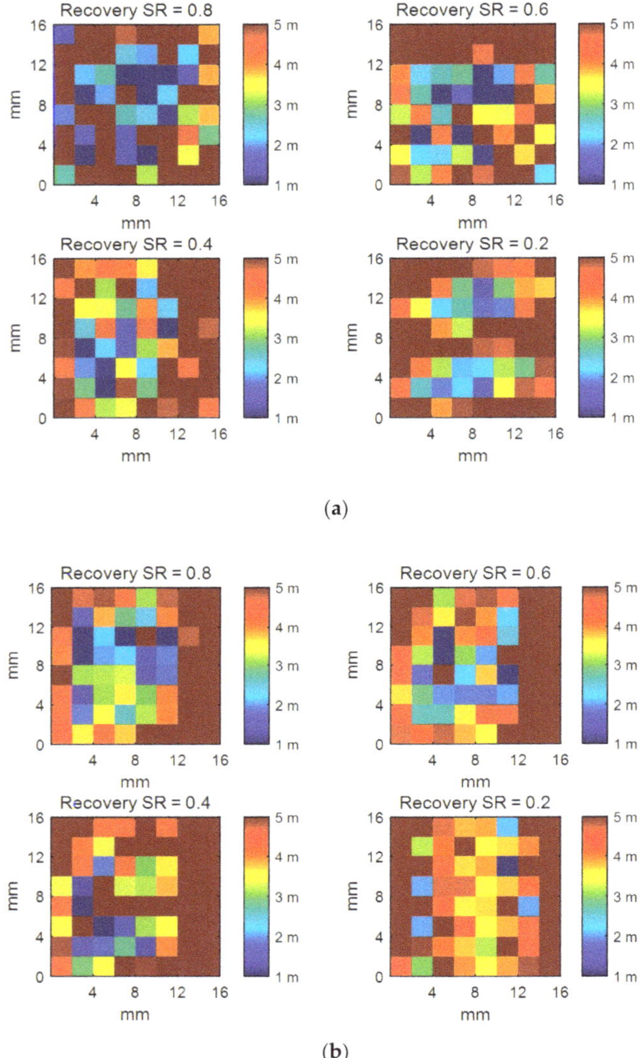

(a)

(b)

Figure 11. *Cont.*

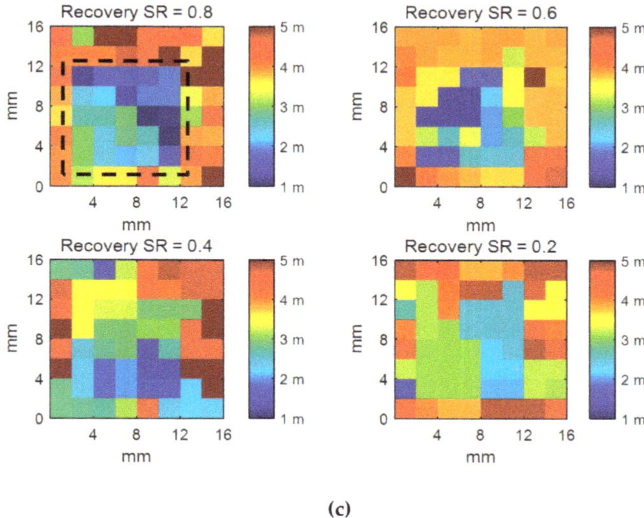

(c)

Figure 11. Three-dimensional image reconstructed by three reconstruction algorithms when the sampling rates are 0.8, 0.6, 0.4, and 0.2. (**a**) BP algorithm; (**b**) OMP algorithm; (**c**) TV algorithm.

Figure 10 shows that the phase RMSE of the three reconstruction algorithms decreases as the sampling rate increases, which indicates a better 3D imaging effect. The TV reconstruction algorithm has the best reconstruction effect, followed by BP, with OMP being the worst, which is consistent with the simulation results of Section 3.2. From the experimental 3D image in Figure 11, it can be observed that when the TV algorithm has a sampling rate of 0.8, the imaging effect of the square target can maintain the basic contour shape (dashed box). When the sampling rate is reduced to 0.6, the edge information of the 3D imaging is partially missing, and the basic contour shape is not obvious. When the sampling rate is 0.4 or 0.2, the noise is large, and the target shape contour can no longer be discerned at all. BP and OMP cannot display the basic shape of the target well for any sampling rate. The experimental results verify the effectiveness of the single-pixel detector non-scanning 3D imaging system, and they further verify that the TV reconstruction algorithm is far better than the BP and OMP algorithms.

In the experiment, the imaging results are blurry. Limited by the existing laboratory conditions, we set the DMD resolution to 8×8, which resulted in a low lateral resolution of the imaging. A weak reflected laser from the diffuse target and a limited laser modulation frequency result in low longitudinal accuracy. In future work, we will optimize the optical path system to reduce the laser power loss in the optical system and increase the laser modulation frequency to improve the longitudinal accuracy. Then, a high imaging resolution can be obtained by setting the DMD to high resolution.

5. Conclusions

In this paper, a single-pixel non-scanning 3D imaging system was designed, which combines compressed sensing technology with phase method laser ranging technology. It can use a single-pixel detector to achieve 3D imaging of a target. Parametric simulation studies and actual imaging experiments prove that the Hadamard measurement matrix and the TV reconstruction algorithm produce better imaging results at the same sampling rate. In the future, we will attempt to increase the laser modulation frequency and imaging resolution to further improve the measurement accuracy and imaging resolution. This system will have even wider application prospects in the field of 3D imaging.

Author Contributions: G.S. conceived the method, designed the experiments, and revised the manuscript; L.Z. performed the simulations and experiments, processed the data and wrote the original manuscript; W.W. provided the experimental funds and reviewed the manuscript; K.L. provided assistance to the experiments and proofread the manuscript. All authors have read and agreed to the published version of the manuscript.

Funding: This work was supported by the National Natural Science Foundation of China under Grant No. 51505113, the Zhejiang Provincial Natural Science Foundation of China under Grant No. LZ16E050001, and the State Key Laboratory of Precision Measuring Technology and Instruments Project under Grant No. PIL1601.

Conflicts of Interest: The authors declare no conflict of interest.

References

1. Anthes, J.P.; Garcia, P.; Pierce, J.T.; Dressendorfer, P.V. Nonscanned ladar imaging and applications. *Proc. SPIE.* **1993**, *1936*, 11–22. [CrossRef]
2. Wang, H.; Liu, Z. 3-D laser imaging technology and applications. *Electron. Des. Eng.* **2012**, *12*, 160–163, 168.
3. Albota, M.A.; Heinrichs, R.M.; Kocher, D.G.; Fouche, D.G.; Player, B.E.; O'Brien, M.E.; Aull, B.F.; Zayhowski, J.J.; Mooney, J.; Willard, B.C.; et al. Three-dimensional imaging laser radar with a photon-counting avalanche photodiode array and microchip laser. *Appl. Opt.* **2002**, *41*, 7671–7678. [CrossRef] [PubMed]
4. Li, L.; Hu, Y.; Zhao, N.; He, M. Application of Three-Dimensional Laser Imaging Technology. *Laser Optoelectron. Prog.* **2009**, *46*, 66–71. [CrossRef]
5. Aull, B.F.; Schuette, D.R.; Young, D.J.; Craig, D.M.; Felton, B.J.; Warner, K. A study of crosstalk in a 256 × 256 photon counting imager based on silicon Geiger-mode avalanche photodiodes. *IEEE Sens. J.* **2015**, *15*, 2123–2132. [CrossRef]
6. Aull, B.F. Silicon Geiger-mode avalanche photodiode arrays for photon-starved imaging. *Proc. SPIE* **2015**, *9492*, 94920. [CrossRef]
7. Chen, N. Review of 3D laser imaging technology. *Laser Infrared* **2015**, *10*, 14–18.
8. Donoho, D.L. Compressive sensing. *IEEE Trans. Inf. Theory* **2006**, *52*, 1289–1306. [CrossRef]
9. Candès, E.J.; Tao, T. Near-Optimal Signal Recovery From Random Projections: Universal Encoding Strategies? *IEEE Trans. Inf. Theory* **2006**, *52*, 5406–5425. [CrossRef]
10. Candes, E.; Romberg, J.; Tao, T. Robust uncertainty principles: Exact signal reconstruction from highly incomplete frequency information. *IEEE Trans. Inf. Theory* **2006**, *52*, 489–509. [CrossRef]
11. Duarte, M.; Davenport, M.A.; Takbar, D.; Laska, J.N.; Sun, T.; Kelly, K.; Baraniuk, R.G. Single-pixel imaging via compressive sampling. *IEEE Signal Process. Mag.* **2008**, *25*, 83–91. [CrossRef]
12. Howland, G.; Dixon, P.B.; Howell, J.C. Photon-counting compressive sensing laser radar for 3D imaging. *Appl. Opt.* **2011**, *50*, 5917–5920. [CrossRef]
13. Guo, B. Three-Dimensional Compressed Sensing Imaging Using Phase-Shift Laser Range Finding. Master's Thesis, Beijing Institute of Technology, Beijing, China, 2014.
14. Sun, M.-J.; Edgar, M.P.; Gibson, G.M.; Sun, B.; Radwell, N.; Lamb, R.; Padgett, M.J. Single-pixel three-dimensional imaging with time-based depth resolution. *Nat. Commun.* **2016**, *7*, 12010. [CrossRef] [PubMed]
15. Eldar, Y.C.; Kutyniok, G. *Compressed Sensing: Theory and Applications*; Cambridge University Press: Cambridge, UK, 2012.
16. Hedayat, A.; Wallis, W.D. Hadamard Matrices and Their Applications. *Ann. Stat.* **1978**, *6*, 1184–1238. [CrossRef]
17. Yu, L.; Barbot, J.P.; Zheng, G.; Sun, H. Compressive Sensing With Chaotic Sequence. *IEEE Signal Process. Lett.* **2010**, *17*, 731–734. [CrossRef]
18. Chen, S.S.; Donoho, D.L.; Saunders, M.A. Atomic Decomposition by Basis Pursuit. *SIAM Rev.* **2001**, *43*, 129–159. [CrossRef]
19. Foucart, S.; Rauhut, H. *A Mathematical Introduction to Compressive Sensing*; Springer: New York, NY, USA, 2013.
20. Beck, A.; Teboulle, M. Fast Gradient-Based Algorithms for Constrained Total Variation Image Denoising and Deblurring Problems. *IEEE Trans. Image Process.* **2009**, *18*, 2419–2434. [CrossRef] [PubMed]
21. Bah, B.; Tanner, J. Improved Bounds on Restricted Isometry Constants for Gaussian Matrices. *SIAM J. Matrix Anal. Appl.* **2010**, *31*, 2882–2898. [CrossRef]

Article

An Improved Circumferential Fourier Fit (CFF) Method for Blade Tip Timing Measurements

Zhibo Liu, Fajie Duan *, Guangyue Niu, Ling Ma, Jiajia Jiang and Xiao Fu

State Key Laboratory of Precision Measuring Technology and Instruments, Tianjin University, Weijin Road, Tianjin 300072, China; zhiboliu@tju.edu.cn (Z.L.); niuguangyue@tju.edu.cn (G.N.); tinama_17116@tju.edu.cn (L.M.); jiajiajiang@tju.edu.cn (J.J.); fuxiao215@tju.edu.cn (X.F.)

* Correspondence: fjduan@tju.edu.cn; Tel.: +86-131-3205-3038

Received: 7 May 2020; Accepted: 20 May 2020; Published: 26 May 2020

Abstract: Rotating blade vibration measurements are very important for any turbomachinery research and development program. The blade tip timing (BTT) technique uses the time of arrival (ToA) of the blade tip passing the casing mounted probes to give the blade vibration. As a non-contact technique, BTT is necessary for rotating blade vibration measurements. The higher accuracy of amplitude and vibration frequency identification has been pursued since the development of BTT. An improved circumferential Fourier fit (ICFF) method is proposed. In this method, the ToA is not only dependent on the rotating speed and monitoring position, but also on blade vibration. Compared with the traditional circumferential Fourier fit (TCFF) method, this improvement is more consistent with reality. A 12-blade assembly simulator and experimental data were used to evaluate the ICFF performance. The simulated results showed that the ICFF performance is comparable to TCFF in terms of EO identification, except the lower PSR or more number probes that have a more negative effect on ICFF. Besides, the accuracy of amplitude identification is higher for ICFF than TCFF on all test conditions. Meanwhile, the higher accuracy of the reconstruction of ICFF was further verified in all measurement resonance analysis.

Keywords: blade tip timing; circumferential Fourier fit; synchronous vibration

1. Introduction

Rotating blade vibration measurements are of great significance for any turbomachinery research and development program [1–4]. As a non-contact measurement technique, the blade tip timing (BTT) technique has been widely used in blade vibration measurements, such as in aero-engines, turbines, and compressors [5–9]. This technique uses the time of arrival (ToA) of the blade tip passing the casing-mounted probes to give the blade vibration. Compared with traditional strain gauges [10], BTT has the advantages of low equipment cost, easy installation, and monitoring vibration of each blade that passes through the probe [11]. However, BTT signals are typically under-sampled. It brings difficulty for blade vibration analysis.

In BTT analysis, the character of the blade response is usually grouped into two distinct classes, namely, asynchronous vibration and synchronous vibration. Asynchronous vibration mainly occurs in the abnormal vibrations, such as rotating stall, surge, flutter, and bearing vibration [12–14]. In these cases, the blade vibration frequency is a non-integer multiple of the rotating frequency and the phase of the response can be arbitrary, which is shown in Figure 1a. Synchronous vibration is excited by the multiples of the rotating frequency [15,16]. At constant speed, the probe can only monitor the blade response at a fixed point since the phase of the response remains fixed relative to a stationary datum at a given speed. The under-sampling of synchronous vibration is shown in Figure 1b. Therefore, the degree of under-sampling is much higher for synchronous vibration, which leads to more difficultly in identifying the blade vibration parameters from raw data.

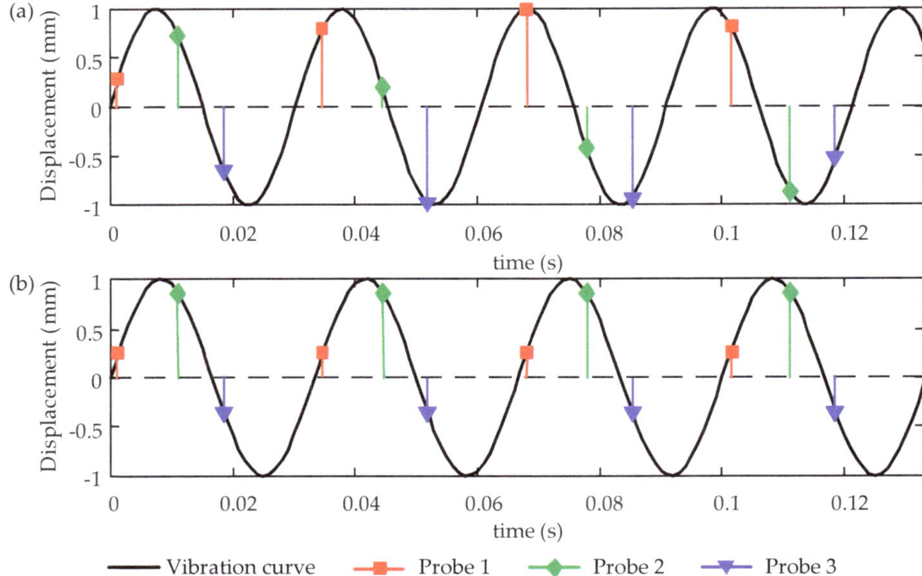

Figure 1. (**a**) Under-sampling of asynchronous vibration and (**b**) under-sampling of synchronous vibration.

Up to now, the identification methods for BTT signals are divided into two categories; one is spectrum analysis and the other is curve fitting. The spectrum analysis method generally maps BTT signals to frequency domain space to obtain information about blade vibration. This type of method usually requires that the sampling frequency is kept constant, that is, the signal is sampled on a constant rotating speed. Spectrum analysis methods mainly include a traveling wave analysis (TW) [17], the "5 + 2" method [18], the minimum variance spectrum estimation (MVSE) [19], non-uniform discrete Fourier transform (NUDFT) [20], cross-spectrum estimation (CSE) [21], sparse reconstruction (SR) method [22], etc. The spectrum method mainly is used to analyze asynchronous vibration. However, the results of spectrum analysis are always corrupted with aliases and replicas of the true frequency components. Consequently, the real spectrum recovery needs to further make use of the knowledge of the blade's dynamic properties, which is usually obtained based on the finite element method (FEM) [23]. The curve fitting method usually uses sine functions to fit the blade vibration displacements, which are obtained by speed sweeping through the resonance region. The curve fitting method mainly used to analyze synchronous vibration. These methods mainly include the single-parameter method [24], the two-parameter plot method [25], autoregressive (AR) method [26,27], the circumferential Fourier fit (CFF) method [28], the method without once per revolution [29,30], and so on. The accuracy of these curve fitting methods is often related to the amount of fitted data and the inter-blade coupling of the mistuned blade.

The CFF method is highly recommended for synchronous vibration analysis by Hood Technology Corporation, which is a commercial vendor of BTT systems [28]. This method can identify the amplitude and phase of vibration on the conditions that the engine order (EO) is known. Tao Ouyang proposed a traversed EO method based on CFF, which is no longer limited to known EO [31]. In this paper, we refer to the Ouyang CFF as the traditional CFF (TCFF). An improved formulation for the calculation of the TOA was proposed by S. Heath and M. Imregun, who proved the BTT analysis technique has limitations regarding synchronous response, in that the blade vibration was not considered in the TOA [32]. In this paper, an improved CFF (ICFF) method is proposed, which considers the influence of blade vibration when measuring the ToA of the blade tip. The ICFF is more robust with more probes employed to complete the parameter identification rather than the single-parameters or the improved single-parameters method that just used one probe. Compared to the AR method that requests the probes are equally spaced installed, the ICFF method has not got so strict limitations for probe layout. Moreover, the ICFF introduced into the EO traversed thought that proposed by the literature [31] and a more precision calculation method proposed by the literature [32]. These improvements are more consistent with reality. A 12-blade assembly simulator and experimental data were used to evaluate the performance of ICFF. In both simulated and experimental tests, the ICFF performed better than TCFF.

2. The Improved Circumferential Fourier Fit (ICFF) Method

2.1. A Brief Introduction to TCFF

The theory of TCFF is consistent with the CFF method. It uses multiple sensors (usually four) distributed at different positions in the circumferential direction of the casing to monitor blade vibration. On the condition that the EO is known, TCFF is equivalent to the CFF method, and the blade vibration parameters can be directly calculated according to the CFF method. In the case that the EO is unknown, the blade amplitude, vibration phase, direct current offset, and fitting residual error are calculated using the CFF method with each possible EO, which is selected within a certain range. After the traversal is completed, according to the principle of least squares, the EO corresponding to the minimum fitting residual error is the true EO, and the blade vibration parameters calculated using this EO are the true vibration parameters. More details about TCFF are provided by the literature [31]. The TCFF method introduces the idea of EO traversal to overcome the shortcomings of CFF, which must rely on known EO to identify blade vibration. Although TCFF did not make any changes to the CFF theoretical method, except to introduce EO traversal, to a certain extent, TCFF has wider engineering applicability than the CFF method. However, neither TCFF nor the CFF method takes into account the influence of blade vibration on the ToA. The ICFF method proposed in this paper further considers the effect of blade vibration on the ToA to achieve a higher accuracy of blade amplitude identification. Considering that TCFF can perform EO traversal, which overcomes the disadvantage that CFF must rely on known EO, this paper uses TCFF as a comparison object to illustrate the performance improvement of ICFF.

2.2. The Theoretical Basis of ICFF

Assuming the blade response is of single-frequency vibration:

$$y = A\sin(\omega t + \varphi) + d \tag{1}$$

where y represents the vibration displacement of the blade tip, A represents the amplitude, ω represents the angular frequency, t represents the ToA, ϕ represents the phase, and d represents the direct current offset. For the synchronous vibration, there is:

$$\omega = \mathrm{EO}\cdot\omega_v \tag{2}$$

where ω_v represents the rotor rotating frequency. This is based on the BTT measurement principle, assuming that the blade rotates to the monitoring position β at time t. The geometry relationship between the rotation angular, monitoring position, and blade vibration is shown in Figure 2. The equivalence relationship can be given by:

$$\omega_v t = \int_0^t \overline{\omega}_v dt = 2\pi n + \beta + \varepsilon - y/R \tag{3}$$

where $\overline{\omega}_v$ is the average rotating speed, and n is the number of revolution. In the Equation (3), it can be seen that the time t (that is ToA) is not only dependent on the rotating speed and monitoring position, but also on the blade vibration.

Let:

$$\widetilde{\beta} = \beta + \varepsilon - y/R \tag{4}$$

According to Equations (1)–(4), the blade vibration displacement monitored by the probe mounted at the position β can be given:

$$y = A\sin(EO\widetilde{\beta} + \varphi) + d. \tag{5}$$

Equation (5) is the mathematical model for monitoring the vibration displacement of the blade tip under the influence of blade vibration that is considered. According to the literature [32], in the case of large vibration displacement or mistuning, the improved single-parameter method that considers the influence of blade vibration has higher accuracy than the traditional method. ICFF is a combination of the theory of the improved single-parameter method and TCFF.

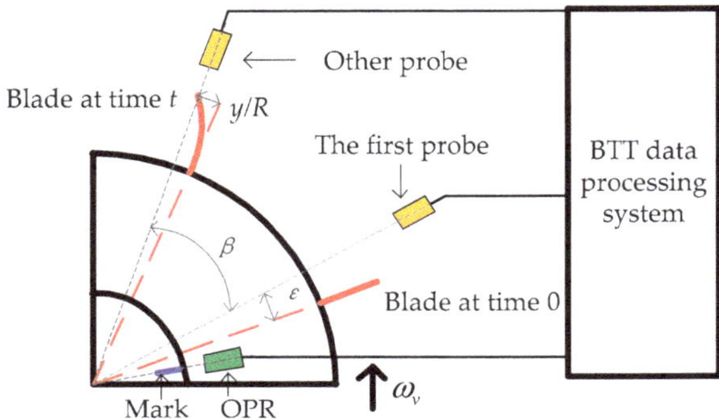

Figure 2. The schematic of the blade tip timing (BTT) measurement. The once per revolution (OPR) sensor monitors the rotating speed through the mark on the shaft and provides a timing reference for the BTT measurement system. The probe is responsible for monitoring the ToA. R represents the distance from the blade tip to the center of the rotor, and y/R represents the circumferential angle of the blade vibration. β represents the monitoring position, and ε represents the angle of the blade from the first probe when the mark passes the OPR sensor.

2.3. The Derivation of ICFF

Equation (5) is transformed by triangle formulas:

$$y = A\sin(\mathrm{EO}\widetilde{\beta})\cos(\varphi) + A\cos(\mathrm{EO}\widetilde{\beta})\sin(\varphi) + d \tag{6}$$

TCFF usually needs at least four probes to monitor blade vibration. In this section, assuming that there are g probes, the blade vibration displacement sequence can be represented according to Equation (6):

$$\begin{bmatrix} y_1 \\ y_2 \\ \vdots \\ y_g \end{bmatrix} = \begin{bmatrix} \sin(\mathrm{EO}\widetilde{\beta}_1) & \cos(\mathrm{EO}\widetilde{\beta}_1) & 1 \\ \sin(\mathrm{EO}\widetilde{\beta}_2) & \cos(\mathrm{EO}\widetilde{\beta}_2) & 1 \\ \vdots & \vdots & \vdots \\ \sin(\mathrm{EO}\widetilde{\beta}_g) & \cos(\mathrm{EO}\widetilde{\beta}_g) & 1 \end{bmatrix} \begin{bmatrix} A\cos(\varphi) \\ A\sin(\varphi) \\ d \end{bmatrix} \tag{7}$$

That is:

$$\mathbf{Y} = \mathbf{BX} \tag{8}$$

where **Y** is the displacement vector, **B** is the coefficient matrix, and **X** is the parameter matrix. The main difference between ICFF and TCFF is the constituent parameters of the coefficient matrix **B**, although Equation (7) in this paper is the same in form as TCFF. For the TCFF method, the coefficient matrix **B** is completely determined by the probe mounted position β and the initial angle ε, that is, once the probe installation scheme is determined, the coefficient matrix **B** is also determined. This means that the coefficient matrix **B** is a constant in the TCFF method. The same applies to the CFF method. However, the coefficient matrix **B** in ICFF is a function of three parameters: the probe installation angle β, the initial angle ε, and the blade vibration displacement y/R. According to the derivation in Section 2.2, blade vibration displacement is also one of the most important parameters affecting the ToA. In this case, the coefficient matrix **B** is no longer a constant but can be adjusted in real-time according to the actual vibration displacement of the blade.

Based on the principle of least squares and EO traversal, when the possible engine order EO_k ($k = 1, 2, 3 \ldots$) is selected, there is:

$$\mathbf{X}_k = \left(\mathbf{B}_k^T \mathbf{B}_k\right)^{-1} \mathbf{B}_k^T \mathbf{Y}. \tag{9}$$

The condition number is the indicator of the quality of the matrix **B**, which influence the accuracy of **X** directly. The larger the condition number, the more ill-condition matrix **B** is, and the more sensitive of the calculated results are to the measurement error. An effective way to improve the quality of the matrix **B** is to adjust the sensor layout. Therefore, based on the minimization the condition number of matrix **B**, a method about determining the sensor layout can be derived. The condition number mentioned above and the probe spacing of the resonance (PSR) used later in the paper as an indicator of sensor distribution both are methods essentially to determine how the sampled interval of periodic signals are arranged. However, PSR is more common.

Define the corresponding fitted residuals as S_k:

$$S_k = \frac{\|\mathbf{B}_k \mathbf{X}_k - \mathbf{Y}\|_2}{\sqrt{g-1}} \tag{10}$$

For the number of revolution n that contains the blade synchronous resonance, S_k can be instead by $\widetilde{S_k}$:

$$\widetilde{S_k} = \frac{\sum S_k}{n} \tag{11}$$

The EO_k (k = 1, 2, 3 ...) that minimizes $\widetilde{S_k}$ is the true EO. After the true EO and parameter matrix **X** are determined, the amplitude, phase, and direct current offset can be further obtained by the following formula:

$$\begin{aligned} A &= \sqrt{\mathbf{X}(1)^2 + \mathbf{X}(2)^2} \\ \varphi &= \begin{cases} \arctan\left(\frac{\mathbf{X}(2)}{\mathbf{X}(1)}\right) & \mathbf{X}(1) > 0 \\ \arctan\left(\frac{\mathbf{X}(2)}{\mathbf{X}(1)}\right) + \pi & \mathbf{X}(1) < 0 \end{cases} \\ d &= \mathbf{X}(3) \end{aligned} \tag{12}$$

where $\mathbf{X}(i)$ (i = 1, 2, 3) represents the ith row element of the matrix **X**. For n revolutions, after obtaining each revolution of A, ϕ, and d through Equation (12), it is necessary to calculate the respective averages to get the final value.

3. Method Evaluation Using Simulated Data

3.1. Simulated Model

For the actual blade vibration measurement, even if the blade dynamic behavior is fully understood, it is still difficult to eliminate uncertainties in the measurement data [33,34], such as speed fluctuations, system random errors, etc. These uncertainties will affect the evaluation accuracy, therefore, it is better to use simulated data without measurement uncertainties to evaluate the accuracy of the reconstruction methods. A 12-blade assembly simulator was used to evaluate the ICFF performance. The schematic of the simulated model was shown in Figure 3. A mass-spring-damper system represents each blade, which is coupled to its two neighbors through two further spring–damper assemblies. A mathematical model was employed to simulate the forced vibration of the rotating bladed assembly. This mathematical model ignored the effects of the centrifugal force and temperature on the blade stiffness.

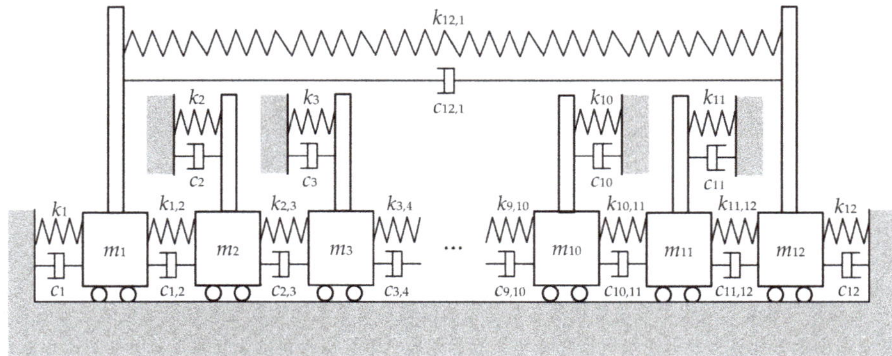

Figure 3. The simulated model. m_i, k_i, and c_i (i = 1, 2, 3... 12) represent the mass, stiffness, and damping coefficient of blade i, respectively. $k_{i,\,j}$ and $c_{i,\,j}$ ($|i-j| = 1$) represent the coupling stiffness and coupling damping coefficient between adjacent blades.

The mathematical model of the bladed assembly is given by:

$$\mathbf{M}\ddot{y} + \mathbf{C}\dot{y} + \mathbf{K}y = \mathbf{F}(t) \tag{13}$$

where **M**, **C**, **K**, and **F**(t) represent the mass matrix, damping matrix, stiffness matrix, and excitation force vector, respectively. Let **W**(i, j) ($1 \le i \le 12$ and $1 \le j \le 12$) represents the element at the ith row and jth column of the matrix **W**, and above matrixes can be expressed as:

$$\begin{aligned} \mathbf{M}(i,j) &= \begin{cases} m_i & i = j \\ 0 & others \end{cases} \\ \mathbf{C}(i,j) &= \begin{cases} c_{i,j-1} + c_j + c_{i,j+1} & i = j \\ -c_{i,j} & |i-j| = 1 \\ 0 & others \end{cases} \\ \mathbf{K}(i,j) &= \begin{cases} k_{i,j-1} + k_j + k_{i,j+1} & i = j \\ -k_{i,j} & |i-j| = 1 \\ 0 & others \end{cases} \end{aligned} \tag{14}$$

In Equation (14), use 1 for subscripts greater than 12 and 12 for subscripts less than 1. The excitation force f_i (t) on blade i is given by:

$$f_i(t) = F_i \sin(\mathrm{EO}\omega_v t + \frac{2\pi \mathrm{EO}}{N} i), for \quad i = 0, 1, \ldots N \tag{15}$$

where F_i is the amplitude of the excitation force, and N represents the count of the blade. N was set to 12 in this paper.

The quality of probe distribution is represented by the probe spacing on the resonance (PSR). The PSR is the ratio of the difference between the times of arrival of a blade at the first and last probe to the period of the blade response at resonance [35], i.e.,:

$$\mathrm{PSR} = \omega_n(ToA_{last} - ToA_{first})/(2\pi), \tag{16}$$

where ω_n represents the natural angular frequency.

Further details on the mathematics of the simulator are given by the literature [36]. According to the literature [37,38], the mistuning coefficient Δf_i and the coupling coefficient h_i are defined to characterize the degree of blade mistuning and the coupling between blades. The concrete expressions of these two parameters are:

$$\begin{aligned} \Delta f_i &= \frac{k_i}{k} - 1 \\ h_i^2 &= \frac{k_{i,j}}{k_i}, \quad |i-j| = 1 \end{aligned} \tag{17}$$

where the subscripts i and j represent the blade number, and k represents the blade nominal stiffness. Let the nominal mass m, stiffness k, and damping c be 1 kg, 8.1×10^5 N/m, and 9 N·s/m, respectively. Then, the nominal natural frequency of the blade is 900 rad/s according to Equation (18):

$$\omega_n = \sqrt{k/m} \tag{18}$$

Let the distribution of the mistuning coefficients Δf_i of the 12-blade assembly simulator conform to a Gaussian distribution with a mean of 0 and a standard deviation of 0.04. The distribution of Δf_i is shown in Table 1. Assume that the blade actual mass m_i (i = 1, 2, 3 ... 12) is equal to nominal mass m. The actual natural frequency of the blade was calculated according to Equations (17)–(18) and Table 1, which is shown in Figure 4.

Table 1. The distribution of the mistuning coefficient Δf_i (%).

Δf_1	Δf_2	Δf_3	Δf_4	Δf_5	Δf_6	Δf_7	Δf_8	Δf_9	Δf_{10}	Δf_{11}	Δf_{12}
−0.706	3.165	−5.328	−9.319	−5.796	1.334	1.565	1.806	−0.521	0.734	−1.904	3.448

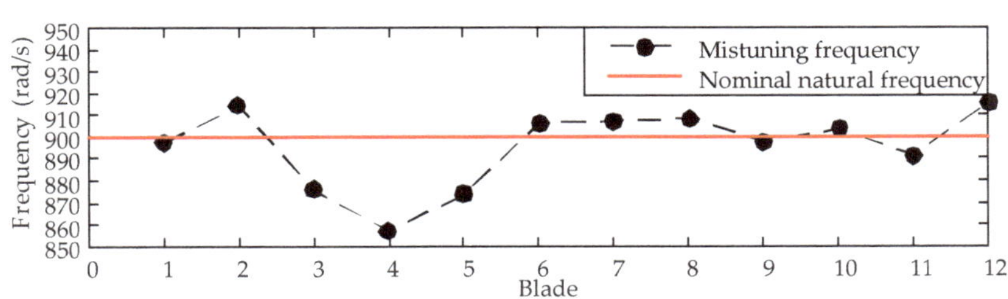

Figure 4. The mistuning frequency of the blade.

3.2. Simulated Test

The object of the simulated test was to assess analysis techniques for determining the EO and amplitude. Several exhaustive tests were performed using BTT data obtained from the simulator to evaluate the accuracy of the reconstruction method and their sensitivity to various test parameters. These parameters were:

(1) Probe spacing on the resonance (PSR);
(2) The number of revolution (n);
(3) Engine order (EO);
(4) Inter-blade coupling of the mistuned blade (ICoMB);
(5) The number of probes (g);
(6) Noise-to-signal ratio (NSR).

The NSR is defined as the ratio of the r.m.s. value of the noise to the value of the 'clean' amplitude of the blade. The NSR was set to 100 points averagely distributed in the range from 0.01 to 0.3. For the same NSR, a white noise generator was used to create 100 distinct noise sequences with which to corrupt the data obtained from the simulator. Each test was repeated using all the white noise sequences for each method on each NSR value. When there is noise in the data, the reconstruction method will contain two types of error [39]:

(1) Bias is where the estimates are incorrect. It is caused by a summated squared noise term in the $\mathbf{B}^T\mathbf{B}$ matrix that does not tend to zero but increases with increasing noise levels. This is an effect of the ICFF model that is used.
(2) Scatter, where there is a range of estimates spread about the mean value, is caused by random variations in the data.

The mean and variance of the recovered EO and amplitude were computed, from which 95 percent confidence intervals were obtained. The best method will exhibit the lowest bias and, preferably, lowest scatter. Six test cases were designed to analyze and compare the performance of the TCFF and ICFF under different test conditions. These are shown in Table 2. Test 1 was specifically chosen to be a very difficult case to analyze. Hence, the other tests were used to investigate whether specific changes in the parameters improve or degrade the performance of the method.

Table 2. The simulated test.

Test	PSR	n	True EO	g	ICoMB
Test 1	98%	30	10	4	No
Test 2	55%	30	10	4	No
Test 3	98%	50	10	4	No
Test 4	98%	30	13	4	No
Test 5	98%	30	10	7	No
Test 6	98%	30	10	4	Yes

The parameters whose background was shaded were the changed parameters relative to Test 1.

The PSR is a key factor that affects the performance of all BTT algorithms. It is also commonly used to determine the quality of probe distribution for blade synchronous vibration measurement. Currently, there are corresponding methods for optimizing the probe installation layout [40]. Combined with the FEM, the optimization of probe layout can be achieved in actual engineering measurement. Therefore, we set the PSR to a high value (98%) in Test 1. In engineering applications, the probe is inevitably affected by the temperature, airflow, etc., and in severe cases, the measurement data will not be available. In this case, where a limited number of probes were used to measure blade vibration, the failed probe could not participate in the blade vibration analysis, which destroyed the optimization of the probe layout and resulted in a decrease of the PSR. For this reason, the PSR was set to a low value (55 percent) in Test 2.

For TCFF, the blade vibration parameters can be calculated based on the single-revolution data obtained by multiple probes (usually four). However, it is impractical that the single-revolution data are used to identify the blade vibration parameters in engineering applications, considering the effects of system measurement errors. Just as the AR method does [39], it is reasonable to select multi-revolution data, including the blade synchronous resonance region, to determine the true vibration parameters based on the average of the calculation results. For any curve fitting algorithm, such as the single-parameter method, the amount of selected data should include the complete blade resonance response region as much as possible, although there are no specific rules on how many data points should be selected. In the simulated tests, 30 (Test 1) and 50 (Test 3) data points were selected for identifying the vibration parameters to demonstrate the effects of the data amounts on the quality of the identification of each method.

BTT technology is sensitive to a mode with a large amplitude at the blade tip, such as the first-order bending mode. In actual rotating machinery operation, due to mechanical structure or airflow disturbance, the same vibration mode of the blade may be excited at different resonance rotating speeds corresponding to different EO. For BTT algorithms, analysis of the blade vibration at different resonance rotating speeds is essentially a process of reconstructing signals with different degrees of under-sampling. The EO (Test 1: EO = 10; Test 4: EO = 13) was used to analyze and compare the sensitivity of TCFF and ICFF to different degrees of under-sampling, and the EO was also an important parameter for the blade vibration analysis.

Based on the CFF theory, at least three probes are needed to monitor blade synchronous vibration. The least-square method is used to calculate the blade synchronous vibration parameters if more than three probes are used. The TCFF usually uses four probes to increase robustness. However, Tao OuYang used seven probes in the experimental verification of TCFF [31]. So far, there is no optimal choice for the number of probes. Test 5 was designed to verify the performance of TCFF and ICFF on the different number of probes.

Mistuning is a universal feature of the objective existence of the blades, and if inter-blade coupling of the mistuned blade exists, it will affect the performance of the BTT algorithms that only analyze single-frequency signals, such as the single-parameter method, the two-parameter plot method, CFF, and TCFF. The same applies to the ICFF method proposed in this paper. Therefore, Test 6 was specially designed to analyze and compare the sensitivity of TCFF and ICFF to ICoMB.

The noise in the measurement data is difficult to eliminate. The "clean" simulated data was polluted by noise in all simulated tests to make the data closer to the real measurements. The anti-noise performance of each method was verified by using simulated data of different pollution degrees that were represented by the NSR value.

The monitor positions of the probes in Test 1, Test 3, and Test 6 were 0°, 13.1°, 25.3°, and 35.3°. The monitor positions of the probes in Test 4 were 0°, 11.3°, 13.3°, and 25.3°. In Test 2, the monitor positions of the probes were 0°, 6.3°, 10.1°, and 19.8°. In Test 5, the monitor positions of the probes were 0°, 6.3°, 13.1°, 19.1°, 25.3°, 30.2°, and 35.3°. The PSR in Table 2 was calculated according to Equation (16). If the option of the ICoMB is "NO", it means that the coupling coefficient is set to zero; if it is "Yes", it means the coupling coefficient is set to nonzero. In this paper, the coupling coefficient h_i (i = 1, 2, 3 ... , 12) was set to 0.1 for the option "Yes". It is must be stressed that the blade vibration does not influence the adjacent blades when the coupling coefficient is set to zero, even under the condition that the blade is mistuned. For this, it is still a single-frequency vibration for the single blade. However, the synchronous resonance of the single blade will be a superposition of multiple-frequency vibration under the condition that the coupling coefficient is set to nonzero, and it is difficult to distinguish. The coupled vibration of the mistuned blade is a serious challenge for TCFF, since it assumes the blade response is of single-frequency vibration. According to the results shown in Figure 4, the natural frequency of blade 7 is close to that of the adjacent blades. To evaluate the performance of ICFF under the condition of blade mistuning and inter-blade coupling, the simulated data of blade 7 were selected in the simulated tests. The rotating speed was set to accelerate uniformly from 300 to 2400 rpm, which included the resonance rotating speed for all simulated tests. Take the resonance rotating speed as the reference and select half n points before and after the reference point to compose the number of revolutions n. The results of each simulated test were analyzed as follows.

- Test 1:

As expected, Test 1 was a very difficult case. As shown in Figure 5a, the EO identification started to produce biased results at about NSR 0.16 for both methods. For tests with correct EO identification, the 95% confidence intervals of the relative error of the amplitude identification were calculated, which was shown in Figure 5b. The intervals increased with the increase of the NSR for both methods. However, the relative error of amplitude identification of ICFF was lower than that of TCFF.

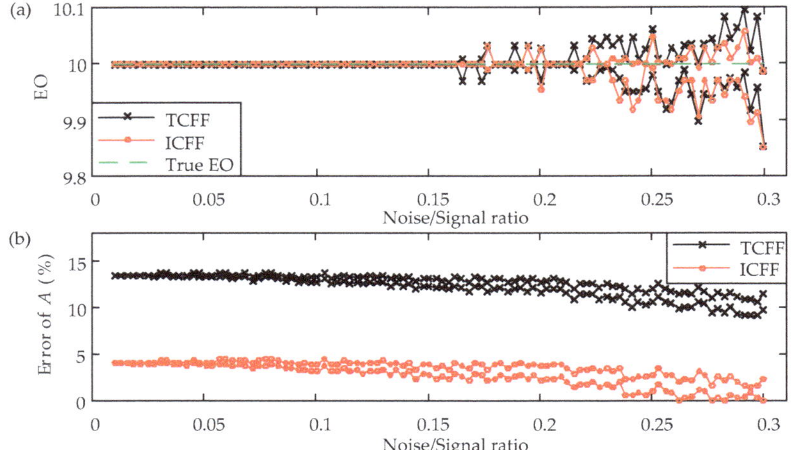

Figure 5. Results of Test 1. (**a**) 95% confidence intervals of the engine order (EO) identification and (**b**) 95% confidence intervals of the relative error of the amplitude identification.

- Test 2:

Reducing the PSR value to 55 percent produced the most dramatic degrade in the quality of the identification. The correct EO identification did not occur for ICFF before NSR 0.15, and the deviation of EO identification was very large for both methods compared with Test 1. In order to compare the two methods synchronously, Figure 6a only showed that the results of the EO identification in the NSR range of 0.15 and 0.30. The relative error of amplitude identification was calculated after NSR 0.15. The scatter of the relative error was higher than that of Test 1. As already discussed by other literature [35,36,39], curve fits of lower PSR data are generally not conducive to parameter identification.

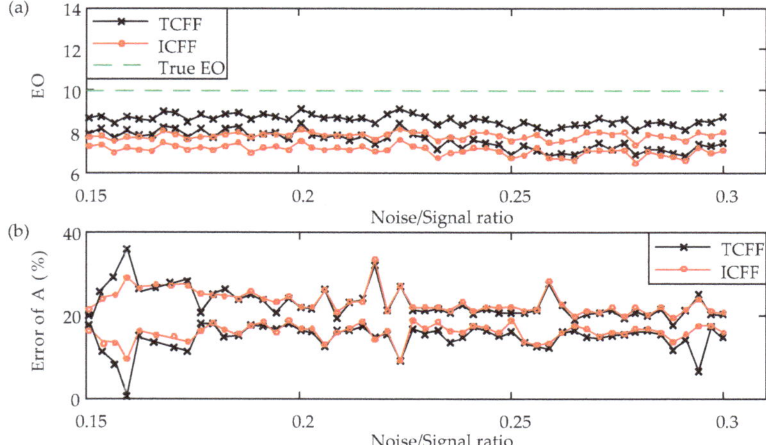

Figure 6. Results of Test 2. (**a**) 95% confidence intervals of the EO identification and (**b**) 95% confidence intervals of the relative error of the amplitude identification.

- Test 3:

Increasing the number of revolutions over which data were used for the identification process can improve the results. The results of EO identification started to be biased when the NSR increase to about 0.18 for both methods, which was shown in Figure 7a. The NSR corresponding to unbiased identification was higher 0.03 than Test 1. However, relative error for amplitude identification increased with increasing the number of data, which was caused by a summated squared noise term in the $\mathbf{B}^T\mathbf{B}$ matrix. This is an effect of the method model that is used. Although the relative error was increased for both methods, the ICFF error increased less than the TCFF.

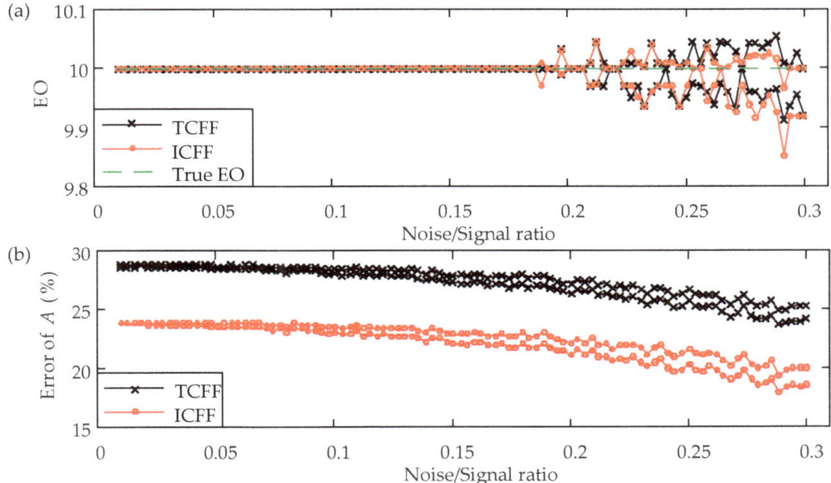

Figure 7. Results of Test 3. (**a**) 95% confidence intervals of the EO identification and (**b**) 95% confidence intervals of the relative error of the amplitude identification.

• Test 4:

The results of Test 4 were shown in Figure 8. Increasing the EO of the tip timing data had no visible effect on the accuracy of the EO identification for ICFF. However, the relative error of amplitude identification was higher than that of Test 1. For the same natural frequency, the larger EO, the higher degree of under-sampling. That means the number of samples was less for a period of vibration of the blade. The amplitude that recovered was more sensitive to the number of samples than the EO identification. Although the relative error was increased for both methods, the quality of ICFF was still higher than TCFF.

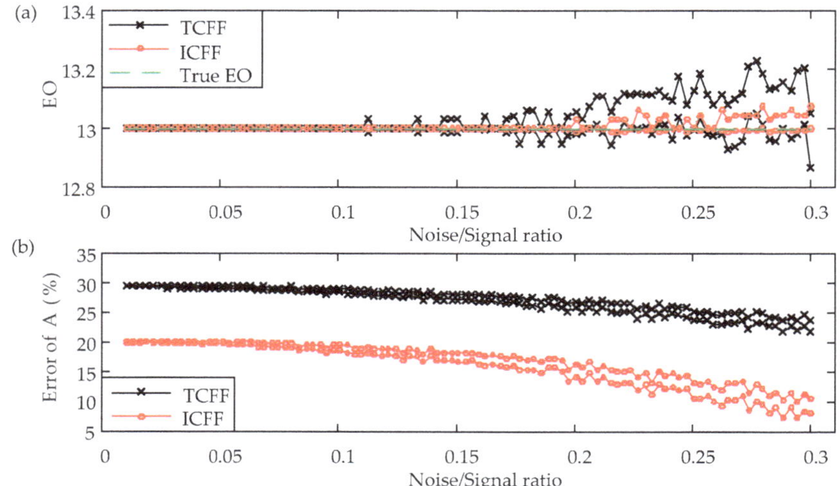

Figure 8. Results of Test 4. (**a**) 95% confidence intervals of the EO identification and (**b**) 95% confidence intervals of the relative error of the amplitude identification.

• Test 5:

Increasing the number of probes made the scatter of EO identification get larger for ICFF, which was shown in Figure 9a. Meanwhile, the relative error of amplitude identification was increased for both methods, which was shown in Figure 9b. The summated squared noise term in the $\mathbf{B}^{\mathrm{T}}\mathbf{B}$ matrix would increase with the number of probes. Then the squared noise term was transferred to amplitude identification. Figure 9b still showed the relative error of ICFF was smaller than TCFF.

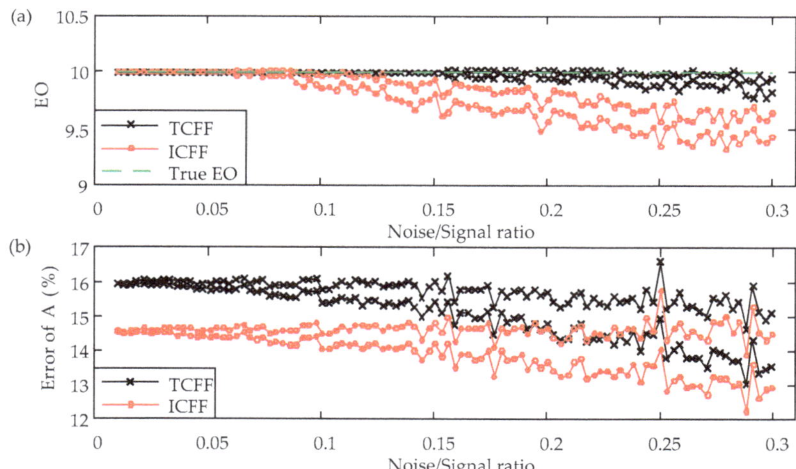

Figure 9. Results of Test 5. (**a**) 95% confidence intervals of the EO identification and (**b**) 95% confidence intervals of the relative error of the amplitude identification.

• Test 6:

The existence of the ICoMB did not have an effect on the EO identification for both methods, which was shown in Figure 10a. BTT data that consists of two modes corresponding to different EO generally had negative effects on the quality of identification, especially for the reconstruction method that processes the single-frequency signals. However, it did not belong to the range considered in this paper. It needs some future work for processing the multi-frequency signals based on ICFF. The ICoMB made the quality of amplitude identification degrade, which was shown in Figure 10b. However, the relative error of ICFF was still lower than that of TCFF.

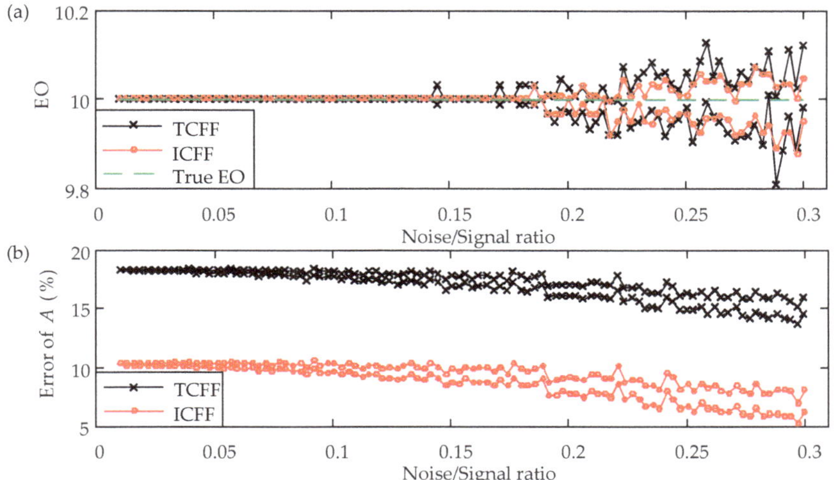

Figure 10. Results of Test 6. (**a**) 95% confidence intervals of the EO identification and (**b**) 95% confidence intervals of the relative error of the amplitude identification.

3.3. Summary

The percentage of correct EO identification and the mean relative error of amplitude identification were summarized in Tables 3 and 4 respectively for all simulated tests at the NSR 0.01, 0.05, 0.10, 0.15, 0.20, 0.25 and 0.30.

Table 3. Percentage of correct EO identification following the noise-to-signal ratio (NSR) values.

NSR	Test 1		Test 2		Test 3		Test 4		Test 5		Test 6	
	TCFF	ICFF	TCFF	ICFF	TCFF	ICFF	TCFF	ICFF	TCFF	ICFF	TCFF	ICFF
0.01	100%	100%	0	0	100%	100%	100%	100%	100%	100%	100%	100%
0.05	100%	100%	1%	0	100%	100%	100%	100%	100%	100%	100%	100%
0.10	100%	100%	14%	0	100%	100%	100%	100%	100%	91%	100%	100%
0.15	100%	100%	19%	2%	100%	100%	99%	100%	100%	87%	100%	100%
0.20	98%	97%	29%	5%	100%	100%	96%	99%	98%	66%	98%	100%
0.25	96%	98%	12%	15%	98%	99%	88%	100%	92%	46%	96%	99%
0.30	88%	88%	18%	11%	96%	96%	78%	96%	89%	55%	87%	94%

The results of the EO identification that were less than 80 percent were marked in red.

Table 4. Mean relative error of the amplitude identification following NSR values.

NSR	Test 1		Test 2		Test 3		Test 4		Test 5		Test 6	
	TCFF	ICFF	TCFF	ICFF	TCFF	ICFF	TCFF	ICFF	TCFF	ICFF	TCFF	ICFF
0.01	13.49%	4.06%	—	—	28.74%	23.82%	29.41%	20.04%	15.97%	14.53%	18.29%	10.32%
0.05	13.42%	4.00%	—	—	28.59%	23.64%	29.29%	19.96%	15.87%	14.46%	18.19%	10.19%
0.10	13.03%	3.56%	—	—	28.29%	23.37%	28.70%	18.97%	15.59%	14.25%	17.97%	10.02%
0.15	12.69%	3.35%	18.63%	19.07%	27.63%	22.48%	27.62%	17.53%	15.72%	14.47%	17.38%	9.30%
0.20	12.23%	2.98%	19.10%	19.38%	26.83%	21.63%	26.39%	15.25%	15.24%	14.16%	16.58%	8.47%
0.25	11.30%	1.92%	18.30%	19.50%	26.22%	20.94%	24.54%	11.78%	15.78%	14.92%	15.56%	7.65%
0.30	10.54%	1.23%	17.32%	18.29%	24.73%	19.32%	22.72%	9.24%	14.35%	13.70%	15.19%	7.17%

The results of amplitude identification that were higher than 20 percent were marked in red.

The results of the EO identification that were less than 80 percent were marked in red in Table 3. Curve fits of lower PSR data are generally not conducive to parameter identification, especially for the EO identification, which was demonstrated by Test 2. Increasing the number of probes would have some negative effect on the EO identification for ICFF, and the degree of influence became significantly larger when the data noise up to 0.20 (Test 5). The higher PSR value and the larger amount of fitted data were conducive to EO identification for both methods. The higher EO value (higher degrees of under-sampling) and coupling between mistuning blades (unconsidering multi-frequency vibration) had little effect on the results of EO identification. Through the comparison from Test 1, Test 3, Test 4 and Test 6, ICFF performance was comparable to TCFF in terms of EO identification.

The relative error for amplitude identification increased with the number of fitted data, which was verified by the results of Test 3. Higher EO value has a more negative effect on the amplitude identification of TCFF than that of ICFF (Test 4). Table 4 showed that the results of amplitude identification of ICFF were more accurate than TCFF, except that the low PSR (Test 2) resulted in the little difference between ICFF and TCFF. The larger number of fitted data and the number of probes would lead to a higher error of amplitude identification for both methods, which was caused by the squared noise term in the $\mathbf{B}^{T}\mathbf{B}$ matrix that increased with the number of fitted data or probes. Compared with TCFF, ICFF considers the influence of blade vibration on the ToA of the blade tip. This improvement is more consistent with reality and leads to more accuracy of the amplitude identification. This was demonstrated through all of the simulated tests.

Conclusions drawn above were based only on the numerical simulations but not based on the analysis of real signals measured during the experimental tests for a real object under loading conditions. In the experimental tests, different factors may affect the accuracy of the reconstruction method. So, the proposed method will be verified by experimental measurements and analysis in the next section.

4. Method Evaluation Using Experimental Data

The experimental data that was provided by the author of the literature [36] was used to further verify the feasibility of the reconstruction method proposed in this paper. The test rig in the literature [36] was shown in Figure 11a. The radius of the rotor is 60 mm, and there are eight blades. The once per revolution (OPR) sensor was installed near the shaft to provide the rotating speed and a timing reference through detecting a marking on the shaft. Seven optic probes were installed on the casing to measure the blade vibration. The installation angles of the probes were 0° (P_0), 18.56° (P_1), 36.17° (P_2), 53.75° (P_3), 72.42° (P_4), 119.82° (P_5), and 239.06° (P_6). The nitrogen shock was used as an exciting force to the rotating blades.

The mode frequency was plotted on the vertical axis in the Campbell diagram of Figure 11b. The diagram plotted the resonant frequency on the vertical axis and the rotating speed on the horizontal axis. The EO lines showed a range from 9 to 15 (blue lines). The seven black dots in the Campbell diagram represent the intersection of the EO lines with the mode frequency. The nominal resonant speeds corresponding to these seven intersections are 7081 rpm (EO = 15), 7613 rpm (EO = 14),

8202rpm (EO = 13), 8907 rpm (EO = 12), 9724 rpm (EO = 11), 10,710 rpm (EO = 10), and 11,890 rpm (EO = 9), respectively.

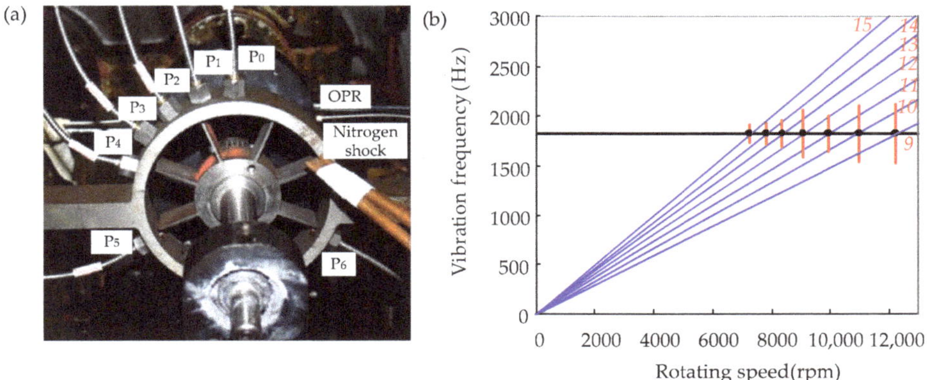

Figure 11. (**a**) The test rig and (**b**) the blade Campbell diagram.

The displacement of blade 7 measured by probe P_0 for rotating speed 6500–12,100 rpm was shown in Figure 12. In the whole speed up, there were seven resonances excited by the nitrogen shock, although the vibration intensity was not completely the same. The EO corresponding to each resonance was predicted through the Campbell diagram (Figure 11b), which was labeled in Figure 12. The similar resonances were also measured by other probes, and here it did not show these vibration waveforms one by one. In the engineering applications, the EO can be predicted through the Campbell diagram, like the experiment in this paper. The EO identification step can be skipped on the condition that the true EO had been known before parameter identification. In this case, the correct EO would be used directly by identification methods to calculate the amplitude, phase, and direct current offset. In this section, the other parameters would be calculated based on the EO that was from prediction rather than identification. Therefore, the reconstruction accuracy was compared for each method base on the EO known.

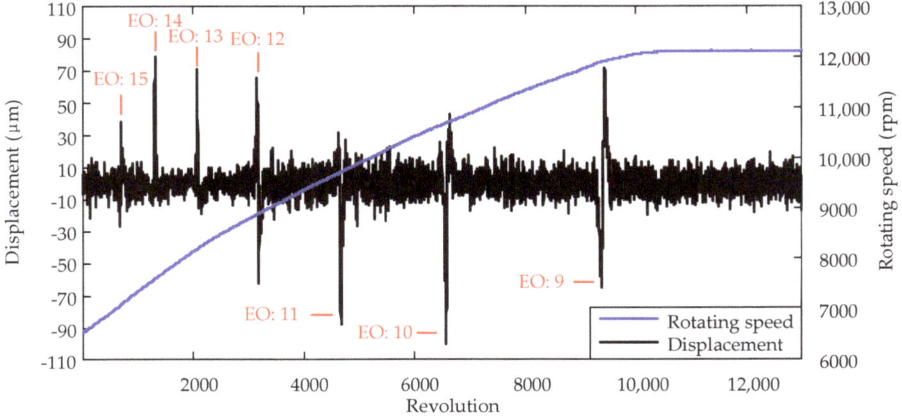

Figure 12. The displacement of blade 7 measured by probe P_0.

Although the EO corresponding to each resonance can be predicted through the Campbell diagram, the corresponding amplitude can not be known exactly, which was not the same as the simulated tests where the true signal was obtained in advance. Even if the strain gauges were applied, it also needs a very complex technique for converting the strain to the amplitude of the blade tip, and the strain measurement error as an integrated part of strain data was also transmitted to the amplitude. Therefore, where the true amplitude is unknown, the relative error of reconstruction (REoR) was used to evaluate the quality of reconstruction. If the REoR is greater than zero, the higher REoR, the higher quality of reconstruction for ICFF. In contrast, if the REoR is negative that means the reconstruction accuracy of ICFF is lower than TCFF. The REoR is given by:

$$\mathrm{REoR} = \frac{\mathrm{STDoE_{TCFF}} - \mathrm{STDoE_{ICFF}}}{\mathrm{STDoE_{TCFF}}} \times 100\% \tag{19}$$

where the STDoE represents the standard deviation of reconstruction error for each method, which is indicated by the subscript. The STDoE is given by:

$$\mathrm{STDoE} = \frac{1}{n}\sum_{j=1}^{n} \sqrt{\frac{1}{g}\sum_{i=0}^{g-1} (y_{i,j} - y^{*}_{i,j})^2} \times 100\% \tag{20}$$

where n represents the number of revolutions, g represents the number of probes, $y_{i,j}$ represents the measurement displacement by probe i at revolution j, and $y^{*}_{i,j}$ represents the reconstruction displacement.

In this experiment, the EO has been determined based on the FEM in advance. So, the true EO can be directly used without identification, which like the CFF does. Based on the results of simulated tests, the more number of probes and fitted data, the higher the reconstruction error for amplitude calculated. Therefore, just four probes (P_0, P_1, P_2, and P_3) instead of all were selected to identify the amplitude. For each resonance analysis, the number of fitted data was 20. The reconstruction data of blade 7 measured by probe P_0 was plotted in Figure 13. Each resonance was divided by the blue lines. The reconstruction resonance waveforms did not display the whole since only 20 data points were selected for each resonance. The results of reconstruction for measured data by the other probe were similar to Figure 13, which were not shown one by one. It was difficult to determine the quality of the identification only through Figure 13. Therefore, the STDoE and REoR were calculated based on the reconstruction data to achieve the purpose of comparing the quality of the identification. The results of identification for each method were summarized in Table 5.

Table 5. The results of the identification.

Resonance	PSR	EO	Identification						REoR
			Amplitude (μm)		Direct Current Offset (μm)		STDoE (μm)		
			TCFF	ICFF	TCFF	ICFF	TCFF	ICFF	
1	23.96%	15	63.01	43.01	4.84	3.83	11.51	9.69	+15.81%
2	9.03%	14	78.48	67.59	7.32	8.67	27.11	23.85	+12.02%
3	94.10%	13	83.55	71.19	13.50	12.70	18.48	16.72	+9.52%
4	79.17%	12	99.28	85.26	2.29	2.23	31.61	29.02	+8.16%
5	64.24%	11	88.47	75.75	3.38	3.77	16.20	14.99	+7.47%
6	49.31%	10	127.63	107.87	2.36	1.73	28.16	24.05	+14.57%
7	34.38%	9	84.92	75.03	4.69	3.77	10.31	9.28	+9.99%

The REoRs were all greater than zero for all resonance analysis, which was shown in Table 5. That means the reconstruction accuracy of ICFF was higher than that of TCFF. Besides, even for lower PSR like the EO 14 corresponding to PSR 9.03 percent, as long as the true EO is known, the results of amplitude identification still was credible for both methods.

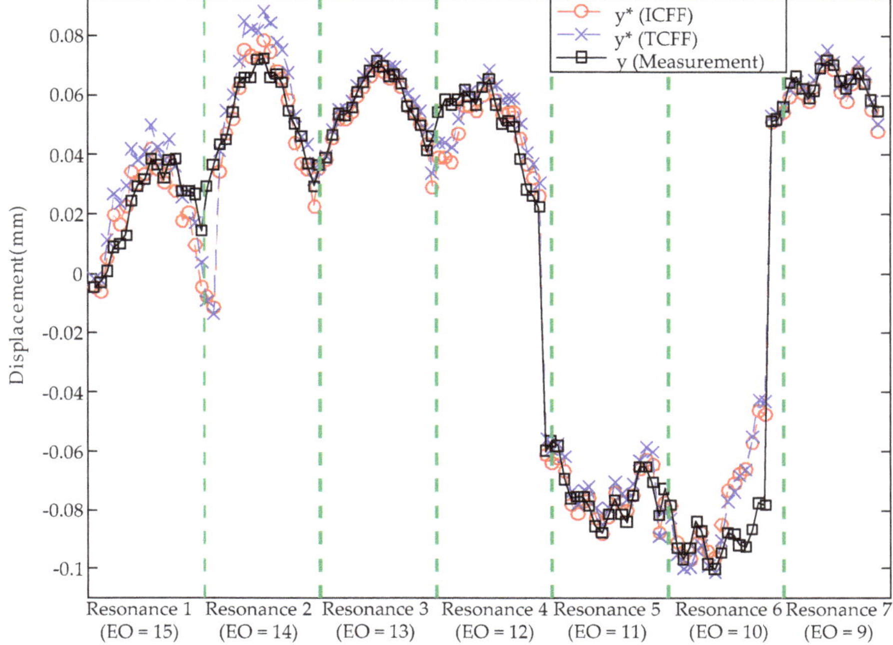

Figure 13. The reconstruction data of blade 7 measured by probe P_0.

5. Conclusions

In this paper, an improved circumferential Fourier fit method named ICFF was proposed, which considers the blade vibration effect on the TOA. In this method, the ToA is not only dependent on the rotating speed and monitoring position, but also on blade vibration. Compared with TCFF, this improvement is more consistent with reality. A 12-blade assembly simulator and experimental data were used to evaluate the ICFF performance. The simulated results showed that the ICFF performance is comparable to TCFF in terms of EO identification, except the lower PSR or more number probes have a more negative effect on ICFF, which results in the poor performance of ICFF in Test 2 and Test 5. However, the accuracy of amplitude identification is higher for ICFF than TCFF on all test conditions. Meanwhile, the higher accuracy of the reconstruction of ICFF was further verified by experimental data, where the quality of reconstruction was higher for ICFF than TCFF in all measurement resonance analysis. The propagation of uncertainty in measurements, the method of sensor layout determination, and the BTT data analysis that consists of two or more modes corresponding to different EOs will be the focus of further research work.

Author Contributions: Conceptualization, Z.L. and F.D.; methodology, Z.L.; validation, F.D.; investigation, Z.L. and G.N.; writing—original draft preparation, Z.L.; writing—review and editing, L.M.; visualization, J.J. and X.F.; funding acquisition, F.D. All authors have read and agreed to the published version of the manuscript.

Funding: This research was funded by Equipment Pre-research Field Fund (Grant Nos. 61400040303 and 61405180505), National Science and Technology Major Project (Grant No. 2017-V-0009), National Natural Science Foundations of China (Grant Nos. 51775377, 61971307 and 61905175), National Key Research and Development Plan (Grant No. 2017YFF0204800), Natural Science Foundations of Tianjin City (Grant No. 17JCQNJC01100), Science and Technology on Underwater Information and Control Laboratory (Grant No. 6142218081811), and 2019 Guangdong Provincial Market Supervision Administration Quality and Safety Supervision Scientific Research Project (Grant No. 2019ZZ02).

Conflicts of Interest: The authors declare no conflict of interest.

Nomenclature

A	the amplitude of blade vibration	N	number of blades
$\mathbf{B}$	coefficient matrix	R	distance from the blade tip to the center of the rotor
c	nominal damping of the blade	REoR	the relative error of reconstruction
c_i	damping of blade i (i = 1, 2, 3 ...)	STDoE	the standard deviation of reconstruction error
$\mathbf{C}$	damping matrix	S_k	fitted residual
$c_{i,j}$	coupling damping	$\widetilde{S_k}$	equivalent fitted residual
d	the direct current offset	t	time of arrival (ToA)
$f_i(t)$	excitation force	ω	angular vibration frequency
Δf_i	mistuning coefficient	ω_n	angular natural frequency
$\mathbf{F}(t)$	excitation force vector	ω_v	angular rotating speed
F_i	the amplitude of the excitation force	$\overline{\omega_v}$	average angular rotating speed
g	number of the probe	$\mathbf{X}$	parameter matrix
h_i	coupling coefficient	y	vibration displacement of the blade
k	nominal stiffness of the blade	$\mathbf{Y}$	displacement vector
k_i	stiffness of blade i (i = 1, 2, 3 ...)	β	installed position of the probe
$\mathbf{K}$	stiffness matrix	$\widetilde{\beta}$	equivalent monitoring position
$k_{i,j}$	coupling stiffness	ε	the angle of the blade from the first probe when the mark passes the OPR sensor
m	the nominal mass of the blade	ϕ	vibration phase
m_i	mass of blade i (i = 1, 2, 3 ...)		
$\mathbf{M}$	mass matrix		
n	number of revolution		

References

1. Kumar, S.; Roy, N.; Ganguli, R. Monitoring Low Cycle Fatigue Damage in Turbine Blade Using Vibration Characteristics. *Mech. Syst. Signal. Process.* **2007**, *21*, 480–501. [CrossRef]
2. Madhavan, S.; Jain, R.; Sujatha, C.; Sekhar, A.S. Vibration Based Damage Detection of Rotor Blades in a Gas Turbine Engine. *Eng. Fail. Anal.* **2014**, *46*, 26–39. [CrossRef]
3. Till, H.; Stefan, B. Axial Fan Blade Vibration Assessment under Inlet Cross-Flow Conditions Using Laser Scanning Vibrometry. *Appl. Sci.* **2017**, *7*, 862. [CrossRef]
4. Presas, A.; Valentin, D.; Valero, C.; Egusquiza Montagut, M.; Egusquiza, E. Experimental Measurements of the Natural Frequencies and Mode Shapes of Rotating Disk-Blades-Disk Assemblies from the Stationary Frame. *Appl. Sci.* **2019**, *9*, 3864. [CrossRef]
5. Battiato, G.; Firrone, C.M.; Berruti, T.M. Forced Response of Rotating Bladed Disks: Blade Tip-Timing Measurements. *Mech. Syst. Signal. Process.* **2017**, *85*, 912–926. [CrossRef]
6. Beauseroy, P.; Lengellé, R. Nonintrusive Turbomachine Blade Vibration Measurement System. *Mech. Syst. Signal. Process.* **2007**, *21*, 1717–1738. [CrossRef]
7. Russhard, P. Development of a Blade Tip Timing Based Engine Health Monitoring System. PhD. Thesis, The University of Manchester, Manchester, UK, 2010.
8. Ye, D.; Duan, F.; Jiang, J.; Cheng, Z.; Niu, G.; Shan, P.; Zhang, J. Synchronous Vibration Measurements for Shrouded Blades Based on Fiber Optical Sensors with Lenses in a Steam Turbine. *Sensors* **2019**, *19*, 2501. [CrossRef]
9. Ye, D.; Duan, F.; Jiang, J.; Niu, G.; Liu, Z.; Li, F. Identification of Vibration Events in Rotating Blades Using a Fiber Optical Tip Timing Sensor. *Sensors* **2019**, *19*, 1482. [CrossRef] [PubMed]
10. Russhard, P. The Rise and Fall of the Rotor Blade Strain Gauge. In *Mechanisms and Machine Science*; Springer: Cham, Switzerland, 2015; pp. 27–37. [CrossRef]
11. Heath, S.; Imregun, M. A Survey of Blade Tip-Timing Measurement Techniques for Turbomachinery Vibration. *J. Eng. Gas. Turbines Power* **1998**, *120*, 784. [CrossRef]
12. Schoenenborn, H.; Breuer, T. Aeroelasticity at Reversed Flow Conditions—Part II: Application to Compressor Surge. *J. Turbomach.* **2011**, *134*, 061031–061035. [CrossRef]
13. Rzadkowski, R.; Rokicki, E.; Piechowski, L.; Szczepanik, R. Analysis of Middle Bearing Failure in Rotor Jet Engine Using Tip-Timing and Tip-Clearance Techniques. In Proceedings of the ASME Turbo Expo 2014, Dusseldorf, Germany, 16–20 June 2014.

14. Krause, C.; Giersch, T.; Stelldinger, M.; Hanschke, B.; Kühhorn, A. Asynchronous Response Analysis of Non-Contact Vibration Measurements on Compressor Rotor Blades. In Proceedings of the ASME Turbo Expo 2017: Turbomachinery Technical Conference and Exposition, Charlotte, NC, USA, 26–30 June 2017.
15. Wang, W.; Ren, S.; Chen, L.; Huang, S. Investigation on the Method of Blade Synchronous Vibration Parameter Identification (Simulation). *J. Vib. Shock* **2017**, *36*, 120–126. [CrossRef]
16. Wang, W.; Ren, S.; Chen, L.; Shao, H. Test for Synchronous Vibration Parametric Identification Method of a Turbine's Blades. *J. Vib. Shock* **2017**, *36*, 127–133. [CrossRef]
17. Watkins, W.; Robinson, W.; Chi, R. Noncontact Engine Blade Vibration Measurements and Analysis. In Proceedings of the AIAA/SAE/ASME/ASEE 21st Joint Propulsion Conference, Monterey, CA, USA, 8–10 July 1985.
18. Zhang, Y.; Duan, F.; Fang, Z.; Ou, Y.; Ye, S. Analysis of Non-Contact Asynchronous Vibration of Rotating Blades Based on Tip-Timing. *Chin. J. Mech. Eng.* **2008**, *44*, 147–150. [CrossRef]
19. Vercoutter, A.; Lardies, J.; Berthillier, M.; Talon, A.; Burgardt, B. Improvement of Compressor Blade Vibrations Spectral Analysis from Tip Timing Data: Aliasing Reduction. In Proceedings of the ASME Turbo Expo 2013: Turbine Technical Conference and Exposition, San Antonio, TX, USA, 7–8 June 2013.
20. Kharyton, V.; Bladh, R. Using Tiptiming and Strain Gauge Data for the Estimation of Consumed Life in a Compressor Blisk Subjected to Stall-Induced Loading. In Proceedings of the ASME Turbo expo 2014: Turbine Technical Conference and Exposition, Duesseldorf, Germany, 16–20 June 2014.
21. Kharyton, V.; Dimitriadis, G.; Defise, C. A Discussion on the Advancement of Blade Tip Timing Data Processing. In Proceedings of the ASME Turbo Expo 2017: Turbomachinery Technical Conference and Exposition, Charlotte, NC, USA, 26–30 June 2017.
22. Lin, J.; Hu, Z.; Chen, Z.-S.; Yang, Y.-M.; Xu, H.-L. Sparse Reconstruction of Blade Tip-Timing Signals for Multi-Mode Blade Vibration Monitoring. *Mech. Syst. Signal. Process.* **2016**, *81*, 250–258. [CrossRef]
23. Ewins, D.J. Modal Analysis for Rotating Machinery. In *Modal Analysis and Testing*; Springer: New York, NY, USA, 1999; pp. 549–568.
24. Zablotskiy, I.Y.; Korostelev, Y.A. *Measurement of Resonance Vibrations of Turbine Blades with the Elura Device*; Foreign Technology Div Wright-Patterson AFB OH: Dayton, OH, USA, 1978.
25. Heath, S. A New Technique for Identifying Synchronous Resonances Using Tip-Timing. *J. Eng. Gas. Turbines Power* **2000**, *122*, 219–225. [CrossRef]
26. Gallego-Garrido, J.; Dimitriadis, G.; Wright, J.R. A Class of Methods for the Analysis of Blade Tip Timing Data from Bladed Assemblies Undergoing Simultaneous Resonances—Part I: Theoretical Development. *Int. J. Rotat. Mach.* **2007**, *2007*. [CrossRef]
27. Gallego-Garrido, J.; Dimitriadis, G.; Carrington, I.B.; Wright, J.R. A Class of Methods for the Analysis of Blade Tip Timing Data from Bladed Assemblies Undergoing Simultaneous Resonances—Part II: Experimental Validation. *Int. J. Rotat. Mach.* **2007**, *2007*. [CrossRef]
28. Joung, K.-K.; Kang, S.-C.; Paeng, K.-S.; Park, N.-G.; Choi, H.-J.; You, Y.-J.; Von Flotow, A. Analysis of Vibration of the Turbine Blades Using Non-Intrusive Stress Measurement System. In Proceedings of the PWR2006, Atlanta, GA, USA, 2–4 May 2006.
29. Guo, H.; Duan, F.; Zhang, J. Blade Resonance Parameter Identification Based on Tip-Timing Method without the Once-Per Revolution Sensor. *Mech. Syst. Signal. Process.* **2016**, *66*, 625–639. [CrossRef]
30. Chen, K.; Wang, W.; Zhang, X.; Zhang, Y. New Step to Improve the Accuracy of Blade Tip Timing Method without Once Per Revolution. *Mech. Syst. Signal. Process.* **2019**, *134*, 106321. [CrossRef]
31. Ou, Y.; Duan, F.; Li, M.; Kong, X. Method for Identifying Rotating Blade Synchronous Vibration at Constant Speed. *J. Tianjin Univ.* **2011**, *44*, 742–746. [CrossRef]
32. Heath, S.; Imregun, M. An Improved Single-Parameter Tip-Timing Method for Turbomachinery Blade Vibration Measurements Using Optical Laser Probes. *Int. J. Mech. Sci.* **1996**, *38*, 1047–1058. [CrossRef]
33. Rossi, G.; Brouckaert, J.-F. Design of Blade Tip Timing Measurement Systems Based on Uncertainty Analysis. In Proceedings of the International Instrumentation Symposium, San Diego, CA, USA, 4–8 June 2012.
34. Russhard, P. Blade Tip Timing (BTT) Uncertainties. In Proceedings of the 12th International A.I.VE.LA. Conference on Vibration Measurements by Laser and Noncontact Techniques, Anona, Italy, 29 June–1 July 2016.
35. Dimitriadis, G.; Carrington, I.B.; Wright, J.R.; Cooper, J.E. Blade-Tip Timing Measurement of Synchronous Vibrations of Rotating Bladed Assemblies. *Mech. Syst. Signal. Process.* **2002**, *16*, 599–622. [CrossRef]

36. Ou, Y. Rotating Blade Vibration Detection and Parameters Identification Technique Using Blade Tip-Timing. PhD. Thesis, Tianjin University, Tianjin, China, 2011.
37. Wei, S.T.; Pierre, C. Localization Phenomena in Mistuned Assemblies with Cyclic Symmetry Part I: Free Vibrations. *J. Vib. Acoust. Stress Reliab. Des.* **1988**, *110*, 429–438. [CrossRef]
38. Wei, S.T.; Pierre, C. Localization Phenomena in Mistuned Assemblies with Cyclic Symmetry Part II: Forced Vibrations. *J. Vib. Acoust. Stress Reliab. Des.* **1988**, *110*, 439–449. [CrossRef]
39. Carrington, I.B.; Wright, J.R.; Cooper, J.E.; Dimiadis, G. A Comparison of Blade Tip Timing Data Analysis Methods. *Proc. Inst. Mech. Eng. Part G J. Aerosp. Eng.* **2001**, *215*, 301–312. [CrossRef]
40. Diamond, D.H.; Stephan Heyns, P. A Novel Method for the Design of Proximity Sensor Configuration for Rotor Blade Tip Timing. *J. Vib. Acoust.* **2018**, *140*. [CrossRef]

Review

Optical Angle Sensor Technology Based on the Optical Frequency Comb Laser

Yuki Shimizu *, Hiraku Matsukuma and Wei Gao

Department of Finemechanics, Tohoku University, 6-6-01, Aramaki Aza Aoba, Aoba-ku, Sendai, Miyagi 980-8579, Japan; hiraku.matsukuma@nano.mech.tohoku.ac.jp (H.M.); gaowei@cc.mech.tohoku.ac.jp (W.G.)

* Correspondence: yuki.shimizu@nano.mech.tohoku.ac.jp; Tel.: +81-22-795-6950

Received: 29 May 2020; Accepted: 8 June 2020; Published: 11 June 2020

Abstract: A mode-locked femtosecond laser, which is often referred to as the optical frequency comb, has increasing applications in various industrial fields, including production engineering, in the last two decades. Many efforts have been made so far to apply the mode-locked femtosecond laser to the absolute distance measurement. In recent years, a mode-locked femtosecond laser has increasing application in angle measurement, where the unique characteristics of the mode-locked femtosecond laser such as the stable optical frequencies, equally-spaced modes in frequency domain, and the ultra-short pulse trains with a high peak power are utilized to achieve precision and stable angle measurement. In this review article, some of the optical angle sensor techniques based on the mode-locked femtosecond laser are introduced. First, the angle scale comb, which can be generated by combining the dispersive characteristic of a scale grating and the discretized modes in a mode-locked femtosecond laser, is introduced. Some of the mode-locked femtosecond laser autocollimators, which have been realized by combining the concept of the angle scale comb with the laser autocollimation, are also explained. Angle measurement techniques based on the absolute distance measurements, lateral chromatic aberration, and second harmonic generation (SHG) are also introduced.

Keywords: optical angle sensor; mode-locked femtosecond laser; optical frequency comb; laser autocollimation; diffraction grating; absolute angle measurement; nonlinear optics; second harmonic generation

1. Introduction

The angle and length are among the most fundamental parameters that determine the form of an object [1,2]. As can be seen in measurements of the angle between two surfaces of parts or assemblies by a protractor or fixed angle gauges, angle measurement has been carried out since ancient times [3]. In the current production engineering, angle sensors play important roles in many types of machine tools, arm robots, and measuring instruments [4]. For measurement of the angular displacement or angular position of an object with a fixed axis of rotation, rotary encoders are often employed. In a rotary encoder, a relative angular displacement between a rotating scale disk coupled coaxially with that of the rotating object and a reading head kept stationary with respect to the axis of rotation can be detected by reading circular graduations on the scale disk in an optical or electromagnetic manner [5–8]. Rotary encoders are capable of carrying out high-precision angle measurement over an angular range of 360° and can be employed as a feedback sensor for the control of the axial position of an object. In the state-of-the-art rotary encoder, a self-calibration method has been established [9,10]. Furthermore, the improvements in the resolution and measurement speed of absolute rotary encoders contribute to achieving further higher positioning accuracy and fabrication throughput in machine tools and robot systems [11].

On the other hand, precision positioning is one of the key technologies for the precise fabrication of a component with a complex surface form such as an aspheric form or a freeform, as well as the fabrication of a device with three-dimensional micrometric or nanometric structures [11]. One-axis precision positioning can be achieved by employing a linear slide equipped with precision displacement sensors such as laser interferometers or linear encoders [12]. In multi-axis machine tools and measuring instruments, multi-dimensional precision positioning is carried out by employing positioning systems composed of such precision linear slides. For further higher precision positioning, it is necessary to evaluate the angular error motions of such linear slides with a high measurement throughput in a non-contact manner [4,13]. The rotary encoders cannot be employed in such applications, since the distance between a reading head and a scale disk cannot be changed during the measurement. For such a purpose, optical autocollimators are often employed [14,15]. By employing a two-dimensional image sensor such as an area sensor or a multi-cell photodiode as the photodetector, two-axis angular displacement can be measured simultaneously [4,16]. Nowadays, optical autocollimators have expanded applications to the surface form measurement of precision optical components [17–21]. In recent years, most of the commercial autocollimators employed a charge-coupled device (CCD) or a complementary metal-oxide semiconductor (CMOS) as their photodetectors for the achievement of a large angle measurement range as well as a high resolution [22–24]. One of the disadvantages of employing a CCD image sensor is its low measurement throughput; this can be overcome by employing a CMOS or a PSD (position-sensitive detector) [24]. The relatively large size of the optical setup is another issue for the conventional optical autocollimator, since its large footprint could prevent the optical autocollimator from being employed in machine tools where the space for such a sensor is limited. For the achievement of high measurement throughput and high resolution in a compact manner, a laser autocollimator based on laser autocollimation [25] has been developed. With the employment of a laser diode and a photodiode as a compact light source and a photodetector, respectively, highly sensitive high-speed angular displacement measurement has been achieved [26,27]. A laser autocollimator can also be applied for surface form measurement of precision components [28]. Furthermore, a three-axis laser autocollimator and six-degree-of-freedom (6-DOF) surface encoder capable of measuring the three-axis angular displacement of a grating reflector with a single measurement laser beam [29–33], as well as a three-axis inclinometer [34] based on the principle of the three-axis laser autocollimator, have been developed.

In the field of dimensional metrology, many types of optical sensors employing a mode-locked femtosecond laser, in which the optical frequency of each mode can be directly linked to a national standard of frequency/time, have been developed in the past two decades since the establishment of the laser source [35–39]. For example, methods for measurement of the absolute distance of an object [40–46] and the absolute thickness of an optical glass [47–49] have been developed with the enhancement of a well-controlled pulse repetition rate of the mode-locked femtosecond laser. The establishment of a fiber-based mode-locked femtosecond laser [50–53] designed in a compact size and capable of being operated stably even in a limited environmental condition has contributed to the growth of these techniques in dimensional metrology. In recent years, the trend of employing a mode-locked femtosecond laser source can also be seen in angle measurement technologies. A new concept of generating an absolute angle scale with the enhancement of the dispersive characteristic of a diffraction grating has been proposed [54], and a femtosecond laser autocollimator based on the concept as well as laser autocollimation, has been established [55,56]. Furthermore, the method has been extended to the measurement of the absolute position of an object [57]. The above-mentioned femtosecond laser autocollimators are based on the well-controlled, equally spaced frequency comb in the spectrum of the mode-locked femtosecond laser. Angle measurement with a mode-locked femtosecond laser source has also been achieved by the phenomenon of chromatic aberration [58], as well as second harmonic generation (SHG) which is a well-known phenomenon in nonlinear optics [59]. Table 1 summarizes the optical methods for angle measurement employing a mode-locked femtosecond laser source. The new angle measurement techniques with a mode-locked femtosecond laser are expected to achieve the performances that cannot be achieved by conventional optical angle sensors with a single-mode laser source or a white light source.

Table 1. Optical angle sensors based on a mode-locked femtosecond laser.

Utilized Characteristics	Physical Phenomenon/Measuring Technique to be Coupled
Stable discrete modes in frequency domain	➢ Diffraction [54] ➢ Diffraction and laser autocollimation [55–57] ➢ Chromatic aberration [58] ➢ Sagnac effect [60,61] (Fiber optic gyroscope)
High pulse energy	➢ Dispersive interferometry [62] (Absolute distance measurement) ➢ Time-of-flight (TOF) counted by dual-comb interferometry with balanced cross-correlation of second harmonics [63] (Absolute distance measurement) ➢ Second-harmonic generation (SHG) [59]

In this review article, optical angle sensors employing a mode-locked femtosecond laser source are investigated. It should be noted that an optical fiber gyro based on the Sagnac effect [64,65] is another method for angle measurement. Many types of optical fiber gyros such as an interferometric Fiber optic gyroscope (I-FOG) [66], a resonator Fiber optic gyroscope (R-FOG) [67] and Brillouin scattering [68] have been developed so far, and some trials have been performed using optical fiber gyros to employ a mode-locked femtosecond laser [60,61]. The measurement principles of these optical fiber gyros are the same as the conventional ones employing a pulsed laser [69]. In addition, optical fiber gyros are not appropriate for applications in production engineering, such as the evaluation of the angular error motion of a precision linear slide or the measurement of the freeform or the aspheric form of an optical component, due to their drift characteristics associated with their principle based on the detection of angular velocity. The optical fiber gyros with a mode-locked femtosecond laser are thus not included in this review article.

2. Angle Measurement Methods Based on the Discrete Modes of a Mode-Locked Femtosecond Laser in Frequency Domain

2.1. A Method Employing the Dispersive Characteristics of a Diffraction Grating

2.1.1. Principle of the Generation of an Angle Scale Comb from an Optical Frequency Comb

The equally-spaced optical modes in a mode-locked femtosecond laser, often referred to as an optical frequency comb, can be employed to generate an "angle scale comb" that can be employed as optical graduations for angle measurement. A schematic of the principle of generating an angle scale comb from a mode-locked femtosecond laser is shown in Figure 1. The angle scale comb is generated with the enhancement of the dispersive characteristics of a diffraction grating. A mode-locked femtosecond laser from a light source is constructed incident to a reflective-type diffraction grating. The ultra-short pulse train of a mode-locked femtosecond laser in the time domain is a series of equally-spaced optical modes with a pulse repetition rate ν_{rep} in the frequency domain. A reflective-type diffraction grating generates a series of first-order diffracted beams, in which each optical mode has a different angle of diffraction from those of the others. In the case where a two-dimensional reflective-type diffraction grating is employed as shown in the figure, groups of first-order diffracted beams will be generated in the two directions. It should be noted that only the groups of the positive first-order diffracted beams in the X- and Y-directions are indicated in the figure for the sake of simplicity. The angle of diffraction of the ith optical mode θ_i in the group of the positive first-order diffracted beams can be expressed as follows [54]:

$$\theta_i = \arcsin\left(\frac{c}{n_{air} g \nu_i}\right) (i = 1, 2, 3, \ldots, n), \quad (1)$$

where c, n_{air}, g, and ν_i are the speed of light in vacuum, the refractive index in air, the pitch of the diffraction grating, and the optical frequency of the ith mode, respectively. As can be seen in the Equation, the discrete optical frequency ν_i in the spectrum of the mode-locked femtosecond laser and the angle of diffraction θ_i of the ith mode are in a one-to-one relationship, and thus the group of the diffracted beams can be treated as optical graduations for angle measurement: angle scale comb.

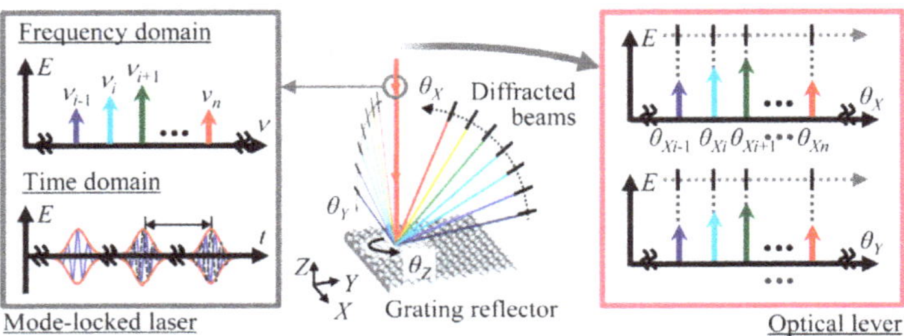

Figure 1. Angle scale comb generated from the equally-spaced optical modes of a mode-locked femtosecond laser in the frequency domain with the enhancement of the dispersive characteristics of a reflective-type diffraction grating [54].

The angle scale comb generated from a mode-locked femtosecond laser can be employed for angle measurement. Figure 2a shows one of the examples of applying the angle scale comb for measurement of a rotary table on which a grating reflector is mounted. Since the modes in the angle scale comb pass through the detector one after another with the rotation of the grating reflector, the angular displacement of the rotary table can be detected by monitoring the reading output of the detector. Figure 2b shows another example of the application of the angle scale comb; measurement of the free-form surface of an optical component. In this case, due to the local slope $\Delta\theta$ at a certain position on the free-form surface, the angle of incidence of the reflected beam from the optical component experiences a change in its propagation direction of $2\Delta\theta$. The angle scale comb emanating from the grating reflector also experiences a change in its angle of diffraction of $2\Delta\theta$. A local slope mapped over the surface of the optical component can thus be obtained by detecting the change in the angle of diffraction of the angle scale comb during the lateral scanning of the optical component. Through the integral calculation of the obtained local slope map, the surface form of the optical component can be reconstructed.

A mode-locked femtosecond laser has superior characteristics as a high-precision, highly stable laser source that can be directly traceable to the national standard of frequency and time [70]. The angle scale comb takes over these characteristics of a mode-locked femtosecond laser. In an ideal case, an angular distance between neighboring comb modes and the angular range in the angle scale comb correspond to the pulse repetition rate and the spectral bandwidth of the mode-locked femtosecond laser, respectively, from which the angle scale comb is generated. In the general case, the pulse repetition rate and the spectral bandwidth of an optical frequency comb are on the order of 100 MHz and 10 THz, respectively; this means that the dynamic range of the angle scale comb (the ratio of the angular range to the angular distance of neighboring modes) thus becomes large.

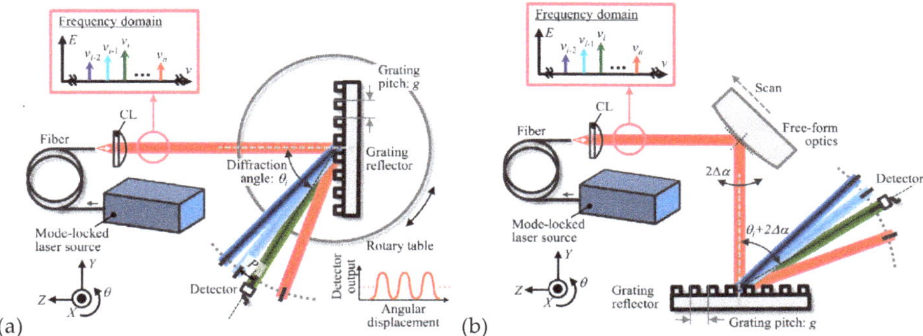

Figure 2. Applications of the angle scale comb [54]: (**a**) Measurement of the angular displacement of a rotary table; (**b**) Measurement of the surface form of an optical component.

2.1.2. Light Intensity Detecting-Type Mode-Locked Femtosecond Laser Autocollimator

In the optical setups shown in Section 2.1.1., the resolution of the angle measurement can be improved by increasing the distance between the grating reflector and the detector. However, in most of the applications, the space available for the optical setup of the angle scale comb is limited; this fact means that there is a trade-off relationship between the resolution of angle measurement and the size of the optical setup. This issue can be addressed by introducing the laser autocollimation into the optical frequency comb.

Figure 3a shows an example of the optical setup of a conventional laser autocollimator, which is based on the laser autocollimation [25]. A single-mode laser emitted from a laser source such as a laser diode (LD) is collimated by a collimating lens (CL) and is then made incident to a plane mirror reflector through a polarizing beam splitter (PBS) and a quarter-wave plate (QWP). The reflected beam from the plane mirror reflector goes through the QWP again, is reflected by the PBS, and is then captured by an autocollimation unit composed of a collimator objective (CO) and a photodiode (PD). In the autocollimation unit, the active cell of the PD is placed at the back focal plane of the CO so that the laser beam made incident to the CO can be focused on the PD active cell. The displacement Δd of the focused laser beam on the PD active cell due to the angular displacement $\Delta\theta$ of the plane mirror reflector can be expressed by the following Equation [4]:

$$\Delta\theta = \arctan\left(\frac{\Delta d}{2f}\right) \tag{2}$$

In the case where a single-cell photodiode is employed as the photodetector to detect the spot displacement Δd as shown in Figure 3a, the measuring range of the photodetector is defined by the diameter D of the focused laser beam on the PD active cell, and the corresponding measuring range of the angular displacement becomes $\pm$ arctan($D/4f$). The detection sensitivity of Δd by the PD is inversely proportional to the focused spot diameter D [4]. The decrease of D for the achievement of the highly sensitive measurement of the angular displacement is thus required; namely, there is a trade-off relationship between the sensitivity and the measuring range in the conventional laser autocollimator with a photodiode and a single-mode laser source.

A femtosecond laser autocollimator, which can be realized by combining the angle scale comb with the conventional laser autocollimation, can overcome the aforementioned problem. Figure 3b shows a schematic of the femtosecond laser autocollimator in which a mode-locked femtosecond laser source is employed as the light source. In the optical setup of an angle scale comb shown in Figure 2a, the laser autocollimation unit composed of a collimator objective and a single-cell photodiode is newly employed instead of the sole photodetector. As can be seen in Figure 3b, each of the first-order diffracted beams emanating from the grating reflector is focused onto the focal plane of the collimator objective. As a result, a series of focused diffracted beams aligned in a line can be obtained. The relative position of each of the focused beams is determined by Equations (1) and (2). Since all the focused diffracted beams experience the translational displacement on the PD active cell associated with the angular displacement of the grating reflector, a continuous reading output with a cycle corresponding to the relative position of each of the focused diffracted beams can be obtained. Denoting the number of first-order diffracted beams in the angle scale comb as N, a theoretical measurement range of the femtosecond laser autocollimator becomes N times that of the conventional laser autocollimator employing a single-mode laser source and with a single focused spot on the PD active cell. It should be noted that the detection of the displacement of a focused beam in the femtosecond laser autocollimator shown in Figure 3b is based on the photocurrent output from the PD, the amount of which is associated with the intensity of the light rays captured by the active cell on the PD. Therefore, the femtosecond autocollimator with a PD is referred to as the light intensity detecting-type femtosecond laser autocollimator.

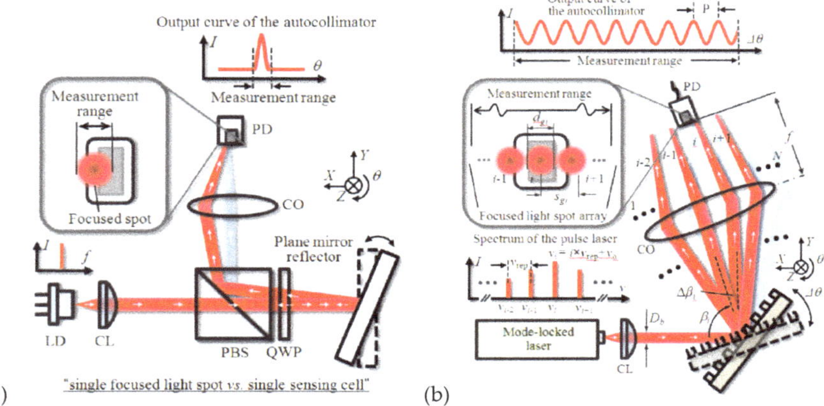

Figure 3. A comparison between the conventional laser autocollimator and the femtosecond laser autocollimator [55]: (**a**) A laser autocollimator with a single-mode laser source; (**b**) A femtosecond laser autocollimator with a mode-locked femtosecond laser source.

Figure 4a,b show a schematic and a photograph of the optical setup for the light intensity detecting-type femtosecond laser autocollimator, respectively. In the setup, a Fabry–Pérot etalon with a free spectral range (FSR) of 770 GHz was employed as an optical bandpass filter to enlarge the distance between the neighboring modes in the spectrum of the optical frequency comb.

Figure 5 shows the obtained photocurrent from the PD converted into the voltage output by a trans-impedance amplifier. As can be seen in the figure, the cycle of the obtained voltage output agreed well with the FSR of the etalon. These results demonstrated that the light intensity detecting-type femtosecond laser autocollimator has a measurement range greater than 11,000 arc-seconds (3.06), which is much larger than that of the conventional laser autocollimator with a single-mode laser source [4,27].

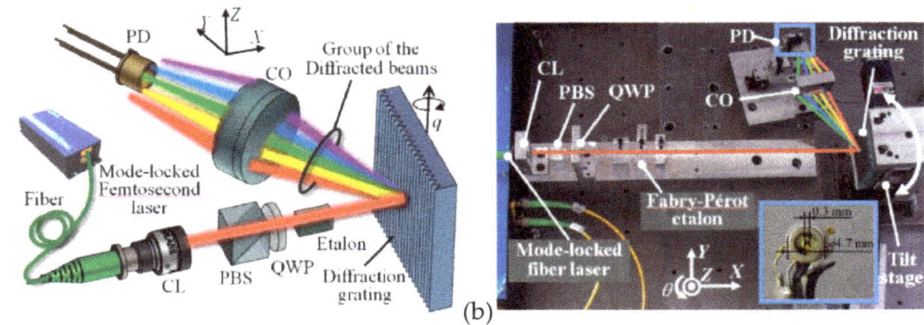

Figure 4. Optical setup of the mode-locked femtosecond laser autocollimator [55]: (**a**) A schematic of the optical setup; (**b**) A photograph of the optical setup.

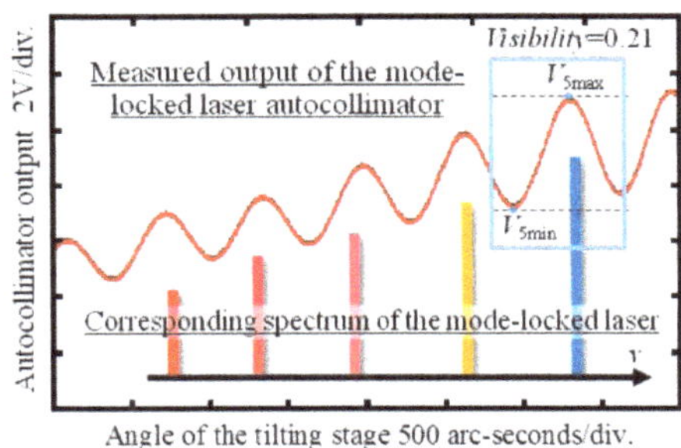

Figure 5. Variation of the reading output of the femtosecond laser autocollimator due to the angular displacement of the grating reflector [55].

2.1.3. An Optical Frequency Domain Angle Measurement Method Associated with a Mode-Locked Femtosecond Laser Autocollimator

Most of the commercial fiber-based mode-locked femtosecond lasers have a pulse repetition rate on the order of 100 MHz. When such a laser source is employed in the light intensity detecting-type femtosecond laser autocollimator, the distance between the neighboring spots in the focal plane of the collimator objective reaches the order of several nanometers, and the spots cannot be distinguished separately. The pulse repetition rate is thus required to be extended by a Fabry–Pérot etalon in the femtosecond laser autocollimator described in Section 2.1.2. Due to the diffraction limit, it is unavoidable for the neighboring spots in the focal plane of the collimator objective to overlap with each other in most of the cases. This results in the degradation of the signal quality (visibility) of the reading output of the light intensity detecting-type femtosecond laser autocollimator. As can be seen in the result shown in Figure 5, maximum visibility that can be achieved by the femtosecond laser is approximately 0.5; this prevents the light intensity detecting-type femtosecond laser autocollimator to achieve further higher sensitivity.

To address the issue, an optical frequency-domain femtosecond laser autocollimator has been established [56]. Figure 6 shows a schematic of the detection of the angle scale comb in optical frequency domain. A detector unit composed of a collimator objective, a single-mode fiber, and a spectrometer is newly employed. By observing the spectrum of the first-order diffracted beams captured by the single-mode fiber, the angle scale comb can be detected with a visibility of 100%. Furthermore, this technique realizes the direct conversion of the optical frequency comb to the angle scale comb, which contributes to achieving further precision angle measurement.

Figure 7 shows the developed optical frequency-domain femtosecond laser autocollimator. A collimated mode-locked femtosecond laser is made to pass through a Fabry–Pérot etalon with a free spectral range (FSR) of 100 GHz and is then made incident to a grating reflector, and a part of the first-order diffracted beams emanating from the grating surface is captured by the single-mode fiber detector through the collimator objective. The spectrum of the captured laser beam is analyzed by an optical spectrum analyzer. It should be noted that the etalon is employed in this setup to identify each mode in the angle scale comb in optical frequency domain by using a spectrometer with a limited frequency resolution.

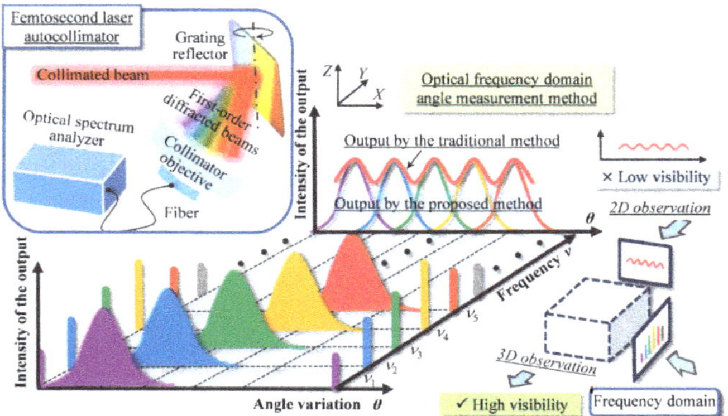

Figure 6. Mode-locked femtosecond laser autocollimator based on the optical frequency-domain angle measurement [56].

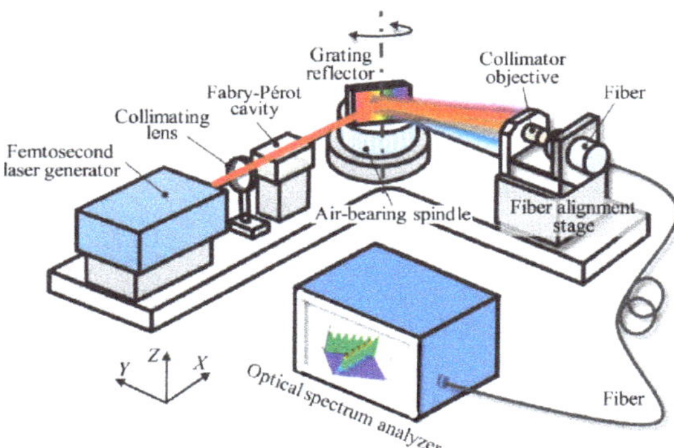

Figure 7. Femtosecond laser autocollimator based on the principle of the optical frequency-domain angle measurement [56].

Figure 8 shows an example of the angle scale comb observed in optical frequency domain. As can be seen in the figure, each of the comb modes is successfully identified in the optical frequency domain. Although a limited result over an angle range of approximately 400 arc-seconds is indicated in the figure, each of the comb modes is verified to be identified over an angular range of approximately 6°, which is limited by the spectral width of the mode-locked femtosecond laser, as well as the resolving power of the diffraction grating employed in the setup. The expansion of the spectral range of a mode-locked femtosecond laser with a supercontinuum [71] is expected to achieve a further wider measurement range. In addition, the results have demonstrated that a high resolution of 0.03 arc-seconds can be achieved by interpolating the intensity variation of each of the comb modes.

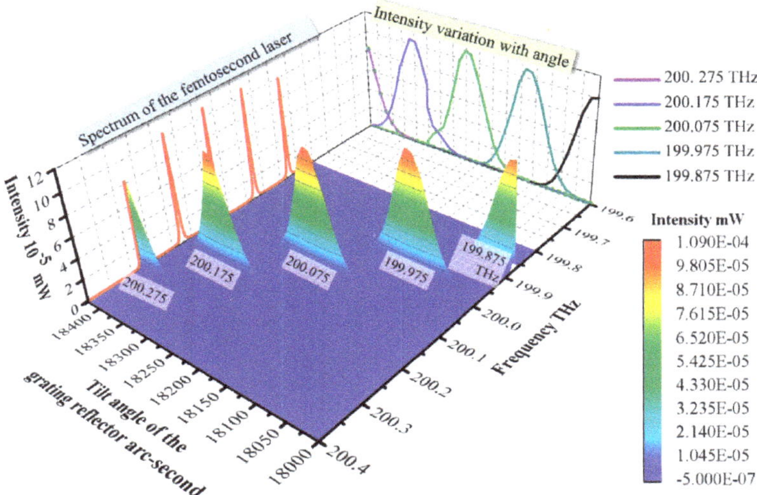

Figure 8. Reading output of the mode-locked femtosecond laser autocollimator with a visibility of 100% observed in the optical frequency domain [56].

Measurement of the absolute angular position of an object is a difficult task for the conventional autocollimator [17–24] and the laser autocollimator with a single-mode laser source [25–28], since it is difficult to verify the absolute angular position of a target reflector with respect to the optical axis of the autocollimation unit in a space; namely, the determination of the "zero-angle" position. Although a method employing a retroreflector is available for the zero-angle position adjustment [72], this method can only be applied to the applications where the propagation directions of the incident beam and the reflected beams become parallel to each other. Furthermore, the accuracy of the zero-angle position adjustment is dominated by the accuracy of the retroreflector. This issue can be addressed by the optical frequency-domain femtosecond laser autocollimator with the enhancement of an assisted angle sensor [57]. Figure 9 shows a schematic of the high-precision zero-angle position adjustment in the optical frequency-domain femtosecond laser autocollimator. As can be seen in the figure, $\Delta\theta$, which is the difference of the angles of diffraction of the $(i + 1)$th and ith first-order diffracted beams with mode frequencies of ν_{i+1} and ν_i, respectively, is verified by measuring the angular displacement of the grating reflector by the assisted angle sensor such as a conventional autocollimator or a rotary encoder embedded to the rotary table on which the grating reflector is mounted. By using the obtained $\Delta\theta$, the absolute angle Φ between the femtosecond laser beam and the optical axis of the laser autocollimation unit can be obtained. It should be noted that the final measurand of the method is not Φ but θ, the angle of the normal of the grating reflector with respect to the incident femtosecond laser.

The above mentioned zero-angle position adjustment can be applied to the optical setup where the propagation directions of the incident beam and the reflected beam are not parallel to each other. Once the angle Φ is confirmed during the fabrication process of the optical head of the mode-locked femtosecond laser autocollimator based on the principle of the optical frequency-domain angle measurement, absolute angle position of the grating reflector with respect to the incident femtosecond laser beam can be obtained with the following measurements.

Figure 10 shows one of the examples of the absolute angle measurement based on the proposed method. As can be seen in the figure, the absolute angular position of the grating reflector in a step of 180 arc-seconds was well distinguished over the absolute angle range from −37.805 degrees to −37.354 degrees in the frequency domain; this means that the optical frequency comb can be directly converted into the angle scale comb for absolute angular position measurement.

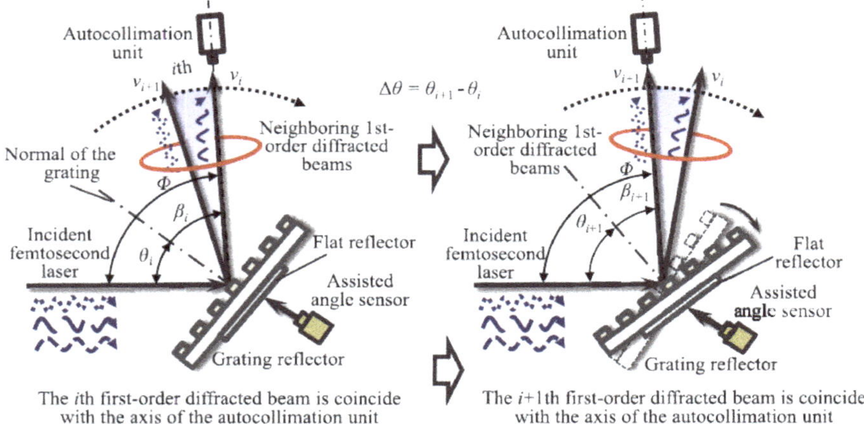

Figure 9. Absolute angular position measurement by the mode-locked femtosecond laser autocollimator based on the principle of the optical frequency-domain angle measurement with the enhancement of an assisted angle sensor [57]. In the setup, the angle of the normal of the grating reflector with respect to the incident femtosecond laser is a final measurand.

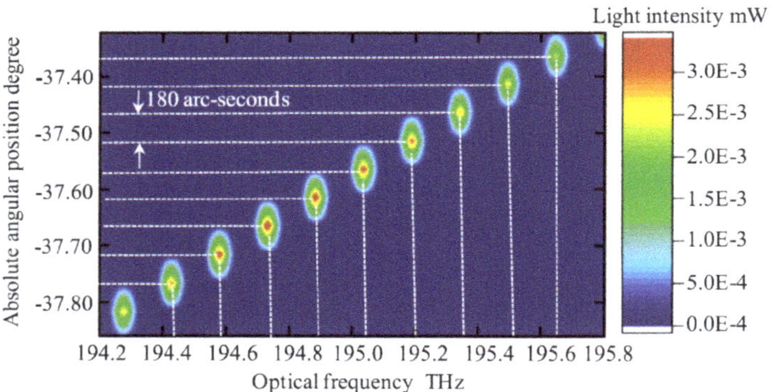

Figure 10. An example of the absolute angular position measurement by the optical frequency-domain femtosecond laser autocollimator [57].

2.2. Methods Based on the Chromatic Aberrations of a Simple Lens

According to the thin-lens equation, the focal length of a simple lens depends on the light wavelength of an incident light via the lens refractive index [73]. When a collimated mode-locked femtosecond laser is made incident to a simple lens in such a way that the laser axis is aligned to be coaxial with respect to the optical axis of the lens, the optical modes in the femtosecond laser are focused at different points on the optical axis of the lens. The light ray from an off-axis point will arrive at a different height above the optical axis; namely, the frequency-dependent lens focal length causes a frequency dependence of the transverse magnification as well [73]. This characteristic, which is referred to as the lateral chromatic aberration, can be employed for measurement of the small angular displacement of an object with the enhancement of the laser autocollimation [25,58].

Figure 11 shows the optical setup for measurement of the small angular displacement of an object based on the chromatic aberrations of a simple lens, where the principle of the laser autocollimation is integrated. A mode-locked femtosecond laser is employed as the light source, while the laser autocollimation unit composed of a collimator objective and a single-mode fiber connected to a spectrometer is employed as the detector for angle measurement. The mode-locked femtosecond laser from the light source is collimated by a collimating lens and is then made incident to a target reflector. The reflected beam is then made incident to the laser autocollimation unit. The fiber detector in the laser autocollimation unit is placed at an off-axis position d_{Fiber} ($\neq 0$) with respect to the optical axis of the collimator objective; namely, the optical axis of the laser autocollimation unit has an angle with respect to the optical axis of the collimator objective. This arrangement enables the optical setup to detect the small angular displacement of the target reflector with the effect of the lateral chromatic aberration of the lens.

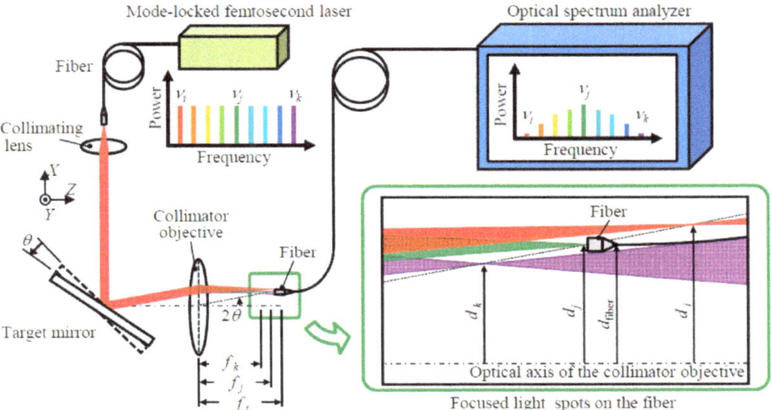

Figure 11. Optical setup for measurement of the small angular displacement of an object based on the chromatic aberrations of a simple lens [58].

Denoting the focal length of the simple lens for the ith mode in the mode-locked femtosecond laser as f_i, the lateral displacement of the focused mode d_i with respect to the optical axis of the simple lens due to the angular displacement θ of the target reflector can be expressed as follows [58]:

$$d_i = f_i \tan 2\theta \tag{3}$$

On the assumption that the mode-locked femtosecond laser has a uniform spectrum over its spectral range, and the optical frequency of the jth mode is the peak frequency in the spectrum of the light rays captured by the fiber detector at the condition where $\theta = \theta_0$ ($\neq 0$), the following relationship should be satisfied [58]:

$$d_{\text{fiber}} = d_j = f_j \tan 2\theta_0 \tag{4}$$

In the same manner, the following Equation should be satisfied in the case with the angular displacement $\theta_0 + \Delta\theta$ and the corresponding lateral displacement d_k of the kth mode at the peak in the spectrum of the captured light rays [58]:

$$d_{\text{fiber}} = d_k = f_k \tan 2(\theta_0 + \Delta\theta) \tag{5}$$

From Equations (4) and (5), the angular displacement of the target reflector $\Delta\theta$ can be obtained as follows [58]:

$$\Delta\theta = \frac{1}{2}\tan^{-1}\left(\frac{f_j}{f_k}\tan 2\theta_0\right) - \theta_0 \tag{6}$$

Since θ_0 is known as the design parameter in the optical setup, $\Delta\theta$ can be obtained based on Equation (6) by calculating f_j and f_k based on the lens equation by using the detected peak frequencies

ν_j and ν_k of the jth and kth modes, respectively. It should be noted that the setup shown in Figure 11 is similar to that shown in Figure 7 based on the dispersive characteristics of a grating reflector. One of the advantages of the optical setup shown in Figure 11 is that a reflective-type grating reflector is not required for measurement; this contributes to building up the optical setup to a compact size. In addition, an arbitrary reflective surface on an object can be measured by the setup shown in Figure 11. It should also be noted that the angle sensor with the diffraction grating shown in Figure 7 can be employed for the evaluation of an object with a reflective surface without grating pattern structures by modifying the optical setup as shown in Figure 2b. However, in this case, attention should be paid to misalignments of the optical components, as well as the alignment of an object under inspection.

Figure 12a shows the change in the light intensity of each mode captured by the fiber detector analyzed in optical frequency domain. In the figure, only several modes from 185 THz to 200 THz in a frequency difference of 5 THz are plotted for the sake of clarity. As can be seen in the figure, the light intensity of each mode has been changed by the angular displacement of the mirror reflector. Figure 12b shows the peak frequencies observed at each angular position of the mirror reflector. By detecting the peak frequency in the spectrum of the light rays captured by the fiber detector, the angular displacement of the mirror reflector can thus be detected.

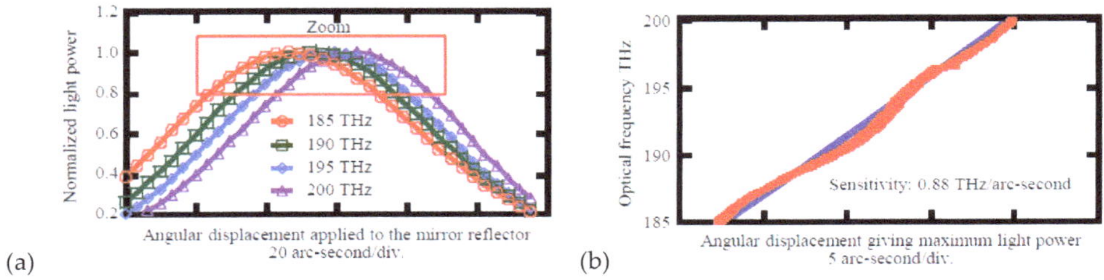

Figure 12. Optical modes captured in the setup based on the chromatic aberrations of a simple lens [58]: (**a**) Variation of the light intensities of optical modes captured by the fiber detector; (**b**) Peak optical frequency detected at each angular position of the mirror reflector.

3. Methods Based on the Absolute Distance Measurement

The angular displacement of an object can be measured by multilateration or triangulation [11,74,75] in which the displacements of several points on its surface are measured by interferometric methods [13,76] or non-interferometric methods such as the time-of-flight (TOF) method [77]. By employing a method with a mode-locked femtosecond laser source capable of measuring the absolute distance of a measuring point [40–46,62,63], absolute angular position measurement can be realized.

Figure 13 shows an example of the absolute angular position measurement based on the absolute distance measurement by the TOF method combined with the second harmonic generation (SHG) [63]. In the proposed method, a pair of the mode-locked femtosecond laser sources synchronized with the same reference clock but with slightly different pulse repetition rates is employed; one of the mode-locked femtosecond lasers is employed as a signal laser, while the other is employed as a local laser for down-conversion. In the setup, the femtosecond laser beam from the light source is at first collimated by a collimating lens (CL) and is then divided into four sub-beams by using a diffractive optical element (DOE). The sub-beams are made incident to the target surface where four mirrors are placed in an angular distance of 90° along the circumference direction to reflect the sub-beams to the DOE. It should be noted that the four mirrors are arranged in such a way that the optical path lengths of the sub-beams become different from each other. The reflected sub-beams are then combined at the DOE and are made to pass through the circulator and to superimpose with the local femtosecond laser in free space. The reflected sub-beams are then converted into electric signals by the balanced cross-correlator based on the principle of the nonlinear optical cross-correlation for measurement of the absolute distance of the four mirrors on the object with respect to the DOE. The novel optical configuration with the DOE and the four mirror reflectors makes it possible to carry out simultaneous measurement of the absolute distances of the four points on the object by a single balanced cross-correlation (BCC) electrical signal in which the four corresponding pulse trains can be observed independently. By using the measured absolute distances d_i ($i = 1, 2, 3, 4$) of the mirror reflectors with respect to the DOE, absolute angular positions can be obtained through a simple calculation based on the geometric relationship of the optical components; the X- and Y-directional absolute normal angles of the surface under inspection θ_x and θ_y, respectively, with respect to the femtosecond laser beam incident to the DOE can be calculated as $\theta_x = \sin^{-1}[(d_1 - d_3)/A]$ and $\theta_y = \sin^{-1}[(d_2 - d_4)/A]$, where A is the distance between the two corresponding mirror reflectors.

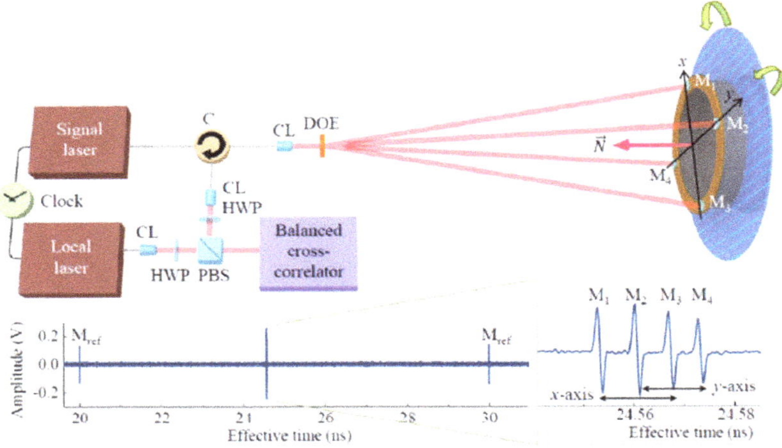

Figure 13. An example of the angle measurement based on the absolute distance measurement [63].

Figure 14 shows the detected angular motion of an object placed 3.7 m away from the DOE. In the experiment, angular motion with a frequency of 1 Hz was given to the object by using a piezoelectric tilt stage. Figure 14a shows the absolute displacements measured at the four positions, and Figure 14b shows the angular motions about the X- and Y-axes calculated from the obtained absolute displacements. As can be seen in the figures, the absolute distances at the four positions were successfully detected simultaneously, and the angular motions about the X- and Y-axes were successfully reconstructed from the obtained absolute distances.

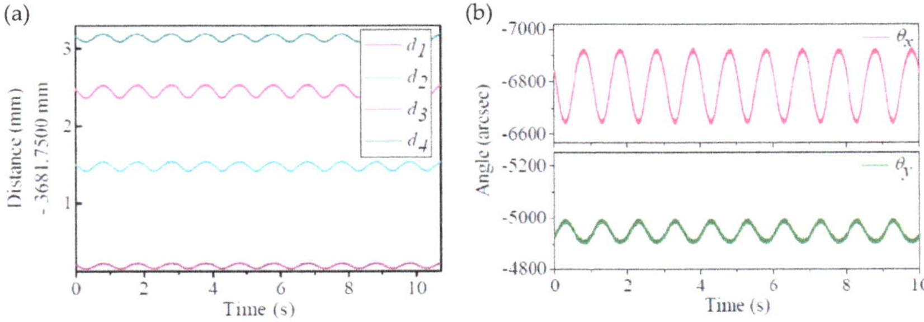

Figure 14. Dynamic measurement of the absolute angular position of an object by a dual-comb based absolute distance measurement method [63]: (**a**) Measured absolute distances of the four measurement points; (**b**) Variations of the absolute angular positions calculated from the measured absolute distances.

Figure 15a shows another example of applying the displacement measurement based on the mode-locked femtosecond laser [62]. The setup is based on a Michaelson interferometer designed to have a single reference arm and two measurement arms by splitting the collimated mode-locked femtosecond laser beam into three beams with beam splitters (BS1 and BS2). Through the demodulation of the interference spectrum of the three beams from the reference arm and the measurement arms based on a Fourier transform method [78], optical path differences among the three beams can be obtained simultaneously, as shown in Figure 15b.

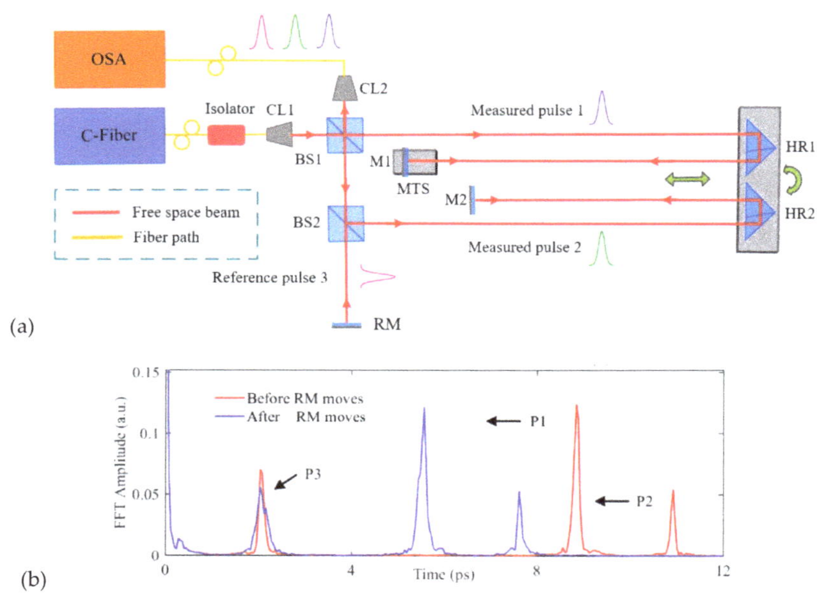

Figure 15. Angle measurement based on the absolute distance measurement by a Michaelson interferometer with a mode-locked femtosecond laser source [62]: (**a**) Optical configuration with two measurement arms; (**b**) Spectrum of the combined beams.

It should be noted that environmental fluctuations (the refractive index of air) could affect the absolute distance measurement [79] in the same manner as the technique based on the single-mode laser source. Angle measurement based on the absolute distance measurement with a mode-locked femtosecond laser source could also be affected by the environmental fluctuations, and attention should be paid to the control of environmental parameters such as temperature, pressure, and humidity.

4. Angle Detection Based on the High Pulse Energy of a Mode-Locked Femtosecond Laser

The small angular displacement of an object can also be measured by utilizing the angle-dependency of the second harmonic generation (SHG) in nonlinear optics, where the second harmonic light with the doubled optical frequency $\nu_2 = 2\nu_1$ of the fundamental light ν_1 incident to a nonlinear optical component is generated. It is well known that an efficient SHG can be accomplished by the procedure referred to as index matching [73], where the intensity of the second harmonic light depends on the angle of incidence of the fundamental light; this angle dependency can be employed for small angular displacement measurement. Owing to the characteristic of a mode-locked femtosecond laser with high pulse energy, effective SHG can be achieved by focusing a mode-locked femtosecond laser beam in a nonlinear crystal. On the assumption that the Rayleigh length b of the focused fundamental light lay with wavelength λ_1 is much longer than the length of the negative uniaxial crystal with refractive indices of n_e and n_o for extraordinary and ordinary rays, respectively, the intensity of the second harmonic light I_2 with wavelength λ_2 to be generated by SHG can be expressed as follows [59]:

$$I_2 = \frac{8\pi^2 d_{\text{eff}}^2 L^2}{n_1^2 n_2 \varepsilon_0 c \lambda_1^2} I_1^2 \sin c^2(\Delta k L/2) \quad (7)$$

where I_1 is the light intensity of the fundamental light ray, d_{eff} is the effective nonlinear coefficient [32], ε_0 is the vacuum permittivity, c is the speed of light in a vacuum. In the above Equation, Δk is a θ-dependent phase mismatching that can be represented as $\Delta k = 4\pi[n_o(\lambda_1) - n_e(\theta,\lambda_2)]/\lambda_1$. The angle $\theta = \theta_m$ satisfying the sinc-term in the above Equation to be the maximum value of 1 (namely, $n_e = n_o$ and Δk thus becomes zero) is referred to as the matching angle (Figure 16) that can be obtained by the following Equation [59]:

$$\theta_m = \arcsin\left(\sqrt{\frac{[n_O(\lambda_1)]^{-2} - [n_O(\lambda_2)]^{-2}}{[n_e(\lambda_1)]^{-2} - [n_O(\lambda_2)]^{-2}}}\right) \quad (8)$$

According to the above Equations, the intensity of the second-harmonic light I_2 decreases rapidly with the small angular displacement of the nonlinear crystal from the matching angle; this characteristic of the SHG can be employed for angle measurement.

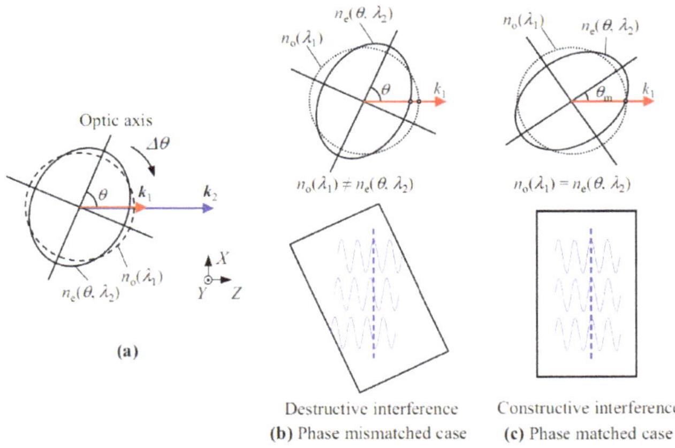

Figure 16. The matching angle in the second harmonic generation (SHG) [59]: (**a**) Refractive index ellipse of a negative uniaxial crystal; (**b**) Phase mismatched case; (**c**) Phase matched case.

Figure 17a shows an example of how to apply the characteristic of the matching angle in SHG for measurement of the small angular displacement of an object [59]. In the setup, a nonlinear crystal is mounted on a rotary table. For angle measurement, the optic axis of the nonlinear crystal is aligned to have the angle θ, which is almost equal to the matching angle θ_m, with respect to the propagating direction of the fundamental light wave as shown in Figure 18b so that the second harmonic light with enough power can be obtained. A small angular displacement of the rotary table can be measured by detecting the angular displacement of the nonlinear crystal by monitoring the change in the intensity of the second harmonic light with a photodetector.

The feasibility of the method described above has been verified in experiments. Figure 18a,b show a schematic and a photograph of the developed setup. An Erbium-doped fiber-based mode-locked femtosecond laser with the spectrum ranging from 1480 nm to 1640 nm has been employed as the light source. The laser beam introduced into the setup is at first collimated by a collimating lens and is then made incident to a nonlinear crystal mounted on a rotary table. As can be seen in Equation (8), the intensity of the second harmonic light to be generated in a nonlinear crystal depends on the crystal material, as well as the light wavelength. In the setup, barium borate (BBO) crystal with a similar matching angle for each mode in the femtosecond laser has thus been employed as the nonlinear crystal. The optic axis of the nonlinear crystal is adjusted to be approximately 20 degrees. It should be noted that, regarding Equation (7), the fundamental light is made to focus in the BBO crystal so that the second harmonic light can be generated effectively. The second harmonic light generated by SHG is then condensed onto a photodiode by a condenser lens, and the photocurrent from the photodiode is converted into voltage signal through a trans-impedance amplifier to monitor the signal by an oscilloscope. It should be noted that not only the second harmonic light but also the remaining fundamental light would come out from the BBO crystal. In the setup, a polarizer is placed in front of the photodiode so that the fundamental light will not be detected. For the verification of the angular displacement of the BBO crystal, a commercial laser autocollimator is also employed as a reference sensor.

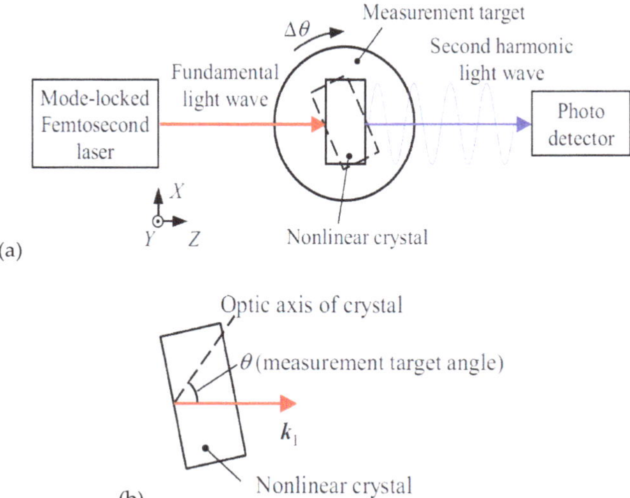

Figure 17. Measurement of the angular displacement of an object based on the characteristics of the matching angle in the second harmonic generation (SHG) [59]: (**a**) Schematic of the optical setup; (**b**) The angle to be measured by the method.

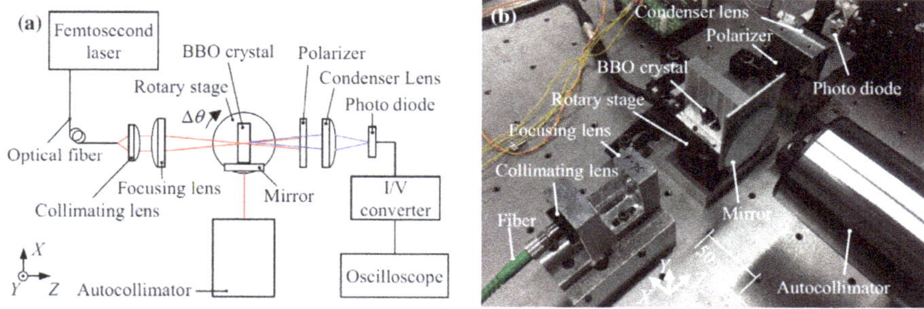

Figure 18. Measurement of the angular displacement of a rotary table based on SHG: (**a**) Optical configuration of the setup; (**b**) A photograph of the setup [59].

Figure 19a shows the change in the intensity of the second harmonic light as the change in the angular displacement of the BBO crystal observed in experiments, where three types of focusing lens with focal lengths of 40 mm, 75 mm and 150 mm are employed. As can be seen in the figure, a peak can be observed at the matching angle in each of the plots. Figure 19b shows the variation of the intensity of second-harmonic light as the change in the angular position of the BBO crystal. As can be seen in the figure, the angular displacement in a step of 0.4 arc-seconds has clearly been distinguished. These results demonstrate the feasibility of measuring small angular displacement by the SHG. It has also been verified that the shorter focal length of the focusing lens contributes to obtaining second-harmonic light with larger intensity, since the power of the fundamental beam can be further concentrated at the beam waist of the focused beam in the BBO crystal.

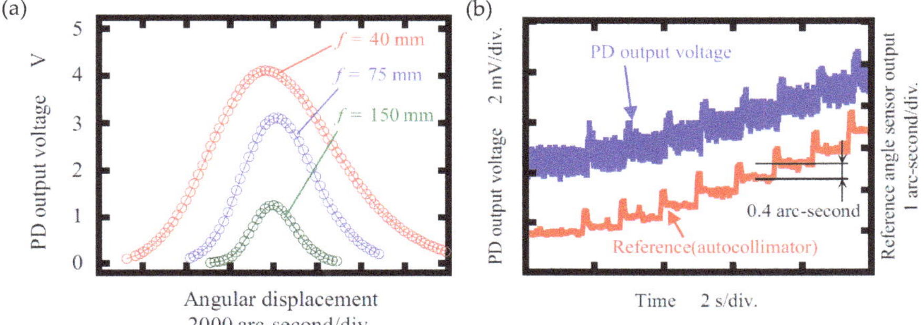

Figure 19. Variation of the light intensity of generated second harmonic light observed in experiments [59]: (**a**) Angle dependency of the second harmonic generation; (**b**) Variation of the light intensity of second harmonic light due to the angular displacement of the BBO crystal in a step of 0.4 arc-seconds.

5. Conclusions

A mode-locked femtosecond laser has been employed in a variety of applications in dimensional metrology for production engineering. In recent years, mode-locked femtosecond lasers have expanded application to precision angle measurement, which is also an important activity in production engineering. In this article, some angle measurement techniques employing a mode-locked femtosecond laser have been reviewed. With the employment of the dispersive characteristic of diffraction grating, equally-spaced modes of the mode-locked femtosecond laser in frequency domain can be converted into the "angle scale comb", which is a series of scale graduations for angle measurement. With the enhancement of the highly-stabilized optical modes of a mode-locked femtosecond laser over a wide spectral range, highly stable optical angle measurement can be achieved over a wide measuring range. In addition, by combining the angle scale comb with the laser autocollimation, a mode-locked femtosecond laser autocollimator with a high resolution over a wide angular range can be realized. The mode-locked femtosecond laser autocollimator has the possibility of realizing a direct link of angle measurement to the national standard of time/frequency by observing the reading output in the frequency domain. A mode-lock femtosecond laser can also be employed for angle measurement by utilizing physical phenomena such as the second harmonic generation of a nonlinear crystal or lateral chromatic aberration of a single lens. In addition, novel optical setups for the absolute distance measurement with a mode-locked femtosecond laser source can also achieve the measurement of angular displacement/absolute angular position of an object. Table 2 summarizes the features of the angle measurement techniques reviewed in this paper, including the achieved resolutions and measuring ranges.

Table 2. Comparison of the techniques reviewed in this paper.

Type of the Optical Angle Sensor	Resolution and Measuring Range	Features (Advantages and Disadvantages)
Light intensity detecting-type mode-locked femtosecond laser autocollimator (in Section 2.1.2) [55]	Resolution: Sub-arc-second Range: >3°	➢ High measurement throughput with a high-speed photodetector ➢ A grating reflector required ➢ Low signal visibility ➢ The resolution is mainly limited by the wavelength resolution of the spectrometer in the detector unit
Frequency-domain mode-locked femtosecond laser autocollimator (in Section 2.1.3) [56,57]	Resolution: 0.03 arc-seconds Range: >6°	➢ High signal visibility with a spectrometer that enables measurement of an object with low surface reflectivity ➢ A grating reflector required ➢ Low measurement throughput ➢ The resolution mainly limited by the wavelength resolution of the spectrometer in the detector unit
A method based on the chromatic aberration of a simple lens (in Section 2.2) [58]	Resolution: 0.23 arc-seconds Range: >100 arc-seconds	➢ Simple optical configuration without a grating reflector ➢ High signal visibility with a spectrometer ➢ The resolution mainly limited by the chromatic lens and the wavelength resolution of the spectrometer in the detector unit
Methods based on the absolute distance measurement (in Section 3) [63]	Resolution: 0.073 arc-seconds Range: >300 arc-seconds	➢ Long working distance ➢ High measurement throughput with a high-speed photodetector ➢ Several mirror reflectors are required ➢ The resolution mainly affected by environmental fluctuations (temperature, humidity, air pressure)
A method based on the second harmonic generation (in Section 4) [59]	Resolution: 0.4 arc-seconds Range: >3.3°	➢ High measurement throughput with a high-speed photodetector ➢ An expensive nonlinear crystal required ➢ Limited applications (need to mount a nonlinear crystal for angle measurement)

One of the disadvantages of the above-mentioned techniques is the high cost of the mode-locked femtosecond laser source. With the decrease of the cost of the femtosecond laser source, the angle measurement techniques explained in this review article are expected to be employed in many industrial applications in the near future where the traceability of measurement becomes a more important issue to be addressed.

Author Contributions: Conceptualization, W.G. and Y.S.; methodology, W.G. and Y.S.; software, W.G. and Y.S.; validation, W.G. and Y.S.; formal analysis, W.G. and Y.S.; investigation, Y.S. and H.M.; resources, W.G. and Y.S.; data curation, W.G., H.M. and Y.S.; writing—original draft preparation, Y.S.; writing—review and editing, W.G. and Y.S.; visualization, Y.S. and W.G.; supervision, W.G.; project administration, W.G.; funding acquisition, W.G., Y.S. and H.M. All authors have read and agreed to the published version of the manuscript.

Funding: A part of this research is supported by the Japan Society for the Promotion of Science (JSPS) 15H05759 and 18H01345.

Acknowledgments: The authors would like to thank all our laboratory members who joined the related projects at Tohoku University for their contributions to the achievements described in this paper.

Conflicts of Interest: The authors declare no conflict of interest. The founding sponsors had no role in the design of the study; in the collection, analyses, or interpretation of data; in the writing of the manuscript, and in the decision to publish the results.

References

1. ISO. *Geometrical Product Specifications (GPS)—Dimensional tolerancing—Part 2: Dimensions Other than Linear Sizes*; ISO 14405-2:2011; ISO: Genava, Switzerland, 2012.
2. Gao, W.; Haitjema, H.; Fang, F.Z.; Leach, R.K.; Cheung, C.F.; Savio, E.; Linares, J.M. On-machine and in-process surface metrology for precision manufacturing. *CIRP Ann.* **2019**, *68*, 843–866. [CrossRef]

3. Wallis, D.A. History of Angle Measurement. In Proceedings of the FIG Working Week 2005 and the 8th International Conference for Global Spatial Data Infrastructure, Cairo, Egypt, 16–21 April 2005; pp. 1–17.
4. Gao, W. *Precision Nanometrology—Sensors and Measuring Systems for Nanomanufacturing*; Springer: Berlin/Heidelberg, Germany, 2010.
5. Just, A.; Krause, M.; Probst, R.; Bosse, H.; Haunerdinger, H.; Spaeth, C.; Metz, G.; Israel, W. Comparison of angle standards with the aid of a high-resolution angle encoder. *Precis. Eng.* **2009**, *33*, 530–533. [CrossRef]
6. Renishaw plc. *RESOLUTE Absolute Optical Encoder with Biss Serial Communications*; Renishaw plc: Wotton-under-Edge, UK, 2013; pp. 1–8.
7. Magnescale, C. Feedback Scale General Catalog. Available online: http://www.magnescale.com/mgs/product/catalog/FeedbackScale_en.pdf (accessed on 29 May 2020).
8. Miyashita, K.; Takahashi, T.; Yamanaka, M. Features of a magnetic rotary encoder. *IEEE Trans. Magn.* **1987**, *23*, 2182–2184. [CrossRef]
9. Watanabe, T.; Kon, M.; Nabeshima, N.; Taniguchi, K. An angle encoder for super-high resolution and super-high accuracy using SelfA. *Meas. Sci. Technol.* **2014**, *25*, 065002. [CrossRef]
10. Watanabe, T.; Fujimoto, H.; Masuda, T. Self-calibratable rotary encoder. *J. Phys. Conf. Ser.* **2005**, *13*, 240–245. [CrossRef]
11. Gao, W.; Kim, S.W.; Bosse, H.; Haitjema, H.; Chen, Y.L.; Lu, X.D.; Knapp, W.; Weckenmann, A.; Estler, W.T.; Kunzmann, H. Measurement technologies for precision positioning. *CIRP Ann. Manuf. Technol.* **2015**, *64*, 773–796. [CrossRef]
12. Kunzmann, H.; Pfeifer, T.; Flügge, J. Scales vs. Laser Interferometers Performance and Comparison of Two Measuring Systems. *CIRP Ann. Manuf. Technol.* **1993**, *42*, 753–767. [CrossRef]
13. Cheng, F.; Fan, K.C. High-resolution angle measurement based on Michelson interferometry. *Phys. Procedia* **2011**, *19*, 3–8. [CrossRef]
14. Moore, W.R. *Foundations of Mechanical Accuracy*; The Moore Special Tool Company: Bridgeport, CT, USA, 1970.
15. Luther, G.G.; Deslattes, R.D.; Towler, W.R. Single axis photoelectronic autocollimator. *Rev. Sci. Instrum.* **1984**, *55*, 747–750. [CrossRef]
16. Gao, W.; Ohnuma, T.; Satoh, H.; Shimizu, H.; Kiyono, S.; Makino, H. A precision angle sensor using a multi-cell photodiode array. *CIRP Ann. Manuf. Technol.* **2004**, *53*, 425–428. [CrossRef]
17. Geckeler, R.D.; Just, A.; Krause, M.; Yashchuk, V.V. Autocollimators for deflectometry: Current status and future progress. *Nucl. Instrum. Methods Phys. Res. Sect. A Accel. Spectrometers Detect. Assoc. Equip.* **2010**, *616*, 140–146. [CrossRef]
18. Geckeler, R.D.; Weingaertner, I.; Just, A.; Probst, R. Use and treaceable calibration of autocollimators for ultra-precise measurement of slope and topography. *Recent Dev. Traceable Dimens. Meas.* **2001**, *4401*, 184–195.
19. Qian, S.; Geckeler, R.D.; Just, A.; Idir, M.; Wu, X. Approaching sub-50 nanoradian measurements by reducing the saw-tooth deviation of the autocollimator in the Nano-Optic-Measuring Machine. *Nucl. Instrum. Methods Phys. Res. Sect. A Accel. Spectrometers Detect. Assoc. Equip.* **2015**, *785*, 206–212. [CrossRef]
20. Bitou, Y.; Kondo, Y. Scanning deflectometric profiler for measurement of transparent parallel plates. *Appl. Opt.* **2016**, *55*, 9282–9287. [CrossRef]
21. Chen, M.; Xie, S.; Wu, H.; Takahashi, S.; Takamasu, K. Three-dimensional surface profile measurement of a cylindrical surface using a multi-beam angle sensor. *Precis. Eng.* **2020**, *62*, 62–70. [CrossRef]
22. MÖLLER-WEDEL OPTICAL Electroninc Autocollimators. Available online: www.moeller-wedel-optical.com (accessed on 29 May 2020).
23. Nikon Corporation Autocollimators 6B-LED/6D-LED. Available online: https://www.nikon.com/products/industrial-metrology/support/download/brochures/pdf/2ce-iwqh-4.pdf (accessed on 29 May 2020).
24. Trioptics GmbH OptiTest (R) a Complete Range of Optical Instrument. Available online: http://www.trioptics.com (accessed on 29 May 2020).
25. Ennos, A.E.; Virdee, M.S. High accuracy profile measurement of quasi-conical mirror surfaces by laser autocollimation. *Precis. Eng.* **1982**, *4*, 5–8. [CrossRef]
26. Saito, Y.; Gao, W.; Kiyono, S. A single lens micro-angle sensor. *Int. J. Precis. Eng.* **2007**, *8*, 14–18.
27. Shimizu, Y.; Tan, S.L.; Murata, D.; Maruyama, T.; Ito, S.; Chen, Y.-L.; Gao, W. Ultra-sensitive angle sensor based on laser autocollimation for measurement of stage tilt motions. *Opt. Express* **2016**, *24*, 2788–2805. [CrossRef]

28. Gao, W.; Huang, P.S.; Yamada, T.; Kiyono, S. A compact and sensitive two-dimensional angle probe for flatness measurement of large silicon wafers. *Precis. Eng.* **2002**, *26*, 396–404. [CrossRef]
29. Gao, W.; Saito, Y.; Muto, H.; Arai, Y.; Shimizu, Y. A three-axis autocollimator for detection of angular error motions of a precision stage. *CIRP Ann. Manuf. Technol.* **2011**, *60*, 515–518. [CrossRef]
30. Li, X.; Gao, W.; Muto, H.; Shimizu, Y.; Ito, S.; Dian, S. A six-degree-of-freedom surface encoder for precision positioning of a planar motion stage. *Precis. Eng.* **2013**, *37*, 771–781. [CrossRef]
31. Shimizu, Y.; Matsukuma, H.; Gao, W. Optical sensors for multi-axis angle and displacement measurement using grating reflectors. *Sensors* **2019**, *19*, 5289. [CrossRef] [PubMed]
32. Matsukuma, H.; Ishizuka, R.; Furuta, M.; Li, X.; Shimizu, Y.; Gao, W. Reduction in Cross-Talk Errors in a Six-Degree-of-Freedom Surface Encoder. *Nanomanuf. Metrol.* **2019**, *2*, 111–123. [CrossRef]
33. Gao, W.; Araki, T.; Kiyono, S.; Okazaki, Y.; Yamanaka, M. Precision nano-fabrication and evaluation of a large area sinusoidal grid surface for a surface encoder. *Precis. Eng.* **2003**, *27*, 289–298. [CrossRef]
34. Shimizu, Y.; Kataoka, S.; Ishikawa, T.; Chen, Y.L.; Chen, X.; Matsukuma, H.; Gao, W. A liquid-surface-based three-axis inclination sensor for measurement of stage tilt motions. *Sensors* **2018**, *18*, 398. [CrossRef]
35. Jones, D.J.; Cundiff, S.T.; Fortier, T.M.; Hall, J.L.; Ye, J. Carrier–Envelope Phase Stabilization of Single and Multiple Femtosecond Lasers. In *Few-Cycle Laser Pulse Generation and Its Applications. Topics in Applied Physics*; Kärtner, F.X., Ed.; Springer: Berlin/Heidelberg, Germany, 2012; Volume 95, pp. 317–343.
36. Jones, D.J.; Diddams, S.A.; Ranka, J.K.; Stentz, A.; Windeler, R.S.; Hall, J.L.; Cundiff, S.T. Carrier-envelope phase control of femtosecond mode-locked lasers and direct optical frequency synthesis. *Science* **2000**, *288*, 635–640. [CrossRef]
37. Udem, T.; Holzwarth, R.; Hänsch, T.W. Optical frequency metrology. *Nature* **2002**, *416*, 233–237. [CrossRef]
38. Udem, T.; Reichert, J.; Holzwarth, R.; Hänsch, T.W. Accurate measurement of large optical frequency differences with a mode-locked laser. *Opt. Lett.* **1999**, *24*, 881–883. [CrossRef]
39. Jang, Y.-S.; Kim, S.-W. Distance Measurements Using Mode-Locked Lasers: A Review. *Nanomanuf. Metrol.* **2018**, *1*, 131–147. [CrossRef]
40. Minoshima, K.; Matsumoto, H. High-accuracy measurement of 240-m distance in an optical tunnel by use of a compact femtosecond laser. *Appl. Opt.* **2000**, *39*, 5512–5517. [CrossRef]
41. Bitou, Y.; Schibli, T.R.; Minoshima, K. Accurate wide-range displacement measurement using tunable diode laser and optical frequency comb generator. *Opt. Express* **2006**, *14*, 644–654. [CrossRef]
42. Kajima, M.; Matsumoto, H. Picometer positioning system based on a zooming interferometer using a femtosecond optical comb. *Opt. Express* **2008**, *16*, 1497–1506. [CrossRef]
43. Joo, K.-N.; Kim, Y.; Kim, S.-W. Distance measurements by combined method based on a femtosecond pulse laser. *Opt. Express* **2008**, *16*, 19799–19806. [CrossRef]
44. Bitou, Y. High-accuracy displacement metrology and control using a dual Fabry-Perot cavity with an optical frequency comb generator. *Precis. Eng.* **2009**, *33*, 187–193. [CrossRef]
45. Coddington, I.; Swann, W.C.; Nenadovic, L.; Newbury, N.R. Rapid and precise absolute distance measurements at long range. *Nat. Photonics* **2009**, *3*, 351–356. [CrossRef]
46. Kim, S.W. Metrology: Combs rule. *Nat. Photonics* **2009**, *3*, 313–314. [CrossRef]
47. Bae, J.; Park, J.; Ahn, H.; Jin, J. Total physical thickness measurement of a multi-layered wafer using a spectral-domain interferometer with an optical comb. *Opt. Express* **2017**, *25*, 12689–12697. [CrossRef]
48. Park, J.; Jin, J.; Wan Kim, J.; Kim, J.A. Measurement of thickness profile and refractive index variation of a silicon wafer using the optical comb of a femtosecond pulse laser. *Opt. Commun.* **2013**, *305*, 170–174. [CrossRef]
49. Jin, J.; Kim, J.W.; Kim, J.A.; Eom, T.B. Thickness and refractive index measurement of a wafer based on the optical comb. *Opt. InfoBase Conf. Pap.* **2010**, *18*, 18339–18346.
50. Kubina, P.; Adel, P.; Adler, F.; Grosche, G.; Hänsch, T.W.; Holzwarth, R.; Leitenstorfer, A.; Lipphardt, B.; Schnatz, H. Long term comparison of two fiber based frequency comb systems. *Opt. Express* **2005**, *13*, 904–909. [CrossRef]
51. Adler, F.; Moutzouris, K.; Leitenstorfer, A.; Schnatz, H.; Lipphardt, B.; Grosche, G.; Tauser, F. Phase-locked two-branch erbium-doped fiber laser system for long-term precision measurements of optical frequencies. *Opt. Express* **2004**, *12*, 5872–5880. [CrossRef] [PubMed]
52. Hundertmark, H.; Wandt, D.; Fallnich, C.; Haverkamp, N.; Telle, H.R. Phase-locked carrier-envelope-offset frequency at 1560 nm. *Opt. Express* **2004**, *12*, 770–775. [CrossRef] [PubMed]

53. Inaba, H.; Daimon, Y.; Hong, F.-L.; Onae, A.; Minoshima, K.; Schibli, T.R.; Matsumoto, H.; Hirano, M.; Okuno, T.; Onishi, M.; et al. Long-term measurement of optical frequencies using a simple, robust and low-noise fiber based frequency comb. *Opt. Express* **2006**, *14*, 5223–5231. [CrossRef] [PubMed]
54. Shimizu, Y.; Kudo, Y.; Chen, Y.L.; Ito, S.; Gao, W. An optical lever by using a mode-locked laser for angle measurement. *Precis. Eng.* **2017**, *47*, 72–80. [CrossRef]
55. Chen, Y.-L.; Shimitzu, Y.; Kudo, Y.; Ito, S.; Gao, W. Mode-locked laser autocollimator with an expanded measurement range. *Opt. Express* **2016**, *24*, 425–428. [CrossRef] [PubMed]
56. Chen, Y.; Shimizu, Y.; Tamada, J.; Kudo, Y.; Madokoro, S.; Nakamura, K.; Gao, W.; Gao, W.; Kim, S.W.; Bosse, H.; et al. Optical frequency domain angle measurement in a femtosecond laser autocollimator. *Opt. Express* **2017**, *25*, 16725–16738. [CrossRef] [PubMed]
57. Chen, Y.L.; Shimizu, Y.; Tamada, J.; Nakamura, K.; Matsukuma, H.; Chen, X.; Gao, W. Laser autocollimation based on an optical frequency comb for absolute angular position measurement. *Precis. Eng.* **2018**, *54*, 284–293. [CrossRef]
58. Shimizu, Y.; Madokoro, S.; Matsukuma, H.; Gao, W. An optical angle sensor based on chromatic dispersion with a mode-locked laser source. *J. Adv. Mech. Des. Syst. Manuf.* **2018**, *12*. [CrossRef]
59. Matsukuma, H.; Madokoro, S.; Dwi, W.; Yuki, A.; Wei, S. A New Optical Angle Measurement Method Based on Second Harmonic Generation with a Mode—Locked Femtosecond Laser. *Nanomanuf. Metrol.* **2019**, *2*, 187–198. [CrossRef]
60. Dennis, M.L.; Diels, J.-C.M.; Lai, M. Femtosecond ring dye laser: A potential new laser gyro. *Opt. Lett.* **1991**, *16*, 529–531. [CrossRef]
61. Diddams, S.; Atherton, B.; Diels, J.C. Frequency locking and unlocking in a femtosecond ring laser with application to intracavity phase measurements. *Appl. Phys. B Lasers Opt.* **1996**, *63*, 473–480. [CrossRef]
62. Liang, X.; Lin, J.; Yang, L.; Wu, T.; Liu, Y.; Zhu, J. Simultaneous Measurement of Absolute Distance and Angle Based on Dispersive Interferometry. *IEEE Photonics Technol. Lett.* **2020**, *32*, 449–452. [CrossRef]
63. Han, S.; Kim, Y.-J.; Kim, S.-W. Parallel determination of absolute distances to multiple targets by time-of-flight measurement using femtosecond light pulses. *Opt. Express* **2015**, *23*, 25874–25882. [CrossRef] [PubMed]
64. Post, E.J. Sagnac effect. *Rev. Mod. Phys.* **1967**, *39*, 475–493. [CrossRef]
65. Bergh, R.A.; Lefevre, H.C.; Shaw, H.J. An Overview of Fiber-optic Gyroscopes. *J. Light. Technol.* **1984**, *2*, 91–107. [CrossRef]
66. Culshaw, B. Fibre optic gyroscopes. *Phys. Technol.* **1982**, *13*, 79–80. [CrossRef]
67. Ezekiel, S.; Balsamo, S.R. Passive ring resonator laser gyroscope. *Appl. Phys. Lett.* **1977**, *30*, 478–480. [CrossRef]
68. Kadiwar, R.K.; Giles, I.P. Optical Fibre Brillouin Ring Laser Gyroscope. *Electron. Lett.* **1989**, *25*, 1729–1731. [CrossRef]
69. Rosker, M.J.; Christian, W.R.; Mcmichael, I.C.; Oaks, T. Picosecond-pulsed diode ring laser gyroscope. *Proc. SPIE* **1994**, *2116*, 365–373.
70. Fortier, T.; Baumann, E. 20 Years of Developments in Optical Frequency Comb Technology and Applications. *Commun. Phys.* **2019**, *2*, 1–16. [CrossRef]
71. Tausenev, A.V.; Kryukov, P.G.; Bubnov, M.M.; Likhachev, M.E.; Romanova, E.Y.; Yashkov, M.V.; Khopin, V.F.; Salganskii, M.Y. Efficient source of femtosecond pulses and its use for broadband supercontinuum generation. *Kvantovaya Elektron.* **2005**, *35*, 581–585. [CrossRef]
72. Tamada, J.; Kudo, Y.; Chen, Y.-L.; Shimizu, Y.; Gao, W. Determination of the zero-position for an optical angle sensor. *J. Adv. Mech. Des. Syst. Manuf.* **2016**, *10*. [CrossRef]
73. Hecht, E. *Optics*, 5th ed.; Pearson: London, UK, 2017; ISBN 9780133977226.
74. Acosta, D.; Albajez, J.A.; Yagüe-Fabra, J.A.; Velázquez, J. Verification of Machine Tools Using Multilateration and a Geometrical Approach. *Nanomanuf. Metrol.* **2018**, *1*, 39–44. [CrossRef]
75. Shiou, F.J.; Liu, M.X. Development of a novel scattered triangulation laser probe with six linear charge-coupled devices (CCDs). *Opt. Lasers Eng.* **2009**, *47*, 7–18. [CrossRef]
76. Fan, K.C.; Chen, M.J. 6-Degree-of-freedom measurement system for the accuracy of X-Y stages. *Precis. Eng.* **2000**, *24*, 15–23. [CrossRef]
77. Pellegrini, S.; Buller, G.S.; Smith, J.M.; Wallace, A.M.; Cova, S. Laser-based distance measurement using picosecond resolution time-correlated single-photon counting. *Meas. Sci. Technol.* **2000**, *11*, 712–716. [CrossRef]

78. Cui, M.; Zeitouny, M.G.; Bhattacharya, N.; van den Berg, S.A.; Urbach, H.P. Long distance measurement with femtosecond pulses using a dispersive interferometer. *Opt. Express* **2011**, *19*, 6549–6562. [CrossRef] [PubMed]
79. Joo, K.N.; Kim, S.W. Absolute distance measurement by dispersive interferometry using a femtosecond pulse laser. *Opt. Express* **2006**, *14*, 5954–5960. [CrossRef]

Article

A Form-Free and High-Precision Metrological Method for the Twist of Aeroengine Blade

Xuezhe Li [1] and Zhaoyao Shi [2,*]

1 Hebei Key Laboratory of Safety Monitoring of Mining Equipment, College of Mechanical and Electrical Engineering, North China Institute of Science and Technology, Sanhe 065201, China; 200800897lxz@ncist.edu.cn
2 Beijing Engineering Research Center of Precision Measurement Technology and Instruments, College of Mechanical Engineering and Applied Electronics Technology, Beijing University of Technology, Beijing 100124, China
* Correspondence: shizhaoyao@bjut.edu.cn; Tel.: +86-139-1030-7299

Received: 16 May 2020; Accepted: 13 June 2020; Published: 16 June 2020

Abstract: In order to solve the problems in the accuracy and adaptability of the existing methods for blade twist measurement, a high-precision and form-free metrological method of blade twist based on the parameter evaluation of twist angular position and twist angle is proposed in the study, and the theoretical model, the measurement principle and the key technologies of the method are discussed in detail. Three key issues of the twist metrology of a blade are solved based on technologies of calibration, a priori planning and geometric analysis: aeroengine axis matching, high-precision coordinate acquisition of the leading edge and the trailing edge, and extraction of twist angular position of the profile. The measurement path planning, sampling strategy optimization and high-precision coordinate collection are executed automatically without theoretical model data of the measured blade, thus the form-free and high-precision metrology of the blade twist is achieved. The research results show that the metrological method of blade twist presented in this study is effective, and that its measurement uncertainty is less than 0.01°. This method is form-free, efficient and accurate, and can solve the problems of high-precision measurement and evaluation for the twist of aeroengine blade primely.

Keywords: aeroengine blade; blade twist; measurement and evaluation; a priori planning; geometric analysis

---

1. Introduction

In order to maintain the optimal distribution of airflow in the cascade passage, avoid the separation of airflow and reduce the blade loss, the aeroengine blade must be twisted and stacked by profiles along the airfoil height direction [1]. The twist of the blade plays an important role in optimizing and improving the aerodynamic performance of an aeroengine, and its measurement and evaluation are of great significance. Gao et al. [2,3] studied the influence law of blade machining error on the aerodynamic performance of a compressor and pointed out that the torsional error of the blade is an important factor affecting the loss of the blade profile. Zhang et al. [4] analyzed the influence of blade profile deviation on turbine performance by numerical simulation, and concluded that the deviation of blade profile changes the field structure and the acceleration law of the airflow in the turbine passage, thus affecting the aerodynamic performance of the turbine. Cheng et al. [5] researched the effects of six typical blade parameters on compressor performance and drew a conclusion that blade torsion error has an important influence on overall pressure ratio, efficiency, flow rate, etc.

At present, the measuring technologies of two-dimensional profile parameters such as the maximum thickness of the profile and the radius of the leading and trailing edges are relatively perfect, and the extraction algorithms of parameters are more mature [6–8]. However, the measurement and evaluation of blade twist, which characterizes the torsion law of a blade in three-dimensional space, remains a challenge. Difficulties in the measurement of the blade twist are mainly manifested in the following aspects: (1) The twist is a three-dimensional geometric parameter of the blade. The measuring model is complex and the characterization and evaluation are difficult. (2) The parameter of twist angular position is defined by the angle between the profile chord and the axial line of engine. It is another difficult problem to locate the axial line of the engine fast and accurately, and to establish the measurement datum. (3) It is hard to achieve a high-precision measurement of blade twist. Many error factors such as the coordinate acquisition accuracy of the leading and trailing edges and the extraction algorithms of the profile chord line have a great influence on the accuracy of blade twist measurement.

At present, some progress has been made in blade twist measurement, and some mature commercial products in the market can be found to meet the basic requirements of blade twist measurement, such as CMM (Coordinate Measuring Machine) of the Global Series from HEAXGON and CORE DS from WENZEL. The existing measuring methods of blade twist mainly involve two categories: the contact measurement method of CMM and the non-contact optical measurement method [9].

Contact measurement technology is the first applied blade twist measurement technology. Su et al. [10] proposed a two-point measurement method for blade torsional deformation. This method can be used to estimate the rough deformation state of the blade, but measurements with just two points may lose important information and cannot evaluate the blade twist angle with high accuracy. Shi et al. [11] studied a contact measuring method for blades based on CMM and proposed a variable-speed scanning method based on curvature recognition. The method improves the accuracy of measurement by optimizing the measurement path, but has a poor efficiency. Bu et al. [12] analyzed the influence of CMM sampling points on the measurement accuracy of characteristic parameters for turbine blades, and improved the measurement accuracy of parameters such as twist angle by optimizing the sampling strategy. There are disadvantages and constraints in the measurement of blade twist based on contact measurement technology: (1) The theoretical model of the measured blade is necessary for path planning and result analysis, which limits measurement adaptability greatly. (2) The measurement efficiency is low, and the high-speed and full-information measurement cannot be achieved. (3) Challenges arise, due to the effect of the radius of the probe ball, the cosine error, serious fluctuations or even the loss of measured data.

With the development of technology, non-contact measurement technology has become a new hotspot in blade measurement research. Sun et al. [13] presented a method for non-contact measurement and evaluation of the aeroengine blade. The coordinates are collected by scanning the blade profile using the laser sensor, the analysis and evaluation of the blade parameters are realized by the comparison of measurement model and design model of the blade. Li et al. [14] studied a mathematical model of inclination error compensation, where the coordinate acquisition accuracy of non-contact measurement is improved based on the proposed error-compensation model. The non-contact optical measuring technology improves the measurement efficiency significantly, but it still has some problems in optical adaptability and measurement accuracy due to the limitation of the measurement principle.

Aiming at the problems in the accuracy and adaptability of blade twist measurements, a high-precision and form-free measurement method of blade twist based on evaluations of twist angular position and twist angle is proposed, and the theoretical model, the measurement principle and the key technologies of this method are discussed in detail. The high-precision acquisition of coordinates and the extraction of chord angles without a theoretical model of the measured blade are key issues and difficulties of this study. The method proposed in this study significantly improves the accuracy and adaptability of blade twist measurement, and provides a new technical solution for the measurement and evaluation of the twist of aeroengine blade.

2. Characterization and Measurement of Blade Twist

2.1. Definition of Twist Angular Position of the Profile

The profile is a closed two-dimensional blade contour with a special aerodynamic performance, which consists of a convex profile, a concave profile, a leading edge arc and a trailing edge arc [15,16]. The twist angular position of the profile characterizes the absolute torsion level of profile relative to the axial line of engine in the space, and its geometric definition is shown in Figure 1.

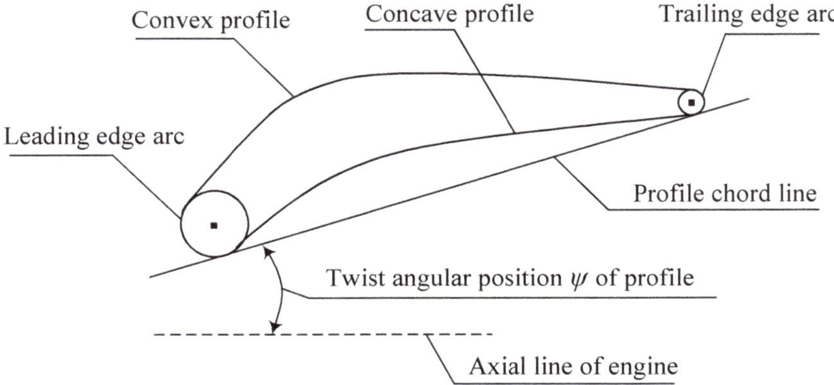

Figure 1. Definition diagram of the twist angular position of profile.

In Figure 1, the profile chord line refers to the common tangent of the leading and trailing edges of the profile, and the twist angular position ψ of the profile is defined by the angle between the profile chord and the axial line of engine. As can be seen from Figure 1, the registration of the engine axis and extraction of the profile chord line are key issues for measuring the twist angular position of profile.

2.2. Characterization of Blade Twist

The blade is composed of a series of profiles which are twisted and stacked according to certain rules. The selection of the test profile and the extraction of the twist characteristic parameters are two key problems to be solved in the characterization of the blade twist. In order to evaluate the spatial torsion characteristics of blade efficiently and scientifically, two characteristic parameters of twist angular position $\overline{\psi}$ of blade and twist angle η of blade are proposed in this study, which characterize and evaluate the twist level of the blade comprehensively from two aspects of position and form.

The method of characterization and evaluation of the blade twist is described as follows:

(1) As shown in Figure 2, two test profiles I and II are planned on the blade airfoil. Test profile I is set at 3 mm below the minimum radius of the blade tip and test profile II is set at 4 mm above the maximum radius of the blade root.

(2) The twist angular positions ψ_{I} and ψ_{II} of two test profiles are measured respectively based on technologies of a priori planning and geometric analysis.

(3) The characteristic parameters $\overline{\psi}$ and η of the blade twist are extracted and calculated. In order to reveal the torsional characteristics of blade comprehensively and intuitively, two evaluation parameters are defined in the study: The twist angular position $\overline{\psi}$ of blade characterizes the absolute torsion level of blade relative to the axial line of the engine, which can be determined by average twist angle position of the blade; the twist angle η of blade characterizes the relative torsion degree of the blade itself, which can be

determined by the difference value between maximum twist angle position and minimum twist angle position. The mathematical models of parameters $\overline{\psi}$ and η are defined respectively as follows:

$$\overline{\psi} = \frac{\psi_{\mathrm{I}} + \psi_{\mathrm{II}}}{2} \tag{1}$$

$$\eta = \left|\psi_{\mathrm{I}} - \psi_{\mathrm{II}}\right| \tag{2}$$

By optimizing the number and location of the test profiles, the measuring process of the blade twist is effectively simplified and the evaluation efficiency is significantly improved. Based on the parameter evaluations of the twist angular position $\overline{\psi}$ and twist angle η, the spatial torsion state of the blade is characterized intuitively and comprehensively. The method proposed in this study has the advantage of being a simple model that delivers a comprehensive evaluation, which authentically reveals the torsion rule of the blade.

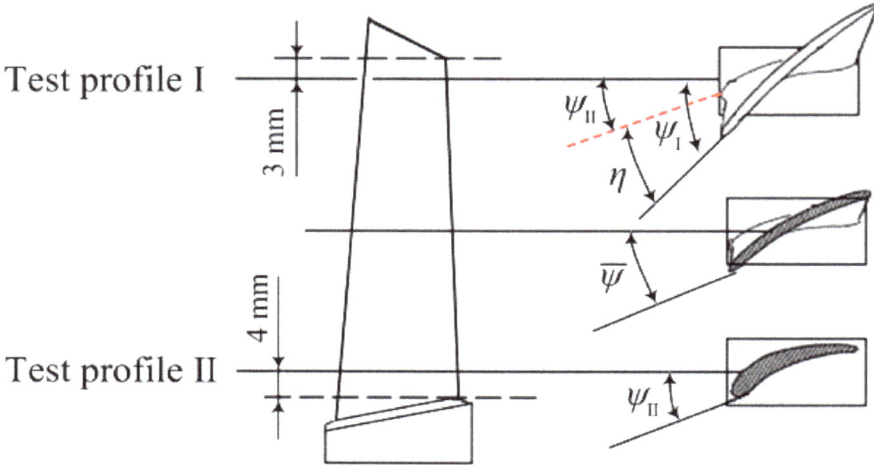

Figure 2. Model of characterization and evaluation for blade twist.

2.3. Measurement Method of Blade Twist

In this study, a high-precision special machine for blade measurement shown in Figure 3 is applied to the measurement research of the blade twist. The measuring machine is essentially a four-coordinate laser measurement system, and consists of a four-axis motion platform, a fixture system and a high-precision laser probe. The four-axis motion platform is made up of three linear shaftings of X, Y, and Z with a resolution of 0.1 μm, a rotary shafting of C with a resolution of 0.0002° and a precision CNC (Computer Numerical Control) system [17,18]. The fixture system solves the problems of clamping, positioning and measurement calibration. The high-precision laser probe is designed with a conoscopic holography sensor CP-3 from Optimet (Israel), with a measurement range of −1 mm to +1 mm and a resolution of 0.1 μm. The measuring machine provides a precise hardware platform for the research of the measurement method, and the subsequent experimental verification is also implemented on the high-precision special machine. Coordinate measurements and chord angle extraction with high precision are two key issues to be solved in blade twist measurement.

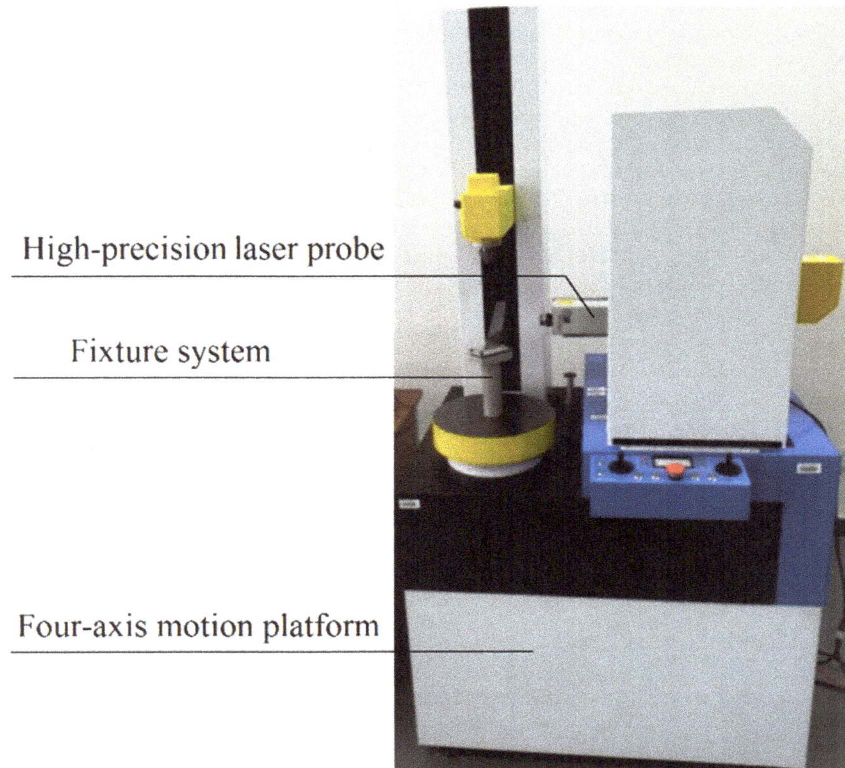

Figure 3. High-precision special machine for blade measurement.

In order to improve the accuracy and adaptability of the coordinate measurement, an a priori planning measurement method is proposed in this study. Firstly, the test profiles of the blade are measured without theoretical model data by the laser probe, and the theoretical model of the blade is self-constructed based on feature recognition. Then, the measuring path is planned and the sampling strategy is optimized based on the theoretical model solved of blade, and the high-precision acquisition of the blade coordinates is achieved. All coordinates are collected in the positions near the reference distance of the laser probe based on a priori planning technology, thus the depth of measurement approaches 0 mm and the measurement error is no more than 10 μm. In addition, this method is a form-free measurement method and does not need the theoretical model of the blade, which improves the adaptability of the measurement. By optimizing the measurement method, the accuracy level can meet the measuring requirements of aero-turbine blades with a first precision grade. The a priori planning measurement method is an innovation of this study.

In order to improve the measurement accuracy of the chord angle, an algorithm for edge extraction based on sampling optimization and least squares fitting is proposed, which can provide the authentic measurement data for the extraction of the profile chord. The statistical uncertainty of the method is less than 3 μm.

The principle and the process of the blade twist measurement are summarized as follows:

Step 1. Establishment of workpiece coordinate system and registration of engine axis. The workpiece coordinate system is established by scanning the selected section of the mounting column of the fixture, and the axial line of the engine is matched and aligned with the X axis by measuring the side of the base platform of the fixture. In this study, the least squares fitting algorithm is used to process the measured data, which greatly improves the datum accuracy of the twist measurement.

Step 2. Acquisition of coordinates of the test profiles. The high-precision coordinate data is obtained by scanning and measuring test profiles I and II with measurement technology of a priori planning.

Step 3. Feature recognition and profile fitting. By feature recognition and piecewise fitting for the collected coordinate data, the test profiles I and II are extracted, which provides authentic and accurate measurement data for subsequent parameter calculation and evaluation.

Step 4. Basic parameters calculation of test profiles. Based on the fitted profiles I and II, combined with the mathematical model of each parameter, the basic parameters of blade profile, such as leading edge radius, trailing edge radius, leading edge center and trailing edge center, are calculated.

Step 5. Calculation of twist angular positions of two test profiles. Based on the basic parameters of the profiles calculated in Step 4, the twist angular positions ψ_{I} and ψ_{II} are analyzed and calculated with the technology of geometric analysis.

Step 6. Evaluation of blade twist. Based on the solutions of ψ_{I} and ψ_{II} obtained in Step 5, the characteristic parameters of blade twist $\overline{\psi}$ and η are calculated using Equations (1) and (2). Thus, the spatial torsion state of blade is characterized and evaluated comprehensively with twist angular position $\overline{\psi}$ and twist angle η.

The measuring process of blade twist on the high-precision special machine for blade measurement is shown in Figure 4.

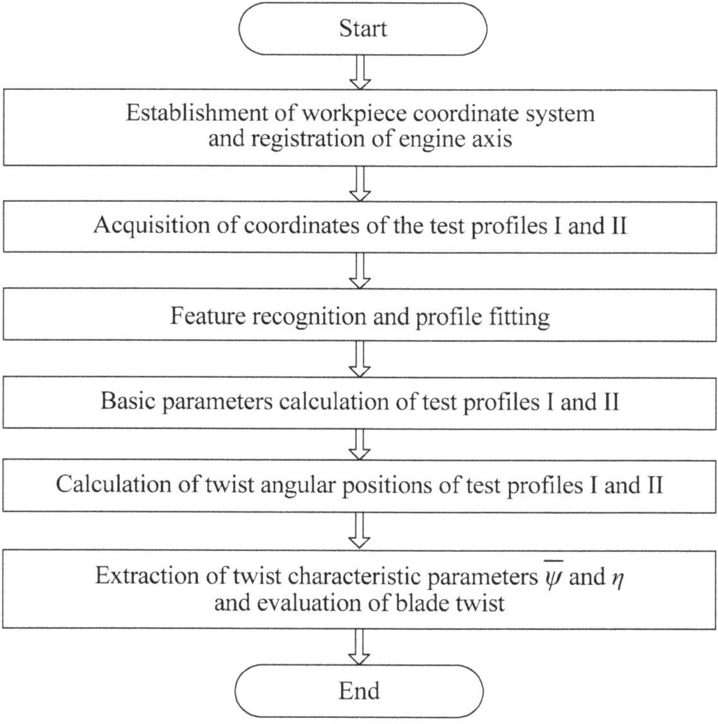

Figure 4. Flowchart of blade twist measurement.

3. Analysis of Key Technologies

3.1. Registration Method of Axial Line of Engine Based on Least Squares Principle

In order to locate the axial line of engine quickly and accurately, a fixture for blade clamping as shown in Figure 5 is designed in this study.

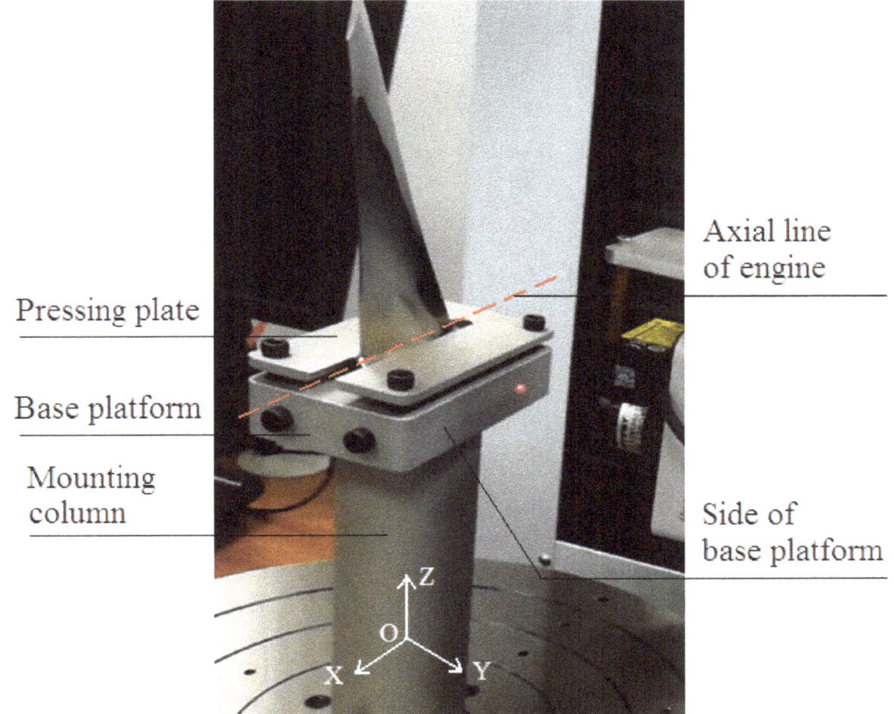

Figure 5. Fixture for blade clamping.

The fixture is composed of three exact centrosymmetric components: the mounting column, the base platform and the pressing plate. The high-precision positioning and clamping of blade is achieved using an adjustable dovetail groove structure formed by sides of the pressing plate and base platform. In addition, the dowel datum planes of the measured blade are closely clamped with the sides of pressing plate, which guarantees the parallel relation between axial line of engine and side of base platform. Therefore, the precise adjustment of the engine axis can be realized by measuring and rotating the side of base platform.

The registration method of axial line of engine based on the fixture for blade clamping shown in Figure 5 is described as follows:

(1) Establishment of workpiece coordinate system. Firstly, a set of coordinates are obtained by measuring the selected profile of the mounting column with the probe system along the X-axis direction. Then, the center coordinates and radius of the mounting column are solved based on the least squares circle fitting for the collected coordinate data. Finally, through motion control, the workpiece coordinate system O-XYZ is established on the axis of the mounting column as shown in Figure 5.

(2) Registration of the engine axis. As shown in Figure 6, the axial line of the engine is matched and aligned with the X axis in three steps. Firstly, a line with a length of 50 mm on the side of base

platform in X-axis direction is selected, and a set of coordinates are collected by measuring the planned profile with the probe system in spacing of 2 mm. Then, the slope and intercept are calculated based on least squares linear fitting for the measured data and the registration angle α between the normal line on the side of the base platform and Y axis is obtained. Finally, the angle α is adjusted to zero by the rotary motion of the measuring machine to realize the registration of engine axis.

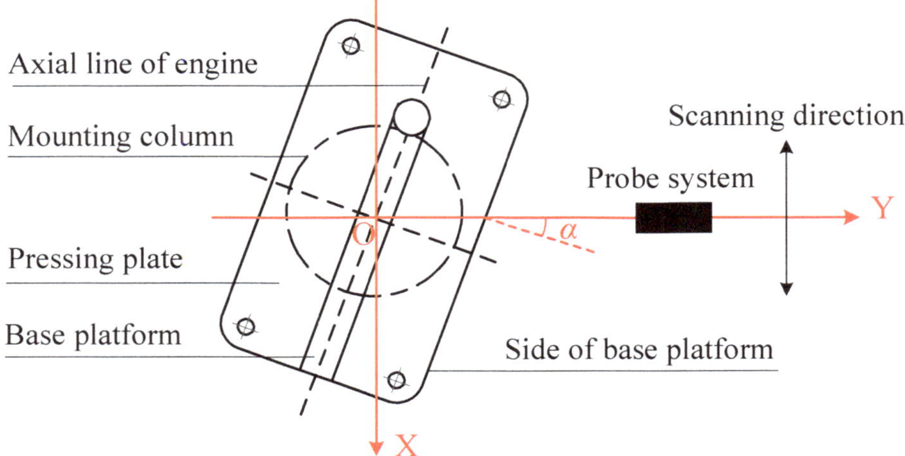

Figure 6. Registration diagram of aeroengine axis.

3.2. Extraction Method of the Leading and Trailing Edges Based on A Priori Planning

The high-precision extraction of the leading and trailing edges is the primary and key issue to be solved of blade twist measurement. The leading and trailing edges of the blade have characteristics of high geometric accuracy, ultra-thin shape and large curvature fluctuation, which cause great difficulties with the machining and measurement of blade edges [19,20]. Influenced by the probe accuracy and system scanning resolution, the existing measuring method of blade edges has disadvantages of large coordinate error, serious fluctuation or even loss of measured data.

In order to solve the technical problem of the blade edge measurement, a new measuring method for leading and trailing edges based on a priori planning is proposed in this study. The blade edges are measured in three steps using the method proposed in this study. Firstly, the test profiles are scanned by the probe system in the absence of the theoretical model of measured blade [21,22], and a set of a priori coordinates of $P_{si}(x_{si}, y_{si}, z_{si})$ for planning are obtained, $i = 1, 2, \ldots, N$. Then, the number and location of measurement points are optimized based on the feature recognition and sampling strategy analysis for the a priori coordinates collected [23,24]. Finally, according to the optimization results, the probe system is controlled to collect data at the planned positions, and the precise coordinates of $P_{mi}(x_{mi}, y_{mi}, z_{mi})$ of each measuring point are obtained.

The curves of the leading and trailing edges measured by priori planning technology are shown in Figure 7. It can be concluded from Figure 7 that the sampling strategy is optimized automatically without the theoretical model of the measured blade, and the edge features of the blade are extracted effectively by planning measuring points densely at the edge parts with larger curvature variation. Based on the least squares fitting for the precision coordinates collected of the leading and trailing edges, parameters of center coordinates and radius of the blade edge can be solved quickly and accurately. The high-precision measurement of the leading and trailing edges based on the technologies of a priori planning and least squares fitting is a key issue and innovation, which overcomes the influence of the error factors such as the movement accuracy of the machine tool and provides the authentic measurement data for the extraction of the profile chord and subsequently the calculation of twist angular position.

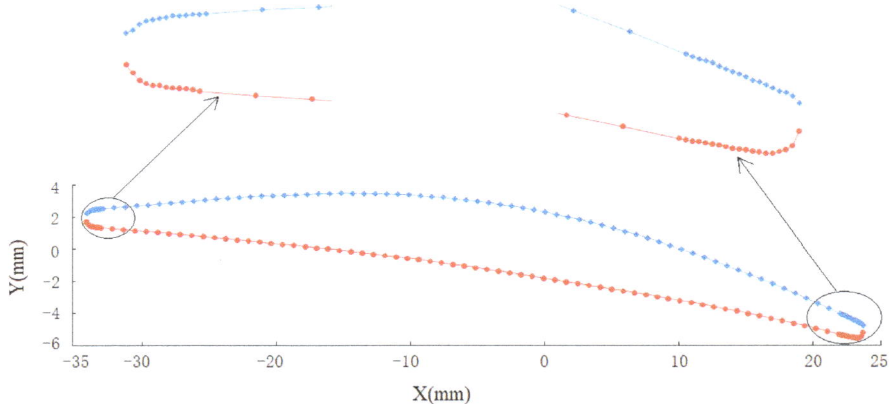

Figure 7. Curves of the leading and trailing edges extracted based on a priori planning technique.

The adaptive optimization of sampling strategy is achieved through form-free and a priori planning measurement of the measured blade in this method. In addition, coordinates of $P_{mi}(x_{mi}, y_{mi}, z_{mi})$ are collected at the planned positions, where the depth of field is close to zero, thus the inclination error is effectively reduced and the measurement accuracy is improved greatly [25,26]. The research results show that the measuring method based on a priori planning has the characteristics of being high precision and form-free. The comprehensive measurement accuracy reaches the level of 10 μm, which can meet the measuring requirements of the leading and trailing edges of ultra-thin blade.

3.3. Extraction Algorithm of Twist Angular Position of the Profile Based on Geometric Analysis

As shown in Figure 8, $O_1(x_1, y_1)$ and $O_2(x_2, y_2)$ represent the centers of the leading and trailing edges respectively, R_q and R_h represent the radius of the leading and trailing edges respectively, and the line of EF is the profile chord. The axial line of engine is parallel to X axis after registration, thus the mathematical model of twist angular position of the profile can be defined as follows:

$$\psi = \beta + \theta \tag{3}$$

where β is the angle between the center line O_1O_2 of the leading and trailing edges and X axis, θ is the angle between the profile chord and the center line O_1O_2. By geometric analysis, we can get the mathematical models of parameters θ and β as follows:

$$\theta = \arcsin \frac{|R_q - R_h|}{\sqrt{(x_2 - x_1)^2 + (y_2 - y_1)^2}} \tag{4}$$

$$\beta = \arctan \frac{y_2 - y_1}{x_2 - x_1} \tag{5}$$

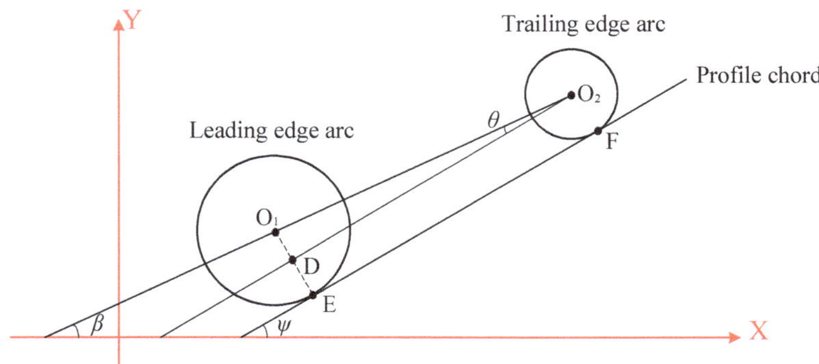

Figure 8. Analysis diagram of the mathematical model for the twist angular position of the profile.

Based on the least squares fitting technology, the center coordinates and radius of the blade edges are solved. Then the twist angular position of the profile can be calculated using Equations (3)–(5).

4. Experiments and Analysis

Taking an aeroengine blade as the measured object, the method for the blade twist measurement proposed in the paper has been experimented and studied on the measuring machine of the blade shown in Figure 3. The experiment is carried out according to the measurement process shown in Figure 4.

4.1. Measurement and Analysis of Test Profiles

The test profiles have been measured based on the technology of a priori planning measurements, and some of the experimental data is shown in Table 1. In Table 1, all coordinates are obtained in workpiece coordinate system after registration with engine axis. The measurement curves of the test profiles are shown in Figure 9. As can be concluded from Figure 9, the closed contour curves of the test profiles are achieved based on the adaptive adjustment of the distribution of measuring points and the depth of field, the edge features of the blade are extracted effectively by planning measuring points densely at edges with larger curvature variation, and the exact form and position relationship of the test profiles are intuitively represented. The experimental results show that the a priori planning measurement method proposed in the study improves the adaptability of the coordinate measurement by self-constructing the theoretical model of the blade, and provides accurate data for the calculation of the blade edge parameters and the extraction of the twist characteristic quantity.

Table 1. Measurement coordinates of the test profiles (unit: mm).

	Test Profile I				Test Profile II			
	Leading Edge		Trailing Edge		Leading Edge		Trailing Edge	
ID	X	Y	X	Y	X	Y	X	Y
1	−48.6	11.236	7.1	−11.077	−33.4	2.493	23.0	−4.426
2	−48.7	11.265	7.2	−11.142	−33.5	2.496	23.1	−4.445
3	−48.8	11.286	7.3	−11.192	−33.6	2.470	23.2	−4.494
4	−48.9	11.298	7.4	−11.263	−33.7	2.453	23.3	−4.539
5	−49.0	11.283	7.5	−11.336	−33.8	2.433	23.4	−4.588
6	−49.1	11.250	7.6	−11.409	−33.9	2.418	23.5	−4.607
7	−49.2	11.184	7.7	−11.486	−34.0	2.368	23.6	−4.662
8	−49.3	10.739	7.8	−11.852	−34.1	2.279	23.7	−4.771
9	−49.2	10.663	7.7	−11.891	−34.1	1.758	23.7	−5.424
10	−49.1	10.572	7.6	−11.946	−34.0	1.630	23.6	−5.484

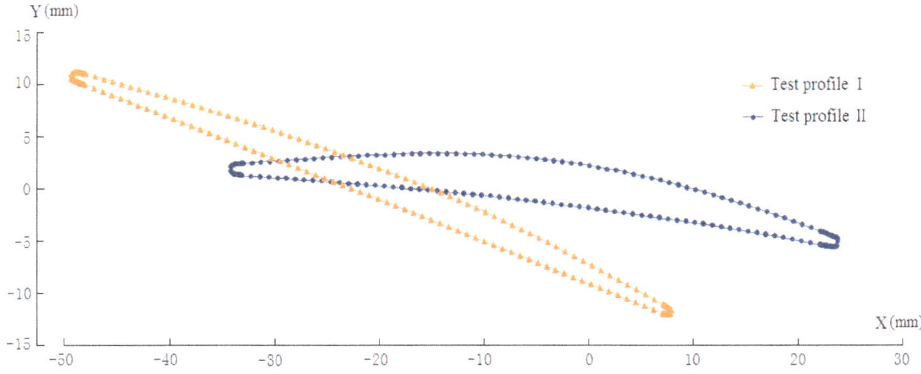

Figure 9. Measurement curves of test profiles.

4.2. Measurement and Evaluation of Parameters of Leading and Trailing Edges

The leading and trailing edges of the blade studied in this paper are constructed by circular arc, and its mathematical model is as follows:

$$x^2 + y^2 + ax + by + c = 0 \tag{6}$$

where a, b, c are the model coefficients, which can be solved by the least squares fitting algorithm based on coordinates collected of the leading and trailing edges [27,28]. Then the parameters of leading and trailing edges can be determined by the following equations:

$$\begin{cases} x_0 = -a/2 \\ y_0 = -b/2 \\ r = \sqrt{a^2 + b^2 - 4c}/2 \end{cases} \tag{7}$$

where (x_0, y_0) are the center coordinates of the leading and trailing edges, and r is the radius.

The test profiles I and II have been measured five times using the method of the a priori planning measurement, and the coordinates collected of blade profiles are used for the accuracy analysis of the leading and trailing edges. Based on the five sets of coordinates collected and the least squares circle fitting algorithm, the parameters and measurement standard deviations of the leading and trailing edges are solved as shown in Table 2. In Table 2, R_q is the leading edge radius, (x_1, y_1) are the center coordinates of the leading edge, R_h is the trailing edge radius, (x_2, y_2) are the center coordinates of the trailing edge, $m_1 \sim m_5$ represent five sets of parameters solved of blade edges, μ represents the average value of measurements, and σ represents the standard deviation of measurements.

Table 2. Measurement results of the leading and trailing edges (unit: mm).

	Test Profile I						**Test Profile II**					
	Leading Edge			**Trailing Edge**			**Leading Edge**			**Trailing Edge**		
	R_q	x_1	y_1	R_h	x_2	y_2	R_q	x_1	y_1	R_h	x_2	y_2
m_1	0.458	−48.810	10.853	0.417	7.336	−11.585	0.537	−33.611	1.964	0.542	23.228	−5.024
m_2	0.462	−48.805	10.852	0.416	7.339	−11.586	0.538	−33.609	1.965	0.543	23.222	−5.024
m_3	0.456	−48.812	10.853	0.419	7.336	−11.583	0.534	−33.616	1.967	0.541	23.229	−5.023
m_4	0.457	−48.810	10.850	0.415	7.335	−11.587	0.534	−33.614	1.966	0.539	23.231	−5.022
m_5	0.461	−48.806	10.853	0.418	7.335	−11.583	0.535	−33.613	1.966	0.541	23.229	−5.027
μ	0.459	−48.809	10.852	0.417	7.336	−11.585	0.536	−33.613	1.966	0.541	23.228	−5.024
σ	0.003	0.003	0.002	0.002	0.002	0.002	0.002	0.003	0.002	0.002	0.003	0.002

The measurement results show that the algorithm for edge extraction based on sampling optimization and least squares fitting presented in the study is effective, the parameters of the blade edges are extracted exactly, and the statistical uncertainty of the measurements is as follows. The accurate measurement of the leading and trailing edges provides the authentic data for the extraction of profile chord.

$$\begin{cases} u_A(x_0) = 0.003 \text{ mm} \\ u_A(y_0) = 0.002 \text{ mm} \\ u_A(r) = 0.003 \text{ mm} \end{cases} \tag{8}$$

4.3. Measurement and Evaluation of Blade Twist

Based on the form and position parameters solved of the leading and trailing edges, the characteristic quantities $\overline{\psi}$ and η of blade twist are calculated using Equations (1)–(5) as follows:

$R_q = 0.459$, $x_1 = -48.809$, $y_1 = 10.852$; $R_h = 0.417$, $x_2 = 7.336$, $y_2 = -11.585$ → Formulas (4) (5) → $\theta_1 = 0.04°$, $\beta_1 = -21.783°$ → Formula (3) → $\psi_{\text{I}} = -21.743°$

$R_q = 0.536$, $x_1 = -33.613$, $y_1 = 1.966$; $R_h = 0.541$, $x_2 = 23.228$, $y_2 = -5.024$ → Formulas (4) (5) → $\theta_2 = 0.005°$, $\beta_2 = -7.011°$ → Formula (3) → $\psi_{\text{II}} = -7.006°$

ψ_{I}, ψ_{II} → Formulas (1) (2) → $\overline{\psi} = -14.375°$, $\eta = 14.737°$

The uncertainty of the measurement of blade twist is analysed as follows:

(1) Uncertainty of parameter β:

The mathematical model of combined standard uncertainty of parameter β can be derived from the Equation (5):

$$u_C(\beta) = \frac{1}{1+M^2} \times \left[\frac{1}{A^2} \times u^2(y_2) + \frac{1}{A^2} \times u^2(y_1) + \left(\frac{-B}{A^2}\right)^2 \times u^2(x_2) + \left(\frac{-B}{A^2}\right)^2 \times u^2(x_1) \right]^{\frac{1}{2}} \tag{9}$$

where $A = x_2 - x_1$, $B = y_2 - y_1$, $M = B/A$.

The parameters of blade edges and their statistical uncertainties shown in Table 2 are brought into Equation (9), then the result of uncertainty evaluation of β is as follows:

$$\begin{cases} u_C(\beta_1) = 5.065 \times 10^{-5} \text{deg} \\ u_C(\beta_2) = 4.985 \times 10^{-5} \text{deg} \end{cases} \tag{10}$$

(2) Uncertainty of parameter θ:

The mathematical model of combined standard uncertainty of parameter θ can be derived from the Equation (4):

$$u_C(\theta) = K_1 \times \left[K_2 \times u^2(R_q) + K_2 \times u^2(R_h) + K_3 \times u^2(y_2) + K_3 \times u^2(y_1) + K_4 \times u^2(x_2) + K_4 \times u^2(x_1) \right]^{\frac{1}{2}} \tag{11}$$

In Equation (11), the coefficients $K_1 \sim K_4$ are determined by the following equations:

$$\begin{cases} K_1 = \left[1 - \frac{(R_q - R_h)^2}{(x_2-x_1)^2 + (y_2-y_1)^2} \right]^{-\frac{1}{2}} \\ K_2 = \frac{1}{(x_2-x_1)^2 + (y_2-y_1)^2} \\ K_3 = \frac{(R_q-R_h)^2 \times (y_2-y_1)^2}{\left[(x_2-x_1)^2 + (y_2-y_1)^2\right]^3} \\ K_4 = \frac{(R_q-R_h)^2 \times (x_2-x_1)^2}{\left[(x_2-x_1)^2 + (y_2-y_1)^2\right]^3} \end{cases} \tag{12}$$

The parameters of blade edges and their statistical uncertainties shown in Table 2 are brought into Equations (11) and (12), then the result of uncertainty evaluation of θ is as follows:

$$\begin{cases} u_C(\theta_1) = 7.016 \times 10^{-5} \text{ deg} \\ u_C(\theta_2) = 7.408 \times 10^{-5} \text{ deg} \end{cases} \quad (13)$$

(3) Uncertainty of blade twist:

The mathematical models of combined standard uncertainties of parameters $\overline{\psi}$ and η can be derived respectively from Equations (1)–(3):

$$u_C(\overline{\psi}) = \frac{1}{2}[u_C{}^2(\beta_1) + u_C{}^2(\theta_1) + u_C{}^2(\beta_2) + u_C{}^2(\theta_2)]^{\frac{1}{2}} \quad (14)$$

$$u_C(\eta) = [u_C{}^2(\beta_1) + u_C{}^2(\theta_1) + u_C{}^2(\beta_2) + u_C{}^2(\theta_2)]^{\frac{1}{2}} \quad (15)$$

The uncertainty analysis results of parameters β and θ are brought into Equations (14) and (15), then the uncertainties of parameters $\overline{\psi}$ and η are obtained as follows:

$$\begin{cases} u_C(\overline{\psi}) = 0.004° \\ u_C(\eta) = 0.008° \end{cases} \quad (16)$$

The experimental results show that the absolute twist angular position $\overline{\psi}$ of blade relative to the engine axis is $-14.375°$ and its uncertainty of measurement is 0.004°, the relative twist angle η of blade itself is 14.737° and the measurement uncertainty is 0.008°. The measurement method of blade twist presented in this study is effective and accurate, the problems of measurement and evaluation of the blade twist are solved commendably using technologies of calibration, a priori planning and geometric analysis.

4.4. Deviation Assessment of the Method

In order to verify the reliability of the measurement results, the verification measurement of the test profiles is carried out on a high-precision four-coordinate measuring machine of JE42, whose overall measurement accuracy is less than 2 μm. The four-coordinate measuring machine of JE42 collects coordinate data with a high-precision contact probe of GT31 from TESA Switzerland, and its technical parameters are provided in Table 3. After coordinate transformation and registration of the measured data, the measurement curves of test profiles are shown in Figure 10.

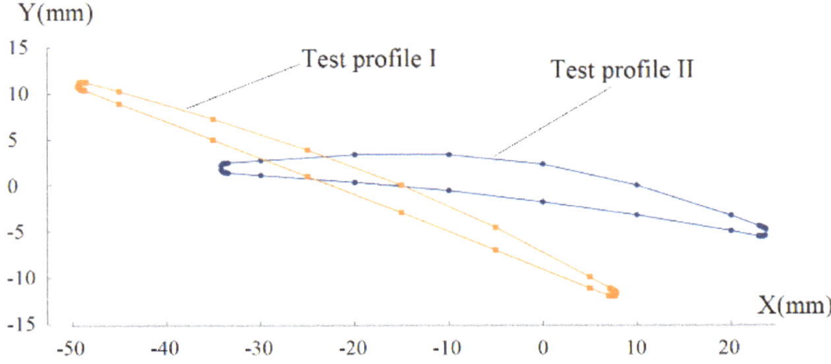

Figure 10. Measurement curves of test profiles on JE42.

Table 3. Parameters of four-coordinate measuring machine.

Parameters	Specifications
Range	X: 400 mm, Y: 500 mm, Z: 650 mm, C: 360°
Resolution	X, Y, Z: 0.1 µm, C: 0.0002°
Total accuracy	≤2 µm

The experimental data collected by JE42 is processed based on least squares fitting algorithm, and the form and position parameters of the leading and trailing edges for test profiles I and II are solved as shown in Table 4.

Table 4. Parameters of the leading and trailing edges measured by JE42 (unit: mm).

	Leading Edge			Trailing Edge		
	R_q	x_1	y_1	R_h	x_2	y_2
Test profile I	0.461	−48.858	10.848	0.422	7.27	−11.58
Test profile II	0.527	−33.699	1.951	0.54	23.176	−5.013

Based on the parameters solved of the leading and trailing edges and evaluation model of blade twist proposed in the study, the results are as follows:

$$\begin{cases} \overline{\psi}_0 = -14.356° \\ \eta_0 = 14.776° \end{cases} \tag{17}$$

where $\overline{\psi}_0$ and η_0 represents respectively the twist angular position and twist angle measured by JE42.

Taking the measurement results of JE42 as the agreed true value, the measurement deviation of twist angular position is less than 0.02°, and the measurement deviation of twist angle is less than 0.04°. The experiments show that the method proposed in this study can achieve high-precision and form-free measurement of blade twist.

4.5. Analysis and Comparison of Measurement Methods

The common methods for blade twist measurement are analyzed and compared as shown in Table 5. It can be concluded as follows:

Table 5. Analysis and comparison of common methods for blade twist measurement.

	CMM of Global Series from HEAXGON	CORE DS from WENZEL	Form-Free and High-Precision Method
Methods of detection	Contact measurement	Non-contact optical measurement	Non-contact and form-free measurement
Adaptability	Poor adaptability, theoretical model is necessary	Poor adaptability, theoretical model is necessary	Good adaptability, self-constructing theoretical model
Efficiency	Low efficiency	Higher efficiency due to optical scanning	The efficiency is further improved by optimizing measurement method
Accuracy for edge detection	Serious fluctuation or even loss of measured data	10 µm level	Less than 10 µm due to a priori planning measurement method

(1) CMM collects blade coordinates point by point using a contact probe and the measurement accuracy can be improved through path planning and sampling optimization, but is limited by the measuring principle and radius of probe ball. The CMM method is insufficient in adaptability, efficiency and edge measurement accuracy.

(2) CORE DS collects coordinates efficiently by scanning the blade profile using the laser sensor and the measurement accuracy can reach 10 μm level through inclination error compensation. However, it still has some problems in adaptability, evaluation algorithm and measurement accuracy.

(3) The method for the blade twist measurement proposed in the study is form-free, efficient and accurate, and can solve the problems of high-precision measurement and evaluation for the twist of aeroengine blade primely.

5. Conclusions

In this paper, the precision measurement method of the twist of the aeroengine blade is studied, and a comprehensive measuring method of the blade twist based on evaluations of twist angular position and twist angle is proposed. Three key problems for the twist metrology of a blade are solved based on technologies of calibration, a priori planning and geometric analysis: aeroengine axis matching, high-precision coordinate acquisition of the leading and trailing edges, and extraction of twist angular position of the profile. The contents and conclusions of the innovative research are as follows:

(1) A simple and efficient method for registration of the engine axis based on the special fixture and least squares algorithm is studied, which achieves the fast establishment of the measurement benchmark of blade twist and improves the datum positioning accuracy significantly.

(2) A high-precision method for the measurement of the leading and trailing edges based on a priori planning is proposed in this study, and the technical difficulties of blade edge measurement are solved. By optimizing the sampling strategy automatically, the measurement accuracy is improved significantly and the edge features of blade are extracted effectively, which provides authentic measurement data for the twist evaluation.

(3) A high-precision extraction method for blade edges based on sampling optimization and least squares fitting is presented, which can overcome the fluctuation error of measured data and extract the parameters of blade edges exactly. The statistical uncertainty of the method is less than 3 μm.

(4) An algorithm for parameters calculation of the blade twist based on geometric analysis is analyzed in this study, which solves the problem of characterization and evaluation of aeroengine blade twist.

(5) Taking an aeroengine blade as the test object, the measuring method is studied experimentally. The results show that the metrological method of blade twist presented in this study is effective and its measurement uncertainty is less than 0.01°. This method is form-free, efficient, accurate, and can solve the problems of high-precision measurement and evaluation for the twist of aeroengine blade primely.

Author Contributions: X.L. and Z.S.; methodology, X.L.; software, X.L.; validation, X.L.; formal analysis, X.L.; investigation, X.L.; resources, Z.S.; data curation, Z.S.; writing—original draft preparation, X.L.; writing—review and editing, X.L.; visualization, X.L.; supervision, Z.S.; project administration, Z.S.; funding acquisition, Z.S. All authors have read and agreed to the published version of the manuscript.

Funding: This research was funded by the Key Project of National Natural Science Foundation of China, grant number 51635001. The APC was funded by 51635001.

Acknowledgments: This work was supported by the Key Project of National Natural Science Foundation of China (Grant No. 51635001) and the Fundamental Research Funds for the Central Universities of China (Grant No. 3142018054 and No. 3142019055).

Conflicts of Interest: The authors declare no conflict of interest.

Nomenclature

ψ	twist angular position of profile, deg
$\overline{\psi}$	twist angular position of blade, deg
η	twist angle of blade, deg
ψ_{I}	twist angular position of test profile I, deg

ψ_{II}	twist angular position of test profile II, deg
α	registration angle of engine axis, deg
$P_{si}(x_{si}, y_{si}, z_{si})$	A priori coordinates, mm
$P_{mi}(x_{mi}, y_{mi}, z_{mi})$	precise coordinates, mm
x_1, y_1	center coordinates of leading edge, mm
x_2, y_2	center coordinates of trailing edge, mm
R_q	radius of leading edge, mm
R_h	radius of trailing edge, mm
β	angle between the center line O_1O_2 of the leading and trailing edges and X axis, deg
θ	angle between the profile chord and the center line O_1O_2, deg
X	abscissas of measuring points on test profile, mm
Y	ordinates of measuring points on test profile, mm
μ	average value of measurements, mm
σ	standard deviation of measurements, mm

References

1. Wang, H.S.; Yuan, X.; Yue, G.Q.; Zhong, J.J.; Wang, Z.Q. Effects of Curved Blade Stacking Line on the Aerodynamic Performance of Compressor Cascade. *J. Aerosp. Power* **2002**, *17*, 327–331.
2. Gao, L.M.; Cai, Y.T.; Zeng, R.H.; Tian, L.C. Effects of Blade Machining Error on Compressor Cascade Aerodynamic Performance. *J. Propuls. Technol.* **2017**, *38*, 525–531.
3. Gao, L.M.; Cai, Y.T.; Hao, Y.P.; Zhao, X.; Tian, L.C. Experimental Investigation on Aerodynamic Performance of Compressor Blade Considering Manufacturing Error. *J. Propuls. Technol.* **2017**, *38*, 1761–1766.
4. Zhang, W.H.; Zou, Z.P.; Li, W.; Luo, J.Q.; Pan, S.N. Unsteady Numerical Simulation Investigation of Effect of Blade Profile Deviation on Turbine Performance. *Acta Aeronaut. Astronaut. Sin.* **2010**, *31*, 2130–2138.
5. Cheng, C.; Wu, B.H.; Zheng, H.; Gao, L.M. Effect of blade machining errors on compressor performance. *Acta Aeronaut. Astronaut. Sin.* **2020**, *41*, 623237.
6. Yu, X.L.; Ye, P.Q. Parameters Identification of the Vane Based on MATLAB. *Aviat. Maint. Eng.* **2009**, *4*, 56–58.
7. Peng, Z.G.; Li, W.L. Feature Parameters Extraction of the Blade Surface based on the Improved Convex Hull Algorithm. *Equip. Manuf. Technol.* **2012**, *1*, 37–43.
8. Mao, C.L. *Inspection of Aero Engine Blade Cross-Sectional Feature Parameters*; Tianjin University: Tianjin, China, 2015.
9. Shi, X.Q.; Wu, B.H.; Zhang, D.H. Development Tendency of Inspecting Technology for Aeroengine Blade. *Aeronaut. Manuf. Technol.* **2015**, *12*, 80–84.
10. Su, J.; An, Z.Y.; Yu, Y.F.; Bai, J. A New Synchronous Measurement Method for Torsion and Bending Deformation of Engine Blade. *J. Exp. Mech.* **2017**, *2*, 279–285.
11. Shi, J.H.; Liu, P. High Efficiency Measurement Method for Large-size Aeroengine Blade Profile. *Acta Metrol. Sin.* **2018**, *5*, 605–608.
12. Bu, K.; Zhang, X.; Ren, S.; Qiu, F.; Tian, G. Research on Influence of CMM Sampling Points on Detection of Feature Parameters for Turbine Blade. *J. Northwest. Polytech. Univ.* **2019**, *4*, 767–773. [CrossRef]
13. Sun, B.; Li, H.Z.; He, D.F.; Liu, H.T.; Wang, J.H. Profile Error Detection and Evaluation of Aero-engine Blade. *Tool Eng.* **2019**, *4*, 103–106.
14. Li, B.; Li, F.; Liu, H. A measurement strategy and an error-compensation model for the on-machine laser measurement of large-scale free-form surfaces. *Meas. Sci. Technol.* **2014**, *1*, 5204–5214. [CrossRef]
15. Aviation Industry Standard of People's Republic of China. *Label Tolerance and Surface Roughness of Blade Profile HB 5647-98*; Aviation Industry Corporation of China: Beijing, China, 1999.
16. National Military Standard of People's Republic of China. *Aircraft Gas Turbine Powerplant Terminology and Symbols GJB 2103A-97*; Commission of Science, Technology and Industry for National Defense: Beijing, China, 1998.
17. Shao, W.; Guo, H.R.; Wu, Y.; Zhou, A.W.; Peng, P. Measurement method for aeroengine blade based on large reflection angle noncontact sensing technology. *Opt. Eng.* **2018**, *57*, 054115. [CrossRef]
18. Li, X.Z.; Shi, Z.Y.; Chen, H.F.; Lin, J.C. Current Status and Trends of Aeroengine Blade Profile Metrology. *J. Beijing Univ. Technol.* **2017**, *43*, 557–565.

19. Liu, G.D.; Pu, Z.B.; Zhang, Z.; Sun, Y.B. New method for the measurement of aeroengine blade edge. *Proc. SPIE* **2002**, *4929*, 475–480.
20. Li, Y. The Aero-Engine Blade Edge Detection based on Color Image Wavelet Sub-pixel Method. *Aeronaut. Sci. Technol.* **2009**, *1*, 25–27.
21. Shi, Z.Y.; Zhang, B.; Lin, J.C. Design of measuring machine for complex geometry based on form-free measurement mode. *Chin. J. Sci. Instrum.* **2012**, *33*, 1377–1384.
22. Shi, Z.Y.; Zhang, B.; Li, X.M. Form-Free Measurement Mode in Precision Engineering. *Nanotechnol. Precis. Eng.* **2012**, *10*, 132–136.
23. Li, X.Z.; Shi, Z.Y.; Li, K.; Li, Y.K. A measuring method for the leading and trailing edges of blade based on feature modelling. In Proceedings of the IEEE the 14th International Conference on Electronic Measurement & Instruments, Changsha, China, 1–3 November 2019; Volume 11, pp. 680–685.
24. Zhang, H.; Shi, Z.Y.; Zhang, B. Study on 3D Geometric Shape Discriminant and Form Error Evaluation. *Chin. J. Sci. Instrum.* **2014**, *35*, 1217–1222.
25. Sun, B.; Li, B. Laser Displacement Sensor in the Application of Aero-Engine Blade Measurement. *IEEE Sens. J.* **2016**, *16*, 1377–1384. [CrossRef]
26. Li, X.Z.; Shi, Z.Y.; Li, Y.K.; Lin, J.C. Form-free high-precision probe system of blade based on the synchronization of planning and measurement. *Chin. J. Sci. Instrum.* **2018**, *39*, 9–17.
27. Li, Q.; Huang, X.; Li, S.G. A laser scanning posture optimization method to reduce the measurement uncertainty of large complex surface parts. *Meas. Sci. Technol.* **2019**, *30*, 105203. [CrossRef]
28. Li, X.Z.; Shi, Z.Y.; Li, Y.K.; Lin, J.C. A High-precision Form-free Metrological Method of Aeroengine Blades. *Int. J. Precis. Eng. Manuf.* **2019**, *20*, 2061–2076.

Article

Development of a Rapid Optical Measurement System for Circular Workpieces with Irregular Tooth Contours after Broaching Process

Yu-Liang Chen, Xuan-Qi Liang, Zi-Rong Ye and Quang-Cherng Hsu *

Department of Mechanical Engineering, National Kaohsiung University of Science and Technology, 415 Chien-Kung Road, Kaohsiung City 80778, Taiwan; frankchen2341@gmail.com (Y.-L.C.); st0597pjz@gmail.com (X.-Q.L.); a0965349877@gmail.com (Z.-R.Y.)

* Correspondence: hsuqc@nkust.edu.tw

Received: 9 June 2020; Accepted: 24 June 2020; Published: 27 June 2020

Abstract: During a manufacturing process, it is essential to quickly identify whether a tool needs to be replaced or adjusted, to ensure that production quality is not compromised. Therefore, the re-inspection of the product or first article inspection is an important process. Reducing the inspection time can reduce the time spent waiting for a product in the production line. This research aimed to design a system that can automatically and rapidly measure the dimensions of irregular tooth contours in the broaching process, to ensure cutting tools are replaced when necessary. This study developed an automatic machine for measuring the irregular tooth contours of large ring parts; the tooth root, tooth height, and tooth thickness of the workpiece are measured. The measurement diameter is approximately 200 mm, and the radial inspection accuracy is within ±20 μm; we aimed to reduce the detection time considerably. An optical micrometer and an automatic rotating platform were used in the measurement system. The workpieces to be measured were easy to install, and the eccentricity was automatically corrected by the system, thus saving time that would be taken to correct Abbe errors. This research successfully developed a rapid optical measurement system that can reduce the inspection time from 30 min to 60 s. Moreover, the maximum radial measurement error is −0.02 mm, which means that the measurement accuracy is within ±20 μm (total: 40 μm).

Keywords: automated optical inspection; precision measurement; circular contour; edge detection

1. Introduction

Automatic optical inspection (AOI) is commonly used in industries. System stability is essential and must be considered when designing a suitable metrology system for use in manufacturing processes. The stability of a system depends on certain factors involved in the system design, such as the type of method/system used for object identification (e.g., line or area scanning system) and the type of filter employed.

There are many methods to improve the precision of AOI; these methods help save time, increase safety, and increase the reliability of the measurement performance. For example, Ali et al. [1] proposed a camera system with precision measurement that adaptively controls the threshold values to identify the gear profile. Therefore, the gear profile can be measured with safety and reliability. Camelio et al. [2] proposed a measurement system that quantifies broaching tool wear based on the overall worn area. This method uses automated image cropping and digital imaging processing tools to determine the affected area, without any manual intervention. To remove irrelevant noise, the desired portion of the unprocessed image is manually cropped into a region of interest from the original image, after which image processing is performed.

To improve the processing accuracy of products in automated manufacturing, the state of tool wear is monitored [3]. This approach extracts the tool edge by decomposing the original tool-wear image, to reduce the influence of texture and noise in edge measurement. Adamo et al. [4] proposed a low-cost, high-performance glass surface inspection system. The system uses two low-cost complementary metal-oxide-semiconductor (CMOS) (1.3 M pixels) USB cameras to form an array, and as the glass width increases, more cameras can be added to increase the number of images that can be acquired. Defects such as scratches and bubbles can be identified by using the Canny edge detection method.

Xiong et al. [5] proposed a large-scale spatial-positioning laser-scanning system. They proposed a single transmitter station calibration method based on a photoelectric-scanning multi-angle resection positioning model that combines photoelectric-scanning angle measurements and spatial resection with an external receiver array. The experiment results demonstrated that the method could achieve millimeter-level measurement accuracy.

Heikkila et al. identified errors occurring during the entire geometric camera calibration process, such as those occurring during the picking of pictures, model fitting, and image correction, to control point steps. In that paper, they present a four-step calibration procedure, which is an extension of the two-step method. In one of the additional steps, the distortion caused by circular features is compensated for, and in the other additional step, the distorted image coordinates are corrected [6].

Wan et al. [7] presented the concept of using the latitude and longitude coordinate system for spherical correction; various basic matrix formulas were used to perform automatic correction of the dual PTZ (pan-tilt-zoom) camera system. Ng [8] presented an improvement of Otsu's method. In the improved method, image segmentation is better when the grayscale histogram of the image has a double peak; if the grayscale histogram has only a single peak, the threshold will be close to the peak, which increases the possibility of poor image segmentation.

Samopa et al. [9] combined Canny edge detection and the fixed percentage method, to propose a method of setting binarization. The experiment results showed that the binarization threshold is set on the edge detection, and the performance of this method was better than Otsu's method. Jung et al. [10] created a 3D image through reprojection without an external light source, and the scanning time was shorter than that of a conventional stereo vision forming system. Previous studies have used the KEYENCE LS-7030 optical measurement meter to measure the extrudate swell of high-density polyethylenes in capillary flow [11,12]. This line scan device was also used in this proposed study.

Sun et al. [13] proposed a method for controlling autonomous underwater vehicles (AUVs); a Butterworth filter was used in the accelerometer and gravity sensor of the attitude compass. After the signal was filtered, the attitude control of the AUV was more precise, rendering AUVs more suitable for the exploration and rescue of underwater vehicles. It shows the Butterworth filter has a very flat frequency response in the passband.

The signal data of elevator safety pliers are produced by an accelerometer sensor. However, noise interferes with the signal of the accelerometer. This affects the measurement data and hampers the smooth braking of the pliers, or causes a delay in the application of the brakes. This study thus employed a Butterworth filter method to obtain smoother measurement data [14]. Unfortunately, the reliability of acceleration data is severely compromised by measurement noise; this noise needs to be suppressed to the maximum extent possible, to be able to use the data for comfort assessment. Pinto et al. [15] presented a fifth-order Butterworth low-pass filter; it is based on the fully differential difference transconductance amplifier (FDDTA) building blocks. That is, it is fifth-order transfer function according to the Butterworth theory. The Butterworth low-pass filter is a suitable smoothing method for electrical signal processing [13–15].

Ye et al. [16] proposed an in situ deflectometric measurement strategy for the robotic polishing of optical components where the accuracy is comparable to that of a coordinate measuring machine (CMM). In order to enhance the machining accuracy of curved contour on large thin-walled skin, Bi et al. [17] developed a method of isometric-mapping-based adaptive machining where the measurement system is a laser-scanner-based on-machine measurement (OMM) system to obtain 3D real geometry deformed surface. There are several 3D measurement methods [18,19] devoted to the assessment of manufacturing processes or the evolution of surface texture.

Area scanning measurement can quickly obtain images, but a postprocessing step is required for photo distortion correction [6,10], and a graphic processing step is required to obtain the contours [1–4]. Moreover, the requirements of ambient light and a larger workpiece mean that the charge-coupled device (CCD) requires a higher resolution to maintain measurement accuracy; a high-resolution CCD is particularly expensive. The advantage of a line scanning system is that, when measuring a large workpiece, because of 1D linear image data, the sensor (camera) pixel requirements are much lower than those of an area scanning system [11,12]. Moreover, there are no strict requirements for ambient light. However, in the workpiece platform control system, signal processing is more complex, and the workpiece moving speed needs to be adequate to obtain the correct superimposed image.

In this study, we focused on measuring large workpieces of approximately 200 mm. To achieve high-speed detection, the following criteria should be met: easy installation of the workpiece, operation of the system should be simple, the system should be stable, measurement should be quick and precise, and the cost should be low. A general line scan method is applied for 2D measurements in plane, such as width, height, or length. The novel approach of this study is by integrating a rotational table and precise line scan device, to measure large circular workpiece. The micro-vibration of the transmission gears in the rotational table causes high-frequency vibration noise, which affects the measurement result; thus, this study used the Butterworth filter for low-pass filter processing.

2. Proposed Metrology System

In this study, we selected the line scanning system, where the workpiece can be placed directly on the rotating platform for measurement. The system is mainly divided into three parts: measurement unit (CCD and LS-7030), transmission mechanism (automatic detection rotating platform), and data processing (PLC and visual C# program).

A prototype system and a final rapid inspection system were investigated. The difference between these two systems lies in the speed employed and the data transfer interfaces. The prototype used the RS-232 interface to transfer data to the PC; the platform rotation speed for the prototype was 0.25 rpm (AC servo motor of 20 rpm). The platform rotation speed for the final rapid system was 2.5 rpm (AC servo motor of 20 rpm). Because of a 10-fold increase in speed, to improve data acquisition and transmission, we added an expansion input/output unit (KV-SIR32XT) module to the system. After the addition of this module, the Ethernet/IP interface could be used to transfer data to a PC.

In the program function, we needed to determine the coordinates of the rotation platform axis by using the CCD first; these coordinates were then used as the reference center of the subsequent image program. Next, we needed to convert the upper and lower limits (E1 and E2) of the LS-7030 measurement light band to the upper and lower limits of the image coordinates, to convert the measured results to the radius of the image coordinates.

Because the platform and workpiece are not on the same axis, an eccentricity error may occur during rotation; this error causes the measurement result to oscillate like a sine waveform. In this study, two different methods were proposed to determine the center of the measured object, and the eccentricity value was calculated by a program. Finally, the axis eccentricity was corrected by the eccentricity curve, and the measured value without eccentricity was obtained.

The center of a round workpiece, such as a ring gauge, can be determined by using four edge points at every 90°. However, after broaching, because some portions of the contours change, a more accurate method to determine the center of a round workpiece is required. To this end, we propose a method of circular regression from the un-machined surface points.

The control software uses self-developed programs, employing Visual Studio C# with Emgu CV as image processing libraries and the Butterworth filter for low-pass filtering. In this study, numerical calculations were performed on the tooth root, tooth height, and tooth thickness. In addition, before beginning the measurement, the LS-7030 system needs to be switched on for 30 min, in order to ensure stable measurement.

2.1. System Design

Figure 1 shows the schematic of the automatic optical measurement system. Figure 1a shows the direction in which the sample is loaded on the rotary table in the top view. Figure 1b shows the equipment; the equipment includes CCD components (IDS 500 million CCD/UI-3580CP-C-HQ_REV 2), a Computar 2/3″ 8 mm f1.4 locking iris and focus (M0814-MP2) lens, an LS-7030 CCD micrometer (LS-7001 micrometer display), a programmable logic controller (PLC) of KV-7500, a rotary table, an XY linear micrometer manual stage (TSD XY-TS100AR), a harmonic reducer (GTC 1:80), a servo motor (SME-L04030SAB, AC400W) and servo motor driver, and a control box.

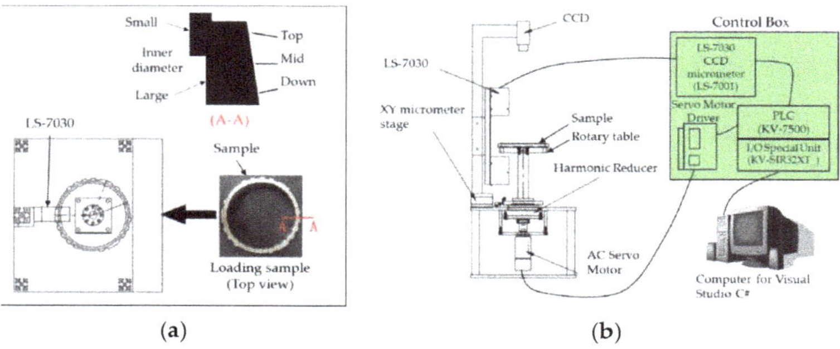

Figure 1. Schematic of the automatic optical measurement system. (**a**) Top view. (**b**) Front view and equipment.

In the rapid measurement system, an I/O Special Unit (KV-SIR32XT) was added; this unit uses the Ethernet/IP interface to communicate with a PC, to facilitate high-speed data transfer.

The upper CCD is used to determine the rotation center of the machine rotating platform. The measurement sample is directly placed on the platform, using the outer diameter of the platform for alignment/fixation. There are two diameters within the inner diameter, as shown in the cross-section A-A of Figure 1a: a large diameter and a small diameter. The large diameter is the outer diameter of the platform and is used for alignment and fixation, and the small diameter is used to stop the sample on the platform. Thus, the workpieces to be measured can be simply and stably placed on the platform; however, a gap typically exists between the platform and the workpiece, resulting in a slight eccentricity error on the workpiece.

The rotary table consists of a harmonic reducer (1:80) and an AC servo motor. In addition, there is a pointer for positioning the machine to the 0° position. An inductive limit switch is used to return the machine to the 0° position.

This study used the KEYENCE line-scanning optical measurement system LS-7030, which is an all-in-one optical measurement instrument; the measurement range is 0.3 to 30 mm, the accuracy is ± 2 μm, and the repeatability is ± 0.15 μm. The LS-7030 consists of an integrated parallel beam system and an imaging system (high-brightness GaN light-emitting diode (LED), telecentric lens, and HL-CCD. The HL-CCD facilitates continuous exposure measurement (reaching 2400 times/s), so it can be used for high-speed continuous detection.

The following are the advantages of a harmonic reducer: (1) Compared with other reducers, the harmonic reducer has a smaller volume and simpler structure, and it generally consists of three basic components (wave generator, flexible wheel, and rigid wheel) [20]. (2) The harmonic reducer can be used with large loads. Because a large number of teeth mesh when the harmonic gear rotates and the flexible wheel meshes with the rigid wheel after elastic deformation, the teeth are in surface contact with other teeth. Therefore, the backlash is small, and meshing is possible even without side backlash, which is suitable for reverse rotation during movement and maintaining output accuracy.

Other control buttons are on the outer cover of the control box, such as the start button, which triggers the rotating table to rotate once, the home position return button, which triggers the rotating table to return to the zero point position of the system, and the emergency stop button, which is required in case the measurement needs to be stopped.

In the prototype system, the rotation speed was only 0.25 rpm, and the measurement time was approximately 4 min. To achieve a measurement time of within 1 min, the rotation speed must be increased. According to Reference [21], the harmonic reducer has different vibration values for different rotation speeds. Due to the elastic deformation of the flexible wheel and the large number of teeth in contact with each other during the transmission, vibration causes measurement errors, and thus, different speeds have to be tested to determine the ideal measurement speed.

2.2. Calculation Method of Rotation Axis Center

First, physical objects must be converted into coordinates of the virtual system. Therefore, the conversion coordinates of the rotation axis center and the measurement of the LS-7030 light band must be obtained. These two factors form the basis for the subsequent software calculation. We calculated the center coordinate C by using the area method.

As shown in Figure 2, the CCD was used to take photos of the correction block at three different positions; the positions of the three correction blocks do not overlap, and the total rotation angle is greater than 180°. Then, image processing is used to identify the coordinates of the large and small circles on the calibration block, and the center of platform is C (x_c, y_c). A triangle (denoted in red) is formed from the center points of two circles and platform, as shown in the Figure 2. The area of the triangle at the three rotation positions is equal because the distance between the big circle and the small circle is fixed, as well as the distance to the center of point C.

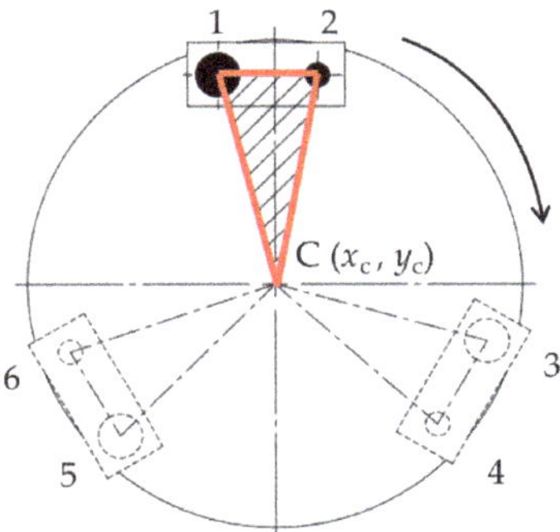

Figure 2. Determining the rotation center by the triangular areas. The big circle locations are at 1, 3, and 5. The small circle locations are at 2, 4, and 6. C is the rotary table center with coordinates (x_c, y_c).

As shown in Figure 2, the positions of the big circles are numbered 1, 3, and 5, and the positions of the small circles are numbered 2, 4, and 6; the coordinates of the 6 points are written as (x_i, y_i), i = 1–6. Equation (1) shows the area calculation formulas for the three triangles, Δ12C, Δ34C, and Δ56C; the areas of the three triangles are equal ($A_0 = A_1 = A_2$) to solve C.

The three triangle-area formulas are as follows:

$$\mathrm{A0} = \frac{1}{2}\begin{vmatrix} x_1 & y_1 & 1 \\ x_2 & y_2 & 1 \\ x_c & y_c & 1 \end{vmatrix};\ \mathrm{A1} = \frac{1}{2}\begin{vmatrix} x_3 & y_3 & 1 \\ x_4 & y_4 & 1 \\ x_c & y_c & 1 \end{vmatrix};\ \mathrm{A2} = \frac{1}{2}\begin{vmatrix} x_5 & y_5 & 1 \\ x_6 & y_6 & 1 \\ x_c & y_c & 1 \end{vmatrix} \tag{1}$$

After derivation, the rotation center (x_c, y_c) values are as follows, in Equations (2) and (3):

$$x_c = \frac{(x_2y_1-x_1y_2+x_3y_4-x_4y_3)(x_2-x_1+x_5-x_6)-(x_2y_1-x_1y_2+x_5y_6-x_6y_5)(x_2-x_1+x_3-x_4)}{(y_1-y_2-y_3+y_4)(x_2-x_1+x_5-x_6)-(y_1-y_2-y_5+y_6)(x_2-x_1+x_3-x_4)} \quad (2)$$

$$y_c = \frac{(x_2y_1-x_1y_2+x_3y_4-x_4y_3)(y_1-y_2-y_5+y_6)-(x_2y_1-x_1y_2+x_5y_6-x_6y_5)(y_1-y_2-y_3+y_4)}{(y_1-y_2-y_5+y_6)(x_2-x_1+x_3-x_4)-(y_1-y_2-y_3+y_4)(x_2-x_1+x_5-x_6)} \quad (3)$$

2.3. Conversion of Measured Values to Radius

In this study, LS-7030 was used to simultaneously measure the inner diameter and outer diameter of the ring gauge, and the measured values were the upper and lower dimensions of the micrometer optical belt (Out_2, Out_1).

As shown in Figure 3, the measurement dimensions are the upper and lower dimensions of the light band (Out_2, Out_1). Using the Out_2 and Out_1 values at 0° and 180°, the E_1 and E_2 values can be solved. Finally, the values of outer contour and inner contour radius can be converted, using E_1 and E_2. This study did not measure the inner contour radius, and therefore it is not discussed (Out_2, E_2).

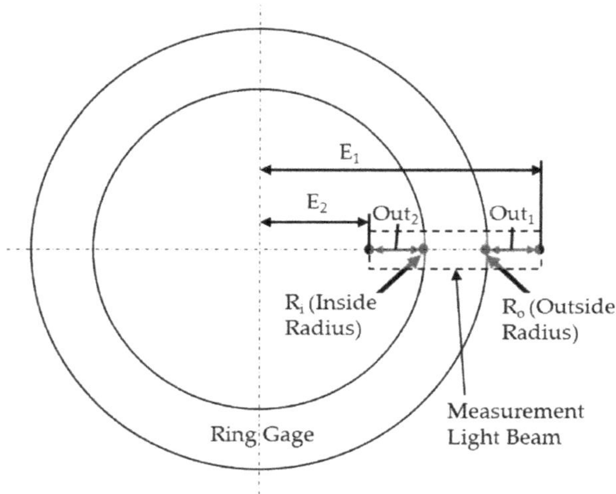

Figure 3. Schematic of ring gauge measurement and symbols used in the calculation.

The ring gauge diameter is used for ensuring accuracy in calculation; the diameter value (2Ro) is calculated by using Equation (4), and Equation (5) is used for the calculation of the inner diameter (2Ri). Then, using these values, we can calculate E_1 and E_2 by using Equations (6) and (7), allowing the proposed system to also measure the inner diameter and contour.

$$2\mathrm{Ro} = E_1-(\mathrm{Out}_1)_{0^\circ}+E_1-(\mathrm{Out}_1)_{180^\circ} \quad (4)$$

$$2\mathrm{Ri} = E_2+(\mathrm{Out}_2)_{0^\circ}+E_2+(\mathrm{Out}_2)_{180^\circ} \quad (5)$$

$$E_1 = \frac{[2\mathrm{Ro}+(\mathrm{Out}_1)_{0^\circ}+(\mathrm{Out}_1)_{180^\circ}]}{2} \quad (6)$$

$$E_2 = \frac{[2\mathrm{Ri}-(\mathrm{Out}_2)_{0^\circ}-(\mathrm{Out}_2)_{180^\circ}]}{2} \quad (7)$$

2.4. Correction Calculation for Workpiece Eccentricity

Because the platform and workpiece are not on the same axis, eccentricity errors may occur during rotation. In the prototype system, the ring gauge coordinate center was obtained by using the ring gauge, and then the correction of the eccentricity value and the measured value were calculated by the C# program. On the rapid measuring machine, using the outer diameter of the ring gauge or the maximum outer diameter of the tooth profile in the workpiece, a circle center was obtained through regression analysis. In this study, the ring gauge diameter is 192.68 mm.

As shown in Figure 4a, the rotating platform and workpiece (or ring gauge) are not on the same axis, and thus, the measurement result X varies due to eccentricity oscillation. The green zone is the LS-7030 measurement zone, C is the rotary table center, r is the sample or the ring gauge radius, r′ is the eccentricity distance ($\overline{CP}$) from the ring gauge center to the rotation center, and θ is the angle between the eccentricity distance ($\overline{CP}$) and the X-axis. In quadrant I, the angle is +θ.

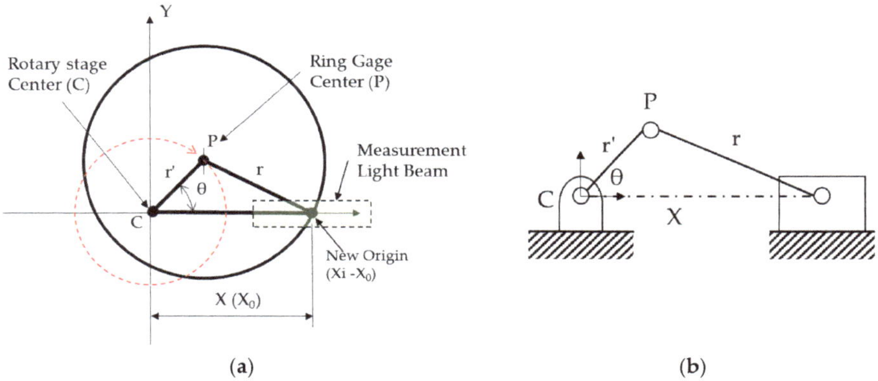

Figure 4. Eccentric correction calculation. (**a**) Eccentricity between rotary table center and ring gauge center. (**b**) Simplification to slider-crank mechanism.

The measurement system moves in a clockwise rotation in the top view, and thus, the progressive rotation angle is −θ. Because the machine rotates in the clockwise direction, the maximum value of X is at 0°; therefore, the measured values of X at the beginning of quadrants I and II gradually increase to the maximum value (θ = 0°) as the platform rotates, after which they gradually decrease. The minimum value of X is at 180°; therefore, the measured values of X at the beginning of quadrants III and IV gradually decrease to the minimum value (θ = 180°) as the platform rotates, after which they gradually increase.

As shown in Figure 4b, all X values can be simplified to a slider-crank mechanism, and the distance X can be obtained by using the slider-crank mechanism, as expressed in Equation (8).

$$X = r' \cos\theta + \sqrt{r^2 - (r' \sin\theta)^2} \tag{8}$$

Equation (8) can be used to calculate the eccentricity sine curve where the input parameter of θ was derived from the cosine theorem in Equation (9) and finally can be obtained by using Equation (10).

When the prototype is used for ring gauge measurement, the eccentricity value (r′) is calculated directly from the measured values; the value is obtained by dividing the difference between the maximum value of X and the minimum value of X by 2, as shown in Equation (11). The ring gauge radius (r) is 96.34 mm.

On the rapid measuring machine, using the outer diameter of the ring gauge or the maximum outer diameter of the tooth profile in the workpiece, a circle profile is obtained through regression analysis.

Then, the circle is used to calculate the center P. Therefore, the eccentricity value r′, the addendum circle radius r, and the X distance can be obtained simultaneously.

$$r^2 = r'^2 + X^2 - 2r'X\cos\theta \tag{9}$$

$$\theta = \cos^{-1}\left(\frac{r'^2 + X^2 - r^2}{2r'X}\right) \tag{10}$$

$$r' = \frac{(Xmax - Xmin)}{2} \tag{11}$$

To determine the error value of the measurement system, we take the result of the first measurement as the reference at origin 0 and then set the 1st value to zero. The method is to deduct the first value of all measured results, and then the origin is translated by subtracting the starting X_0 value from the value of the first measured results.

Figure 5 shows the error graphs obtained from different starting eccentricity angles, when the eccentricity is assumed to be 0.1 mm (r′ = 0.1). As seen in Figure 5a, when the original eccentricity angle is 0°, the X value is the largest when the coordinates of P are (0.1, 0), and the value is 0. Moreover, because the machine rotates clockwise, the resulting error value is less than 0 and has a minimum value at 180°.

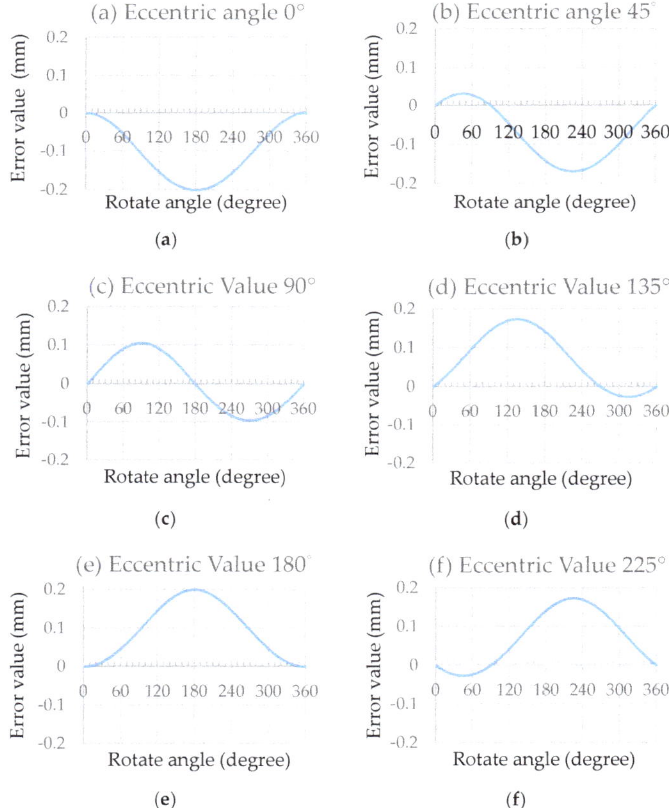

Figure 5. *Cont.*

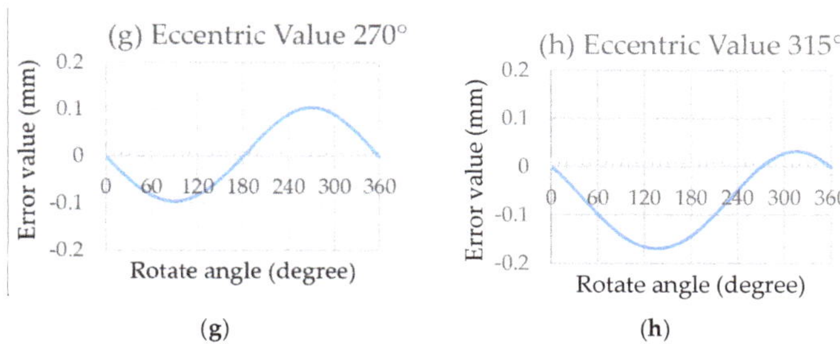

(g) (h)

Figure 5. Error graphs obtained for different starting eccentric angles (**a**–**h**).

As seen in Figure 5b, when the original eccentricity angle is 45°, the coordinates of P are (0.07071, 0.07071), and r′ is in quadrant I. Therefore, when the machine is rotated 45° clockwise, the value of X is the maximum, the error is 0 at 90°, and there is a minimum value at 225°.

As seen in Figure 5c, when the original eccentricity angle is 90°, the coordinates of P are (0, 0.1). Therefore, when the machine is rotated 90° clockwise, the value of X is the maximum, the error is 0 at 180°, and there is a minimum value at 270°.

As seen in Figure 5d, when the original eccentricity angle is 135°, the coordinates of P are (−0.07071, 0.07071). Therefore, when the machine is rotated 135° clockwise, the value of X is the maximum, the error is 0 at 270°, and there is a minimum value at 315°.

As seen in Figure 5e, when the original eccentricity angle is 180°, the X value is the minimum when the coordinates of P are (−0.1, 0). Therefore, when the machine is rotated 180° clockwise, the value of X is the maximum, and the error is 0 at 360°.

As seen in Figure 5f, when the original eccentricity angle is 225°, the coordinates of P are (−0.07071, −0.07071), and r′ is in quadrant III. Therefore, when the machine is rotated 45° clockwise, the value of X is the minimum, the error is 0 at 90°, and there is a maximum value at 225°.

As seen in Figure 5g, when the original eccentricity angle is 270°, the coordinates of P are (0, −0.1). Therefore, when the machine is rotated 90° clockwise, the value of X is the minimum, the error is 0 at 180°, and there is a maximum value at 270°.

As seen in Figure 5h, when the original eccentricity angle is 315°, the coordinates of P are (0.07071, −0.07071), and r′ is in quadrant IV. Therefore, when the machine is rotated 135° clockwise, the value of X is the minimum, the error is 0 at 270°, and there is a maximum value at 315°. As seen in Figure 5, at 180°, the error patterns are reversed.

2.5. Main Size Measurement and Method

In this study, a tooth of a ring object was measured. This object is mainly used for transmission connection, and the tooth shape is similar to that of a rectangular spline.

Because the ring-shaped workpiece with the qualified outer diameter (the maximum outer diameter of the tooth top) is screened and processed before machining of the tooth shape, in this study, the main dimensions considered were the tooth root, tooth height, and tooth thickness.

After the measurement is complete, the software corrects the center of the workpiece and then draws the results, as shown in Figure 6 (showing the contour of the ring-shaped workpiece); the tooth top (Rt), tooth root (Rr), tooth height (Rt − Rr), and tooth thickness (W) are shown in the figure.

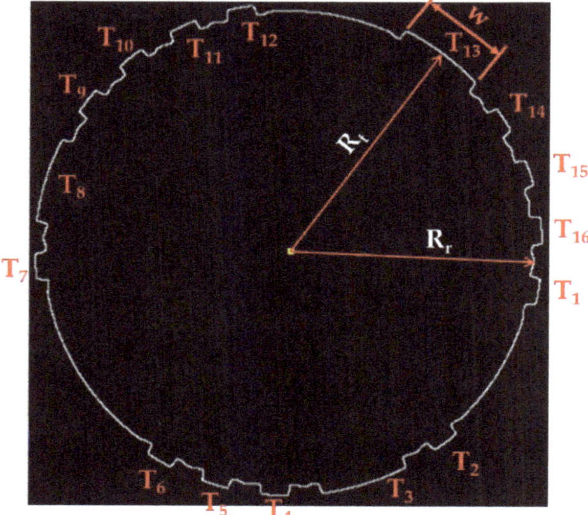

Figure 6. Measurement result for the gear contour (only shows the outside contours of the ring sample). The yellow center point shows the actual sample center.

The tooth height is calculated by subtracting Rr from Rt. The tooth thickness (W) is measured at the top of the tooth, using two endpoints of the tooth. Each tooth is assigned a number (T1–T16), as shown in Figure 6.

2.6. Software and Program Processes

This study used visual C# and Emgu CV as software development programs. Emgu CV is a cross-platform .Net wrapper for the OpenCV image processing library, allowing Open CV functions to be called from .NET compatible languages. Emgu CV can be used to develop real-time image-processing, computer-vision, and pattern-recognition programs.

This study developed a prototype and a rapid system. The program processing flow for each was the same, and only the program for signal processing was different.

2.6.1. Program Processing in the Prototype System

The following steps are involved in the program:

Step 1. Real and virtual conversion: the conversion coordinates of the rotation axis center and the measurement of the LS-7030 light band are obtained first. To comply with the measurement principle, correction needs to be performed, as given in Step 3.

Step 2. Filtering: No filtering is required. Due to the low speed of 0.25 rpm and the RS-232 interface during the prototype development stage of the machine, there is little or no high-frequency noise, and thus, the output result can be used directly.

Step 3. Eccentric correction and calibration: After the eccentricity value is corrected by the ring gauge, the measurement result of the ring gauge can be obtained, and Abbe errors in the system can be reduced by adjusting the XY micrometer stage of the LS-7030 base. Because the ring gauge is circular, the XY micrometer stage is adjusted so that the measured value is the maximum. Then, the measurement center and the rotation center of the object are on the same straight line, almost without any Abbe error. The measured value at this time is equal to the true size of the ring gauge. During the development period, in the eccentric correction of the prototype system, only the eccentricity value of the ring gauge was used.

Step 4. Output: The main measured dimensions are the tooth height, tooth crest, tooth root, and tooth thickness. Except for the tooth thickness, the intermediate position of each tooth is taken as a basis, as shown in Figure 6. Rt and Rr are the intermediate positions of the tooth line. The tooth thickness is the length of the arc of the tooth top.

2.6.2. Program Processing of Rapid Measurement System

Steps 1 and 2 are the same as those in the previous section. Step 2 is performed with a filter. Two types of filters are employed: the Butterworth filter and a simple filter. The Butterworth filter is used in ring gauge calibration, and the simple filter is used in the measurement of the toothed workpieces.

Because the speed was increased to 2.5 rpm and the Ethernet/IP interface was used in the final system, high-frequency noise occurs. A fourth order zero-phase shift low-pass with 1 Hz cutoff frequency of Butterworth filter was used on the ring gauge measurement results.

In the measurement of the tooth-shaped workpieces, a simple filtering method was used for the noise, because the Butterworth filter would distort the tooth shape. In the simple filter program, first, the top and root of the tooth are defined. When the measurement is complete, the addendum radius is considered to be 94.5 mm. The measurement values are between 92.6 and 94.5 mm, and they are considered the tooth root radius. Then, the measurement values are calculated by using the threshold set by the simple filtering method; the current set threshold is 0.1 mm.

We observed a phenomenon from the experimental data: The comparison value of each piece of data with the previous piece of data will not exceed 0.1 mm except for signal noise. This is the origin of the threshold filtering for tooth shaped workpieces.

The simple filtering method compares the current measurement data (the latter data) with the previous data. If the difference is more than 0.1 mm (threshold value), the current data are replaced with the previous data. This threshold is based on the comparison between the measurement results of the study machine and the results of the coordinate measuring machine (CMM) measurement.

Step 3 is eccentric correction and calibration. Two different samples are used: a ring gauge and a workpiece. To achieve further accuracy in the measurement of rectangular spline workpieces, different centers of circles are used for the eccentric correction in rapid measurement systems. For the ring gauge and workpieces, their centers are used for eccentric correction. Circular regression analysis is performed on both of them to determine the center of the circle. Because the arc of the tooth top is the outer diameter of the workpiece and the workpiece is manufactured using a precision lathe, the measurement of the concentricity of the workpiece is accurate.

3. Results and Discussion

In this section, measurements using the prototype system and the rapid system are discussed. Furthermore, the rotation speed in the rapid measurement system is discussed.

The four steps of the program were described in Section 2.6: (1) real and virtual conversion, (2) filtering, (3) eccentric correction and calibration, and (4) output. The program uses the first step of real and virtual conversion to perform the calculation. The filtering requirements are discussed later in the measurement results. In the prototype system, the measurement results are used directly, without filtering. In the rapid measurement system, noise must be filtered out from the measurement results in order to achieve a more accurate eccentric correction. Experiments were conducted to verify the stability, measurement repeatability, and accuracy of the system.

In this study, an optical projector and CMM were used to verify the measurement results. The CMM measurement data are provided by the QC department in the factory, and the CMM measurement results are used for comparison and adjustment in the final rapid measurement system. Therefore, this system can be transferred directly to the production line.

3.1. Prototype Development System

3.1.1. Eccentric Correction

To determine the eccentricity value between the rotation platform and samples, the measurement results must be first converted to X_0 relative offset values.

The X_0 relative offset value (δ) is obtained by deducting all the measured values from the first measured value. In the expression $\delta(i) = X_i - X_0$, $i = 0{\sim}N$, where N is the total number of measurement points for the entire circle, and $\delta(0) = 0$ is the first offset value.

Figure 4a shows the origin translated by subtracting the starting X_0 value, meaning that the first measurement value $\delta(0)$ is transferred to the 0 value as $\delta(0) = 0$.

In the standard ring gauge measurement, the conversion result of the X_0 relative offset shows the effect of eccentricity. The maximum and minimum values of δ equal X max and X min in Equation (11); Equation (11) is used to calculate the eccentricity value r′, and r′ is substituted into Equation (8), to determine the eccentric correction curve.

In the eccentric correction of the prototype system, only the ring gauge eccentricity value was used.

Figure 7 shows the measurement results of the ring gauge that have been converted into X_0 relative offset values. In addition to the sinusoidal oscillation trend, this X_0 relative offset curve also includes the high-frequency error caused by the reducer vibration. The prototype system did not filter the measurement results.

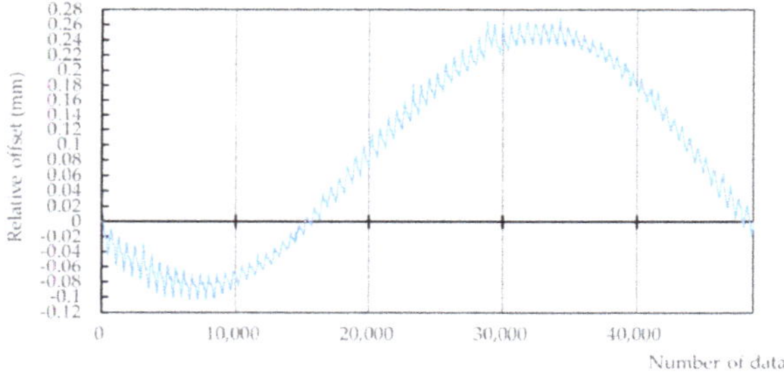

Figure 7. Ring gauge measurement results translated to X_0 relative offset data (0.25 rpm with RS-232).

All results are compared with the relative offset converted to X_0, as shown in Figure 8. As shown in Figure 8a, after calculation of the eccentricity value r′, using Equation (11), the r′ value is used to obtain the eccentric correction curve, using Equation (8). Figure 8b shows the result of subtracting the eccentric correction curve. If the curve is a horizontal line and all values are 0 mm, this means that all the values are the same as the first set of data; values closer to 0 imply that the error is small.

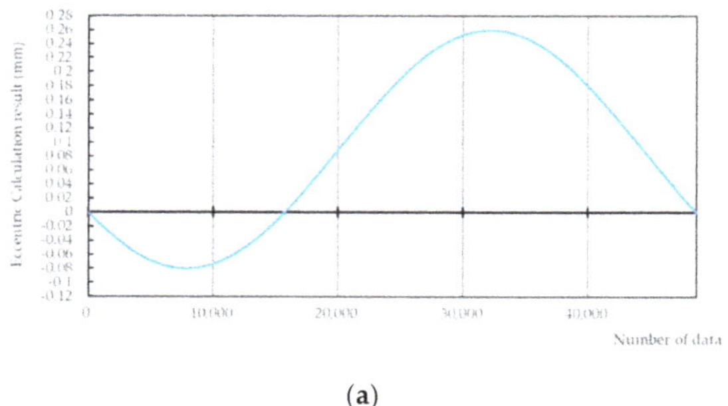

(a)

Figure 8. *Cont.*

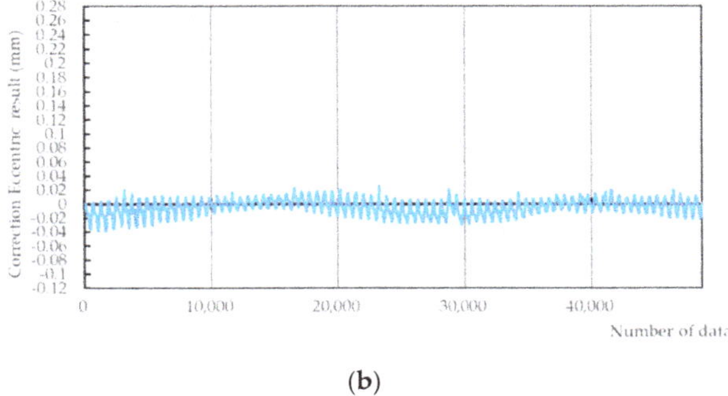

(b)

Figure 8. (**a**) Calculation result of ring gauge eccentricity curve. (**b**) Eccentric correction results.

Figure 8b shows that there are 80 periods of the wavelet phenomenon, and the number of small noise peaks is 80, which is same as the reducer ratio; this indicates the occurrence of loading deformation caused by the meshing of the flexible reducer and the fixed wheel on the harmonic reducer [22]. Because the speed of the prototype system is only 0.25 rpm and the RS-232 has almost no signal transmission noise, the reducer ratio of 1:80 can be determined clearly.

3.1.2. Test for Virtual Measurement System Repeatability and Accuracy

To test the stability of the measurement system, the measurements were repeated thrice with the unloading test. During the unloading test, the sample is not removed or repositioned.

The tooth height and thickness of 16 teeth were measured, as shown in Tables 1 and 2. To easily compare the three sets of results, we subtracted the second and third measurement results from the first measurement results, to obtain the repeatability measurement error. The maximum repeatability measurement error is only 0.004 mm for the tooth height (Table 1), indicating that the measurements were accurate and repeatable.

Table 1. Measurement system repeatability test on tooth height (measurements were repeated thrice).

Measurement System Repeatability Study					
(Unit: mm)	**Tooth Height**			**Repeatability Error**	
No. of Tooth	**Test 1**	**Test 2**	**Test 3**	**2nd–1st**	**3rd–1st**
T1	3.059654	3.058098	3.056252	−0.002	−0.003
T2	2.992897	2.99575	2.995895	0.003	0.003
T3	3.238052	3.23745	3.241447	−0.001	0.003
T4	3.220901	3.22065	3.219147	0.000	−0.002
T5	3.212646	3.210396	3.210106	−0.002	−0.003
T6	3.184547	3.185455	3.185547	0.001	0.001
T7	3.165848	3.166351	3.165047	0.001	−0.001
T8	3.200005	3.2015	3.201103	0.001	0.001
T9	3.181755	3.181557	3.179451	0.000	−0.002
T10	3.164795	3.163551	3.1604	−0.001	−0.004
T11	3.150452	3.151344	3.150154	0.001	0.000
T12	3.212845	3.216354	3.216148	0.004	0.003
T13	3.133095	3.129196	3.131699	−0.004	−0.001
T14	3.108749	3.108849	3.108704	0.000	0.000
T15	3.142197	3.140297	3.14225	−0.002	0.000
T16	3.154251	3.151299	3.152298	−0.003	−0.002

As seen in Table 2, the maximum repeatability error is −0.129 mm for the repeatability measurement results for tooth thickness, and this, too, was an outlier. The tooth thickness error is the sum of the signal noise error and sampling error. The signal noise error is the high-frequency noise in measurement, and the sampling error is determined from the sampling ratio.

Table 2. Measurement system repeatability test on tooth thickness (measurements were repeated thrice).

The Measurement System Repeatability Study					
(Unit: mm)	Tooth Thickness			Repeatability Error	
No. of Tooth	Test 1	Test 2	Test 3	2nd–1st	3rd–1st
T1	10.14783	10.13657	10.16716	−0.011	0.019
T2	9.895499	9.904919	9.946733	0.009	0.051
T3	32.5769	32.52688	32.581	-0.050	0.004
T4	10.05784	10.05223	10.05939	−0.006	0.002
T5	10.20149	10.19728	10.20142	−0.004	0.000
T6	9.865836	9.884316	9.871897	0.018	0.006
T7	9.779203	9.751502	9.73667	−0.028	−0.043
T8	32.75737	32.76112	32.77379	0.004	0.016
T9	10.26547	10.26095	10.24714	−0.005	−0.018
T10	10.40329	10.41216	10.43104	0.009	0.028
T11	10.43783	10.45585	10.40663	0.018	−0.031
T12	10.13186	10.13569	10.13275	0.004	0.001
T13	32.73424	32.60537	32.66642	−0.129	−0.068
T14	10.34993	10.2971	10.29994	−0.053	−0.050
T15	10.262	10.2998	10.29871	0.038	0.037
T16	10.36323	10.33171	10.35833	−0.032	−0.005

For the prototype system, the total number of CCD scan data points was 48,709, and thus, the angle of interval between each data point was 0.00739°. The ring gauge diameter was 192.68 mm, and the sampling interval (0.012 mm sampling blind spot size) was obtained by using Equation (12). In Table 3, it can be seen that the percentage of errors of 0 to 0.01 mm in width are as high as 40.63%. Because the thickness of one tooth spans (pass) two tooth steps, in the worst case, the maximum error is twice the sampling interval: 0.024 mm (approximately 0.03 mm).

Table 3. Repeatability error zone distribution.

Error Zone Distribution Percentage			
Error Max (mm)	Count	Cumulative Number	Cumulative Number %
0~0.01	13	13	40.63%
0.01~0.02	6	19	59.38%
0.02~0.03	2	21	65.63%
0.03~0.04	4	25	78.13%
0.04~0.05	3	28	87.50%
0.05~0.06	2	30	93.75%
0.06~0.07	1	31	96.88%
0.07~0.08	0	31	96.88%
0.08~0.09	0	31	96.88%
0.09~0.10	0	31	96.88%
0.10~0.11	0	31	96.88%
0.11~0.12	0	31	96.88%
0.12~0.13	1	32	100.00%

Figure 9 shows the statistical values in each error range and the percentage distribution of the error area in Table 3. Regarding the error ranges, 87.5% of errors are within 0.05 mm, and 96.88% of errors are within 0.07 mm.

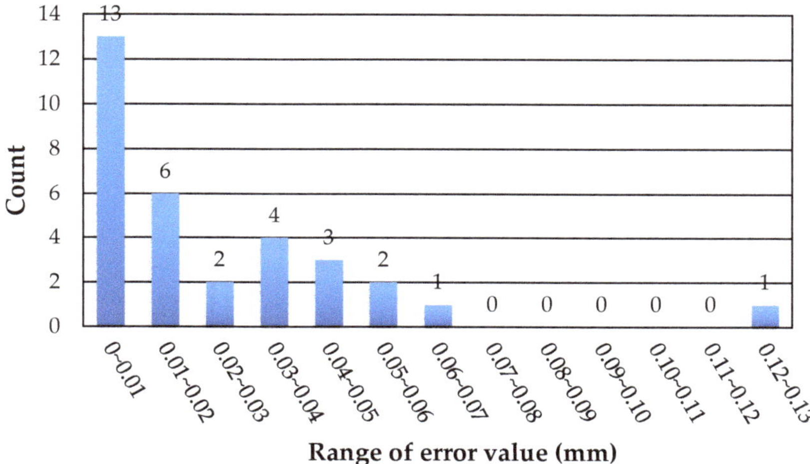

Figure 9. Repeatable error distribution in tooth-thickness tests.

With respect to the influence of the tooth thickness error, the tooth edge positions could not be determined, because of the sampling interval error of the tooth thickness. Without considering the 0.129 mm maximum repeatability error volume (only one), the maximum tooth thickness error was 0.068 mm.

$$\text{Sampling interval} = \pi D \times \frac{\theta}{360} \tag{12}$$

3.1.3. Comparison Between Optical Projector Measurements and Prototype System Measurements

The optical projector can perform the measurement faster than the CMM; thus, we used the optical projector measurements for comparison with the prototype measurements.

Figure 6 shows the position of the teeth, and each tooth is assigned a number. The optical projector measures the tooth height and tooth thickness of T9–T12 of the workpiece, and the average of three measurements was considered in this study and compared with the optical projector measurement results. Table 4 lists the measurement results of the tooth height; the heights, as measured using the optical projector, are 3.192, 3.183, 3.164, and 3.236 mm for T9, T10, T11, and T12, respectively. The heights, as measured using the prototype system measurement, are 3.181, 3.163, 3.151, and 3.215 mm for T9, T10, T11, and T12, respectively. The measurement error between the tooth height values measured using the prototype system and the optical projector is approximately −0.02 mm.

Table 4. Comparison of tooth-height result using optical projector.

	Tooth Height (Unit: mm)		
No. of Tooth	**Avg.**	**Projector**	**Error**
T9	3.181	3.192	−0.011
T10	3.163	3.183	−0.020
T11	3.151	3.164	−0.013
T12	3.215	3.236	−0.021

Table 5 lists the measurement results for tooth thickness; the thicknesses, as measured by using the optical projector, are 10.273, 10.425, 10.475, and 10.155 mm for T9, T10, T11, and T12, respectively. The thicknesses, as measured by using the prototype system, are 10.258, 10.415, 10.433, and 10.133 mm for T9, T10, T11, and T12, respectively. The measurement error between the tooth thickness values measured using the prototype system and the optical projector is approximately −0.04 mm.

Table 5. Comparison of tooth-thickness result using optical projector.

	Tooth Thickness (Unit: mm)		
No. of Tooth	**Avg.**	**Projector**	**Error**
T9	10.258	10.273	−0.015
T10	10.415	10.425	−0.010
T11	10.433	10.475	−0.042
T12	10.133	10.155	−0.022

As in the optical projector comparison result, the developed measurement system is highly precise and accurate, with errors within 0.04 mm.

3.2. Rapid Measurement System

The system proposed in this study can be transferred directly to the production line. In the rapid measurement system, an I/O Special Unit that can use the Ethernet/IP interface to communicate with a PC to facilitate high-speed data transfer was added to the system. The ideal measurement speed was determined first.

As seen in Table 6, when the speed of the prototype system is 0.25 rpm, the sampling rate is estimated to be 203 data points/s. Data transmission through RS-232 is roughly 256 data points/s. In the final rapid measurement system, the rotation speed is 2.5 rpm, and the sampling rate is estimated to be 1326 data points/s, which is higher than the RS-232 transmission value. Therefore, an I/O Special Unit was appropriate to add to the system. This unit can use the Ethernet/IP interface to communicate with a PC to facilitate high-speed data transfer.

The prototype with the RS-232 interface takes approximately 4 min to complete the measurement, whereas the rapid system with the Ethernet/IP interface takes approximately 1 min.

Table 6. Data points estimated in 1 s under different rotary table speeds.

Rotary Table Speed (rpm)	1 Circle Time (s)	Total Number of Data	Total Number of Data in 1 s
0.25	240.0	48709	203
0.32	190	48700	257
2.5	24.0	31817	1326

3.2.1. Effect of Speed on Accuracy

Table 7 indicates that the radial direction error depends on the rotary speed. First, the measurement results must be converted to X_0 relative offset values. The ring gauge is used for error analysis. When the speed changes, the total number of errors also changes. In the radial direction error, the maximum upper limit error is 0.004 mm, and the minimum lower limit error is −0.007 mm (total of 0.011 mm) at a rotary table speed of 2.5 rpm. Thus, increasing the speed helps reduce the deviation in measurement.

Table 7. Radial direction error depends on rotary speed.

Rotary Table Speed (rpm)	Total Data Number	Upper Limit Error (mm)	Lower Limit Error (mm)	Total Error (mm)
0.25	48,709	0.027	−0.038	0.065
0.32	48,700	0.020	−0.012	0.032
2.5	31,817	0.004	−0.007	0.011

Due to the hardware limitations of the PLC, a total of 48,709 data points are captured when the prototype system is employed, but only 31,817 data points are captured when the rapid system is employed.

Table 8 shows the determination of the arc length translated from the sampling interval, divide 360° by the total number of scanned data points to obtain the interval angle between each data point in a 360° rotation, and then calculate the arc length. In the prototype system, the sampling interval is 0.00739°, and in the rapid system, the sampling interval is 0.01131°. The total number of data points affects the sampling interval and precision. The most affected measurement is that of the tooth thickness, as Table 5 shown, the errors in the tooth thickness measurement when the measurement is performed by using an optical projector. Because the start and end positions of the toothed corners may not be measured correctly, the values obtained may be lower than the actual tooth-thickness values.

The maximum sampling interval error is 0.038 mm (0.019 × 2) when the rotary table speed was 2.5 rpm, which is within the tolerance level of 0.05 mm requested by customers; thus, we chose this rotating speed for final use.

Table 8. Translation of sampling interval into arc length.

Rotary Table Speed (rpm)	Total Data Number	Sampling Interval (degree)	Sampling Interval Translates to Arc Length (mm)
0.25	48709	0.00739	0.012
0.32	48700	0.00739	0.012
2.5	31817	0.01131	0.019

3.2.2. Filtering and Eccentric Correction

The rapid measurement system needs to filter the measured results before subsequent program processing; the noise error generated by high-frequency reducer vibration needs to be optimized, as done in Reference [22].

As shown in Figure 10a, all results are converted to X_0 relative offsets values for analysis, and the blue line indicates the X_0 relative offset data of the original measurement result. The orange line represents the results after they have been filtered using a Butterworth filter.

There is considerable signal noise in the blue line, and the maximum upper limit error is 0.0887 mm, whereas the minimum lower limit error is −0.383 mm. The orange line indicates that, when the noise is eliminated, the maximum upper limit error is 0.043 mm, and the minimum lower limit error is −0.211 mm.

Circular regression analysis is used to determine the center of the circle; the value obtained is then input into Equation (8). As shown in Figure 10b, the eccentric correction curves with and without the filter are similar; for the blue line, the maximum upper limit error is 0.037 mm, and the minimum lower limit error is −0.198 mm; and for the orange line, the maximum upper limit error is 0.036 mm, and the minimum lower limit error is −0.197 mm. This indicates that both curves are similar.

Figure 10c shows the eccentric correction results. For the blue line, the ring gauge maximum upper limit value changes from 0.087 mm (without correction) to 0.069 mm (with correction), and the lower limit value changes from −0.383 mm (without correction) to −0.42 mm (with correction)—a total value of 0.488 mm.

For the orange line, the ring gauge maximum upper limit value changes from 0.043 mm (without correction) to 0.026 mm (with correction), and the lower limit value changes from −0.211 mm (without correction) to −0.023 mm (with correction)—a total value of 0.049 mm. The error value is lower than the prototype error value of 0.06 mm.

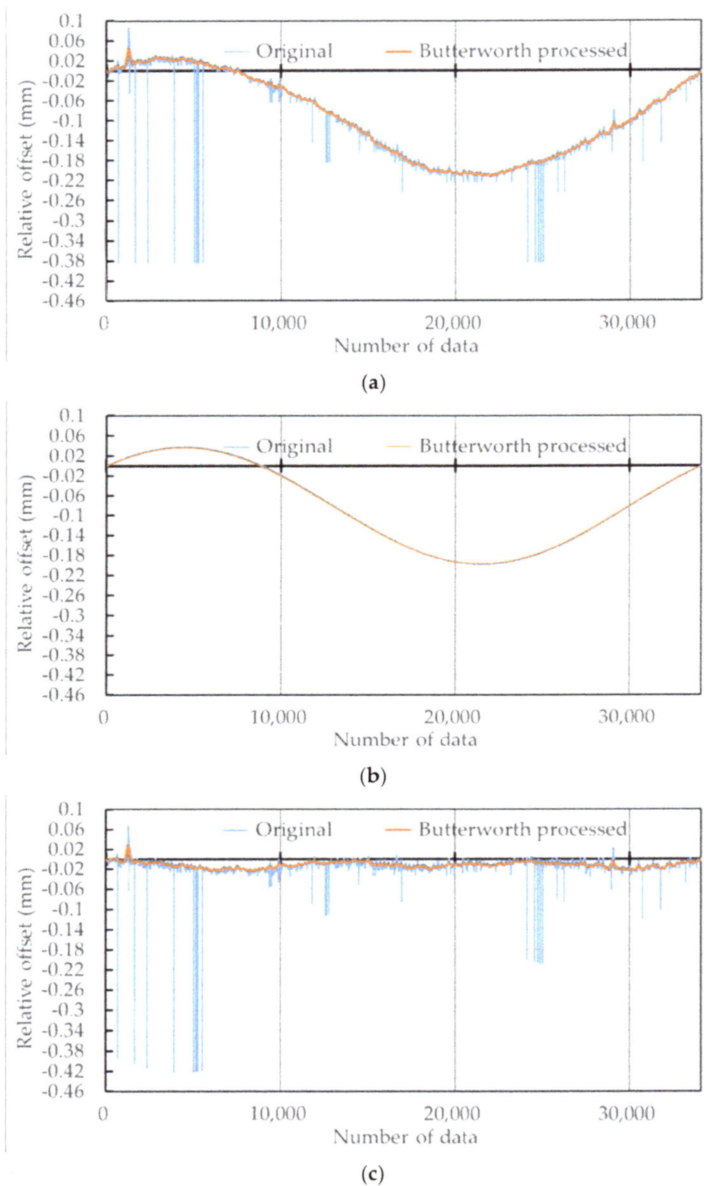

Figure 10. Ring gauge measurement results translated to X_0 relative offset data. The blue line indicates the X_0 relative offset data of the original measurement result. The orange line indicates the data after being filtered using a Butterworth filter. (**a**) Measurement results translated to X_0 relative offset data. (**b**) Calculation result of ring gauge eccentricity curve. (**c**) Eccentric correction results.

3.2.3. Virtual Measurement System Repeatability and Accuracy Test

The results presented in the previous section indicate that the Butterworth filter can improve measurement accuracy and does not affect eccentric correction. In this section, we discuss how both filters affect different parameters. The direct simple filter was used for the tooth-shape workpieces, and the Butterworth filter was used for the ring gauge process.

First, the stability of the rapid system was tested. Table 9 lists the system stability results obtained via the unloading test; the measurement was repeated thrice.

Table 9. Measurement system repeatability test (measurements repeated thrice).

System Repeatability Study					
(Unit: mm)	**Ring Gage Correction Deviation Value**			**Difference**	
Results	**Test 1**	**Test 2**	**Test 3**	**2nd–1st**	**3rd–1st**
Diameter	192.66	192.66	192.66	0.00	0.00
Positive value	0.009	0.002	0.015	−0.007	0.006
Negative value	−0.028	−0.035	−0.028	−0.007	0.000
Total value	0.036	0.037	0.043	0.000	0.006

To easily compare the three measurement results, the first measurement result is subtracted from the second and third measurement results, to obtain the repeatability measurement error.

The repeatability measurement result indicated a maximum difference of 0.007 mm, which demonstrates that the accuracy and repeatability of the radial measurement are highly satisfactory in our system. This result is slightly higher than that of the prototype system (0.004 mm). This implies that the Butterworth filter is effective and can achieve the desired goal.

After comparing with the ring gauge diameter of 192.68 mm, we find that the diameter error value obtained from using this system is −0.02 mm, and the tooth height error value is within −0.02 mm, which is comparable to the optical projector result. Both comparison results show the same error for the radial measurement.

When the radial results shown in Tables 4 and 9 are combined, the difference between the measurement system and the actual size is seen to be −0.02 mm, which means that the system accuracy and repeatability of the diameter measurement are good.

In the tooth-shape workpiece, the measurements were conducted thrice in an unloading test. The tooth height and tooth thickness of 16 teeth were measured, as shown in Tables 10 and 11. To easily compare the three measurement results, the first measurement result was subtracted from the second and third measurement results, to obtain the repeatability measurement error.

As seen in Table 10, the maximum repeatability measurement error was only 0.023 mm for the tooth height, which indicates that the accuracy and repeatability of the radial measurement are highly satisfactory.

Table 11 lists the repeatability measurement results for tooth thickness; the maximum repeatability error is 0.479 mm because the total number of data points is reduced to 31,817.

As mentioned in Section 3.2.1 and indicated in Table 8, in the worst case, the sampling interval translates into an arc length of 0.019 mm (sampling blind spot size). The maximum sampling error for tooth thickness is twice the sampling interval, and thus, it is twice the arc length; therefore, the maximum sampling error is 0.038 mm (0.019 × 2).

Table 10. Measurement system repeatability test on tooth height (measurements were repeated thrice).

Measurement System Repeatability Study					
(Unit: mm)	Tooth Height			Repeatability Error	
No. of Tooth	1st	2nd	3rd	2nd–1st	3rd–1st
T1	3.044182	3.038063	3.065773	−0.006	0.022
T2	3.208313	3.196518	3.200348	−0.012	−0.008
T3	3.204491	3.202126	3.20713	−0.002	0.003
T4	3.146439	3.148987	3.146469	0.003	0.000
T5	3.147575	3.146599	3.150719	−0.001	0.003
T6	3.109627	3.107903	3.112473	−0.002	0.003
T7	3.154678	3.17173	3.153351	0.017	−0.001
T8	3.135544	3.136116	3.147705	0.001	0.012
T9	3.122688	3.111023	3.118095	−0.012	−0.005
T10	3.099098	3.108871	3.097031	0.010	−0.002
T11	3.184891	3.193581	3.190315	0.009	0.005
T12	3.112991	3.117012	3.114571	0.004	0.002
T13	3.076508	3.099327	3.08419	0.023	0.008
T14	3.1362	3.134392	3.132622	−0.002	−0.004
T15	3.141754	3.148598	3.155876	0.007	0.014
T16	3.082542	3.08683	3.084183	0.004	0.002

Table 11. Measurement system repeatability test on tooth thickness (measurements were repeated thrice).

Measurement System Repeatability Study					
(Unit: mm)	Tooth Thickness			Repeatability Error	
No. of Tooth	Test 1	Test 2	Test 3	2nd–1st	3rd–1st
T1	10.41844	10.73286	10.83447	0.314	0.416
T2	32.7338	32.82142	32.87395	0.088	0.140
T3	10.84539	10.62988	10.93286	−0.216	0.087
T4	10.6523	10.76262	10.52345	0.110	−0.129
T5	10.2906	10.74687	10.76949	0.456	0.479
T6	10.30208	10.09921	10.58485	−0.203	0.283
T7	32.69563	32.38203	32.86719	−0.314	0.172
T8	10.70767	10.78356	10.26262	0.076	−0.445
T9	11.22797	11.04815	11.16596	−0.180	−0.062
T10	10.26523	10.04958	10.28927	−0.216	0.024
T11	11.40674	11.15106	10.95964	−0.256	−0.447
T12	32.87467	32.6995	33.04661	−0.175	0.172
T13	10.82223	10.86615	10.42624	0.044	−0.396
T14	10.80893	10.70416	10.53584	−0.105	−0.273
T15	10.68709	10.71829	11.07741	0.031	0.390
T16	10.67616	10.53528	10.7095	−0.141	0.033

Figure 11 shows the thickness error distribution. Because of the sampling interval effect, the count is 0 in the 0–0.01 mm range of error. The count for thickness errors more than 0.2 mm is 15, which is 47% of the total count. The count for thickness errors more than 0.35 mm is 7, which is 22% of the total count.

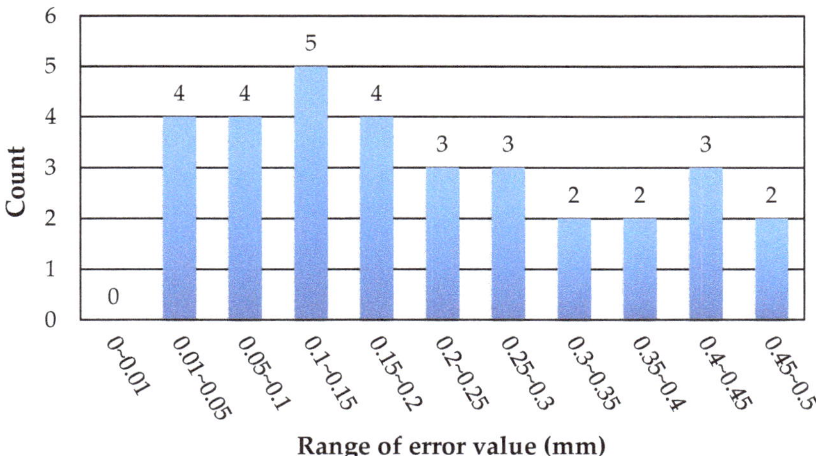

Figure 11. Histogram showing thickness error distribution.

The results show the impact of the decrease in sampling volume. The simple filtering method is only effective for measurement of the radial direction, and the increase in noise also increases the error in the measurement of tooth thickness.

3.2.4. Comparison between CMM and Actual Rapid System Measurements

The measurement results obtained from using the proposed system were compared with results obtained in an actual situation. To reduce the time taken to adjust the parameters when the machine is eventually moved to the factory, the factory provided CMM measurement data to use as the basis for machine adjustment/comparison.

This study successfully developed a rapid optical measurement system that can reduce the inspection time from the 30 min of CMM to 60 s.

Figure 12 shows the CMM measurement positions. Because the workpiece tooth groove is processed by a broach, the groove is tapered toward the top of its face. The CMM measurement is based on the middle position. In this study and in the projector measurement, the maximum size was measured because the optical measurement system can only measure the outermost shape projection. This section discusses the difference in dimensions obtained with the optical method and the CMM.

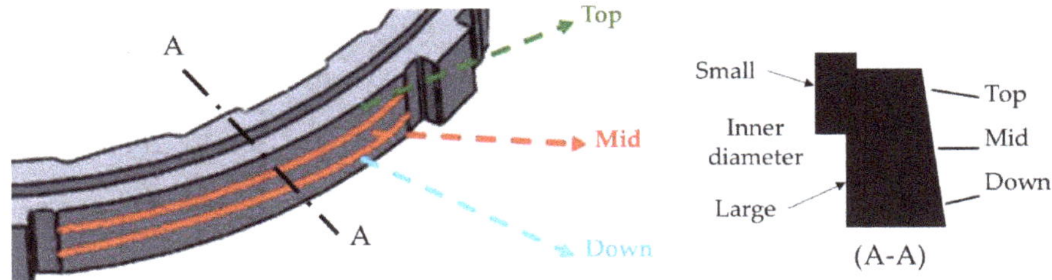

Figure 12. Coordinate measuring machine (CMM) measurement positions and face taper.

Figure 6 shows the tooth positions, which are numbered. The key measurement point of the CMM is the maximum size of 16 roots (Rr) because this is the dimension where interference occurs during product assembly. As in Section 2.6.2, a simple direct filter was added to the data-processing step. The system accuracy and repeatability of the diameter measurement were highly satisfactory. The radial direction error was within 0.007 mm, as shown in Table 9.

As shown in Table 12, the results of using two different eccentric corrections for the measurement of the same tooth-shaped workpiece were compared with the results of the CMM. When a wrong correction value for the ring gauge was used, large deviations between values occurred (Figure 13); two measurement values were lower than those obtained with the CMM. The minimum error value is −0.044 mm, the error range is between −0.044 and 0.121 mm, and the total error value is 0.165 mm. The measurement result indicates that the actual size of the workpiece is larger than the measured size, and therefore, interference may occur during actual assembly and work.

Table 12. Comparison results of three measurement systems.

Comparison Results of Different Rotation Centers					
(Unit: mm)	The Radius of Tooth Root			Error	
No. of Tooth	Ring Gage Center	Sample Center	CMM	Ring Gage Center	Sample Center
T1	93.144	93.105	93.100	0.044	0.005
T2	93.219	93.205	93.171	0.048	0.034
T3	93.146	93.158	93.110	0.036	0.048
T4	93.053	93.127	93.097	−0.044	0.030
T5	93.116	93.157	93.120	−0.004	0.037
T6	93.151	93.199	93.125	0.026	0.074
T7	93.065	93.115	93.065	0.000	0.050
T8	93.183	93.230	93.159	0.024	0.071
T9	93.206	93.239	93.160	0.046	0.079
T10	93.160	93.173	93.145	0.015	0.028
T11	93.149	93.150	93.101	0.048	0.049
T12	93.238	93.225	93.174	0.064	0.051
T13	93.285	93.255	93.208	0.077	0.047
T14	93.140	93.090	93.031	0.109	0.059
T15	93.107	93.060	93.033	0.074	0.027
T16	93.163	93.116	93.042	0.121	0.074
Eccentric of Ring Gage	X	0.02493	Max	0.121	0.079
	Y	0.07299	Min	−0.044	0.005
Eccentric of Sample	X	0.02536	AVG	0.043	0.048
	Y	0.04408	Total	0.165	0.075

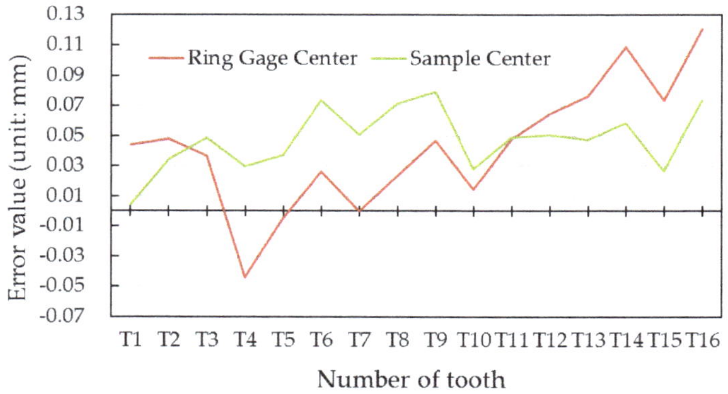

Figure 13. Tooth root error value comparison between the developed system and CMM.

When the right correction value was used, the curve in Figure 13 is almost horizontal, and the oscillation is small; all measurement values were lower than those obtained with the CMM. The minimum error value is 0.005 mm, the error range is between 0.079 and 0.005 mm, and the total error value decreased to 0.075 mm.

The measurement result indicates that the actual size of the workpiece is smaller than the measured size, and therefore, interference may not occur during actual assembly and work.

The proposed optical measurement system can only measure the outermost shape projection of tapered face of the workpiece. Therefore, it is reasonable that the measurement results obtained from using our system will be bigger than the results obtained from using the CMM. Figure 13 shows the error between the correction results of two different circle centers and the CMM measurement result. When using the correct sample center, the error curve is smoother and approaches the horizontal, which indicates consistency in the trend of each error value; moreover, the total error is also small. When using an incorrect center correction, the curve oscillates like a sinusoid.

The CMM measures the middle of the sample, and the dimensions measured by the system and the projector are the outermost contour dimensions of the workpiece. Therefore, differences in dimension and measured results occur. Results obtained with the CMM were compared with the results acquired from using optical projectors; the measurement errors are the same (−0.02 mm) for the ring gauge and tooth shape samples. This means that our measurement system is accurate and stable. The result is satisfactory because the difference between the CMM measurement data and the results obtained with our system is 0.079 mm when using only the direct filter method.

4. Conclusions

We successfully developed a rapid optical measurement system that can reduce inspection time from 30 min to 30 s for 360° measuring time (total 60 s including output the result). This system can quickly identify the cutting tools that need adjustment or replacement, to maintain product dimension accuracy and yield.

Measurements were first conducted by using a prototype system in the laboratory. The rotation speed was slow, and the measurement time was long. Moreover, the amount of data was large, and the resulting tooth thickness measurement error was small. The signal stability of the system was good; no glitches were observed. A wavelet periodic signal with 80 noise peaks was observed; the number of noise peaks is the same as the harmonic reducer ratio. Therefore, the prototype system can be used to observe the extent of vibration generated when the harmonic reducer is in operation.

Measurements were then conducted by using a rapid stable system. The radial dimension measurement results of the prototype and rapid system were similar when using the ring gauge. The difference between these results and that obtained with the optical projector is −0.02 mm. Although the signal of the rapid machine had considerable noise, it could be reduced by using a Butterworth filter, and the total eccentricity error is controlled to within 0.04 mm.

In the rapid measurement machine, due to the need for a larger sampling rate/data, the I/O unit was replaced, leading to errors in the signal noise and the sampling interval. The maximum error in tooth thickness measurement is 0.48 mm because the signal was only filtered by a simple direct filter. We will attempt to resolve this issue in our future work.

Whether we used the prototype or rapid system, the radial measurement dimension error of the ring gauge is within −0.02 mm. This means that the system is highly stable, accurate, and yields reproducible results.

The novel rapid measurement system was compared with the CMM used in the production line. The difference in measurement between the two systems is 0.08 mm, and thus we determined that our system can be directly transferred to the production line after completion. Another good benefit is on the equipment prices: CMM is approximately USD 50,000, and the proposed system is around USD 20,000.

In the future, to solve the problem of reducer-generated noise, the drive mechanism of the rotation platform may be changed to a direct drive motor.

Author Contributions: Conceptualization, Q.-C.H.; methodology, Q.-C.H.; software, Y.-L.C., X.-Q.L. and Z.-R.Y.; validation, Y.-L.C. and Z.-R.Y.; investigation, Q.-C.H.; writing—original draft preparation, Y.-L.C.; writing—review and editing, Q.-C.H.; visualization, Y.-L.C., X.-Q.L. and Z.-R.Y.; supervision, Q.-C.H.; project administration, Q.-C.H.; funding acquisition, Q.-C.H. All authors have read and agreed to the published version of the manuscript.

Funding: This research was partially funded by the Ministry of Science and Technology, Taiwan, under grant number MOST 106-2622-E-151-017-CC3. The authors would like to thank Linesoon Industrial Co., Ltd. (Taiwan) for providing the measurement specimens and for valuable advice.

Conflicts of Interest: The authors declare no conflict of interest.

References

1. Ali, M.H.; Kurokawa, S.; Uesugi, K. Application of machine vision in improving safety and reliability for gear profile measurement. *Mach. Vis. Appl.* **2014**, *25*, 1549–1559. [CrossRef]
2. Loizou, J.; Tian, W.; Robertson, J.; Camelio, J. Automated wear characterization for broaching tools based on machine vision systems. *J. Manuf. Syst.* **2015**, *37*, 558–563. [CrossRef]
3. Yu, X.; Lin, X.; Dai, Y.; Zhu, K. Image edge detection based tool condition monitoring with morphological component analysis. *ISA Trans.* **2017**, *69*, 315–322. [CrossRef] [PubMed]
4. Adamo, F.; Attivissimo, F.; DiNisio, A.; Savino, M. A low-cost inspection system for online defects assessment in satin glass. Meas. *Measurement* **2009**, *42*, 1304–1311. [CrossRef]
5. Xiong, C.; Bai, H. Calibration of Large-Scale Spatial Positioning Systems Based on Photoelectric Scanning Angle Measurements and Spatial Resection in Conjunction with an External Receiver Array. *Appl. Sci.* **2020**, *10*, 925. [CrossRef]
6. Heikkila, J.; Silven, O. A four-step camera calibration procedure with implicit image correction. In Proceedings of the IEEE Computer Society Conference on Computer Vision and Pattern Recognition, San Juan, PR, USA, 17–19 June 1997; pp. 1106–1112.
7. Wan, D.; Zhou, J. Stereo vision using two PTZ cameras. Comput. *Vis. Image Underst.* **2008**, *112*, 184–194. [CrossRef]
8. Ng, H.F. Automatic thresholding for defect detection. *Pattern Recognit. Lett.* **2006**, *27*, 1644–1649. [CrossRef]
9. Samopa, F.; Asano, A. Hybrid Image Thresholding Method using Edge Detection. *J. Comput. Sci.* **2009**, *9*, 292–299.
10. Jung, K.; Kim, S.; Im, S.; Choi, T.; Chang, M. A photometric stereo using re-projected images for active stereo vision system. *Appl. Sci.* **2017**, *7*, 1058. [CrossRef]
11. Behzadfar, E.; Ansari, M.; Konaganti, V.K.; Hatzikiriakos, S.G. Extrudate swell of HDPE melts: I. Experimental. *J. Nonnewton. Fluid Mech.* **2015**, *225*, 86–93. [CrossRef]
12. Konaganti, V.K.; Ansari, M.; Mitsoulis, E.; Hatzikiriakos, S.G. The effect of damping function on extrudate swell. J. Nonnewton. *Fluid Mech.* **2016**, *236*, 73–82.
13. Sun, W.; Xu, A.; Gao, Y. Strapdown gyrocompass algorithm for AUV attitude determination using a digital filter. *Measurement* **2013**, *46*, 815–822. [CrossRef]
14. Pulecchi, T.; Manes, A.; Lisignoli, M.; Giglio, M. Digital filtering of acceleration data acquired during the intervention of a lift safety gears. *Measurement* **2010**, *43*, 455–468. [CrossRef]
15. Pinto, P.M.; Ferreira, L.H.C.; Colletta, G.D.; Braga, R.A.S. A 0.25-V fifth-order Butterworth low-pass filter based on fully differential difference transconductance amplifier architecture. *Microelectron. J.* **2019**, *92*, 104606. [CrossRef]
16. Ye, J.; Niu, Z.; Zhang, X.; Wang, W.; Xu, M. In-Situ deflectometic measurement of transparent optics in precision robotic polishing. *Precis. Eng.* **2020**, *64*, 63–69. [CrossRef]
17. Bi, Q.; Huang, N.; Zhang, S.; Shuai, C.; Wang, Y. Adaptive machining for curved contour on deformed large skin based on on-machine measurement and isometric mapping. *Int. J. Mach. Tools Manuf.* **2019**, *136*, 34–44. [CrossRef]
18. Bednarczyk, J.; Sioma, A. Application of a visual measurement technique to the assessment of electrodynamic stamping. In *Solid State Phenom*; Trans Tech Publications Ltd.: Stafa-Zurich, Switzerland, 2011; pp. 1–9.

19. Sioma, A.; Socha, J.; Klamerus-Iwan, A. A new method for characterizing bark microrelief using 3D vision systems. *Forests* **2018**, *9*, 30. [CrossRef]
20. Masoumi, M.; Alimohammadi, H. An investigation into the vibration of harmonic drive systems. *Front. Mech. Eng.* **2013**, *8*, 409–419. [CrossRef]
21. Ghorbel, F.H.; Gandhi, P.S.; Alpeter, F. On the kinematic error in harmonic drive gears. *J. Mech. Des. Trans. ASME* **2001**, *123*, 90–97. [CrossRef]
22. Nye, T.W.; Kraml, R.P. *Harmonic Drive Gear Error: Characterization and Compensation for Precision Pointing and Tracking*; Trans Tech Publications Ltd.: Stafa-Zurich, Switzerland, 1991; pp. 237–252.

Article

Interlaboratory Empirical Reproducibility Study Based on a GD&T Benchmark

Ali Aidibe *, Souheil Antoine Tahan and Mojtaba Kamali Nejad

Mechanical Engineering Department, École de Technologie Supérieure (ÉTS), Montreal, QC H3C 1K3, Canada; antoine.tahan@etsmtl.ca (S.A.T.); mojtaba.kamalinejad@etsmtl.ca (M.K.N.)

* Correspondence: ali.aidibe@etsmtl.ca

Received: 2 June 2020; Accepted: 6 July 2020; Published: 8 July 2020

Abstract: The ASME Y14.5 geometric dimensioning and tolerancing (GD&T) and ISO-GPS (geometrical product specifications) standards define tolerances that can be added to components to achieve the necessary functionality and performance. The zone that each feature must lie within is defined in each tolerance. Measurement processes, including planning, programming, data collection (with contact or without contact), and data processing, check the compliance of the part with these specifications (tolerances). Over the last two decades, many works have been realized by the metrology community to investigate the accuracy, the measuring methods, and, specifically, the measurement errors of fixed and portable coordinate measuring machines (CMMs). A review of the literature showed the progression of CMMs in terms of accuracy and repeatability. However, discrepancies were observed between measurements using different CMMs or operators. This paper proposed a GD&T-based benchmark for the evaluation of the performance of different CMM operators in computer-aided inspection (CAI), considering different criteria related to the dimensional and geometrical features. An artifact was designed using basic geometries (cylinder and plane) and free-form surfaces. The results obtained from the interlaboratory comparison study showed significant performance variability for complex GD&T, such as in the composite profile and localization. This, in turn, emphasized the importance of GD&T training and certification in order to ensure a uniform understanding among different operators, combined with a fully automated inspection code generator for GD&T purposes.

Keywords: measurement system analysis; coordinate measuring machine; reproducibility; GD&T; quality; metrology; measurement uncertainty

1. Introduction

Geometric dimensioning and tolerancing (GD&T) (or geometrical product specifications (GPS)) is a language of symbols widely used in engineering drawings and computer-generated models to describe, communicate, and determine feature geometry permissible deviations. GD&T is an efficient and unambiguous way of communicating the measurement conditions and specifications of a part. This language accompanies the entire process chain and helps communicate the part intent and function through the design, manufacture, and inspection. As well, it provides a more precise depiction of part features and focuses on the feature-to-feature relationships.

Standards, such as ASME Y14.5-2009 [1] and ISO-GPS [2–8] are comprised of a library of symbols, definitions, rules, and conventions that describe a part in terms of tolerances based on the size, form, orientation, and location. The main and necessary steps needed to derive GD&T results begin with nominal information that describes a specific feature. The manufactured part is inspected using a measurement device (such as a coordinate measuring machine (CMM)) and compared with the nominal definition (e.g., a computer-aided design (CAD) file) in order to verify the dimensional and geometric feature specifications (the actual size and tolerance). The deviations are then computed and displayed. Figure 1 presents the inspection process definition model.

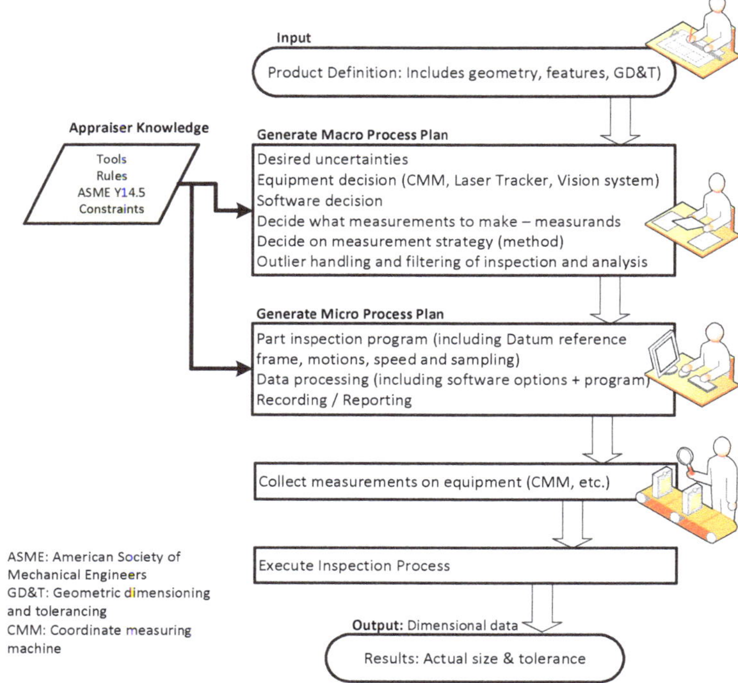

Figure 1. The inspection process definition activity model.

1.1. Measurement Uncertainties—Overview

In metrology science, the true values (ideal quantities) may never be known, and all measurements could potentially have some degree of uncertainty, which is often a function of several variables (sources). The difference between the true and measured values is known as an error. Uncertainty can, as defined by the Guide to the Expression of Uncertainty in Measurement (GUM) [9], be considered as a *'parameter, associated with the result of a measurement that characterizes the dispersion of the values that could reasonably be attributed to the measurand'*. Thus, the estimated value y of the measurand Y is generally calculated using the relationship presented in Equation (1):

$$y = f(x_1, x_2, \ldots, x_n) \tag{1}$$

where x_i is the estimation for each input variable X_i that could potentially have a significant influence on the measurement result (y). The function f can be known and explicit. However, in some cases, the measurement function is unknown or very complex, and no analytic expression is available.

If the function f is explicit and the input quantities are not correlated, the law of propagation of uncertainties given in [9] generally represents the combined standard uncertainty on the estimated value $u(y)$ by:

$$u(y) = \sqrt{\sum_{i=1}^{n} \left(\frac{\partial f}{\partial x_i} \right)^2 u^2(x_i)} \tag{2}$$

where $u(x_i)$ is the standard uncertainty of each input variable x_i In practice, the expanded uncertainty $U(y)$ corresponds to the combined standard uncertainty multiplied by a coverage factor k, where k is chosen, for a prior confidence interval (1−α) to be the $t_{1-\alpha/2,\nu}$ critical value from the t-table, with ν degrees of freedom (Section 6, [9]).

Monte Carlo simulations are typically used to approximate the statistical behavior of the measured value in situations where the measurement function cannot be found directly. To determine the output, the input variables are generated randomly for each simulation within their respective uncertainty ranges. The output probability density function (PDF) is then used for evaluating the uncertainty [10]. Finally, in cases where the measurement function is very complex (or unknown), an empirical estimation can be established using certain assumptions and simplification hypotheses, as proposed in the Measurement System Analysis (MSA) guide from the Automotive Industry Action Group [11].

The MSA consists of a specifically designed experiment aimed at determining the components of variation in the measurement (e.g., the reproducibility, repeatability, bias, etc.). Indeed, the process of obtaining measurements (and defect level estimation) may have variations and produce uncertainty. The analysis tools proposed by the MSA (e.g., the gage repeatability and reproducibility (R&R)) evaluate the uncertainty on a direct measure ($f(x) = x$), such as the thickness measurement from a micrometer. The aim of the whole process is to guarantee the integrity of the data used for quality analysis and to consider the consequences of a measurement error for decisions taken on the product. The reader is referred to [11] for more details.

1.2. Measurement Uncertainties Associated with Dimensional and Geometric Measurement Using CMM

During the last three decades, the coordinate measuring machine (CMM) saw progress in terms of accuracy and repeatability, which, in turn, resulted in productivity improvements. Currently, CMM plays a major role in GD&T standards, such as [1–8], which call for crucial measuring equipment needed for manufacturing quality control [12]. Notwithstanding such improvements, however, uncertainty can be induced not only by the equipment used, but also by the algorithmic choices and the measurement methodology adopted [13–16].

Measurement uncertainty evaluation (quantification) is a crucial step in characterizing and certifying the consistency of the inspection results [17,18]. Measurement uncertainty evaluation must be carried out to ensure advances in measurement science. CMM measurement uncertainty evaluation has become a key focus area for research by many institutions around the world. The Physikalisch-Technische Bundesanstalt (PTB) in Germany, for instance, suggested an expert system scheme for CMM uncertainty evaluation and investigated the impact of the measurement strategy on the overall CMM uncertainty [19].

The National Physical Laboratory (NPL) in the UK standardized the measurement strategies for CMM in order to ensure that the measurement results are reliable [20]. A few authors have employed the design of experiment techniques to estimate the CMM measurement uncertainty. The factorial design of experiments was applied by Feng et al. [21] to study the measurement uncertainty of the position of a hole measured by CMM. They analyzed the effect of variables and their interactions on the uncertainty, while complying with the fundamental rules of the Guide to the Expression of Uncertainty (GUM) [9].

Kritikos et al. [22] designed and implemented a random factorial design of experiments in order to analyze and quantify the influence of different factors (stylus diameter, step width and speed) and their interactions on the CMM measurements' uncertainty of the variable's parallelism, angularity, roundness, diameter, and distance. Other authors, such as Kruth et al. [23] and Sladek et al. [24], proposed methods to determine uncertainties using the Monte Carlo method for feature measurements on CMM. Hongli et al. [25] proposed the Simplified Virtual Coordinate Measuring Machine (SVCMM) method, which makes full use of the CMM acceptance or reinspection report and the Monte Carlo simulation method.

For dimensional metrology with CMM measurements, a task-specific uncertainty estimation was suggested by Haitjema [26], and can be extended to other measurement types as well as linear dimensions, forms (flatness, cylindricity, etc.) and roughnesses. Beaman and Morse [27] performed an experimental evaluation of the software estimation of the task-specific measurement uncertainty for CMMs. Jakubiec et al. [28] addressed this topic and proposed an evaluation of CMM uncertainty, not by studying each axis of the machine, but by proceeding based directly on key specifications expressed in the GD&T standard. Jbira et al. [29] suggested a benchmark including several geometrical and dimensional features for the algorithm efficiency comparison of different Computer-Aided Inspection (CAI) software applications. A comprehensive review of different methods, techniques, and various artifacts for monitoring CMM performance can be found in the research work conducted by [30–33].

In coordinate metrology, Weckenmann et al. [34] mentioned the main contributors to uncertainty, which they subdivided into five groups: measuring devices, environment, workpieces, software, operators, and measurement strategy. A great deal of work has been carried out by the metrology community in terms of investigating the measuring devices, environment, and workpiece components. Although no common understanding of software validation procedures currently exists, the reader is referred to [35], as well as to the European Metrology Research Project (EMRP) under the denomination 'Traceability for computationally-intensive metrology (TraCIM)' [36–38] for research performed on software validation in the field of metrology.

In this paper, we aimed to analyze the measurement uncertainty from an empirical (experimental) perspective. From a review of the literature on the subject, the collective impact of the operator (training, skills, certification, GD&T decoding and interpretation, etc.), the measurement strategy (amount of data, samples, number of measurements, etc.), and the software employed (algorithms used, filtering or removal of outliers, optimization of the stability of the algorithm, layout handling, etc.) had been surprisingly overlooked. We proposed a new GD&T-based benchmark (test artifact) for evaluating (comparing) the performance of measurement systems in different measurement organizations (e.g., industry, schools, and metrology service companies) by considering the uncertainty that can be induced by the operator, the measurement strategy, and the software used.

Under the conditions proposed by the equipment manufacturer, the current hardware is accurate enough to perform the "good" measurements to capture the actual position of a measuring point in the 3D space. In other words, the uncertainty induced by the measuring device is significantly less than that induced by the operator choices, the software options, and the measurement strategies. This means that the performance of a measurement system represents an estimation of the combined variation of the measurement errors (systematic and random errors), which include equipment (hardware) errors, algorithmic errors (software), and operator errors. In this paper, software and operator errors were combined into one, as they can be strongly correlated. According to the MSA approach, this was strictly a reproducibility study [11].

The basic concepts of metrology and related terms that conform to the International Vocabulary of Metrology (VIM) [39] were employed in the present work. According to VIM, reproducibility is the *'closeness of the agreement between the results of measurements of the same quantity, where the individual measurements are made: by different methods, with different measuring instruments, by different observers, in different laboratories, after intervals of time quite long compared with the duration of the single measurement, under different normal conditions of use of the instruments employed'* [39]. According to the Automotive Industry Action Group (AIAG) Measurement System Analysis (MSA) reference manual [11], reproducibility is traditionally referred to as the 'between appraisers' variability. Typically, the term is defined as the average of measurements made by different appraisers using the same measuring instrument when measuring the identical characteristic of the same part.

For the remainder of this paper, MSA terminology will be used [11]. EV stands for Equipment Variation, which is the variation due to repeatability, and AV stands for Appraiser Variation, which is the variation due to reproducibility.

To allow validation of this hypothesis, a GD&T-based artifact was designed using common geometric features (plane, cylinder, etc.) and free-form surfaces. A total of five parts were created, one without any intentional defect (part #1), and four others with a predefined number of intentional dimensional and geometrical defects (parts #2 to #5). The artifacts were intended for use in assessing the performance of many measurement institutes (interlaboratory comparison) in accordance with the dimensional and geometrical tolerance criteria.

The remainder of this paper is structured as follows: In Section 2, we outline the proposed test artifact model, followed by an experimental procedure. A comprehensive metrological and statistical analysis, followed by a general discussion of the results, is presented in Sections 3 and 4. Finally, a summary is provided and future works are described.

2. Materials and Methods

A new GD&T-based test artifact is presented in this section. The model is designed for interlaboratory comparisons of CMMs. Figure 2 provides a visual representation and a description of the proposed artifact, as well as its sub-elements. To ensure the measurement of different shapes and geometrical tolerances, the artifact included basic geometric features (primitives), such as rectangular, planar, cylindrical, and conical surfaces; bore and hole patterns; and free-form surfaces.

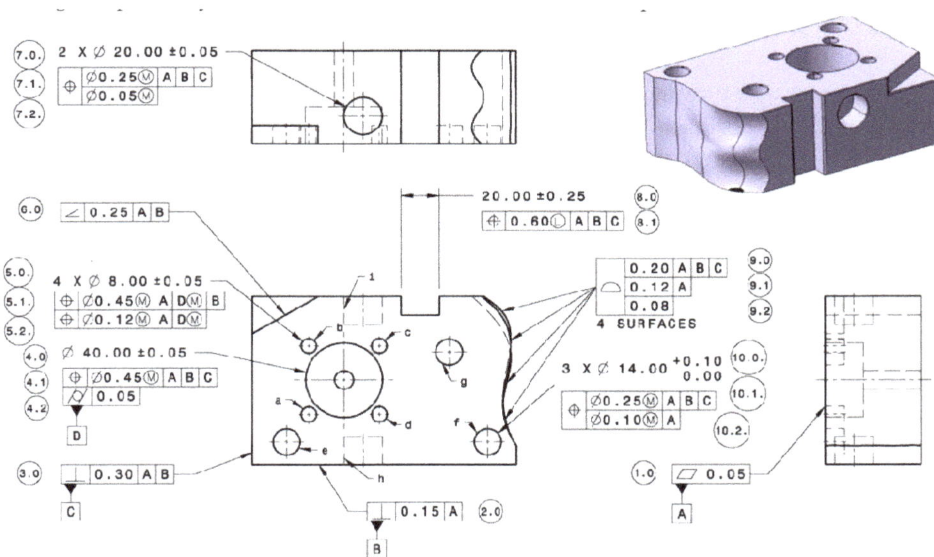

Figure 2. Description of the proposed geometric dimensioning and tolerancing (GD&T)-based artifact.

As shown in Figure 2 and Table 1, a total of ten different features (items) were selected in the artifact, and five main categories related to GD&T were proposed to be characterized and controlled based on ASME Y14.5 (2009) [1]:

- The size tolerances on slab features and cylinder bore/hole diameters.
- The form tolerances, which control the shape of surfaces, such as the flatness of datum plane A, and the cylindricity of cylinder bores.
- The orientation tolerances, which control the tilt of the surfaces and axes for size and non-size features, such as the perpendicularity of datum plane B related to datum plane A.

- The location tolerances presented by the position-locating zones of the bores/holes and the position-relating zone tolerances of the hole patterns. This category locates the center points, axes, and median planes for size features. This category also controls the orientations.
- The profile tolerances, which locate and control the size, form, and orientation of surfaces based on datum references. This is presented by the composite profile-locating, profile-orienting, and profile-form zones.

Table 1. Predefined computer-aided design (CAD) geometrical defects (all dimensions are in mm).

			Defect's CAD Parts				
Item	**Sub Item**	**Description**	**#1**	**#2**	**#3**	**#4**	**#5**
1	1.0	⏥ 0.05	0	0	0	0	0
2	2.0	⊥ 0.15 A	0	0.12	0	0.13	0
3	3.0	⊥ 0.3 A B	0	0.20	0	0.25	0
4	4.0	Ø40 ± 0.05	Ø40	40.00	40.05	40.00	40.14
	4.1	⌖ Ø0.45 Ⓜ A B C	0	0.250	0.11	0.25	0.19
	4.2	⌭ 0.05	0	0	0.17	0	0.18
5	5.0.a	Ø8 ± 0.05	Ø8	8.00	8.00	8.00	8.00
	5.0.b		Ø8	8.00	8.00	8.00	8.00
	5.0.c		Ø8	8.00	8.00	8.00	8.00
	5.0.d		Ø8	8.00	8.00	8.00	8.00
	5.1.a	⌖ Ø0.45 Ⓜ A D Ⓜ B	0	0.45	0	0.10	0
	5.1.b		0	0.45	0	0.09	0
	5.1.c		0	0.45	0	0.16	0
	5.1.d		0	0.45	0	0.11	0
	5.2.a	⌖ Ø0.12 Ⓜ A D Ⓜ	0	0	0	0.10	0
	5.2.b		0	0	0	0.09	0
	5.2.c		0	0	0	0.16	0
	5.2.d		0	0	0	0.11	0
6	6.0	∠ 0.25 A B	0	0	0.200	0.21	0
7	7.0.a	Ø20 ± 0.05	Ø20	20.00	20.00	20.00	20.00
	7.0.b		Ø20	20.00	20.00	20.00	20.00
	7.1.a	⌖ Ø0.25 Ⓜ A B C	0	0.03	0	0.25	0.10
	7.1.b		0	0.03	0.180	0.25	0.10
	7.2.a	⌖ Ø0.05 Ⓜ	0	0	0	0	0
	7.2.b		0	0	0	0	0
8	8.0	20 ± 0.25	20	20.00	20.260	20.00	20.00
	8.1	⌖ 0.6 Ⓛ A B C	0	0.03	0.430	0.250	0.60
9	9.0	⌓ 0.20 A B C	0	1.95	2.15	1.79	2.04
	9.1	⌓ 0.12 A	0	1.09	0.90	1.09	2.04
	9.2	⌓ 0.08	0	0	0	0	0
10	10.0.a	Ø14–14.10	Ø14	14.00	14.00	14.00	14.00
	10.0.b		Ø14	14.00	14.00	14.00	14.00
	10.0.c		Ø14	14.00	14.00	14.00	14.00
	10.1.a	⌖ Ø0.25 Ⓜ A B C	0	0.500	0	0.250	0
	10.1.b		0	0.250	0	0.250	0
	10.1.c		0	0.250	0	0.250	0
	10.2.a	⌖ Ø0.1 Ⓜ A	0	0	0	0	0
	10.2.b		0	0	0	0	0
	10.2.c		0	0	0	0	0

The overall dimensions of the artifacts were 138 × 90 × 50 mm. They were conveniently transportable and could fit into small CMM metrology systems. The artifacts were made from aluminum: Part #1, with no intentional defects, and parts #2 to #5, with some predefined and intentional geometrical imperfections. The geometrical imperfections were performed in accordance with the procedure described in Table 2.

Table 2. Creation of the defects (all dimensions are in mm).

Type	Description
Size	In the case of item #4.0 (Ø40), the hole was machined by slightly modifying the circular path. In the case of items #5.0 (Ø8), #7.0 (Ø20) and #10.0 (Ø14), a drill bit was used for drilling the holes. For the slab feature (item #8.0), size defects were created in CATIA® V5 by slightly modifying the distance between the corresponding parallel planes.
Form	No flatness and straightness defects were created for plane surfaces (item #1.0) and cylindrical features. In the case of item #4.0 (Ø40), cylindricity defects were created while machining the hole by slightly modifying the circular path.
Orientation	Orientation defects were created in CATIA® V5 by rotating the plane surfaces (items #2.0 and #3.0).
Location	Location defects were created in CATIA® V5 by imposing translations and rotations of the axes of cylindrical features (e.g., items # 4.1, 5.1, 5.2, 7.1, 7.2, 10.1, and 10.2) and slab features (item #8.1).
Profile	Profile defects were created in CATIA® V5 by imposing translations and rotations along the three axes of the surface (items #9.0 and #9.1). For item #9.2, no profile defects were created.

Table 1 presents the predefined defects, which were considered as the reference values (nominal defects). Their respective amplitudes were approximately in the same order of magnitude of tolerance. The final real geometry of the part 'as manufactured' was unknown and the actual values were calculated from the measurement points.

The artifact parts were manufactured on three-axis CNN milling machines at the École de technologie supérieure's Products, Processes, and Systems Engineering Laboratory (P^2SEL). Figure 3 presents one of the five manufactured artifacts.

Figure 3. The proposed GD&T-based artifacts.

A total of 15 fixed and portable CMMs from different and independent industrial and academic collaborators in North America (Canada and the USA) were included in this investigation, and are presented in Figure 4. The CMMs used were named according to ISO 10,360 [40]. The accuracy of the CMMs used (equipment variation) in this study typically ranged between 0.7 and 45 μm (±2 σ level). All the induced defects for artifacts #2 to #5 were significantly higher than the aforementioned accuracies (Table 1).

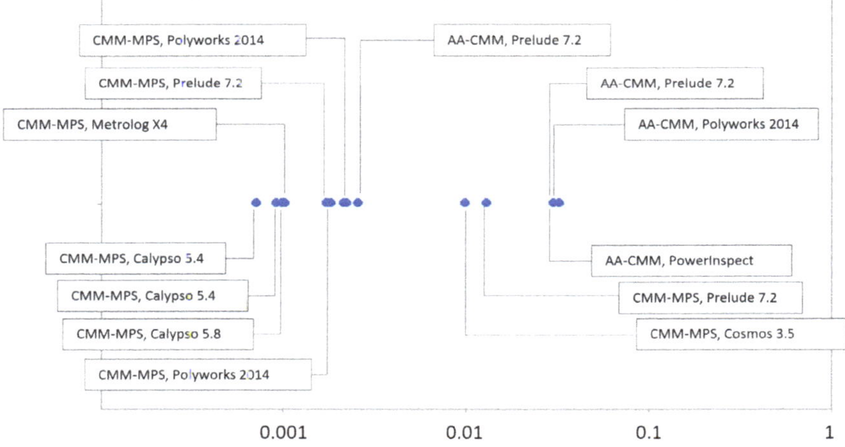

Figure 4. The uncertainty caused by bias and linearity u_E (equipment variation) of the participating fixed and portable coordinate measuring machines (CMMs) (in mm).

The measurements were performed from November 2013 to December 2017. Different institutes and industrial partners (eight industries, three schools, and three companies in the field of dimensional metrology) were asked to measure artifacts #2 to #5 without any particular focus. The aim was to analyze the ordinary measurement performance of each institute. Each artifact received a unique code for each partner (only the coordinator maintained the part-operator-equipment traceability). The circulation of the artifacts was arranged in a circular path, with the evaluation kit forwarded to the next participant and the results sent to the coordinator. Each partner carried out measurements with their own CMM system, which included calibrated equipment, specific software, and an appraiser.

The data were collected through a dynamic pdf form. In this form, each inspection item was mentioned and the operator was asked to: (1) accept or refuse the item and (2) mention the measured value. An online database was connected to this form for fully automatic and secure data collection.

Based on [9,18], the general mathematical model for determining the CMM task-oriented uncertainty is presented in Equation (3):

$$U_c \cong \pm \mathrm{k}\sqrt{u_E^2 + u_{EV}^2 + u_{AV}^2} \tag{3}$$

where u_E is the uncertainty caused by bias and linearity (the equipment variation as provided by the manufacturers); u_{EV} is the uncertainty caused by repeatability as defined in MSA [11]; u_{AV} is the uncertainty caused by reproducibility as defined in MSA [12] (this includes the software used and the measurement strategy); U_c is the expanded combined uncertainty with a coverage factor k (obtained from Student's critical value table); and U_c represents the total error of the inspection process.

Some assumptions were made:

(1) All measurements were done in a controlled environment (metrology laboratory or facilities). Therefore, compared to amplitude defects, uncertainty induced by environmental conditions could be considered negligible in this study.

(2) The uncertainty of the equipment (hardware), including the bias, linearity (u_E), and repeatability (u_{EV}), was much smaller than the amplitude of the induced defects (u_E and u_{EV} << *Defect*). Here as well, the uncertainty of the equipment was much less than that resulting from the reproducibility u_{AV} (variation due to operators–software–measurement strategy error ($u_{EV} << u_{AV}$). Practically, in this study, the equipment uncertainty (u_E) was typically 5 µm for the conventional CMM

(Figure 4). Given the preceding, Equation (2) can be simplified and the measurement variation in this paper can be considered equal to $U_c \approx AV = \pm\, k\, u_{AV}$.

(3) The reproducibility in this study (u_{AV}) represented variations due to algorithmic error, and were mainly due to a programming error (e.g., least squares or minimum zone Chebyshev fit [41]), measurement strategy, or the use of computer programs (how the operator uses all software options, the density and distribution of the measurement points, etc.).

(5) All industrial and academic participants in this study guaranteed a temperature of 20 ± 1 °C in their laboratories. The artifacts, made of aluminum, had overall dimensions of 138 × 90 × 50 mm. The resulting thermal expansion of ±3.4 μm was well below the observed variations.

(4) The confidence representing the 95% interval (error type I = 0.05) was used for the study (we assume $k \approx 1.96$) corresponding to an infinite degree of freedom. Outliers and missing data were not included in the calculations.

3. Results

Table 3 presents the [minimum, median, maximum] geometric and dimensional deviations for items #1 to #10.2, respectively. As illustrated in Figure 5, the results of the investigation are presented on individual value plots with error (interval) bars. For geometrical tolerances with zero target values, the measurements for each part (#2–5) are shown directly. For dimensional tolerances (in this case, the target is the nominal value of CAD), the deviations between the digitized parts (measurement) and the nominal part (CAD) are presented. Figure 6 presents the plots for size tolerances, while Figure 7 presents the plots for form tolerances (items #1 and #4.2) and orientation tolerances (items #3 and #6), Figure 8 presents the plots for location tolerances (items #5.1, #5.2, #7.1, #7.2, #8.1, and #10.2), and Figure 9 presents the plots for profile tolerances (items #9 and #9.1).

Table 3. Results (all dimensions are in mm).

		Results [Minimum, Median, Maximum]			
Sub Item	**Description**	**#2**	**#3**	**#4**	**#5**
1.0	⏥ 0.05	[0.002, 0.014, 0.031]	[0.002, 0.010, 0.029]	[0.001, 0.015, 0.025]	[0.004, 0.0163, 0.025]
2.0	⊥ 0.15 A	[0.015, 0.025, 0.146]	[0.003, 0.030, 0.110]	[0.002, 0.071, 0.158]	[0.011, 0.0386, 0.140]
3.0	⊥ 0.3 A B	[0.016, 0.034, 0.148]	[0.005, 0.028, 0.170]	[0.008, 0.025, 0.279]	[0.009, 0.0381, 0.303]
4.0	Ø40 ± 0.05	[39.84, 39.90, 39.94]	[39.84, 39.91, 40.04]	[39.84, 39.92, 39.96]	[39.84, 39.95, 40.10]
4.1	⌖ Ø0.45 Ⓜ A B C	[0.053, 0.198, 0.360]	[0.063, 0.154, 0.340]	[0.028, 0.200, 0.427]	[0.097, 0.195, 0.454]
4.2	⌭ 0.05	[0.020, 0.040, 0.213]	[0.012, 0.102, 0.163]	[0.017, 0.049, 0.170]	[0.001, 0.1162, 0.216]
5.0.a	Ø8 ± 0.05	[7.845, 7.863, 7.888]	[7.840, 7.910, 7.955]	[7.781, 7.831, 7.941]	[7.789, 7.842, 7.885]
5.0.b		[7.840, 7.853, 7.889]	[7.836, 7.908, 7.960]	[7.786, 7.831, 7.944]	[7.788, 7.840, 7.884]
5.0.c		[7.841, 7.860, 7.894]	[7.840, 7.910, 7.960]	[7.781, 7.829, 7.941]	[7.785, 7.837, 7.886]
5.0.d		[7.833, 7.856, 7.889]	[7.722, 7.888, 7.959]	[7.785, 7.830, 7.947]	[7.785, 7.840, 7.884]
5.1.a	⌖ Ø0.45 Ⓜ A D Ⓜ B	[0.013, 0.442, 0.600]	[0.012, 0.149, 0.451]	[0.018, 0.089, 0.532]	[0.006, 0.186, 0.300]
5.1.b		[0.013, 0.461, 0.699]	[0.008, 0.122, 0.470]	[0.018, 0.077, 0.587]	[0.015, 0.097, 0.380]
5.1.c		[0.020, 0.103, 0.476]	[0.016, 0.150, 0.460]	[0.032, 0.152, 0.767]	[0.016, 0.095, 0.560]
5.1.d		[0.008, 0.266, 0.633]	[0.013, 0.151, 0.464]	[0.031, 0.112, 0.592]	[0.017, 0.106, 0.399]
5.2.a	⌖ Ø0.12 Ⓜ A D Ⓜ	[0.006, 0.020, 0.456]	[0.009, 0.120, 0.451]	[0.012, 0.086, 0.527]	[0.014, 0.116, 0.290]
5.2.b		[0.002, 0.014, 0.452]	[0.006, 0.120, 0.470]	[0.018, 0.087, 0.462]	[0.011, 0.127, 0.316]
5.2.c		[0.011, 0.032, 0.452]	[0.010, 0.120, 0.460]	[0.030, 0.149, 0.764]	[0.004, 0.053, 0.513]
5.2.d		[0.005, 0.032, 0.487]	[0.013, 0.120, 0.464]	[0.026, 0.110, 0.391]	[0.0058, 0.060, 0.399]
6.0	∠ 0.25 A B	[0.001, 0.014, 0.024]	[0.0001, 0.163, 0.215]	[0.022, 0.169, 0.218]	[0.001, 0.0151, 0.180]
7.0.a	Ø20 ± 0.05	[19.89, 19.94, 20.04]	[19.90, 19.94, 20.01]	[19.93, 19.97, 19.99]	[19.97, 19.99, 20.01]
7.0.b		[19.92, 19.96, 20.03]	[19.93, 19.94, 19.99]	[19.94, 19.98, 20.03]	[19.97, 19.99, 20.02]
7.1.a	⌖ Ø0.25 Ⓜ A B C	[0.106, 0.279, 2.777]	[0.072, 0.216, 2.633]	[0.082, 0.238, 2.991]	[0.095, 0.122, 2.667]
7.1.b		[0.096, 0.185, 0.499]	[0.033, 0.225, 0.444]	[0.084, 0.248, 0.734]	[0.038, 0.246, 0.404]
7.2.a	⌖ Ø0.05 Ⓜ	[0.042, 0.142, 0.405]	[0.026, 0.072, 0.346]	[0.021, 0.187, 0.368]	[0.017, 0.087, 0.514]
7.2.b		[0.042, 0.142, 0.405]	[0.020, 0.046, 0.246]	[0.021, 0.144, 0.367]	[0.019, 0.072, 0.404]

Table 3. *Cont.*

Sub Item	Description	Results [Minimum, Median, Maximum] #2	#3	#4	#5
8.0	20 ± 0.25	[19.92, 19.98, 20.01]	[19.93, 19.95, 20.23]	[19.82, 19.97, 20.02]	[19.71, 19.97, 20.34]
8.1	⌖ 0.6 Ⓛ A B C	[0.175, 0.391, 0.735]	[0.16, 0.4657, 0.608]	[0.029, 0.392, 0.539]	[0.025, 0.528, 0.806]
9.0	⌓ 0.20 A B C	[0.333, 1.966, 3.430]	[0.196, 1.464, 3.558]	[0.217, 1.279, 2.369]	[0.36, 2.4498, 3.454]
9.1	⌓ 0.12 A	[0.218, 0.348, 3.362]	[0.201, 0.299, 1.007]	[0.175, 0.261, 0.369]	[0.202, 1.956, 2.369]
9.2	⌓ 0.08	[0.089, 0.481, 3.284]	[0.074, 0.196, 1.521]	[0.065, 0.099, 0.426]	[0.062, 2.341, 2.797]
10.0.a		[13.86, 13.98, 14.03]	[13.82, 13.99, 14.11]	[13.86, 13.95, 14.01]	[13.87, 13.97, 14.12]
10.0.b	Ø14–14.10	[13.87, 13.88, 13.99]	[13.86, 13.99, 14.01]	[13.95, 13.97, 14.01]	[13.95, 13.97, 14.00]
10.0.c		[13.83, 13.88, 13.99]	[13.82, 13.99, 14.01]	[13.94, 13.97, 14.01]	[13.94, 13.98, 14.00]
10.1.a		[0.113, 0.297, 0.708]	[0.034, 0.077, 0.860]	[0.049, 0.171, 0.745]	[0.016, 0.112, 0.508]
10.1.b	⌖ Ø0.25 Ⓜ A B C	[0.042, 0.221, 0.400]	[0.037, 0.058, 0.388]	[0.066, 0.146, 0.434]	[0.012, 0.072, 0.393]
10.1.c		[0.040, 0.195, 0.588]	[0.049, 0.067, 0.388]	[0.062, 0.157, 0.444]	[0.035, 0.082, 0.408]
10.2.a		[0.020, 0.247, 0.583]	[0.002, 0.089, 0.560]	[0.019, 0.171, 0.723]	[0.005, 0.155, 0.418]
10.2.b	⌖ Ø0.1 Ⓜ A	[0.001, 0.027, 0.351]	[0.004, 0.021, 0.332]	[0.003, 0.022, 0.360]	[0.001, 0.038, 0.393]
10.2.c		[0.004, 0.024, 0.684]	[0.002, 0.020, 0.310]	[0.001, 0.020, 0.667]	[0.001, 0.041, 0.408]

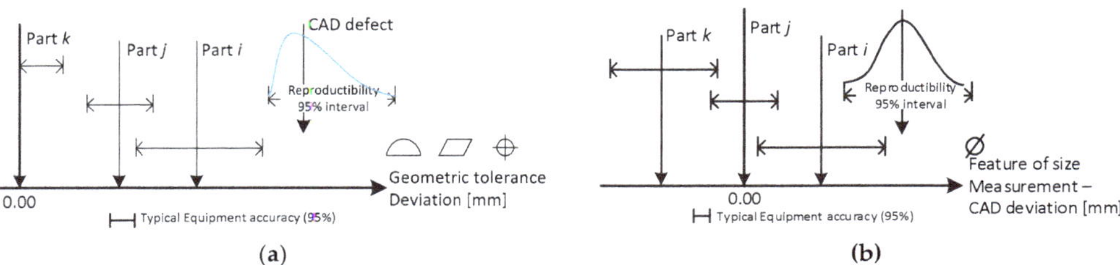

Figure 5. General representation of the results: (**a**) geometric and (**b**) dimensional tolerances.

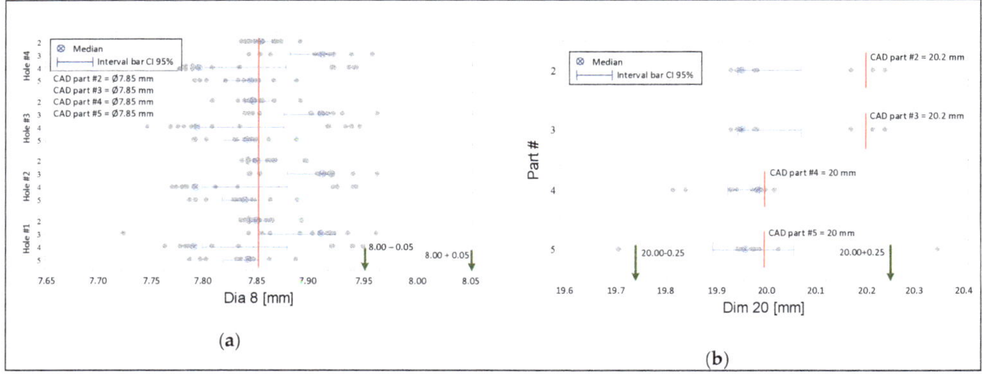

Figure 6. Results of the size tolerance items (**a**) #5 and (**b**) #8.

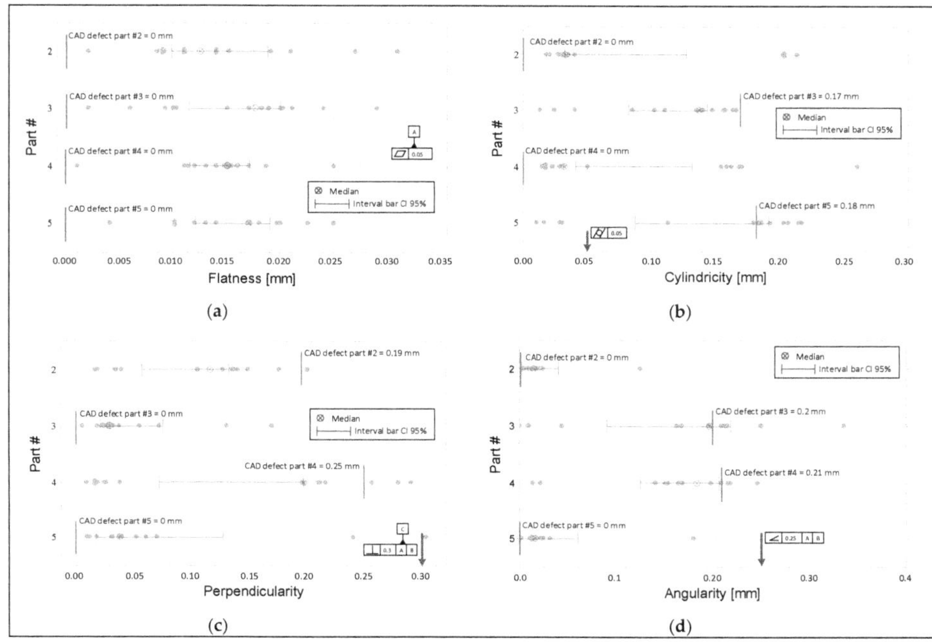

Figure 7. Results of the form tolerance items (**a**) #1 and (**b**) #4.2 and angularity tolerance items (**c**) #3 and (**d**) #6.

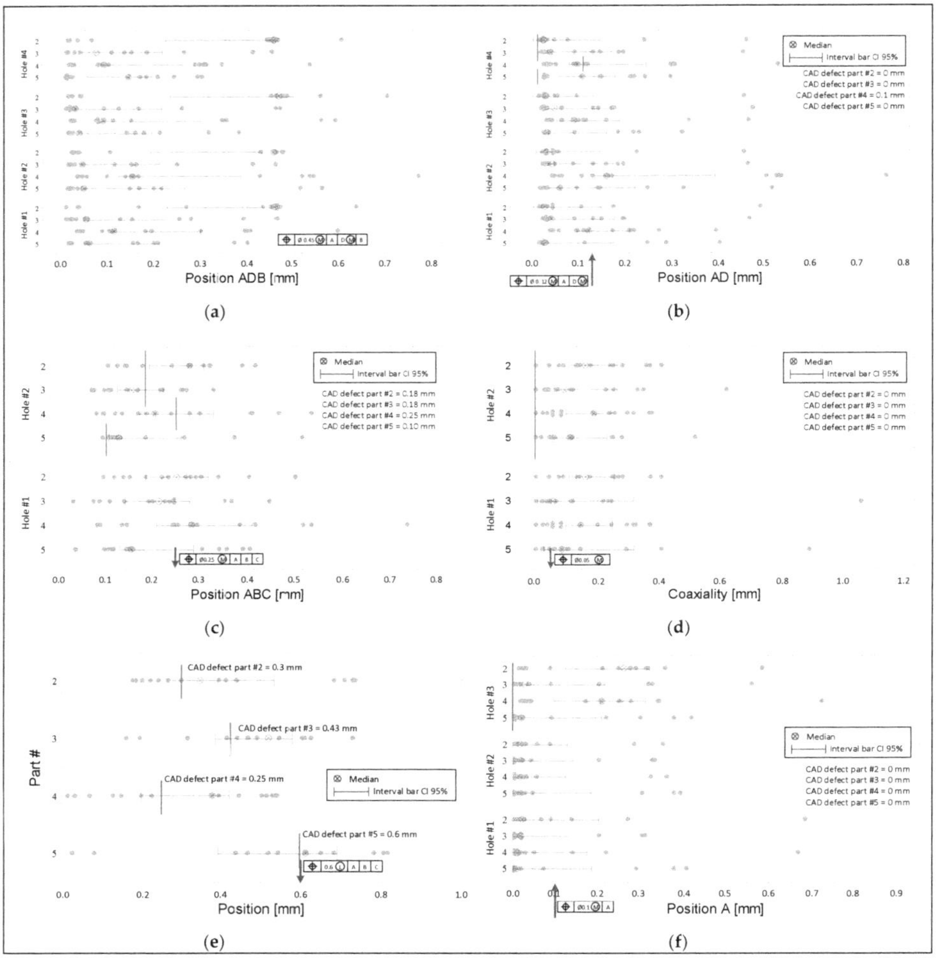

Figure 8. Results of the location tolerance items (**a**) #5.1, (**b**) #5.2, (**c**) #7.1, (**d**) #7.2, (**e**) #8.1, and (**f**) #10.2.

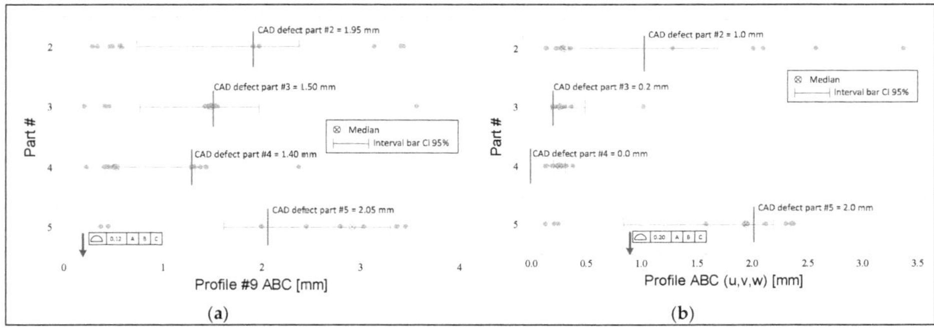

Figure 9. Results of the profile tolerance items (**a**) #9 and (**b**) #9.1.

4. Discussion

This investigation revealed the presence of varying degrees of uncertainty in measurement reproducibility while operating CMMs in different laboratories and institutions. Differing amounts of appraiser variation (*AV*) were present when identical parts were measured by different operators on different (but similar) CMMs of approximately similar designs. Based on the results of the different analyses:

- A lower level of measurement uncertainty was observed on non-defective parts. This observation seems trivial, but deserves to be underlined. Indeed, in the absence of form error, the algorithmic error factor and the number of measurement points had no impact on the 'measurand', except for the perpendicularity tolerance (Table 3 and Figure 7c, parts #3 and #5) and the angularity (Table 3 and Figure 7d, parts #2 and #5).
- For simple requirements (e.g., the flatness (Figure 7) and diameter tolerance (Table 3 and Figure 6)), the range of variation was relatively small and very close to the inherent variation of the equipment.
- On the other hand, a greater presence of measurement variation was observed for more sophisticated and complex GD&Ts (e.g., in the composite profile (Table 3 and Figure 9) and localization tolerances (Figures 8 and 10a)). The combination of different factors, such as the logistics and measurement strategy, the operator type, the set-up type, the size of the point clouds, the choice of the inspection algorithm, etc., appeared to be the source of this overall high measurement uncertainty.
- For the composite profile tolerance (Figure 9), many partners gave the same result for different features. In the specific instance of profile tolerance with all degrees of freedom (Table 3 and Figure 10b), the range of variation was significantly larger than in other cases. In addition to the inadequate choices that the operators can make during inspection operations, the registration (*bestfit*) induced an additional source of uncertainty.
- Although not generalized, several coaxial tolerance results (Figure 8d) (with coaxial tolerance being a specific case of position tolerance) clearly indicated an interpretation error (or manipulation) because the values were larger than in the pattern-locating tolerance zone framework (PLTZF) located in the ABC datum reference frame (Table 3 and Figure 8c).

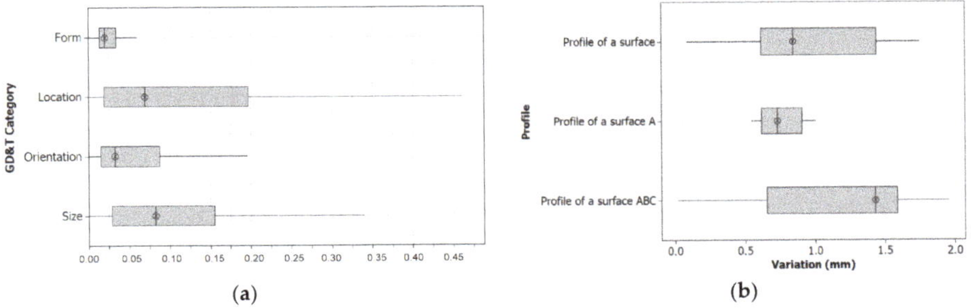

Figure 10. This boxplot of variation = |measured-nominal| amplitude for different GD&T categories; (**a**) form, location, orientation, and size tolerances; (**b**) profile tolerances.

Overall, items without induced defects presented low variability (measurement uncertainty), while those with complex GD&T (e.g., composite features), as well as those recently added to the standard, presented high variability. The combination of different factors, such as the logistics and measurement strategy, the operator type, the set-up type, the size of the point clouds, the choice of the inspection algorithm, etc., appeared to be the source of this overall high measurement uncertainty.

These experimental findings may be applied to technical industrial practice to ensure the quality of the measurement results. They may also serve as an inspiration for proposing solutions to reduce the measurement uncertainty. These solutions may include GD&T training and certification to recognize proficiency in the application and understanding of the GD&T principles expressed in the standards. This would ensure a uniform understanding of the drawings prepared using the GD&T language by different operators, as well as a uniform selection and application of geometric controls to drawings. Another such solution could be in the form of innovative combinations of applied methods, such as a fully automated inspection code generator for GD&T purposes.

Author Contributions: Conceptualization, S.A.T.; methodology, A.A., S.A.T. and M.K.N.; formal analysis, A.A., S.A.T. and M.K.N.; investigation, S.A.T. and M.K.N.; data curation, A.A.; writing—original draft preparation, A.A.; writing—review and editing, A.A., S.A.T. and M.K.N.; visualization, A.A. and S.A.T.; supervision, S.A.T.; project administration, S.A.T.; funding acquisition, S.A.T. All authors have read and agreed to the published version of the manuscript.

Funding: This work was supported by the Natural Sciences and Engineering Research Council of Canada (NSERC), grant number RGPIN-2015-05995.

Acknowledgments: The authors would like to thank École de technologie supérieure (Montreal, QC, Canada), the Natural Sciences and Engineering Research Council of Canada (NSERC), as well as all industrial and academic participants in North America for their support and contributions.

Conflicts of Interest: The authors declare no conflict of interest.

References

1. ASME. *ASME Y14.5-2009 Dimensioning and Tolerancing: Engineering Drawing and Related Documentation Practices*; American Society of Mechanical Engineers: New York, NY, USA, 2009.
2. ISO. *ISO 1101:2012 Geometrical Product Specification (GPS)—Geometrical Tolerancing—Tolerances of Form, Orientation, Location, and Runout*; International Organization for Standardization: Geneva, Switzerland, 2012.
3. ISO. *ISO 12781:2011 Geometrical Product Specification (GPS)—Flatness*; International Organization for Standardization: Geneva, Switzerland, 2011.
4. ISO. *ISO 12181:20122 Geometrical Product Specification (GPS)—Roundness*; International Organization for Standardization: Geneva, Switzerland, 2011.
5. ISO. *ISO 5458:1998 Geometrical Product Specification (GPS)—Geometrical Tolerancing—Positional Tolerancing*; International Organization for Standardization: Geneva, Switzerland, 1998.
6. ISO. *ISO 14405-1:2010 Geometrical Product Specification (GPS)—Dimensional Tolerancing—Part 1: Linear Sizes*; International Organization for Standardization: Geneva, Switzerland, 2010.
7. ISO. *ISO 14405-2:2011 Geometrical Product Specification (GPS)—Dimensional Tolerancing—Part 1: Dimensions Other than Linear Sizes*; International Organization for Standardization: Geneva, Switzerland, 2011.
8. ISO. *ISO 14253-1:1998 Geometrical Product Specifications (GPS)—Inspection by Measurement of Workpieces and Measuring Equipment—Part 1: Decision Rules for Proving Conformance or Non-Conformance with Specifications*; International Organization for Standardization: Geneva, Switzerland, 1998.
9. JCGM, BIPM. *JCGM 100:2008 Guide to the Expression of Uncertainty in Measurement, GUM 1995 with Minor Corrections*; The Joint Committee for Guides in Metrology (JCGM) and the Bureau International des Poids et Mesures (BIPM): Paris, France, 2008.
10. Schwenke, H.; Siebert, B.R.L.; Wäldele, F.; Kunzmann, H. Assessment of Uncertainties in Dimensional Metrology by Monte Carlo Simulation: Proposal of a Modular and Visual Software. *CIRP Ann. Manuf. Technol.* **2000**, *49*, 395–398. [CrossRef]
11. Automotive Industry Action Group. *Measurement Systems Analysis Reference Manual*, 4th ed.; Daimler Chrysler Corporation, Ford Motor Company, General Motors Corporation: Southfield, MI, USA, 2010.
12. Thalmann, R.; Meli, F.; Küng, A. State of the art of tactile micro coordinate metrology. *Appl. Sci.* **2016**, *6*, 150. [CrossRef]
13. Savio, E. Uncertainty in testing the metrological performances of coordinate measuring machines. *CIRP Ann. Manuf. Technol.* **2006**, *55*, 535–538. [CrossRef]
14. Trapet, E.; Savio, E.; De Chiffre, L. New advances in traceability of CMMs for almost the entire range of industrial dimensional metrology needs. *CIRP Ann. Manuf. Technol.* **2004**, *53*, 433–438. [CrossRef]

15. Vrba, I.; Palencar, R.; Hadzistevic, M.; Strbac, B.; Spasic-Jokic, V.; Hodolic, J. Different Approaches in Uncertainty Evaluation for Measurement of Complex Surfaces Using Coordinate Measuring Machine. *Meas. Sci. Rev.* **2015**, *15*, 111–118. [CrossRef]
16. Gąska, P.; Gąska, A.; Sładek, J.; Jędrzejewski, J. Simulation model for uncertainty estimation of measurements performed on five-axis measuring systems. *Int. J. Adv. Manuf. Technol.* **2019**, *104*, 4685–4696. [CrossRef]
17. Bich, W. Revision of the 'Guide to the Expression of Uncertainty in Measurement'. Why and how. *Metrologia* **2014**, *51*, 155–158. [CrossRef]
18. Yinbao, C.; Zhongyu, W.; Xiaohuai, C.; Yaru, L.; Hongyang, L.; Hongli, L.; Hanbin, W. Evaluation and Optimization of Task-oriented Measurement Uncertainty for Coordinate Measuring Machines Based on Geometrical Product Specifications. *Appl. Sci.* **2019**, *9*, 6.
19. Weckenmann, A.; Knauer, M.; Kunzmann, H. The Influence of Measurement Strategy on the Uncertainty of CMM-Measurements. *CIRP Ann. Manuf. Technol.* **1998**, *47*, 451–454. [CrossRef]
20. AMT. *BS 7172:1989 Guide to Assessment of Position, Size and Departure from Nominal Form of Geometric Features*; Advanced Manufacturing Technology Standards Policy Committee of Britain (AMT): London, UK, 2010.
21. Feng, C.X.J.; Saal, A.L.; Salsbury, J.G.; Ness, A.R.; Lin, G.C.S. Design and analysis of experiments in CMM measurement uncertainty study. *Precis. Eng.* **2007**, *31*, 94–101. [CrossRef]
22. Kritikos, M.; Concepción Maure, L.; Leyva Céspedes, A.A.; Delgado Sobrino, D.R.; Hrušecký, R. A Random Factorial Design of Experiments Study on the Influence of Key Factors and Their Interactions on the Measurement Uncertainty: A Case Study Using the ZEISS CenterMax. *Appl. Sci.* **2020**, *10*, 37. [CrossRef]
23. Kruth, J.P.; Gestel, N.V.; Bleys, P.; Welkenhuyzen, F. Uncertainty determination for CMMs by Monte Carlo simulation integrating feature form deviations. *CIRP Ann. Manuf. Technol.* **2009**, *58*, 463–466. [CrossRef]
24. Sładek, J.; Gąska, A. Evaluation of coordinate measurement uncertainty with use of virtual machine model based on monte carlo method. *Measurement* **2012**, *45*, 1564–1575. [CrossRef]
25. Li, H.L.; Chen, X.H.; Cheng, Y.B.; Liu, H.D.; Wang, H.B.; Cheng, Z.Y.; Wang, H.T. Uncertainty modeling and evaluation of CMM task oriented measurement based on SVCMM. *Meas. Sci. Rev.* **2017**, *17*, 226–231. [CrossRef]
26. Haitjema, H. Task specific uncertainty estimation in dimensional metrology. *Int. J. Precis. Technol.* **2011**, *2*, 226–245. [CrossRef]
27. Beaman, J.; Morse, E. Experimental evaluation of software estimates of task specific measurement uncertainty for CMMs. *Precis. Eng.* **2010**, *34*, 28–33. [CrossRef]
28. Jakubiec, W.; Płowucha, W. First Coordinate Measurements Uncertainty Evaluation Software Fully Consistent with the GPS Philosophy. *Procedia CIRP* **2013**, *10*, 317–322. [CrossRef]
29. Jbira, I.; Tahan, A.; Bonsaint, S.; Mahjoub, M.A. Reproducibility Experimentation among Computer-Aided Inspection Software from a Single Point Cloud. *J. Control Sci. Eng.* **2019**, *2019*, 9140702. [CrossRef]
30. ISO. *ISO 10360-2:2009 Geometrical Product Specifications (GPS)—Acceptance and Reverification Tests for Coordinate Measuring Machines (CMM)—Part 2: CMMs Used for Measuring Linear Dimensions*; International Organization for Standardization: Geneva, Switzerland, 2009.
31. ISO. *ISO 17450-2:2012 Geometrical Product Specifications (GPS)—General Concepts—Part 2: Basic Tenets, Specifications, Operators, Uncertainties and Ambiguities*; International Organization for Standardization: Geneva, Switzerland, 2012.
32. ISO. *ISO/TS 15530-4:2008Geometrical Product Specifications (GPS). Coordinate Measuring Machines (CMM): Technique for Determining the Uncertainty of Measurement. Part 4: Evaluating Task-specific Measurement Uncertainty Using Simulation*; International Organization for Standardization: Geneva, Switzerland, 2008.
33. Hammad Mian, S.; Al-Ahmari, A. New developments in coordinate measuring machines for manufacturing industries. *Int. J. Metrol. Qual. Eng.* **2014**, *5*, 101. [CrossRef]
34. Weckenmann, A.; Knauer, M.; Killmaier, T. Uncertainty of coordinate measurements on sheet-metal parts in the automotive industry. *J. Mater. Process. Technol.* **2001**, *115*, 9–13. [CrossRef]
35. Greif, N.; Schrepf, H.; Richter, D. Software validation in metrology: A case study for a GUM-supporting software. *Measurement* **2006**, *39*, 849–855. [CrossRef]
36. EURAMET. Report: Traceability for Computationally-Intensive Metrology. Available online: https://www.euramet.org/research-innovation/search-research-projects/details/project/traceability-for-computationally-intensive-metrology/ (accessed on 6 June 2020).

37. Forbes, A.B.; Smith, I.M.; Härtig, F.; Wendt, K. Overview of EMRP Joint Research Project NEW06 "Traceability for computationally-intensive metrology". In *Advanced Mathematical and Computational Tools in Metrology and Testing X*; World Scientific Publishing Co Pte Ltd.: Singapore, 2015; pp. 164–170.
38. Müller, B. Repeatable and Tracable Software Verification for 3D Coordinate Measuring Machines. In Proceedings of the 18th World Multi-Conference on Systemics, Cybernetics and Informatics, Orlando, FL, USA, 15–18 June 2014.
39. JCGM, BIPM. *JCGM 200:2012 International Vocabulary of Metrology: Basic and General Concepts and Associated Terms (VIM)*; The Joint Committee for Guides in Metrology and The Bureau International des Poids et Mesures: Paris, France, 2012.
40. ISO. *ISO 10360:2016 Geometrical Product Specification (GPS)—Geometrical Tolerancing—Acceptance and Reverification Tests for Coordinate Measuring Systems (CMS)*; International Organization for Standardization: Geneva, Switzerland, 2016.
41. Vemulapalli, P.; Shah, J.J.; Davidson, J.K. Reconciling the differences between tolerance specification and measurement methods. In Proceedings of the ASME 2013 International Manufacturing Science and Engineering Conference, MSEC2013, Madison, WI, USA, 10–14 June 2013.

Article

Measurement Uncertainty Analysis of a Stitching Linear-Scan Method for the Evaluation of Roundness of Small Cylinders

Qiaolin Li, Yuki Shimizu *, Toshiki Saito, Hiraku Matsukuma and Wei Gao

Precision Nanometrology Laboratory, Department of Finemechanics, Tohoku University, Sendai 980-8579, Japan; liqiaolin@nano.mech.tohoku.ac.jp (Q.L.); saito@nano.mech.tohoku.ac.jp (T.S.); hiraku.matsukuma@nano.mech.tohoku.ac.jp (H.M.); gaowei@cc.mech.tohoku.ac.jp (W.G.)

* Correspondence: yuki.shimizu@nano.mech.tohoku.ac.jp; Tel.: +81-22-795-6950

Received: 22 June 2020; Accepted: 6 July 2020; Published: 10 July 2020

Abstract: Influences of angular misalignments of a small cylinder on its roundness measurement by the method referred to as the stitching linear scan method are theoretically investigated. To compensate for the influences, a technique for measuring angular misalignments of a small cylinder by utilizing the linear-scan surface form stylus profilometer, which is employed for roundness measurement, is newly proposed. In addition, for roundness measurement, a holder unit capable of compensating for the angular misalignments of a small cylinder is developed, and the feasibility of the proposed technique is verified in experiments. Furthermore, a measurement uncertainty analysis of the stitching linear-scan method is carried out through numerical calculations based on a Monte Carlo method.

Keywords: precision metrology; form measurement; stitching linear-scan method; roundness measurement; measurement uncertainty; Monte Carlo method

---

1. Introduction

Roundness measurement is an important operation to assure the quality of cylinders [1]. Various sizes of small cylinders are employed in mechanical components; for example, a needle roller in a mechanical bearing [2]. Since the roundness of a needle roller affects the performance of the mechanical bearing in which the needle roller is integrated [3], it is necessary to evaluate the roundness of such a small cylinder with high precision [4].

In machine shops, a method with a dial gauge and a V-block is often employed in roundness measurement of a cylinder [5–7]. Meanwhile, the accuracy of roundness measurement in this method strongly depends on the skill of an operator, and it becomes much more difficult to carry out measurement with the decrease of the size of a workpiece. On the other hand, a roundness measuring instrument based on the rotary-scan method, in which a precision spindle and a displacement gauge are used, can carry out precise roundness measurement of a cylinder [8]. Compared with the V-block method described above, a roundness measuring instrument can reduce a deviation in roundness measurement induced by the operator. Meanwhile, even with a roundness measuring instrument, it becomes much more difficult to carry out measurement with the decrease of the size of a workpiece. Especially, the centering alignment of a small cylinder is a challenging task, and the influence of the eccentricity of a workpiece becomes much more significant with the decrease of the length and the diameter of a workpiece. Therefore, it has been difficult to carry out precision roundness measurement of a small cylinder having a diameter of less than a few millimeters [9].

In responding to the background described above, an alternative method referred to as the stitching linear-scan method has been proposed [10]. In the method, a series of arc-profiles around the circumference of a small cylinder is obtained by using a linear-scan surface form stylus profiler [11], which is often employed for measurement of a surface form/roughness [12], and the roundness profile is reconstructed by stitching the obtained arc-profiles. It has been revealed that the proposed method can carry out roundness measurement of a small cylinder having a diameter and a length of 1.5 mm and 5 mm, respectively, which has been a challenging task using the conventional methods. Also, a good agreement can be found between the obtained roundness of a small cylinder with a diameter of 3 mm by the stitching linear-scan method and a conventional roundness measuring instrument. These results have demonstrated the feasibility of the proposed stitching linear-scan method. Meanwhile, in the proposed method, the angular misalignments of a small cylinder could affect the measurement uncertainty. Furthermore, an intensive analysis of the measurement uncertainty of the stitching linear-scan method has remained to be addressed.

To address the aforementioned issues, a new technique for the compensation of angular misalignments of a small cylinder in the stitching linear-scan method is proposed in this paper. The proposed technique is designed in such a way that a linear-scan surface form stylus profilometer for roundness measurement is utilized to detect angular misalignments of a small cylinder in a simple manner. An experimental setup capable of compensating for the angular misalignments is also developed, and the feasibility of the proposed technique is verified in experiments. Furthermore, numerical calculations are carried out based on a Monte Carlo method [7,13] to estimate the measurement uncertainty of the stitching linear-scan method.

2. Detection and Compensation of Angular Misalignments of a Small Cylinder in the Stitching Linear-Scan Method

2.1. Principle of the Stitching Linear-Scan Method

Figure 1 shows a schematic of the setup for measurement of a small cylinder by the stitching linear-scan method [10]. The setup is composed of a linear-scan surface form stylus profilometer, a round magnet and a workpiece holder with a V-groove. A small cylinder to be measured is attached to the round magnet, on whose outer surface indexing marks with an angular interval of 45° are prepared over 360°. Due to the magnetic force of the round magnet, a small cylinder can automatically be aligned to the center of the round magnet. It should be noted that there should be eccentricity of a small cylinder with respect to the round magnet. However, the influence of the eccentricity can be canceled during the following stitching operation; this is one of the advantages of employing the stitching linear-scan method [10]. The small cylinder with the round magnet is placed in the V-groove on the workpiece holder in such a way that the round magnet surface contacts the side of the work holder.

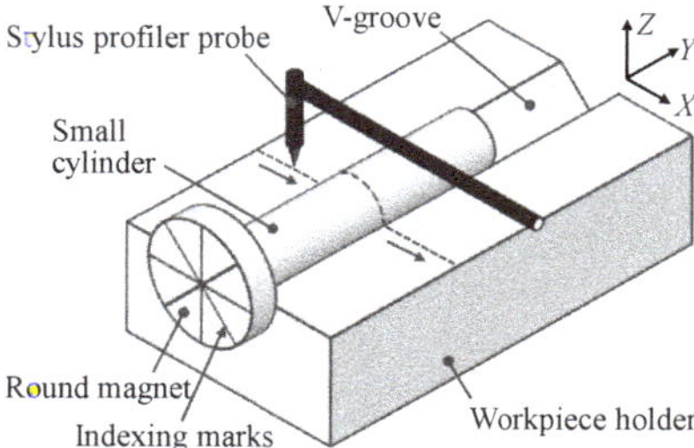

Figure 1. A schematic of the roundness measurement by the stitching linear-scan method.

Measurement of the circumferential profile of the small cylinder is carried out by repeating the 45° rotation of the workpiece and the linear-scan of its arc-profiles, as shown in Figure 2. For the workpiece rotation, n indexing marks (n is a small integer) prepared on the round magnet by a dividing head are utilized. By using the obtained eight arc-profiles of the workpiece, the workpiece profile over 360° can be obtained through the stitching process. From the obtained workpiece profile, evaluations of the roundness and diameter can be carried out. The cylinder-holding mechanism composed of the round magnet and the workpiece holder with the V-groove enables the linear-scan method to evaluate the whole length of a small cylinder, which cannot be achieved by the conventional roundness measuring instruments where a part of the cylinder in its axial direction should be held by a clamping mechanism. Furthermore, the precision positioning of a small cylinder along its axial direction can be achieved in a simple manner by utilizing the round magnet surface as the datum for the positioning.

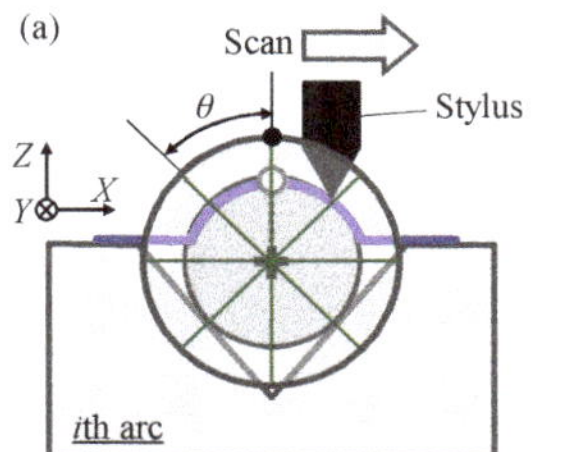

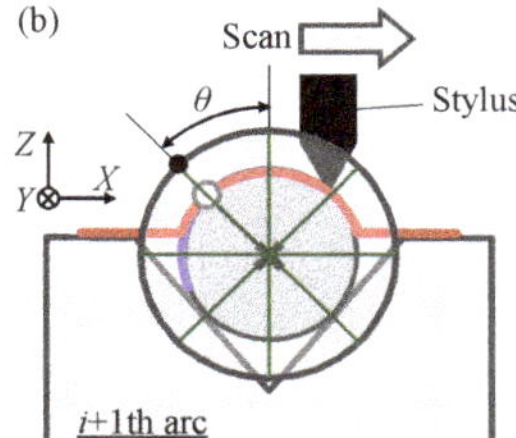

Figure 2. Measurement of the series of arc-profiles and the reconstruction of the circular profile of a workpiece: (**a**) measurement of the *i*th arc profile; (**b**) measurement of the *i* + 1th arc profile.

The stitching process of the obtained eight arcs contains three-step calculations: the coordinate transformation, the radial stitching, and the circumferential stitching. Figure 3 shows a schematic of the stitching procedure. First we repeat the linear-scan and rotation of a small cylinder to obtain a series of arc profiles. After that, radial stitching is carried out for the obtained eight arcs. Each arc-profile datum of a small cylinder is in the Cartesian coordinate system ($x_{i,j}$, $z_{i,j}$), where i is the sampling number in each of the arcs, and j (j = 1, 2, ... , 8) is the measurement order of the arc. By fitting a circle to each of the obtained arcs based on the least-squares method, the center coordinates and radius of the jth arc ($x_{c,j}$, $z_{c,j}$) and r_j, respectively, can be obtained. Then, the coordinates of the eight arcs are shifted based on the following equation to match the center coordinates of the arcs:

$$\begin{pmatrix} x'_{i,j} \\ z'_{i,j} \end{pmatrix} = \begin{pmatrix} x_{i,j} - x_{c,j} \\ z_{i,j} - z_{c,j} \end{pmatrix} \quad (1)$$

Furthermore, each arc profile is rotated by the following equation:

$$\begin{pmatrix} x''_{i,j} \\ z''_{i,j} \end{pmatrix} = \begin{pmatrix} \cos j\theta & \sin j\theta \\ -\sin j\theta & \cos j\theta \end{pmatrix} \begin{pmatrix} x'_{i,j} \\ z'_{i,j} \end{pmatrix} \quad (2)$$

It should be noted that each of the arcs has a different radius r_j. To stitch the arcs in the radial direction, a mean radius $\overline{R}$ is calculated by the following equation:

$$\overline{R} = \frac{1}{8}\sum_{j=1}^{8} r_j \quad (3)$$

By using the obtained $\overline{R}$, each sampling point in each of the arcs is adjusted by the following equation:

$$r'_{i,j} = r_{i,j} - r_j + \overline{R} \quad (4)$$

where $r_{i,j}$ is the distance from the center of the arc to the point ($x''_{i,j}$, $z''_{i,j}$) in the jth arc that can be calculated by the following equation:

$$r_{i,j} = \sqrt{\left(x''_{i,j} - x_{c,j}\right)^2 + \left(z''_{i,j} - z_{c,j}\right)^2} \quad (5)$$

To minimize the deviation of each point from the fitting circle, $x_{c,j}$, $z_{c,j}$, and r_j are determined by minimizing the residual sum of squares S that can be calculated by the following equation [14]:

$$S = \sum_{i=1}^{N} \left(\sqrt{\left(x''_{i,j} - x_{c,j}\right)^2 + \left(z''_{i,j} - z_{c,j}\right)^2} - r_j \right)^2 \tag{6}$$

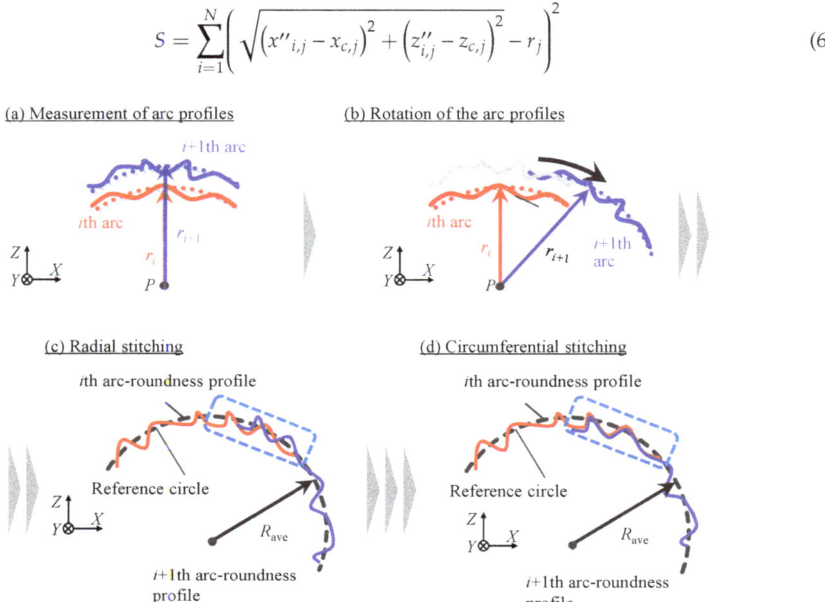

Figure 3. Stitching procedure for the reconstruction of the circular profile of a small cylinder: (**a**) measurement of arc profiles; (**b**) rotation of the obtained arcs; (**c**) radial stitching; (**d**) circumferential stitching.

As the next step of the stitching process, the circumferential stitching is carried out. The angular position $\theta_{i,j}$ of the ith sampled points in the jth arc profile can be calculated by the following equation:

$$\theta_{i,j} = \tan^{-1}\left(\frac{z''_{i,j}}{x''_{i,j}}\right) \tag{7}$$

Since a small cylinder is positioned approximately by using the indexing marks prepared on the outer face of the round magnet, each of the arcs has a certain amount of misalignment in the circumferential direction. Denoting the compensating angle shift for the jth arc as $\Delta\theta_j$, the angular position of the ith point in the jth arc can be calculated by the following equation:

$$\theta'_{i,j} = \theta_{i,j} + \Delta\theta_j \tag{8}$$

For the determination of $\Delta\theta_j$, a cross-correlation function [15,16] can be employed. By stitching the neighboring arcs sequentially from the first arc to the last arc, circumferential stitching can be carried out. For the portions where the neighboring two arcs are overlapping, a mean value of the two radial coordinates can be employed.

By using the profile of the workpiece obtained through the above procedures, the diameter D and roundness of the workpiece can be evaluated. Meanwhile, the roundness can be evaluated by using the obtained profile data through processing it with the procedure defined by ISO [17]. For a small cylinder, the out-of-roundness ΔZ_θ is evaluated based on the least square circle [18]. Denoting the mean distance of the sampling points from the center coordinates as $\overline{r_i}$, ΔZ_θ can be evaluated by the following equation:

$$\Delta Z_\theta = (\overline{r_i})_{\text{Max}} - (\overline{r_i})_{\text{min}} \tag{9}$$

It should be noted that the eccentricity of the workpiece rotation will not affect the measurement in the proposed stitching linear-scan method [10]; this is one of the remarkable advantages compared with the conventional roundness measuring instruments, where the influence of the eccentricity becomes significant with the decrease of the diameter of a workpiece.

2.2. Influences of the Angular Misalignments of a Roll Workpiece

In the stitching linear-scan method, the angular misalignment θ_X of a small cylinder about the X-axis could affect the roundness measurement. Figure 4 shows a schematic of the influence of θ_X on the measured arc profile of a small cylinder. As can be seen in the figure, the radius of the observed arc becomes larger than the real one. Denoting the diameter of an ideal small cylinder as D, the X- and Z-coordinates (x_i, z_i) of the ith sampled point satisfy the following equation:

$$x_i^2 + z_i^2 \cos^2\theta_X = D^2/4 \tag{10}$$

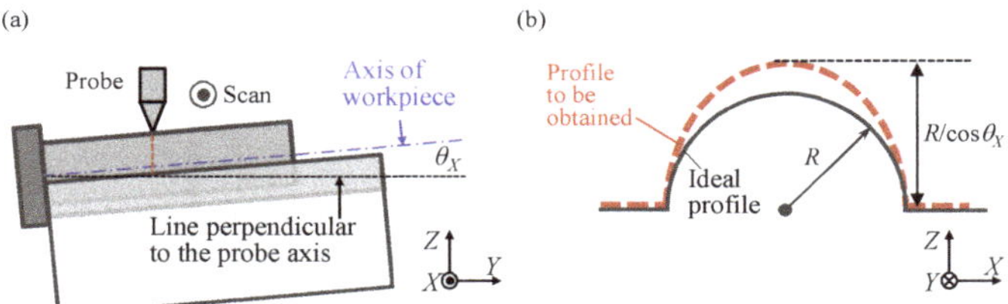

Figure 4. A schematic of the influence of angular misalignment θ_X on the arc-profile to be obtained in measurement: (**a**) definition of θ_X; (**b**) error caused by θ_X.

Denoting the ith sampled point in the polar coordinate system as (r_i,θ_i), the above equation can be modified as follows:

$$r_i(\theta_i) = \frac{D}{2}\left(\cos^2\theta_i + \sin^2\theta_i\cos^2\theta_X\right)^{-\frac{1}{2}} \tag{11}$$

The influence of the angular misalignment of the small cylinder about the X-axis $e_{\theta X}$ can thus be obtained by the following equation:

$$e_{\theta X}(\theta_i) = \frac{D}{2}\left[1 - \left(\cos^2\theta_i + \sin^2\theta_i\cos^2\theta_X\right)^{-\frac{1}{2}}\right] \tag{12}$$

Figure 5a shows the variation of $e_{\theta X}$ over an angular range from θ = 42.5° to θ = 137.5° under the condition of D = 1 mm and θ_X = 1°. The error becomes maximum at the top of the arc (namely, θ = 90°). Numerical calculations are extended to further investigate the influence of θ_X. Figure 5b shows the variation of the maximum $e_{\theta X}$ as the increase of θ_X under the condition of θ_i = 42.5°. Calculations are carried out for the cases with different workpiece diameter D ranging from 1 mm to 6 mm. As can be seen in the figure, the influence of θ_X becomes larger with the increase of D. Meanwhile, the calculation results show that $e_{\theta X}$ can be suppressed to be less than 0.01 μm by reducing θ_X to less than 0.3° for a small cylinder with a diameter D of less than 3 mm, for which it is difficult to carry out roundness measurement by the conventional rotary-scan method.

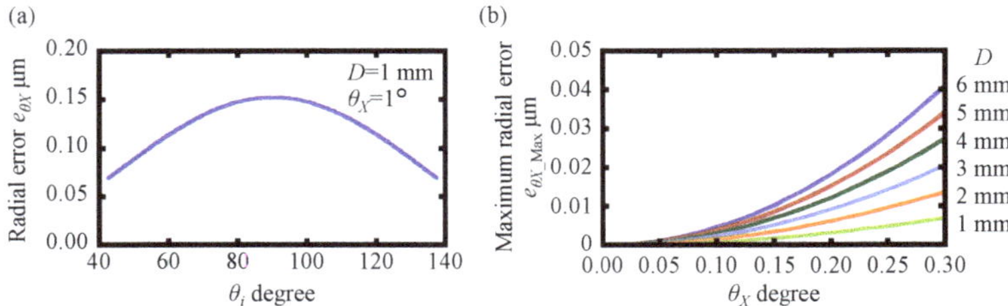

Figure 5. Radial error induced by the angular misalignment θ_X of a cylindrical workpiece: (**a**) radial error at each θ_i; (**b**) variation of the maximum radial error due to the increase of θ_X.

The angular misalignment θ_Z of a small cylinder about the Z-axis could also affect the roundness measurement. Figure 6 shows a schematic of the influence of θ_Z on the measured profile of a small cylinder. As can be seen in the figure, the radius of the observed arc becomes larger than the real one. The influence of the angular misalignment of the small cylinder about the Z-axis $e_{\theta Z}$ can be expressed by the following equation:

$$e_{\theta z}(\theta_i) = \frac{D}{2}\left[1 - \left(\cos^2\theta_i \cos^2\theta_z + \sin^2\theta_i\right)^{-\frac{1}{2}}\right] \quad (13)$$

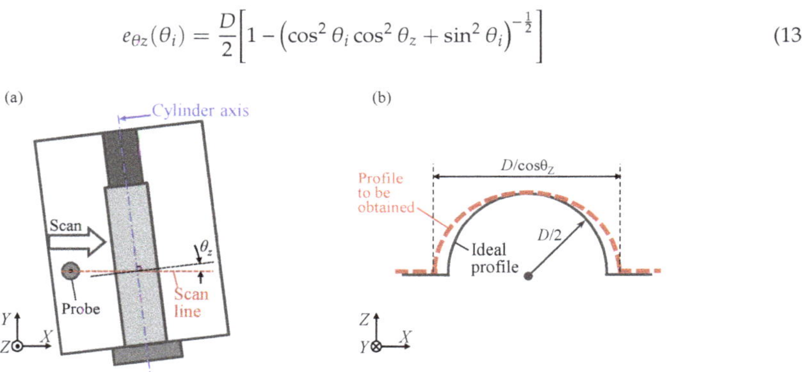

Figure 6. A schematic of the influence of angular misalignment θ_Z on the arc-profile to be obtained in measurement: (**a**) definition of θ_Z; (**b**) error caused by θ_Z.

In the same manner as θ_X, the influence of θ_Z is also investigated through numerical calculations. Figure 7a shows the variation of $e_{\theta Z}$ under the condition of D = 1 mm and θ_Z = 1°. The error becomes maximum at the edges of the arc (namely, θ_i = 42.5° and 137.5°). Figure 7b shows the variation of the maximum $e_{\theta Z}$ as the increase of θ_Z under the condition of θ_i = 42.5°. According to the calculation results, $e_{\theta Z}$ can be suppressed to be less than 0.01 μm by reducing θ_Z to less than 0.2° for a small cylinder with a diameter D of less than 3 mm.

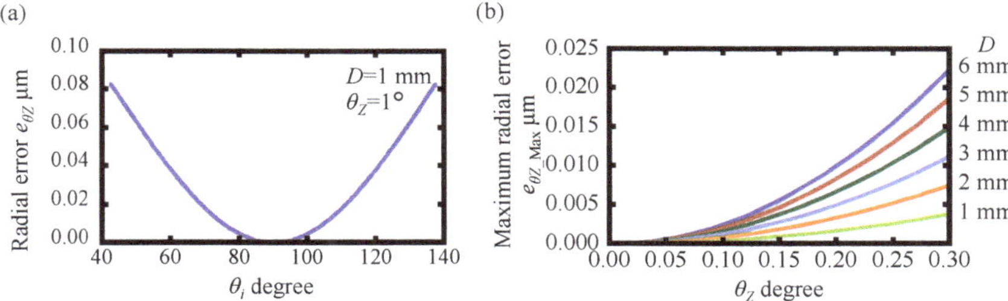

Figure 7. Radial error induced by the angular misalignment θ_Z of a cylindrical workpiece: (**a**) radial error at each θ_i; (**b**) variation of the maximum radial error due to the increase of θ_Z.

2.3. A Technique to Evaluate Angular Misalignments of a Small Cylinder

Regarding the results of numerical calculations described above, angular misalignments θ_X and θ_Z of a small cylinder need to be reduced as much as possible. It is thus necessary to establish a technique to detect θ_X and θ_Z in advance of the roundness measurement of a workpiece. However, regarding the small dimensions of a workpiece to be measured by the stitching linear-scan method, it is not so easy to apply conventional angle sensors such as an optical autocollimator [19] or a multi-axis laser autocollimator [20] for this purpose. To address the issue, attempts are made in this paper to utilize the linear scan surface form stylus profiler, which is employed for roundness measurement in the stitching linear-scan method, for the detection of θ_X and θ_Z.

Now we consider a positioning system shown in Figure 8a composed of a tilt stage, a Z-rotary stage, and a workpiece holder. The tilt stage, the rotary stage, and the workpiece holder are stacked and assembled as a holder unit. Figure 8b shows a photograph of the developed holder unit. The holder unit can be mounted on a linear stage.

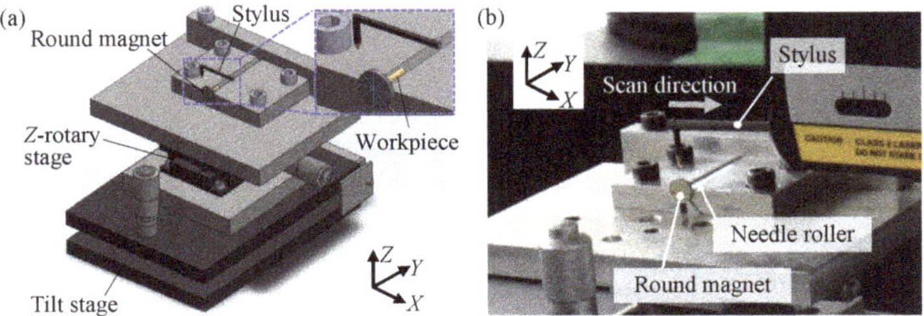

Figure 8. Setup for roundness measurement of a small cylinder with a surface profilometer: (**a**) a schematic of the holder unit; (**b**) a photograph of the holder unit with a surface profilometer.

In the proposed method, θ_X is detected through a two-step procedure. In the first step, the workpiece axis is aligned to be parallel with the X-axis as shown in Figure 9a. By scanning over the workpiece holder surface with the stylus, profile data (x_{i_hold}, z_{i_hold}) of the workpiece holder surface can be obtained. The linear approximation of the obtained profile based on the least squares method provides the inclination angle θ_{Scan_Y} of the workpiece holder surface with respect to the datum surface of the surface form stylus profiler. Denoting the inclination angle of the top surface of the linear stage about the Y-axis with respect to the datum surface of the surface form stylus profiler as θ_{Linear_Y}, and the inclination of the workpiece holder surface about the Y-axis with respect to the top surface of the linear stage as θ_{Holder}, the following relationship should be satisfied:

$$\theta_{Scan\_Y} = \theta_{Linear\_Y} + \theta_{Holder} \tag{14}$$

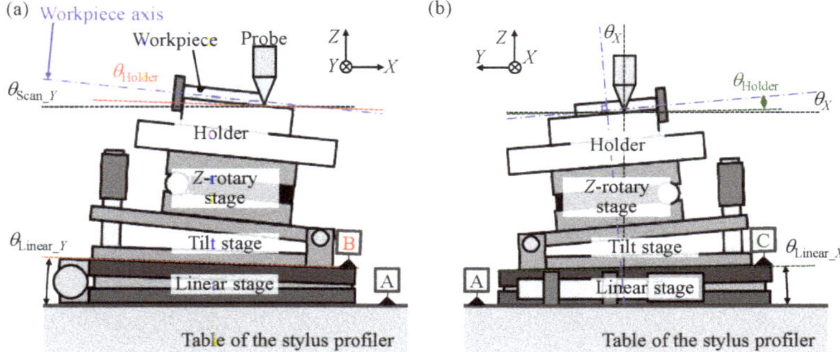

Figure 9. Evaluation of the angular misalignment θ_X of a small cylinder about the X-axis: (**a**) before the rotation of the holder unit about the Z-axis; (**b**) after 90° rotation of the holder unit about the Z-axis.

In the second step, the holder unit is rotated 90° about the Z-axis on the linear stage through the demounting and remounting operations, as shown in Figure 9b. It should be noted that θ_{Holder} does not change during the operations since the components in the holder unit are rotated together without separation. Denoting the inclination angle of the top surface of the linear stage about the X-axis with respect to the datum surface of the surface form stylus profiler as $\theta_{\text{Linear\_X}}$, the angular position of the workpiece holder surface about the X-axis with respect to the datum surface of the surface form stylus profiler θ_X can be expressed by the following equation:

$$\theta_x = \theta_{\text{Linear\_X}} + \theta_{\text{Holoder}} \tag{15}$$

From Equations (14) and (15), the following equation can be obtained:

$$\theta_X = \theta_{\text{Scan\_Y}} + \theta_{\text{Linear\_X}} - \theta_{\text{Linear\_Y}} \tag{16}$$

Since $\theta_{\text{Linear\_X}}$ and $\theta_{\text{Linear\_Y}}$ can be treated as the known parameters through scanning the top surface of the linear stage by the surface form stylus profiler in advance of the roundness measurement, θ_X can be evaluated by measuring $\theta_{\text{Scan\_Y}}$ in the first step.

It should be noted that the parallelism of the cylinder/workpiece with respect to the X-axis was confirmed visually, and that there should exist a certain amount of angular misalignment of the workpiece about the Z-axis (φ_Z), which could affect the measurement. According to the geometric relationship, the error $e_{\theta X}$ in measurement of θ_X due to φ_Z can be expressed by the following equation:

$$e_{\theta X} = \theta_X - \tan^{-1}(\cos\varphi_Z \tan\theta_X) \tag{17}$$

Under the condition of θ_Z = 2.5° and θ_X = 0.1°, $e_{\theta X}$ becomes approximately 0.002°, and is small enough to be neglected. Meanwhile, the influence of the demounting and remounting operations of the holder unit in measurement of θ_X should be evaluated in experiments.

θ_Z can be evaluated by obtaining the arc profiles of a small cylinder at different Y-positions with enough intervals. Figure 10 shows a schematic of the evaluation of θ_Z. First, we obtain the arc-profile of a workpiece at a Y-position of y_1, and detect the X-coordinate of the peak (x_{peak1}) in the obtained arc-profile through fitting a circle based on the least squares method. In the same manner, we detect the X-coordinate of the peak (x_{peak2}) in the arc-profile obtained at a Y-position of y_2. From the obtained peak coordinates, θ_Z can be calculated by the following equation:

$$\theta_z = \tan\left(\frac{x_{\text{peak2}} - x_{\text{peak1}}}{y_2 - y_1}\right) \tag{18}$$

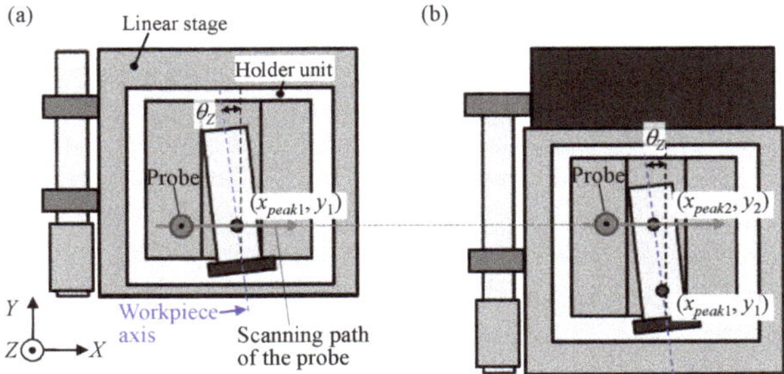

Figure 10. Evaluation of the angular misalignment θ_Z of a small cylindrical workpiece about the Z-axis: (**a**) before shifting; (**b**) after shifting.

It should be noted that the angular misalignment of the workpiece about the X-axis (φ_X) could affect the evaluation of θ_Z. From the geometric relationship, the error $e_{\theta Z}$ in measurement of θ_Z due to φ_X can be expressed by the following equation:

$$e_{\theta Z} = \theta_Z - \tan^{-1}(\cos\varphi_X \tan\theta_Z) \tag{19}$$

In the proposed method, the evaluation of θ_Z is carried out after the measurement and alignment of θ_X. On the assumption that θ_X is aligned to be less than 0.1°, $e_{\theta Z}$ becomes $2.66 \times 10^{-6\circ}$ under the condition of θ_Z = 0.1°; the influence of θ_X can thus be treated as small enough to be neglected.

3. Experiments

3.1. Verification of the Evaluation Method of the Angular Misalignments of a Small Cylinder

Experiments were carried out to verify the feasibility of the proposed technique for evaluation of the angular misalignments of the workpiece in roundness measurement of a small cylinder by the stitching linear-scan method. In the following verification experiments, a calibrated pin gauge with a diameter of 3.000 mm was employed as the measurement specimen. At first, basic characteristics of the proposed technique for the evaluation of θ_X were investigated in experiments. Figure 11 shows a schematic of the experimental setup. Experimental conditions are summarized in Table 1. The axis of a small cylinder was aligned to be parallel with the scanning direction of the surface form stylus profilometer (X-axis in the figure). A commercial autocollimator was employed as the reference angle sensor for measurement of the angular displacement of the workpiece holder. A flat mirror was employed as the target for the commercial autocollimator and was mounted on the workpiece holder. The workpiece holder surface was scanned by the surface form stylus profilometer to obtain the inclination angle θ_{Scan_Y} of the workpiece holder surface with respect to the datum surface of the surface form stylus profiler.

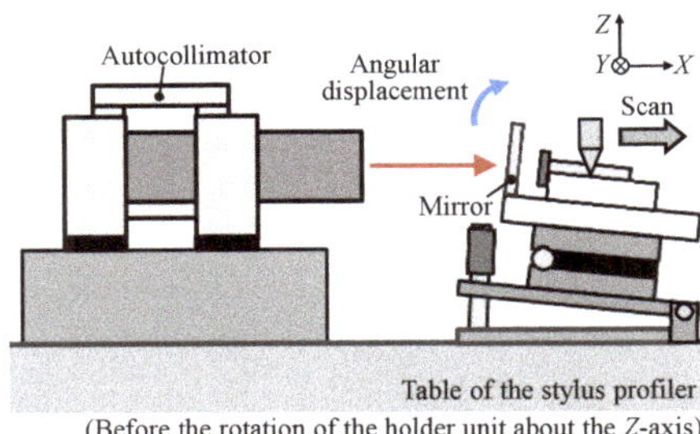

Figure 11. Setup for the verification of the method for measurement of θ_X.

Table 1. Parameters for scanning experiments.

Items	Value	Unit
Tip angle of a stylus	60	degree
Tip radius of a stylus	2	μm
Probe scanning speed	0.1	mm/s
Measuring force	1.0	mN

Figure 12a shows an example of the obtained surface profile of the workpiece holder. From the obtained profile data, the inclination angle θ_{Scan_Y} was obtained through the linear fitting by the least-squares method. Ten repetitive trials, including the demounting and remounting operations of the holder unit at each trial, were made to evaluate the reproducibility. Figure 12b shows the results. The standard deviation of θ_{Scan_Y} was evaluated to be 0.005°.

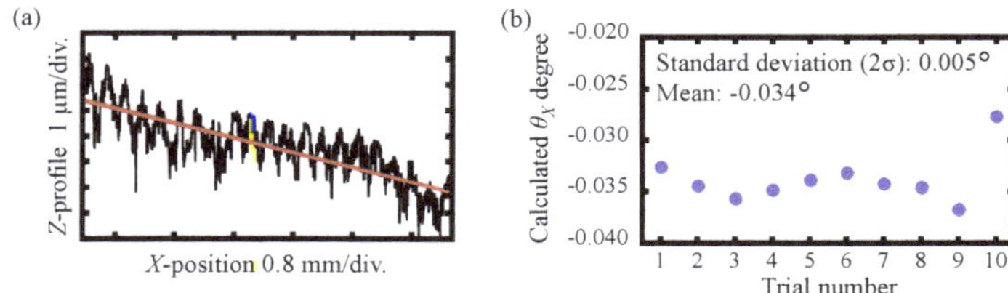

Figure 12. Evaluation of the θ_X and its reproducibility: (**a**) measured profile of the workpiece holder; (**b**) reproducibility of the evaluation of θ_X.

Following the verification of the measurement reproducibility, an attempt was made to detect the angular displacement of the workpiece holder. By using the tilt stage, the workpiece holder was rotated about the Y-axis in steps of approximately 0.005°, and the angular displacement of the workpiece holder was evaluated by both the proposed method and the commercial autocollimator at each step. It should be noted that the scanning length of the stylus profilometer for evaluation of θ_{Scan_Y} was set to be 5 mm. Figure 13 shows the results. As can be seen in the figure, the maximum deviation of the detected θ_{Scan_Y} from the reading of the autocollimator was less than ±0.002°. These experimental results demonstrated the feasibility of the proposed technique of measuring θ_{Scan_Y}.

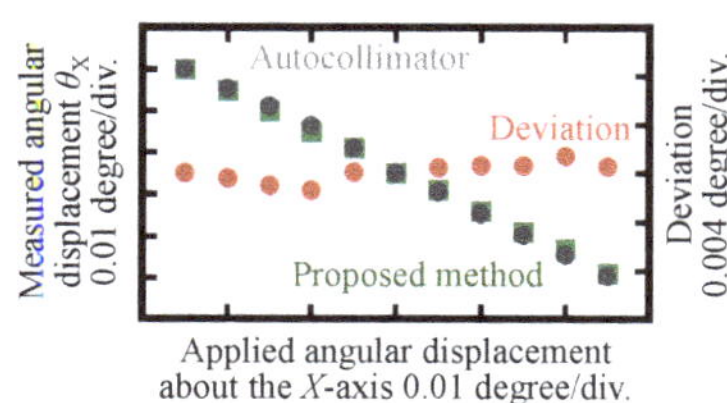

Figure 13. The angular displacement of the workpiece about the X-axis measured by the proposed technique.

Experiments were also carried out to evaluate the basic characteristics of the proposed technique for the evaluation of θ_Z. Figure 14 shows a schematic of the experimental setup. The pin gauge was first attached to a round magnet and was then placed on the V-groove in the workpiece holder. The axis of the workpiece was aligned to be parallel with the Y-axis so that its arc-profiles could be measured by the stylus profilometer.

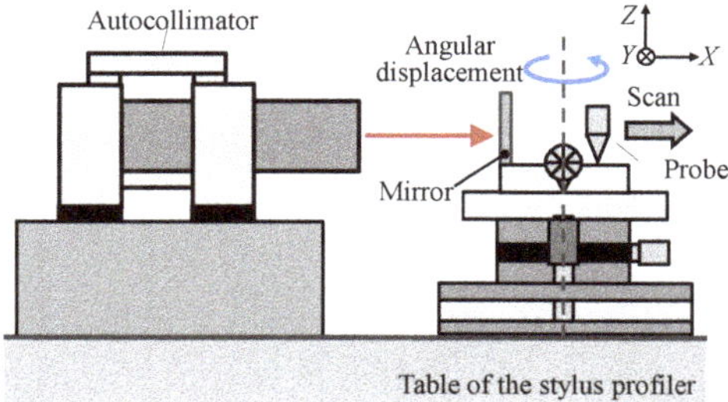

Figure 14. Setup for the verification of the technique for measurement of θ_Z.

By using the setup, the arc-profile of the workpiece was first obtained at a Y-position of y_1. After that, the workpiece was made to travel along the Y-direction together with the holder unit by using the linear stage. After that, another arc-profile was obtained at a Y-position of y_2 (= y_1 + Δy). Figure 15a shows an example of the arc-profile obtained during the experiments. For each of the obtained arc-profiles, a circle was fitted based on the least-squares method to obtain the peak coordinate as shown in Figure 15b. After that, by using the obtained data, θ_Z was evaluated based on Equation (18). In the experiments, Δy was set to 4 mm. Five repetitive trials were made by following the above procedure without the demounting and remounting operations of the holder unit at each trial, and the standard deviation of the measurement of θ_Z was evaluated to be 0.006°, as shown in Figure 15c.

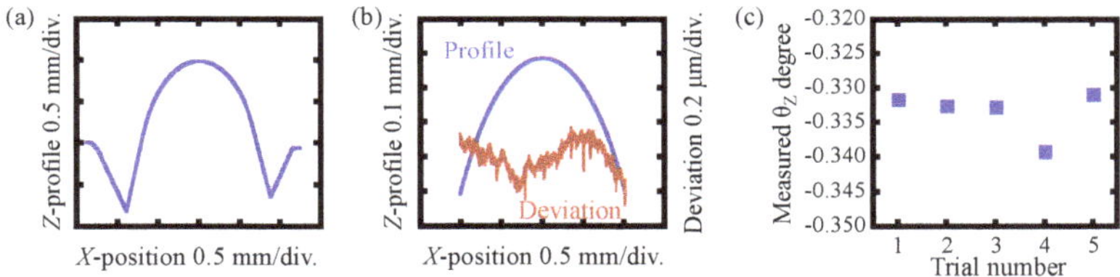

Figure 15. Evaluation of the θ_Z and its repeatability: (**a**) measured profile (**b**) arc-profile cut from the profile in (**a**) and its deviation from a fitting circle; (**c**) repeatability of θ_Z measurement.

Following the verification of the measurement repeatability, an attempt was made to detect the angular displacement θ_Z of the pin gauge. By using the rotary stage, the workpiece was rotated about the Z-axis in steps of approximately 0.023°, corresponding to the resolution of the rotary stage. The angular displacement of the workpiece was evaluated by both the proposed technique and the commercial autocollimator at each step. Figure 16 shows the results. As can be seen in the figure, the maximum deviation of the detected θ_Z from the reading of the autocollimator was less than 0.01°. These experimental results demonstrate the feasibility of the proposed technique for measurement of θ_Z. By using the obtained θ_X and θ_Z, the angular misalignment of the workpiece can thus be compensated.

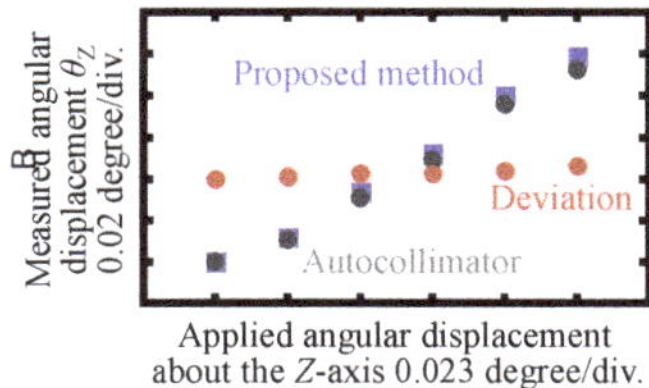

Figure 16. The angular displacement of the workpiece about the Z-axis measured by the proposed method.

3.2. Evaluation of the Roundness Profile of a Small Cylinder with a Diameter of 3 mm

After the compensation of the angular misalignments θ_X and θ_Z of the workpiece by the proposed technique, roundness measurement was carried out. Figure 17a shows the series of arc-profiles obtained in measurements. Figure 17b shows the profile reconstructed from the obtained arc-profiles before the stitching process. To evaluate the effectiveness of the compensation of the angular misalignments, experiments were carried out for the cases of $\theta_Z = 0.004°$ and $\theta_Z = 0.951°$.

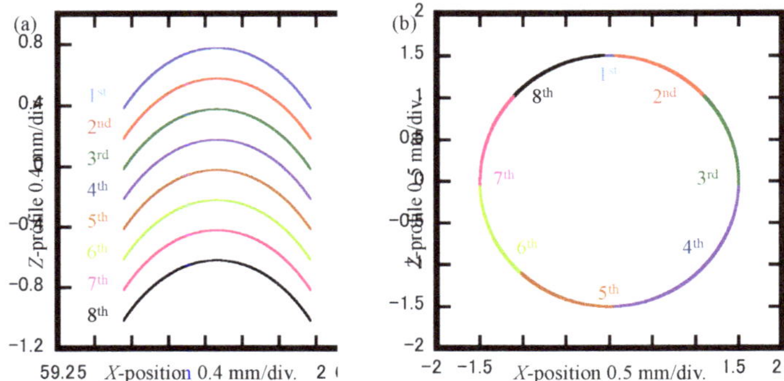

Figure 17. Profiles of a pin gauge with a diameter of 3 mm measured by the stitching linear-scan method: (**a**) measured arc profiles; (**b**) the arc profiles in (**a**) after the rotation of each arc.

Figure 18 shows the roundness profiles of the workpiece before and after the stitching process and after the filtering process. The results are summarized in Table 2. As can be seen in the figure and table, the angular misalignment θ_Z was found to affect the result of roundness measurement in the stitching linear-scan method.

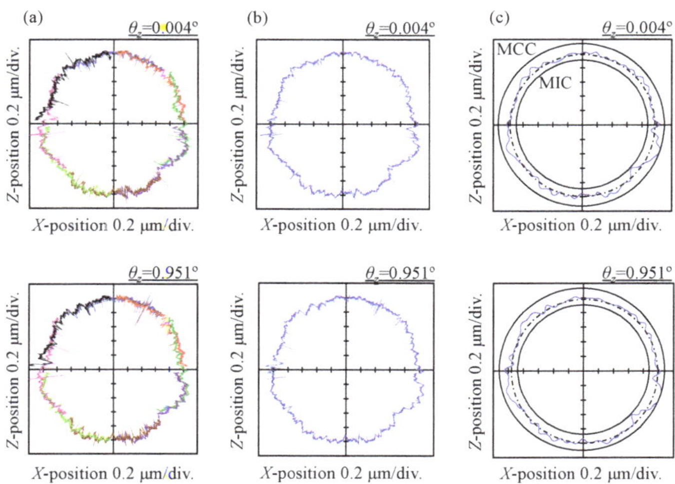

Figure 18. Evaluation of the roundness by the obtained arc profiles: (**a**) roundness profiles after radial stitching; (**b**) roundness profiles after circumferential stitching; (**c**) roundness profiles after filtering.

Table 2. Radius and roundness.

Angular Misalignment of the Pin Gage θ_z		0.004°	0.951°
Diameter	Mean value	2.99963 mm	2.99898 mm
	Standard deviation	0.10 μm	0.13 μm
Roundness	Mean value	0.21 μm	0.19 μm
	Standard deviation	0.03 μm	0.01 μm

4. Measurement Uncertainty Analysis of the Stitching Linear Scan Method

Theoretical investigation of the measurement uncertainty of the stitching linear-scan method was carried out. In the method, several arc-profiles were first measured in the Cartesian coordinate system (x_i,z_i). After that, the obtained data of the arc-profiles were converted into the polar coordinate system (r_i,θ_i) to reconstruct the workpiece profile through the stitching process. Finally, the roundness of the workpiece was evaluated by the reconstructed profile. Since the above procedure contained a fitting process based on the least-squares method, numerical calculations based on the Monte Carlo method were carried out for the estimation of the measurement uncertainty. Since the stitching linear-scan method was developed for roundness measurement of a small cylinder having a radius of a few millimeters, following the previous work by the authors [10], a series of arc-profiles obtained by measuring a needle roller with a diameter of 1.5 mm was employed for the following numerical calculations.

First, the uncertainty of the radius of each sampling point was estimated. On the assumption that the workpiece had a perfect cylindrical profile, the radius r_i at the ith sampled point (x_i,z_i) could be expressed by the following equation:

$$r_i = \sqrt{x_i^2 + z_i^2} \tag{20}$$

The standard uncertainty of r_i (u_{ri}) can thus be obtained by the following equation:

$$u_{ri} = \sqrt{\left(\frac{\partial R_i}{\partial x_i}\right)^2 u_{xi}{}^2 + \left(\frac{\partial R_i}{\partial z_i}\right)^2 u_{zi}{}^2} = \sqrt{\frac{x_i^2}{x_i^2 + z_i^2} u_{xi}{}^2 + \frac{z_i^2}{x_i^2 + z_i^2} u_{zi}{}^2} \tag{21}$$

For the evaluation of u_{ri}, the uncertainties of z_i and x_i need to be obtained. Table 3 summarizes the sources of uncertainty contributing to the uncertainty of z_i. The stylus profilometer employed in the series of experiments was calibrated by using a master sphere artifact in advance of the roundness measurements. According to the specification sheet of the stylus profilometer [11], the uncertainty of the calibration is within ±200 nm. It should be noted that the contribution due to the drift between calibrations is included in this value. Assuming the rectangular probability distribution with a divisor of $\sqrt{3}$, the standard uncertainty of the calibration (u_{Cal}) was evaluated to be 115.47 nm. It should be noted that u_{Cal} contains the influences of the straightness of the linear scan axis, the form errors of the master sphere artifact, and the stylus tip. Also, the uncertainty associated with the resolution of the Z-reading ($u_{\mathrm{res\_z}}$) was evaluated to be 0.92 nm, with regard to the value (3.2 nm) shown in the specification sheet of the stylus profiler. Furthermore, the contribution of repeatability of the Z-reading ($u_{\mathrm{rep\_z}}$) was estimated from a standard deviation in ten repetitive arc-profile measurements (67.66 nm). A standard uncertainty of z_i (u_{zi}) was thus evaluated to be 133.84 nm.

Table 3. Uncertainty of the Z-coordinate of each sampled point.

Sources of Uncertainty	Symbol	Type	Probability Distribution	Standard Uncertainty
Calibration of the probe (including the contribution due to the drift between calibrations)	u_{Cal}	B	Rectangular	115.47 nm
Resolution of probe	$u_{\text{Res\_z}}$	B	Rectangular	0.92 nm
Repeatability	u_{Rep}	A	Gaussian	67.66 nm
Standard uncertainty of z_i	u_{zi}			133.84 nm

The uncertainty of x_i was also estimated. Table 4 summarizes the sources of uncertainty contributing to the uncertainty of x_i. According to the manufacturer of the stylus profilometer, the standard uncertainty of the reading of X-position ($u_{\text{Read\_}x}$) is 150 nm. It should be noted that this value contains the contributions from the repeatability, the drift, and so on. Also, the contribution of the resolution of the x-reading ($u_{\text{Res\_}x}$) was evaluated to be 36.08 nm, with regard to the value (125 nm) shown in the specification sheet of the stylus profiler and the rectangular probability distribution with a divisor of $2\sqrt{3}$. Meanwhile, for the evaluation of u_{xi}, influences of the angular misalignments of a workpiece about the Z- and X-axes need to be considered. Denoting the angular misalignment of a workpiece about the Z-axis as θ_Z, the resultant error in the X-coordinate of each sampled point $\Delta x_{i\_\theta Z}$ can be expressed by the following equation from the geometric relationship:

$$\Delta x_{i\_\theta Z} = R\sin(\varphi/2)\cdot[(1/\cos\theta_Z) - 1] \tag{22}$$

Table 4. Uncertainty of the X-coordinate of each sampled point.

Sources of Uncertainty	Symbol	Type	Probability Distribution	Standard Uncertainty
Reading of x (including the contributions from the repeatability, the drift and so on)	$u_{\text{read\_}x}$	B	Gaussian	150.00 nm
Resolution of x-reading	$u_{\text{Res\_}x}$	B	Rectangular	36.08 nm
Angular misalignment θ_Z	$u_{\theta Z}$	B	Rectangular	0.89 nm
Angular misalignment θ_X	$u_{\theta X}$	B	Rectangular	0.89 nm
Standard uncertainty of x_i	u_{xi}			154.28 nm

The contribution of θ_Z ($u_{\theta Z}$) can thus be evaluated by the following equation:

$$u_{\theta Z} = \sqrt{\left(\frac{\partial \Delta x_{i\_\theta Z}}{\partial \theta_Z}\right)^2 u^2(\theta_Z)} = \sqrt{\left[R\sin\left(\frac{\varphi}{2}\right)\sin\theta_Z/\cos^2\theta_Z\right]^2 \cdot u^2(\theta_Z)} \tag{23}$$

On the assumption that θ_Z was within ±0.1°, by employing a mean radius $\overline{R}$ of 1.5 mm after the stitching process, $u_{\theta Z}$ was evaluated to be 0.89 nm. The contribution of the angular misalignment of a workpiece about the X-axis θ_X was also estimated. According to the geometric relationship, the error in the X-coordinate due to θ_X ($\Delta x_{i\_\theta X}$) can be expressed by the following equation:

$$\Delta x_{i\_\theta X} = r_i\sin(\varphi/2)\cdot[1 - \cos\theta_X] \tag{24}$$

On the assumption that θ_X was within ±0.1°, by employing a mean radius $\overline{R}$ of 1.5 mm after the stitching process, $u_{\theta X}$ was evaluated to be 0.89 nm in the same manner as $u_{\theta Z}$. By combining these contributions, the standard uncertainty of x_i (u_{xi}) was evaluated to be 154.28 nm. Compensations of the angular misalignment of a workpiece are effective in reducing the standard uncertainties of x_i and

z_i, and the contributions from the specifications of the stylus profiler were found to be dominant in these uncertainties.

By combining these contributions based on Equation (21), u_{ri} was evaluated to be 178.75 nm, as summarized in Table 5. It should be noted that the sensitivity coefficients $(\partial r_i/\partial x_i)$ and $(\partial r_i/\partial z_i)$ were treated to have maximum values $(\partial r_i/\partial x_i) = \sin(\varphi/2) = 0.67559$ and $(\partial r_i/\partial z_i) = 1$, respectively, to avoid underestimation.

Table 5. Uncertainty of the radius of each sampled point.

Sources of Uncertainty	Symbol	Value	Sensitivity Coefficient	Standard Uncertainty
Standard uncertainty of z	u_{zi}	133.84 nm	0.67559	90.42 nm
Standard uncertainty of x	u_{xi}	154.28 nm	1	154.28 nm
Standard uncertainty of r_i			u_{ri}	178.75 nm

A standard uncertainty of θ_i ($u_{\theta i}$) was also evaluated. Since θ_i can be calculated as $\theta_i = \arctan(z_i/x_i)$, $u_{\theta i}$ can be calculated by the following equation:

$$u_{\theta i} = \sqrt{\left(\frac{\partial \theta_i}{\partial x_i}\right)^2 u_{xi}{}^2 + \left(\frac{\partial \theta_i}{\partial z_i}\right)^2 u_{zi}{}^2} = \sqrt{\left\{\frac{z_i}{x_i{}^2 + z_i{}^2}\right\}^2 u_{xi}{}^2 + \left\{\frac{x_i}{x_i{}^2 + z_i{}^2}\right\}^2 u_{zi}{}^2} \quad (25)$$

The sensitivity coefficients $(\partial\theta_i/\partial x_i)$ and $(\partial\theta_i/\partial z_i)$ were calculated as $(\partial\theta_i/\partial x_i) = 1/R$ and $(\partial\theta_i/\partial z_i) = (1/R)\sin(\varphi/2)$, respectively, to avoid underestimation. As a result, $u_{\theta i}$ was evaluated to be $u_{\theta i} = 1.1 \times 10^{-4}$ rad (0.0065 degree), as summarized in Table 6.

Table 6. Uncertainty of the angular position of each sampled point.

Uncertainty Sources	Symbol	Value	Sensitivity Coefficient	Standard Uncertainty
Standard uncertainty of z_i	u_{zi}	133.84 nm	666.67 rad/mm	0.0000892 rad
Standard uncertainty of x_i	u_{xi}	154.28 nm	450.39 rad/mm	0.0000695 rad
Standard uncertainty of θ_i	$u_{\theta i}$	0.00011 rad (0.0065 deg.)		

By using the obtained u_{ri} and $u_{\theta i}$, numerical calculations were carried out based on the Monte Carlo method. The procedure of the numerical calculations was as follows:

Step 1: Prepare the data of the series of arc profiles in the polar coordinate system by using the circumferential profile of a small cylinder obtained in experiments after filtering (50 URP).

Step 2: Apply a random value in a Gaussian distribution with a standard deviation of u_θ to the θ-coordinate of each point.

Step 3: Apply a random value in a Gaussian distribution with a standard deviation of u_R to the R-coordinate of each point.

Step 4: Convert the arc-profile data into the Cartesian coordinate system.

Step 5: Carry out the stitching process in the same manner as the experiments.

Figure 19 shows the results of numerical calculations. A trial number of 1×10^5 was used for sufficient numerical stability of the output parameters. Expanded uncertainties of diameter and roundness were thus evaluated to be 0.032 μm and 0.024 μm (k = 2, 95% confidence), respectively.

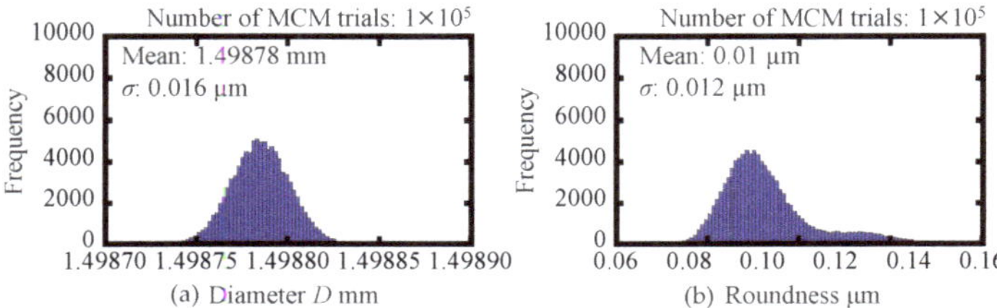

Figure 19. Uncertainties of a radius and roundness estimated by numerical calculations based on the Monte Carlo method: (**a**) radius; (**b**) roundness.

5. Conclusions

A technique to compensate for angular misalignments of a small cylinder in the stitching linear-scan method has been proposed. In the proposed technique, the angular misalignment of a small cylinder about the X-axis was evaluated by measuring the inclination angle of a workpiece holder surface by using the stylus profiler, while the one about the Z-axis was evaluated by using the data of two arc-profiles obtained at different axial positions of the workpiece. Experimental results demonstrated that the proposed method can evaluate the angular misalignments with a repeatability/reproducibility of better than 0.01°, which is enough to suppress the measurement uncertainty of a workpiece diameter as well as roundness. Measurement of a pin gauge having a diameter and a length of 3 mm and 5 mm, respectively, was carried out, and the feasibility of the proposed technique was verified. Furthermore, measurement uncertainty analysis was carried out through numerical calculations based on a Monte Carlo method to theoretically verify the feasibility of the stitching linear-scan method. Expanded uncertainties for workpiece diameter measurement and roundness measurement were evaluated to be 0.032 μm and 0.024 μm, respectively.

Author Contributions: Conceptualization, W.G. and Y.S.; methodology, T.S., Q.L. and W.G.; software, T.S. and Q.L.; validation, Y.S., W.G. and H.M.; formal analysis, T.S., Q.L., Y.S. and W.G.; investigation, T.S., Q.L. and Y.S.; resources, Y.S. and W.G.; data curation, T.S., Y.S. and W.G; writing—original draft preparation, Q.L. and Y.S.; writing—review and editing, W.G. and Y.S.; visualization, W.G. and Y.S.; supervision, W.G.; project administration, W.G.; funding acquisition, W.G., Y.S. and H.M. All authors have read and agreed to the published version of the manuscript.

Funding: This work is supported by the Japan Society for the Promotion of Science (JSPS) 15H05759 and 20H00211.

Acknowledgments: The authors would like to thank Yuki Machida for his help in the preparation of the experimental setup.

Conflicts of Interest: The authors declare no conflict of interest. The funders had no role in the design of the study; in the collection, analyses, or interpretation of data; in the writing of the manuscript, or in the decision to publish the results.

References

1. Uhlmann, E.; Mullany, B.; Biermann, D.; Rajurkar, K.P.; Hausotte, T.; Brinksmeier, E. Process chains for high-precision components with micro-scale features. *CIRP Ann.* **2016**, *65*, 549–572. [CrossRef]
2. ISO. *ISO 1206 Rolling Bearings—Needle Roller Bearings with Machined Rings—Boundary Dimensions, Geometrical Product specIfications (GPS) and Tolerance Values*; ISO: Geneva, Switzerland, 2018.
3. Farhana, N.; Yusof, M.; Ripin, Z.M. Analysis of surface parameters and vibration of roller bearing. *Tribol. Trans.* **2014**, *57*, 715–729.
4. Haitjema, H.; Bosse, H.; Frennberg, M.; Sacconi, A.; Thalmann, R. International comparison of roundness profiles with nanometric accuracy. *Metrologia* **1996**, *33*, 67–73. [CrossRef]
5. Moore, W.R. *Foundations of Mechanical Accuracy*; Moore Special Tool Company, Inc.: Bridgeport, CT, USA, 1970.
6. Oberg, E.; Jones, F.D.; Horton, H.L.; Ryffel, H.H. *Machinery's Handbook*; Industrial Press, Inc.: New York, NY, USA, 2004.

7. Shimizu, Y.; Gao, W.; Matsukuma, H.; Szipka, K.; Archenti, A. On-machine angle measurement of a precision V-groove on a ceramic workpiece. *CIRP Ann.* **2020**, *00*, 2–5. [CrossRef]
8. Talyrond 565/585HS, Taylor Hobson. Available online: www.taylor-hobson.com (accessed on 22 June 2020).
9. Weckenmann, A.; Bruning, J.; Patterson, S.; Knight, P. Grazing incidence interferometry for high precision measurements of cylindrical form deviations. *CIRP Ann.* **2001**, *50*, 381–384. [CrossRef]
10. Chen, Y.L.; Machida, Y.; Shimizu, Y.; Matsukuma, H.; Gao, W. A stitching linear-scan method for roundness measurement of small cylinders. *CIRP Ann.* **2018**, *67*, 535–538. [CrossRef]
11. Form Talysurf PGI, Taylor Hobson. Available online: www.taylor-hobson.com (accessed on 22 June 2020).
12. Gao, W.; Haitjema, H.; Fang, F.Z.; Leach, R.K.; Cheung, C.F.; Savio, E.; Linares, J.M. On-machine and in-process surface metrology for precision manufacturing. *Cirp Ann.* **2019**, *68*, 843–866. [CrossRef]
13. Joint Committee For Guides In Metrology. *Evaluation of Measurement Data—Guide to the Expression of Uncertainty in Measurement*; ISO: Geneva, Switzerland, 2008.
14. Takamasu, K. Least square method for precision measurement. *J. Jpn. Soc. Precis. Eng.* **2010**, *76*, 1130–1133. [CrossRef]
15. Helleseth, T. Some results about the cross-correlation function between two maximal linear sequences. *Discret. Math.* **1976**, *16*, 209–232. [CrossRef]
16. Helleseth, T. A note on the cross-correlation function between two binary maximal length linear sequences. *Discret. Math.* **1978**, *23*, 301–307. [CrossRef]
17. ISO 12181-1. *Geometrical Product Specifications (GPS)-Roundness-Part 1: Vocabulary and Parameters of Roundness*; ISO: Geneva, Switzerland, 2003.
18. Mitutoyo. Quick guide to precision measuring instruments. *J. Prosthet. Dent.* **2003**, *11003*, 2003.
19. MÖLLER-WEDEL OPTICAL Electronic Autocollimators. Available online: www.moeller-wedel-optical.com (accessed on 22 June 2020).
20. Gao, W.; Kim, S.W.; Bosse, H.; Haitjema, H.; Chen, Y.L.; Lu, X.D.; Knapp, W.; Weckenmann, A.; Estler, W.T.; Kunzmann, H. Measurement technologies for precision positioning. *CIRP Ann.* **2015**, *64*, 773–796. [CrossRef]

Article

High-Precision Cutting Edge Radius Measurement of Single Point Diamond Tools Using an Atomic Force Microscope and a Reverse Cutting Edge Artifact

Kai Zhang [1], Yindi Cai [2,*], Yuki Shimizu [1], Hiraku Matsukuma [1] and Wei Gao [1]

1 Precision Nanometrology Laboratory, Department of Finemechanics, Tohoku University, Sendai 980-8579, Japan; zhangkai@nano.mech.tohoku.ac.jp (K.Z.); yuki.shimizu@nano.mech.tohoku.ac.jp (Y.S.); hiraku.matsukuma@nano.mech.tohoku.ac.jp (H.M.); gaowei@nano.mech.tohoku.ac.jp (W.G.)

2 Key Laboratory for Micro/Nano Technology and System of Liaoning Province, Dalian University of Technology, Dalian 116024, China

* Correspondence: caiyd@dlut.edu.cn

Received: 28 May 2020; Accepted: 6 July 2020; Published: 13 July 2020

Abstract: This paper presents a measurement method for high-precision cutting edge radius of single point diamond tools using an atomic force microscope (AFM) and a reverse cutting edge artifact based on the edge reversal method. Reverse cutting edge artifact is fabricated by indenting a diamond tool into a soft metal workpiece with the bisector of the included angle between the tool's rake face and clearance face perpendicular to the workpiece surface on a newly designed nanoindentation system. An AFM is applied to measure the topographies of the actual and the reverse diamond tool cutting edges. With the proposed edge reversal method, a cutting edge radius can be accurately evaluated based on two AFM topographies, from which the convolution effect of the AFM tip can be reduced. The accuracy of the measurement of cutting edge radius is significantly influenced by the geometric accuracy of reverse cutting edge artifact in the proposed measurement method. In the nanoindentation system, the system operation is optimized for achieving high-precision control of the indentation depth of reverse cutting edFigurege artifact. The influence of elastic recovery and the AFM cantilever tip radius on the accuracy of cutting edge radius measurement are investigated. Diamond tools with different nose radii are also measured. The reliability and capability of the proposed measurement method for cutting edge radius and the designed nanoindentation system are demonstrated through a series of experiments.

Keywords: single point diamond tool; cutting edge radius; reversal method; nanoindentation system; elastic recovery

1. Introduction

Ultra-precision diamond cutting, combining a single point diamond tool with an ultra-precision lathe, has been widely employed for the fabrication of microstructure elements, such as microlens arrays [1], compound eye freeform surfaces [2], and sinusoidal grids [3,4]. The achievable machining accuracy of the ultra-precision diamond cutting is significantly affected by the geometry of the diamond tool, including cutting edge contour, cutting edge radius, and tool faces [5–7]. In order to achieve a nanometric machining accuracy, it is essential to conduct a quantitative evaluation of the geometry of the diamond tool, especially cutting edge radius, which determines the minimum depth of cut and the surface finish of the machined microstructures [8,9].

Cutting edge radius of a diamond tool is usually required to be within 10 to 100 nm for ultra-precision diamond cutting [10]. Therefore, the methods for cutting edge radius measurement should have a nanometric measurement accuracy in both lateral and vertical directions [11]. Optical methods can conduct fast and non-contact measurements [12–14]. However, a lateral resolution that is larger than 10 nm is limited by the optical diffraction phenomenon which occurs around the cutting edge when an optical method is used to measure the geometry of the diamond tool. Scanning electron microscopes (SEMs) have a nanometric lateral resolution and a wide view field [15]. However, because SEMs have a limitation of two-dimensional (2D) projection, they can only be used for the qualitative evaluation of diamond tool cutting edge radius and cannot be used for the quantitative evaluation. In addition, the material of diamond tools also influences the measurement accuracy of the diamond tool cutting edge radius using SEMs [16]. Atomic force microscopes (AFMs), which can provide a three-dimensional (3D) measurement with a nanometric accuracy in both lateral and vertical directions, can directly measure the diamond tool cutting edge radius by accurately aligning the AFM cantilever tip with the diamond tool cutting edge within the measurement range of the AFM [17–19]. However, the process for an accurate alignment is extremely time-consuming and the AFM cantilever tip can also be easily damaged due to an inaccurate alignment operation.

An indirect measurement method based on an AFM has been proposed to address the above problems using an AFM to measure cutting edge radius of a diamond tool [20]. A profile of a diamond tool cutting edge is copied on a copper workpiece by indenting the diamond tool into the workpiece on an ultra-precision lathe. Then, the copied profile is scanned by an AFM, from which the diamond tool cutting edge radius can be obtained. Since this indirect measurement method can protect the diamond tool and the AFM cantilever tip from damage, and also shorten the measurement time, it is attractive for practical applications. However, the elastic recovery of the copied profile can affect the measurement accuracy, which is not considered in the indirect measurement method based on the AFM. More importantly, since both the direct and indirect measurement methods are based on an AFM, the obtained profile is a geometric convolution of the AFM tip and the actual and copied diamond tool cutting edges. It is essential to eliminate the convolution effect of the AFM tip whose size is comparable to that of the diamond tool cutting edge, for achieving a high-precision diamond tool cutting edge radius measurement.

There are two approaches for reducing the convolution effect of the AFM tip. One approach is to directly characterize the AFM tip with a form surface measuring instrument. The other is to make an error separation operation on the AFM images for removing the influence of the AFM tip profile. Compared with the direct characterization approach, the error separation approach is more effective because no additional form surface measuring instruments are necessary [21]. An edge reversal error separation method based on an AFM was proposed for measuring cutting edge radius of diamond tool without the effect of the AFM tip radius by the authors of [22]. Firstly, a replicated cutting edge was obtained by indenting a diamond tool into a soft metal material. Then, an AFM was applied to scan the actual and the replicated cutting edges. The cutting edge radius of the diamond tool was obtained by taking the difference between the AFM images of the actual and the replicated cutting edges when the elastic recovery of the replicated cutting edge was small as compared with the cutting edge radius. Molecular dynamics (MD) simulations were carried out to investigate the effect of elastic recovery on cutting edge radius in our previous researches [23,24]. It was verified that when the indentation depth of the replicated cutting edge was set to be larger than 200 nm, the elastic recovery of the replicated cutting edge could be ignored. However, a large indentation depth would cause more measurement uncertainties in the AFM measurement of the replicated tool cutting edge.

Meanwhile, a nanoindentation instrument has been designed for replicating the diamond tool cutting edge onto a soft workpiece surface [25]. The displacement of the diamond tool was monitored by a capacitive sensor of a fast tool servo (FTS) unit, which was employed to drive the diamond tool. The workpiece was pasted on a cantilever whose deflection was detected by another capacitive sensor. The indentation depth of the diamond tool was obtained from the outputs of the two capacitive sensors. However, it was a time-consuming process to replicate the diamond tool cutting edge using this nanoindentation instrument. In addition, a contact damage would be generated on the workpiece surface when the tool-workpiece contact was established before the replication. The contact damage would influence the indentation depth and the profile of the replicated cutting edge.

In this paper, we present high-precision cutting edge radius measurements of single point diamond tools using an AFM and a reverse cutting edge artifact based on the proposed edge reversal method and the designed nanoindentation system. After introducing the measurement principles, the operation of the nanoindentation system is optimized for achieving high-precision control of the indentation depth. The effects of the elastic recovery and the AFM cantilever tip on the measurement accuracy and the measurement uncertainty are investigated. A series of experiments are preformed to verify the reliability and the capability of the proposed method and the designed system.

2. Principle of Measuring Cutting Edge Radius

There are three steps in the process of measuring the diamond tool cutting edge radius using the proposed edge reversal method, as shown in Figure 1.

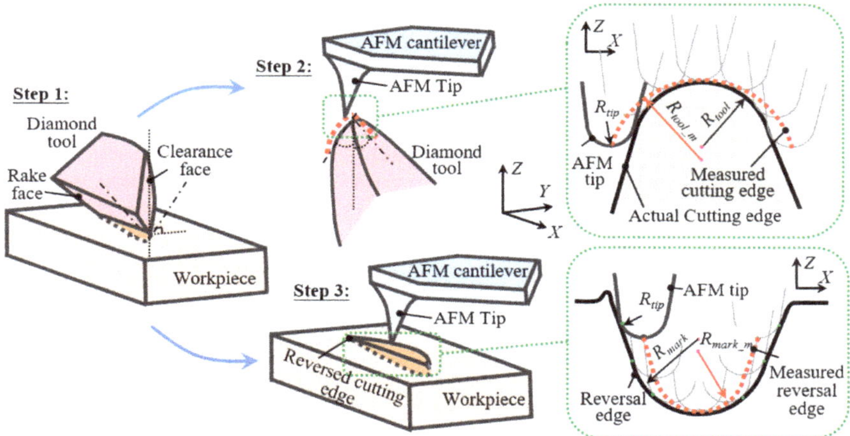

Figure 1. Principle of the proposed edge reversal method.

The first step is related to the cutting-edge replication shown in Figure 1. The single point diamond tool, a piece of copper workpiece, and a nanoindentation system are employed. An indentation mark, referred to as the reverse cutting edge artifact, is fabricated by indenting the diamond tool into the workpiece with the bisector of the included angle between its rake face and clearance face perpendicular to the workpiece surface using the nanoindentation system. The configuration and the principle of the nanoindentation system is introduced later. The reverse cutting edge artifact is, thus, directly transcribed from the geometry of the diamond tool.

The second step is related to the AFM measurement of the actual cutting edge of the diamond tool. The diamond tool, with the cutting edge radius, R_{tool}, is positioned under the AFM cantilever with the bisector of the included angle between its rake face and clearance face along the vertical axis. An AFM cantilever is moved by an XY-stage and a Z-scanner of the AFM along the X-, Y- and Z-directions to align the AFM cantilever tip with the apex of the diamond tool cutting edge. After an accurate alignment, the AFM cantilever is brought to scan across the diamond tool cutting edge. Since the tip radius R_{tip} of the AFM cantilever is comparable to the cutting edge radius, R_{tool}, of the diamond tool, the scan trace of the AFM with an apex radius of $R_{\text{tool\_m}}$ is a convolution of the AFM tip and the diamond tool cutting edge, as shown in Figure 1. Therefore, the following equation can be obtained:

$$R_{\text{tool\_m}} = R_{\text{tool}} + R_{\text{tip}} \tag{1}$$

The third step is related to the AFM measurement of the reverse cutting edge artifact. Then, the reverse cutting edge artifact is measured by the AFM. The apex radius of the reverse cutting edge is defined as R_{mark}, which is also comparable to the AFM tip radius R_{tip}. After the alignment between the AFM cantilever tip and the apex of reverse cutting edge, the AFM tip scans across the reverse cutting edge artifact to map out its topography. The scanned image is also a convolution of the AFM tip and the reverse cutting edge artifact. Therefore, the following equation can be obtained:

$$R_{\text{mark\_m}} = R_{\text{mark}} - R_{\text{tip}} \tag{2}$$

The copper workpiece surface displays a certain degree of elastic recovery after the diamond tool is withdrawn from the surface in the nanoindentation process [23,24]. The relationship between the actual cutting edge radius and reverse cutting edge radius can be expressed as:

$$R_{\text{tool}} = (1 - \xi) R_{\text{mark}} \tag{3}$$

where ξ represents the elastic recovery coefficient of the reverse cutting edge artifact. According to Equations (1)–(3), cutting edge radius of the diamond tool can be evaluated to be:

$$R_{\text{tool}} = \frac{1}{2 - \xi} \left(R_{\text{tool\_m}} + R_{\text{mark\_m}} \right) \tag{4}$$

Therefore, cutting edge radius of the diamond tool can be obtained without being influenced by the AFM tip radius. As can be seen from Equation (4), the achievable accuracy of this method depends on the elastic recovery coefficient ξ. It has been verified that the elastic recovery coefficient, ξ, of the reverse cutting edge would be reduced to 0.012 from 0.068 when the indentation depth was increased to 20 nm from 2.5 nm based on molecular dynamics (MD) simulations [22]. The detail of the MD simulation can be found in [22] and is not repeated in this paper for the sake of clarity. The difference between R_{mark} and R_{tool} was evaluated to be 0.24 nm when the indentation depth was set to be 200 nm. The difference is small and is negligible for the cutting edge radius measurement as compared with the cutting edge radius. However, when the indentation depth is set to be 200 nm, the measurement uncertainty in the tool cutting edge radius measurement becomes large. Therefore, the indentation depth should be controlled within a range of 20 to 200 nm.

A nanoindentation system is, then, designed for achieving the requirement of high-precision indentation depth control of the reverse cutting edge artifact as shown in Figure 2. A single point diamond tool is mounted on the FTS to indent into a prepolished metal workpiece by a linear stage and/or the FTS. When the distance between the diamond tool tip and the workpiece surface is larger than the motion stroke of the FTS, the linear stage is employed to move the diamond tool to approach the workpiece surface by a steeping motor controller. The displacement of the diamond tool driven by the linear stage and the FTS can be measured by the linear encoder of the stage and the displacement sensor of the FTS unit (inside sensor). The workpiece is attached on a cantilever. One end of the cantilever is fixed on a holder and the other end is preloaded by a preload controller. An outside displacement sensor (outside sensor) is used to detect the deflection of the cantilever caused by the indentation. Two *XY*-manual stages (Stage 1 and Stage 2) are used for the alignment between the workpiece surface and the diamond tool tip. The outputs of two sensors are collected by an oscilloscope for further evaluating the indentation depth d_{depth} based on the following equation:

$$d_{\text{depth}} = d_{\text{in}} - d_{\text{out}} \tag{5}$$

where d_{in} and d_{out} represent the outputs of the inside and the outside sensors, respectively.

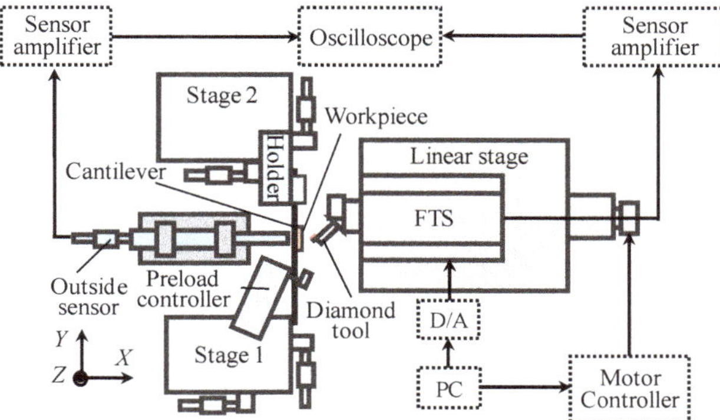

Figure 2. Schematic of the nanoindentation system.

The operation of the nanoindentation system is optimized for achieving high-precision control of the indentation depth during the tool cutting edge replication, as shown in Figure 3. There are three steps for replicating the cutting-edge geometry of a diamond tool into a workpiece surface. In step one, the linear stage is firstly used to move the diamond tool to the initial position shown in Figure 3a, where the distance between the diamond tool tip and the workpiece surface is equal to the motion stroke of the FTS. Then, the diamond tool is actuated by the FTS with a step of 5 nm to avoid any unexpected contact damages on the workpiece surface. The output of the outside sensor does not change in this step, as shown in Figure 3b. In step two, the contact between the diamond tool and the workpiece is established at the contact position where the output of the outside sensor starts to vary due to the deflection of the cantilever. A small indentation mark with a depth less than 5 nm is generated on the workpiece surface, which is small as compared with the reverse cutting edge. Therefore, the effect of the small indentation mark can be ignored. In step three, the diamond tool is indented into the workpiece surface with the command indentation depth, which can be evaluated by substituting the outputs of two sensors at the indentation position into Equation (5). Therefore, a reverse cutting edge artifact with a high-precision depth is fabricated by following these three steps.

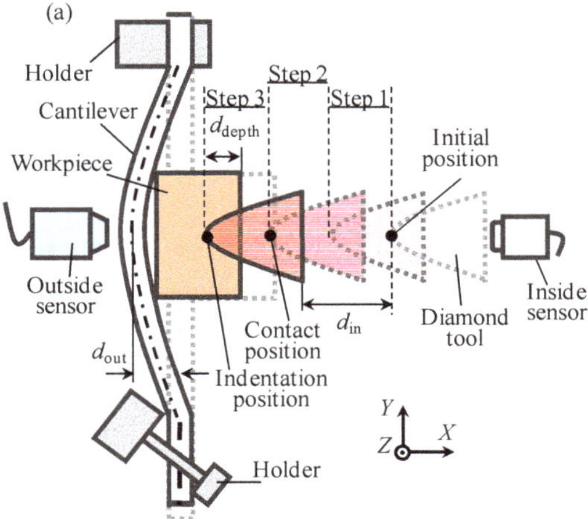

Figure 3. *Cont.*

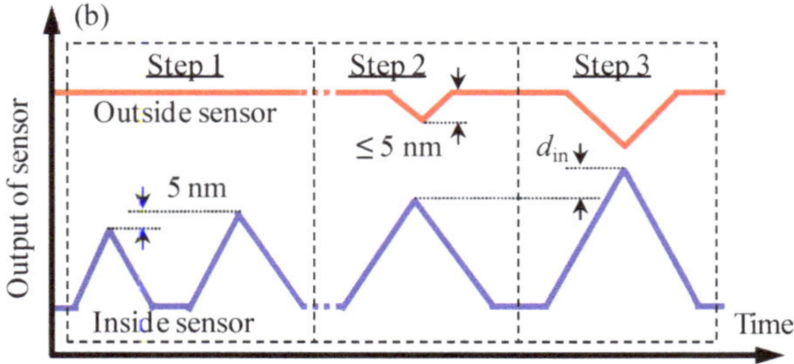

Figure 3. Principle of replicating the tool cutting edge using the proposed nanoindentation system. (**a**) Motion of the diamond tool; (**b**) Outputs of two sensors.

3. Experiment and Discussion

A series of experiments were performed to investigate the effect of elastic recovery on the reverse cutting edge artifact and the AFM cantilever tip on the measurement accuracy of the cutting edge radius as well as to verify the reliability of the proposed edge reversal method and the capability of the designed nanoindentation system.

Since copper has a relatively large Young's modulus, a prepolished copper workpiece with a size of 10 mm (length) × 10 mm (width) × 2 mm (thickness) was selected as the workpiece. Figure 4a shows the topography of the copper workpiece surface measured by a commercial AFM (Innova, Bruker, Billerica, MA, USA). As shown in Figure 4b, the Root Mean Square (RMS) roughness of the section A–A′ of the copper workpiece was evaluated to be 2.43 nm based on a 3D AFM topography. The workpiece surface was smooth enough for replicating the diamond tool cutting edge geometry. The copper workpiece was pasted on an aluminum cantilever with a spring constant of 155 N/m at the contact point. The deformation of the aluminum cantilever was detected by a capacitive sensor (6800, MicroSense, Lowell, MA, USA).

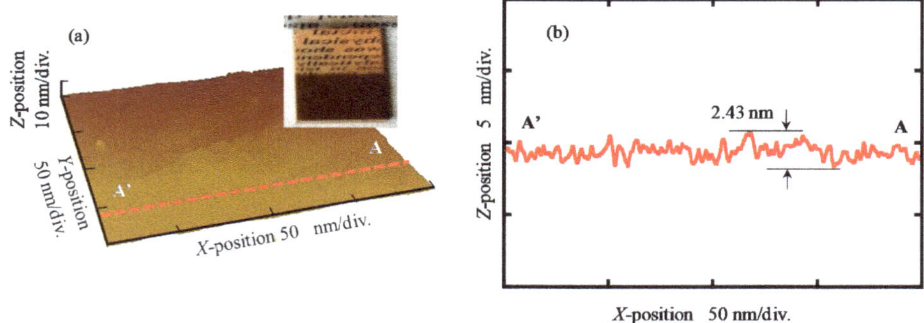

Figure 4. Copper workpiece. (**a**) Atomic force microscope (AFM) topography; (**b**) Roughness of the section A–A′.

3.1. Effect of Elastic Recovery

The effect of elastic recovery of the reverse cutting edge artifact on cutting edge radius measurement accuracy was analyzed. An AFM cantilever (OMCL-AC240TS, Olympus, Tokyo, Japan) with a nominal tip radius of 7 nm and a tip height of 14 µm was selected to measure the topographies of the actual and the reverse diamond tool cutting edge. A single point diamond tool with a nose radius of 1 mm was measured in this test.

Since the indentation depth and the elastic recovery of the reverse cutting edge artifact are related with each other based on the Hertz theory, a group of reverse cutting edge artifacts with various indentation depths from 20 to 180 nm were fabricated on the designed nanoindentation system in order to investigate the effect of the elastic recovery. The interval of indentation depth was set at approximately 20 nm.

Figure 5 shows the outputs of the inside and the outside sensors when the command displacement of the diamond tool actuated by the FTS was set at 50 nm. Only the outputs in step three of Figure 3 are plotted in Figure 5. The output of the inside sensor was 52 nm, which was approximately the same as the command displacement of the diamond tool. The output of the outside sensor was 27 nm. The indentation depth of the reverse cutting edge artifact was calculated to be 25 nm by substituting the outputs of two sensors into Equation (5).

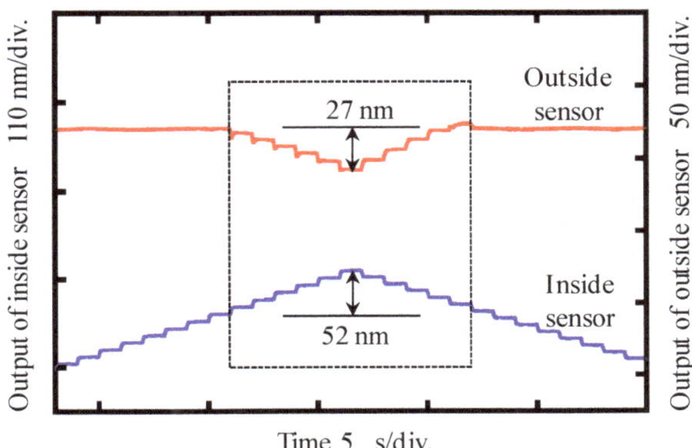

Figure 5. The outputs of the inside and the outside sensors with a command displacement of 50 nm.

The diamond tool was positioned on the Z-scanner of the AFM with the bisector of the included angle between its rake face and clearance face. The Olympus AFM cantilever was moved along the X-, Y- and Z-directions by the XY-stage and the Z-scanner of the AFM to make the alignment between the AFM tip and the apex of the diamond tool cutting edge. The actual cutting edge was, then, scanned by the AFM cantilever. The scan range and the scan rate were set to be 2 μm (X) × 2 μm (Y) and 1.2 μm/s, respectively. The numbers of scanning lines in the X- and Y-positions were 1024 and 1024, respectively. Figure 6a shows a 3D topography of the diamond tool cutting edge. It can be seen from the figure that there was a clear edge between the rake face and the clearance face, which can be employed for quantitative evaluation of the diamond tool cutting edge radius. The cross-sectional profile extracted from the 3D topography is shown in Figure 6b. The measured tool cutting edge radius was obtained from that of the fitted arc, which was evaluated by fitting the points in the apex of the cross-sectional profile of the cutting edge based on the least square method. The radius of the fitted arc, referred to as R_{tool_m}, was evaluated to be approximately 40.32 nm. It should be noted that the fitted arc was a geometric convolution of the AFM tip and the actual diamond tool cutting edge. Therefore, the value of 40.32 nm was not the actual diamond tool cutting edge radius.

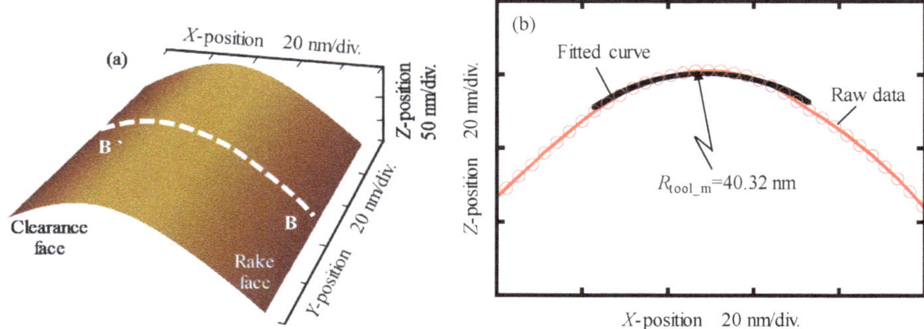

Figure 6. Measured diamond tool cutting edge. (**a**) AFM topography; (**b**) Profile of Section B–B′.

The reverse cutting edge artifact was also placed on the Z-scanner of the AFM. The same AFM cantilever, which had been used to scan the actual cutting edge, was brought to scan the reverse cutting edge. The scanning range, the scan rate, and the number of scanning lines in the X- and Y-position were set to be the same as those in the AFM measurement of the actual cutting edge. Figure 7a,b shows the AFM topography and the cross-sectional profile of the reverse cutting edge artifact, respectively. It can be seen from Figure 7a that the reverse cutting edge was well transcribed from the geometry of the diamond tool. The depth of the reverse cutting edge artifact was evaluated to be 28 nm. It was approximately the same as the indentation depth of 25 nm, which was evaluated based on the outputs of the inside and the outside sensors of the nanoindentation system. The control accuracy of the indentation depth in the designed nanoindentation system was, therefore, verified from the result. The points in the apex of the measured revered cutting edge were fitted using the least square method, as shown in Figure 7b. The radius of the reverse fitted arc, referred to as R_{mark_m}, was evaluated to be 21.04 nm. Similarly, the reverse fitted arc was a geometric convolution of the AFM tip and the reverse cutting edge. The value of 21.04 nm was not the reverse diamond tool cutting edge radius. Assuming ξ is equal to zero, the actual cutting edge radius, R_{tool}, was calculated by substituting the evaluated R_{tool_m} and R_{mark_m} into Equation (4) to obtain 30.68 nm. The difference between R_{tool} and R_{tool_m} was evaluated to be 9.64 nm, which was close to the nominal AFM tip radius of 7 nm. On the basis of the experimental results, the effectiveness of the proposed measurement method for cutting edge radius using an AFM and a reverse cutting edge artifact was verified.

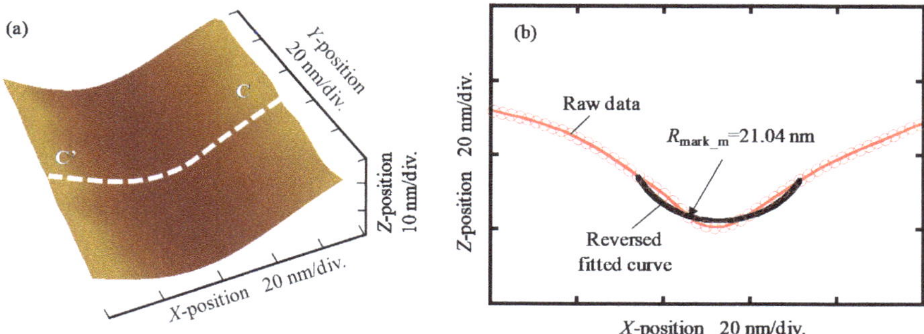

Figure 7. Measured reverse diamond tool cutting edge. (**a**) AFM topography; (**b**) Profile of Section C–C′.

R_{mark_m} of other reverse cutting edges with various indentation depths were also evaluated using the proposed measurement method for cutting edge radius. The evaluated R_{mark_m} and R_{tool} are plotted in Figure 8. As can be seen in the figure, the average diamond tool cutting edge radius was estimated to be 30.38 nm with a standard deviation of 0.31 nm. Therefore, it was verified that the elastic recovery of the reverse cutting edge artifact did not affect the measurement accuracy of cutting edge radius when the indentation depth was within 20 to 200 nm.

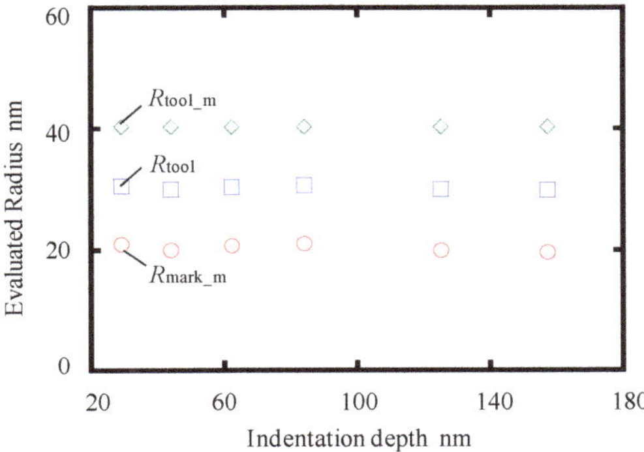

Figure 8. Experimental results under various indentation depths.

In addition, the effect of elastic recovery on measurement uncertainty of cutting edge radius was investigated. The indentation area between the diamond tool cutting edge and the workpiece, shown in Figure 9, can be recognized as a cylinder and a plane surface. According to the Hertz theory [26], for a plane surface and a cylinder with a radius of R_{tool}, the elastic recovery d_{er} of the reverse cutting edge along the X-direction can be expressed by the following equation:

$$d_{er} = \frac{4}{\pi} \cdot \frac{1}{L} \cdot F \cdot \frac{1 - v^2}{E} \quad (6)$$

where F is the applied indentation force between the diamond tool and the workpiece, which can be obtained from the deflection of the cantilever at the indentation position and the spring constants of the cantilever in the designed nanoindentation system. L is the length of cylinder, which can be obtained based on the indentation depth and the nose radius of the diamond tool. E and v represent the Young's modulus and the Poisson's ratio of the diamond tool, respectively.

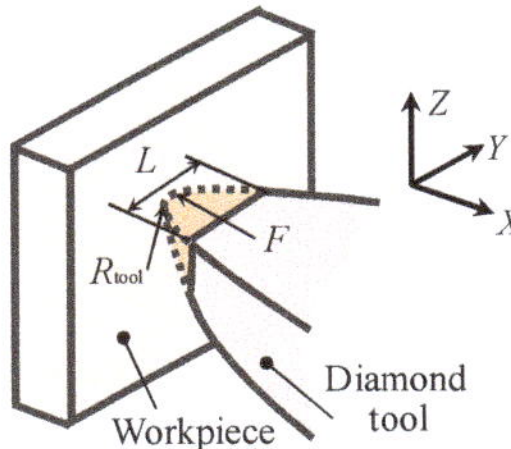

Figure 9. Schematic of the indentation area between the diamond tool and the copper workpiece.

Figure 10 shows the elastic recovery of the reverse cutting edge under various indentation depths. The values were employed to evaluate the measurement uncertainty induced by the elastic recovery. The measurement uncertainty of cutting edge radius of a diamond tool with 1 mm nose radius was evaluated based on the Guide to the Uncertainty in Measurement (GUM) [27].

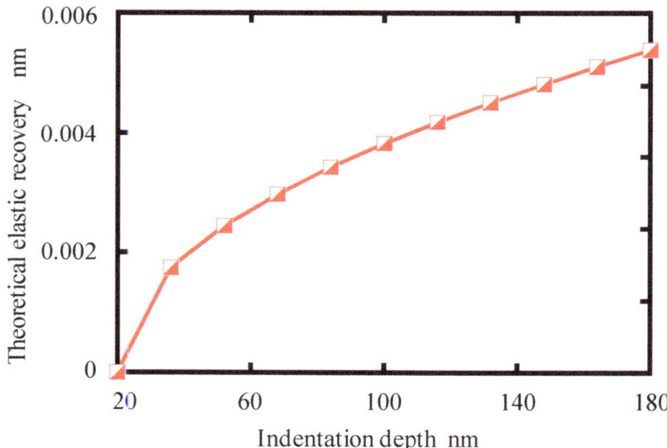

Figure 10. The theoretical elastic recovery under various indentation depths.

Table 1 shows the calculated uncertainties when the indentation depth was equal to 25 nm. The uncertainty in R_{tool_m} and R_{mark_m} was evaluated to be 1.36 and 1.42 nm, respectively. The uncertainty with a coverage factor of $k = 2$ (95% confidence) in the measurement of the diamond tool cutting edge radius was evaluated to be 1.97 nm, which was smaller than that obtained in our previous research [22].

The measurement uncertainties under various indentation depths are shown in Figure 11. The standard deviation of the evaluated uncertainties was calculated to be 0.005 nm. The results further demonstrated that elastic recovery of the reverse cutting edge did not influence the measurement accuracy of the diamond tool cutting edge radius based on the proposed edge reversal method and the designed nanoindentation system.

Table 1. Uncertainty analysis in the measurement of a diamond tool with a nose radius of 1 mm.

Uncertainty Sources	Symbol	Uncertainty Value	Distribution	Standard Uncertainty
Lateral imaging resolution	u_r	0.97 nm	Rectangular	0.56 nm
Lateral positioning resolution	u_{L_n}	1.2 nm	Normal	1.2 nm
Vertical positioning resolution	u_{V_n}	0.2 nm	Normal	0.2 nm
Thermal resolution	u_{t_t}	3.3×10^{-5} nm	Rectangular	1.9×10^{-5} nm
Measurement resolution	u_{t_d}	1.457 nm	-	0.46 nm
Uncertainty in R_{tool_m}	u_{tool_m}	-	-	1.36 nm
Lateral imaging resolution	u_r	0.97 nm	Rectangular	0.56 nm
Lateral positioning resolution	u_{L_n}	1.2 nm	Normal	1.2 nm
Vertical positioning resolution	u_{V_n}	0.2 nm	Normal	0.2 nm
Thermal resolution	u_{m_t}	4.9×10^{-5} nm	Rectangular	2.8×10^{-5} nm
Measurement resolution	u_{m_d}	0.385 nm	-	0.12 nm
Elastic recovery	u_{m_e}	0.358 nm	Rectangular	0.21 nm
Indentation force	u_{m_f}	0.001 nm	Rectangular	5.7×10^{-4} nm
Uncertainty in R_{mark_m}	u_{mark_m}	-	-	1.42 nm

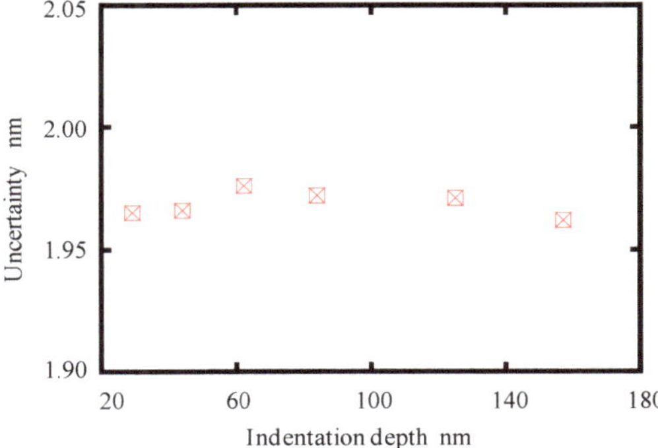

Figure 11. The measurement uncertainties of cutting edge radius of a diamond tool under various indentation depths.

3.2. Effect of AFM Cantilever Tip

The AFM cantilever was one of the crucial factors that affected the evaluation accuracy of the diamond tool cutting edge radius. Therefore, another AFM cantilever (MPP-11100-10, Bruker, Billerica, MA, USA) was installed onto the AFM to scan the actual and the reverse diamond tool cutting edges for investigating the effect of the AFM cantilever tip on the measurement accuracy of the diamond tool cutting edge radius. The Olympus AFM cantilever and the Bruker AFM cantilever were referred to as Cantilever 1 and Cantilever 2, respectively. Differing from Cantilever 1, there was a long base at the top of Cantilever 2. The tip radius of Cantilever 2 was 12 nm, which was larger than that of Cantilever 1.

Then, Cantilever 2 was used to scan the actual and the reverse cutting edges, which had been scanned by Cantilever 1. We confirmed that the AFM topographies of the cutting edges scanned by Cantilever 2 were similar to those measured by Cantilever 1, although the AFM images are not shown here for the sake of clarity. Figure 12 shows the measured diamond tool cutting edge radii using Cantilever 2. The average characterized cutting edge radius of the diamond tool was 29.67 nm, with a standard deviation of 0.16 nm. Characterized cutting edge radius using Cantilever 2 was the same as that using Cantilever 1. The measurement uncertainty using Cantilever 2 was evaluated. The results are also plotted in Figure 12. The standard deviation of the evaluated uncertainty values was calculated to be 0.028 nm. From the above results, it can be seen that the measurement accuracy of the diamond tool cutting edge radius based on the proposed edge reversal method was not affected by the AFM cantilever tip radius. This is consistent with the fact that the effect of the AFM cantilever tip radius can be removed in the proposed method.

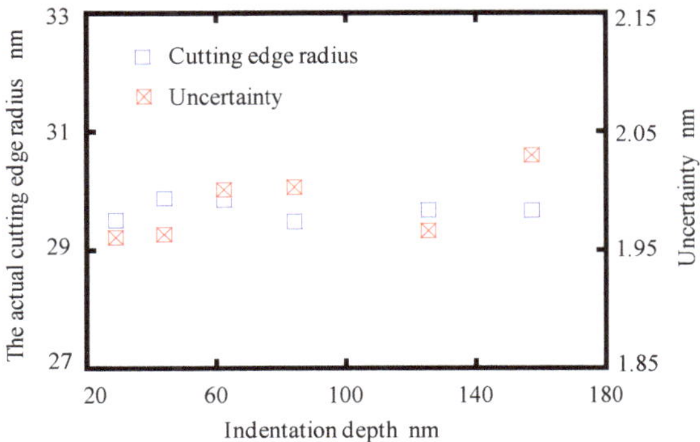

Figure 12. Experimental results using Cantilever 2.

3.3. Reliability of the Proposed Method and the Designed System

To further demonstrate the reliability of the proposed measurement method, cutting edge radius of a diamond tool with a different nose radius of 2 mm, which was referred to as Diamond Tool 2, was also evaluated.

Diamond Tool 2 was mounted on the nanoindentation system to replicate its cutting edge on the copper workpiece with various indentation depths. The AFM with Cantilever 1 was, then, applied to measure the profiles of the actual and replicated cutting edges. The command displacement of Diamond Tool 2 actuated by the FTS was also set from 50 to 300 nm with an interval of 50 nm. However, the corresponding indentation depths were evaluated to be 35, 45, 77, 104, 127, and 163 nm according to the outputs of the inside and the outside sensors, which were different from those by the previous diamond tool with a nose radius of 1 mm.

The measured results of Diamond Tool 2 are summarized in Figure 13. The average radius of the diamond tool cutting edge was evaluated to be 41.29 nm with a standard deviation of 0.66 nm. The standard deviation of the evaluated uncertainty values was calculated to be 0.032 nm. The difference between the actual cutting edge radius by the proposed measurement method (R_{tool}) and the cutting edge radius imaged directly by the AFM (R_{tool_m}) of Diamond Tool 2 was evaluated to be 7.36 nm. This value was also close to the nominal AFM tip radius of 7 nm. The reliability of the proposed method was, thus, verified from the measurement results of Diamond Tool 2.

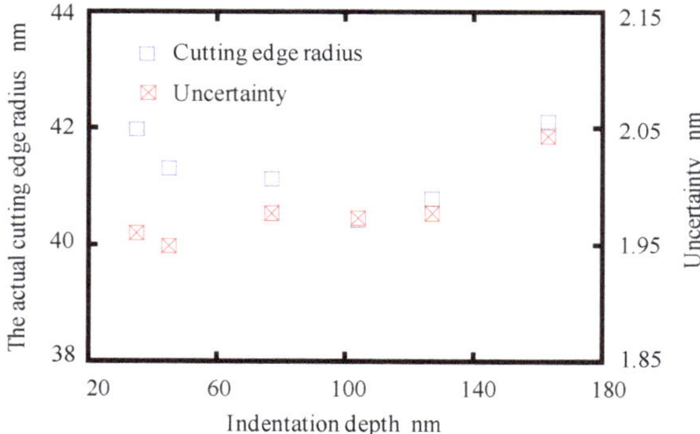

Figure 13. Experimental results of measuring Diamond Tool 2.

4. Summary

In this paper, we presented high-precision cutting edge radius measurement of single point diamond tools using an atomic force microscope and a reverse cutting edge artifact based on the edge reversal method. A reverse cutting-edge artifact with a high-precision depth was fabricated on a workpiece surface using the optimized operation of a newly designed nanoindentation system. Cutting edge radius was evaluated from two AFM images, including an AFM image of the actual cutting edge and an AFM image of the reverse cutting edge, from which the convolution effect of the AFM tip radius was reduced. Cutting edge radii of two diamond tools were evaluated. The first tool had a nose radius of 1 mm and the second tool had a nose radius of 2 mm. The difference between cutting edge radii evaluated by the proposed measurement method and that directly obtained from the AFM image was 9.64 nm for the first tool and 7.36 nm for the second tool. Both values were close to the AFM probe tip nominal radius of 7 nm, from which the feasibility and the reliability of the proposed method were demonstrated. The effect of the elastic recovery was investigated by scanning a group of reverse cutting edge artifacts with various depths from 20 to 180 nm. We confirmed that the elastic recovery of the reverse cutting edge did not affect the measurement accuracy of cutting edge radius when the indentation depth was within 20 to 200 nm. Three-dimensional profile measurements of diamond tool cutting edge based on the proposed measurement method will be carried out in the future work.

Author Contributions: Conceptualization, W.G., Y.S., and Y.C., Data curation, K.Z. and H.M.; Formal analysis, K.Z. and Y.C.; Investigation, Y.S. and Y.C.; Methodology, W.G.; Supervision, W.G.; Visualization, H.M.; Writing—original draft, K.Z.; Writing—review and editing, W.G. and Y.C. All authors have read and agreed to the published version of the manuscript.

Funding: This research is supported by a Japanese Government (MEXT) Scholarship. K.Z. thanks the support from the China Scholarship Council (CSC). In addition, a part of this research is supported by the Japan Society for the Promotion of Science (JSPS) 15H05759 and 20H00211.

Conflicts of Interest: The authors declare no conflict of interest. The funders had no role in the design of the study; in the collection, analyses, or interpretation of data; in the writing of the manuscript, and in the decision to publish the results.

References

1. Zhu, Z.W.; To, S.; Zhang, S.J. Large-scale Fabrication of micro-lens array by novel end-fly-cutting servo diamond machining. *Opt. Express* **2015**, *23*, 20593–20604. [CrossRef] [PubMed]
2. Gong, H.; Wang, Y.; Song, L.; Fang, F.Z. Spiral tool path generation for diamond turning optical freeform surfaces of quasi-revolution. *Comput. Aided Des.* **2015**, *59*, 15–22. [CrossRef]
3. Gao, W.; Araki, T.; Kiyono, S.; Okazaki, Y.; Yamanaka, M. Precision nano-fabrication and evaluation of a large area sinusoidal grid surface for a surface encoder. *Precis. Eng.* **2003**, *27*, 289–298. [CrossRef]
4. Gao, W.; Aoki, J.; Ju, B.F.; Kiyono, S. Surface profile measurement of a sinusoidal grid using an atomic force microscope on a diamond turning machine. *Precis. Eng.* **2007**, *31*, 304–309. [CrossRef]
5. Guo, J.; Zhang, J.; Pan, Y.; Kang, R.; Namba, Y.; Shore, P.; Yue, X.; Wang, B.; Guo, D. A critical review on the chemical wear and wear suppression of diamond tools in diamond cutting of ferrous metals. *Int. J. Extrem. Manuf.* **2020**, *2*, 012001.
6. Brinksmeier, E.; Riemer, O.; Glabe, R.; Lunemann, B.; Kopylow, Y.V.; Dankwart, C.; Meier, A. Submicron functional surface generated by diamond machining. *CIRP Ann. Manuf. Technol.* **2010**, *59*, 535–538. [CrossRef]
7. Wang, J.S.; Fang, F.Z.; Yan, G.P.; Guo, Y.B. Study on diamond cutting of ion implanted tungsten carbide with and without ultrasonic vibration. *Nanomanuf. Metrol.* **2019**, *2*, 177–185. [CrossRef]
8. Shinozaki, A.; Namba, N. Diamond tool wear in the ultra-Precision cutting of large electroless nickel and fabrication of replicated mirrors for a soft X-ray microscope. *JSME Int. J. C Mech. Syst.* **2006**, *49*, 56–62.
9. Gao, W.; Motoki, T.; Kiyono, S. Nanometer edge profile measurement of diamond cutting tools by atomic force microscope with optical alignment sensor. *Precis. Eng.* **2006**, *30*, 396–405. [CrossRef]
10. Gao, W. *Precision Nanometrology–Sensors and Measuring Systems for Nanomanufacturing*; Springer: London, UK, 2010.
11. Gao, W.; Kim, S.W.; Bosse, H.; Haitjema, H.; Chen, Y.L.; Lu, X.D.; Knapp, W.; Weckenmann, A.; Estler, W.T.; Kunzmann, H. Measurement technologies for precision positioning. *CIRP Ann. Manuf. Technol.* **2015**, *64*, 773–796. [CrossRef]
12. Weckenmann, A.; Nalbantic, K. Precision measurement of cutting tools with two matched optical 3D-sensors. *CIRP Ann. Manuf. Technol.* **2003**, *52*, 443–446. [CrossRef]
13. Lim, T.Y.; Ratnam, M.M. Edge detection and measurement of nose radii of cutting tool inserts from scanned 2-D images. *Opt. Lasers Eng.* **2012**, *50*, 1628–1642. [CrossRef]
14. Michihata, M.; Kim, J.; Takahashi, S.; Takamasu, K.; Mizutani, Y.; Takaya, Y. Surface imaging technique by an optically trapped microsphere in air condition. *Nanomanuf. Metrol.* **2018**, *1*, 32–38. [CrossRef]
15. Asai, S.; Taguchi, Y.; Horio, K.; Kasai, T.; Kobayashi, A. Measuring the very small cutting-edge radius for a diamond tool using a new kind of SEM having two detectors. *CIRP Ann. Manuf. Technol.* **1990**, *39*, 85–88. [CrossRef]
16. Denkena, B.; Kohler, J.; Ventura, C.E.H. Customized cutting edge preparation by means of grinding. *Precis. Eng.* **2013**, *37*, 590–598. [CrossRef]
17. Sawano, H.; Ayada, S.; Yoshioka, H.; Shinno, H. A newly developed AFM-Based three dimensional profile measuring system. In Proceedings of the 11th International Conference of the European Society for Precision Engineering and Nanotechnology, Como, Italy, 23–26 May 2011; Volume 1, pp. 108–112.

18. Gao, W.; Asai, T.; Arai, Y. Precision and fast measurement of 3D cutting edge profiles of single point diamond micro-tools. *CIRP Ann. Manuf. Technol.* **2009**, *58*, 451–454. [CrossRef]
19. Li, Z.Q.; Sun, T.; Li, P.; Zhao, X.S.; Dong, S. Measuring the nose roundness of diamond cutting tools based on atomic force microscopy. *J. Vac. Sci. Technol. B* **2009**, *27*, 1394–1398. [CrossRef]
20. Li, X.P.; Rahman, M.; Liu, K.; Neo, K.S.; Chan, C.C. Nano-precision measurement of diamond tool edge radius for wafer fabrication. *J. Mater. Process. Technol.* **2003**, *140*, 358–362. [CrossRef]
21. Gao, W.; Haitjema, H.; Fang, F.Z.; Leach, R.K.; Cheung, C.F.; Savio, E.; Linares, J.M. On-machine and in-process surface metrology for precision manufacturing. *CIRP Ann. Manuf. Technol.* **2019**, *68*, 843–866. [CrossRef]
22. Chen, Y.L.; Cai, Y.; Xu, M.L.; Shimizu, Y.; Ito, S.; Gao, W. An edge reversal method for precision measurement of cutting edge radius of single point diamond tools. *Precis. Eng.* **2017**, *50*, 380–387. [CrossRef]
23. Cai, Y.; Chen, Y.L.; Shimizu, Y.; Ito, S.; Matsukuma, H.; Gao, W.; Zhang, L.C. Molecular dynamics simulation of sub-nanometric tool-workpiece contact on a force sensor-integrated fast tool servo for ultra-precision microcutting. *Appl. Surf. Sci.* **2016**, *369*, 354–365. [CrossRef]
24. Cai, Y.; Chen, Y.L.; Shimizu, Y.; Ito, S.; Gao, W. Molecular dynamics simulation of elastic-plastic deformation associated with tool-workpiece contact in force sensor-integrated fast tool servo. *J. Eng. Manuf.* **2018**, *232*, 1893–1902. [CrossRef]
25. Cai, Y.; Chen, Y.L.; Xu, M.L.; Shimizu, Y.; Ito, S.; Matsukuma, H.; Gao, W. An ultra-precision tool nano-indentation instrument for replication of single point diamond tool cutting edges. *Meas. Sci. Technol.* **2018**, *29*, 054004. [CrossRef]
26. Johnson, K.L. One hundred years of hertz contact. *Proc. Inst. Mech. Eng.* **1982**, *196*, 363–378. [CrossRef]
27. *JCGM 100: 2008 GUM 1995 with Minor Corrections. Evaluation of Measurement Data—Guide to the Expression of Uncertainty in Measurement (GUM)*; BIPM: Sèvres Cedex, France, 2008.

Article

Quantitative Investigation of Surface Charge Distribution and Point Probing Characteristics of Spherical Scattering Electrical Field Probe for Precision Measurement of Miniature Internal Structures with High Aspect Ratios

Xingyuan Bian [1,2], Junning Cui [1,2,*], Yesheng Lu [1,2], Yamin Zhao [1,2], Zhongyi Cheng [1,2] and Jiubin Tan [1,2]

1 Center of Ultra-Precision Optoelectronic Instrument Engineering, Harbin Institute of Technology, Harbin 150080, China; bianxingyuan@hit.edu.cn (X.B.); 13B901016@hit.edu.cn (Y.L.); zhaoyamin@163.com (Y.Z.); 19B901007@stu.hit.edu.cn (Z.C.); jbtan@hit.edu.cn (J.T.)

2 Key Lab of Ultra-Precision Intelligent Instrumentation (Harbin Institute of Technology), Ministry of Industry and Information Technology, Harbin 150080, China

* Correspondence: cuijunning@126.com or cuijunning@hit.edu.cn

Received: 21 June 2020; Accepted: 28 July 2020; Published: 30 July 2020

Abstract: For precision measurement of miniature internal structures with high aspect ratios, a spherical scattering electrical field probe (SSEP) is proposed based on charge signal detection. The characteristics and laws governing surface charge distribution on the probing ball of the SSEP are analyzed, with the spherical scattering electrical field modeled using a 3D seven-point finite difference method. The model is validated with finite element simulation by comparing with the analysis results of typical situations, in which probing balls of different diameters are used to probe a grounded plane with a probing gap of 0.3 μm. Results obtained with the proposed model and finite element method (FEM) simulation indicate that 31% of the total surface charge on a ϕ1 mm probing ball concentrates in an area that occupies 1% of the total probing ball surface. Moreover, this surface charge concentration remains unchanged when the surface being measured varies in geometry, or when the probing gap varies in sensing range. Based on this, the SSEP has realized approximate point probing capability with a virtual "needle" of electrical effect. Together with its non-contact sensing characteristics and 3D isotropy, it can, therefore, be concluded that the SSEP has great potential to be an ideal solution for precision measurement of miniature internal structures with high aspect ratios.

Keywords: surface charge distribution; point probing characteristics; spherical scattering electrical field probe; miniature internal structures; high aspect ratios

---

1. Introduction

Nowadays, parts adopting miniature internal structures with high aspect ratios, such as deep, small holes and grooves [1,2], can transmit working medium, energy, and information over relatively long distances, and are of great significance in achieving high integration and low energy consumption in the aviation, aerospace, and automotive industries. The machining precision of these structures often reaches the sub-micron level; moreover, their aspect ratio is up to several tens and even several hundreds [3–5], posing a great challenge to conventional measurement methods in terms of measurability. Various probing methods have been investigated to measure this particular kind of structure, such as miniaturized ball probes with flexible hinges [6–8], fiber optical probes [9–13], as well as other creative probing methods based on the principle of the tunneling effect [14], optical interference [15,16], pneumatics [17], capacitance [18], atomic force microscope [19], and the piezoelectric effect [20]. However, not having the three features of point probing capability, non-contact probing, and 3D isotropic sensing at the same time makes it difficult to accurately measure high aspect ratio structures with these conventional methods. As part of our efforts to find a possible solution including all of these required capabilities, a spherical scattering electrical field probe (SSEP) was proposed, based on the detection of charge signal on the probing ball, and nanometer resolution displacement sensing, non-contact probing and 3D isotropy sensing capability were achieved simultaneously [1,2]. The only unclear part of the SSEP is point probing capability, on which only qualitative analysis was performed for the time being.

The point probing capability of the SSEP is closely related to surface charge distribution on the probing ball. To analyze the point probing characteristics of the SSEP, a quantitative investigation of the spherical scattering electrical field needs to be conducted to illustrate the law of surface charge distribution. The difficulties are: (1) Conventional theoretical analysis methods [21,22] cannot complete the modeling task with complicated boundary conditions, such as the possibility that the surface being probed could be a plane, spherical, cylindrical, or free geometrical shape; (2) On the other hand, the spherical scattering electrical field is also difficult to model with conventional numerical methods due to the problem of balancing modeling accuracy and computational load. This is because the diameter of the probing ball is on a hundreds of microns to millimeter level, while the probing gap is on a micro and sub-micrometer level, and this creates a multi-scale problem, posing a challenge to gridding and computational accuracy; (3) Experimental methods [23–27] are not applicable here due to the lack of micro-/nano probes with the ability to detect the spherical scattering electrical field in the micro probing gap without introducing violent disturbance into the field. Therefore, an appropriate method of modeling and quantitative analysis of spherical scattering electrical field is the key issue to be solved for the analysis of the surface charge distribution characteristics as well as point probing characteristics of the SSEP.

In this paper, this problem is solved using a proposed 3D seven-point finite difference scheme. In the accordingly built spherical coordinates, the spherical scattering electrical field is modeled by 3D gridding and finite difference computation, solved through iteration calculation, and validated by finite element method (FEM) numerical simulation under special boundary conditions. In this way, the surface charge distribution characteristics and the resulting approximate point probing characteristics of the SSEP are quantificationally analyzed and represented.

2. Principle of SSEP

As shown in Figure 1, when an electrically conductive part is being probed, the surface being probed is grounded while the electric potential of probing ball is set to be a constant such as 1 V, and then a spherical scattering electrical field is formed. As the gap between the probing ball and the surface decreases, the surface charge on the probing ball tends to concentrate in quite a small area around the probing point, which is the closest point on the spherical surface to the surface being probed. The concentration of surface charge on the probing ball becomes obvious when the probing gap δ decreases to a certain level, which is often below the micrometer level. The spherical scattering electrical field and resulting surface charge distribution change drastically when probing gap δ changes, so high-resolution displacement sensing can be achieved by charge signal detection. The SSEP features non-contact, 3D isotropic, and approximate point probing characteristics, and can be fitted in a micro/nano-coordinate measurement machine or an analogous instrument to measure the 3D dimensions and geometry of miniature internal structures with high aspect ratios [2]. This paper focuses on the quantitative investigation of the surface charge distribution characteristics on the probing ball as well as the resulting approximate point probing capacity of the SSEP.

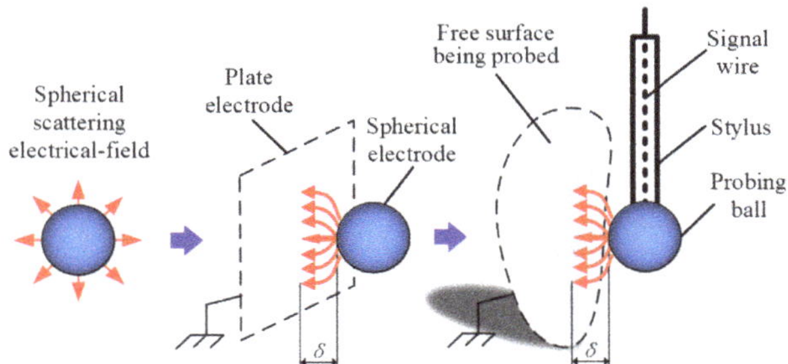

Figure 1. Working principle of spherical scattering electrical field probe (SSEP).

3. Surface Charge Distribution Modeling

Surface charge distribution modeling of the SSEP begins with spherical scattering electrical field analysis. The model of the electrical field is simplified, as shown in Figure 2. The shaft is not shown in Figure 2 because of its negligible effect, due to its delicately designed multi-coaxial-layer active shielding and grounding structure.

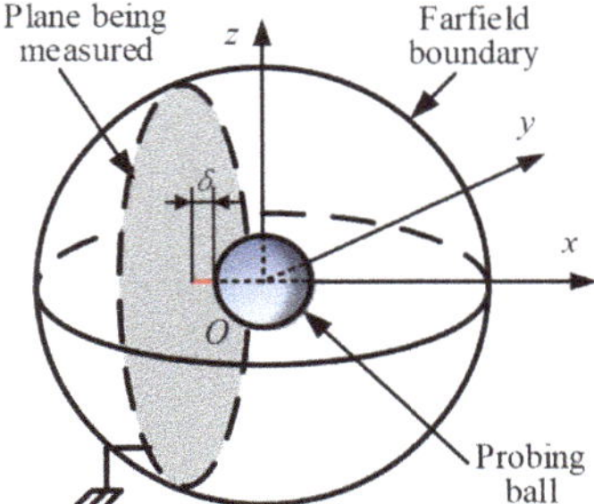

Figure 2. Model for analysis.

The electrical field analysis problem here can be taken as an electrostatic boundary value problem with a Dirichlet boundary condition, in which both the surface being measured and the far-field boundary are grounded, and the electrical potential of the probing ball is known. The electrical potential distribution U in the field can be expressed with Laplace's equation, as shown in Equation (1) below.

$$\nabla^2 U = 0, \tag{1}$$

Laplace's equation is expressed in a spherical coordinate as shown in Equation (2) in order to facilitate the calculation of the surface charge distribution on the probing ball.

$$\nabla^2 U = \frac{1}{r^2}\frac{\partial}{\partial r}\left(r^2\frac{\partial U}{\partial r}\right) + \frac{1}{r^2 \sin\theta}\frac{\partial}{\partial\theta}\left(\sin\theta\frac{\partial}{\partial\theta}\right) + \frac{1}{r^2\sin^2\theta}\frac{\partial^2 U}{\partial\phi^2} = 0, \tag{2}$$

To solve Laplace's equation, a 3D seven-point finite difference scheme is proposed as shown in Equation (3), and the spatial relationship of the seven points is shown in Figure 3. This finite difference scheme is derived from the spherical coordinates with the origin located at the center of the probing ball. Thus, the electrical potential distribution U can be obtained with proper gridding and iteration.

$$U_0{}^{i+1} = \begin{cases} \dfrac{r^2 \sin^2 \theta h_1 h_2 h_3 h_4 h_5 h_6}{2r \sin^2 \theta h_3 h_4 h_5 h_6 (r+h_2-h_1) + \sin^2 \theta h_1 h_2 h_3 h_4 [2+(h_6-h_5)\cot\theta] + 2h_1 h_2 h_5 h_6} \\ \cdot \left[\begin{array}{l} \frac{2(r+h_2)}{rh_1(h_1+h_2)} U_1{}^i + \frac{2(r-h_1)}{rh_2(h_1+h_2)} U_2{}^i + \frac{2}{r^2 \sin^2 \theta h_3 (h_3+h_4)} U_3{}^i \\ + \frac{2}{r^2 \sin^2 \theta h_4 (h_3+h_4)} U_4{}^i + \frac{2+h_6 \cot\theta}{r^2 h_5 (h_5+h_6)} U_5{}^i + \frac{2-h_5\cot\theta}{r^2 h_6 (h_5+h_6)} U_6{}^i \end{array} \right] \Bigg|_{\theta \neq 0 \text{ and } \theta \neq \pi} \\ \dfrac{h_4}{h_3+h_4} U_3{}^i + \dfrac{h_3}{h_3+h_4} U_4{}^i \Big|_{\theta = 0 \text{ or } \theta = \pi} \end{cases} \tag{3}$$

where $I = 1, 2, 3, \ldots$.

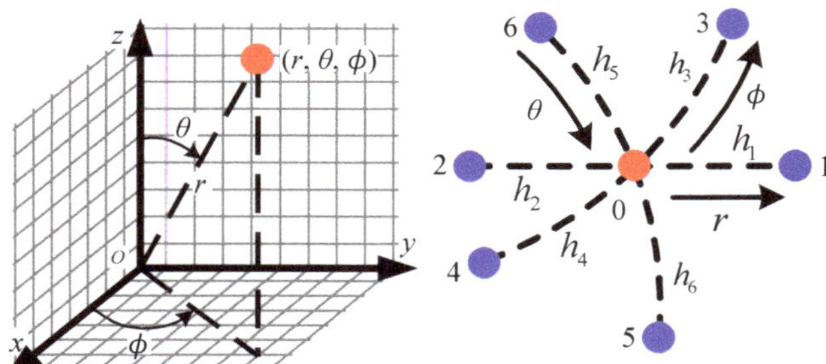

Figure 3. The spatial relationship of 7 points.

Considering the fact that the spherical scattering electrical field is extremely non-uniform, the grids in this method are generated non-uniformly from the spherical coordinates. As shown in Figure 4, fine grids are generated near the probing point while coarse grids are kept away from it. In addition, quick convergence is achieved by combining the over-relaxation iterative method and the non-uniform gridding in this study.

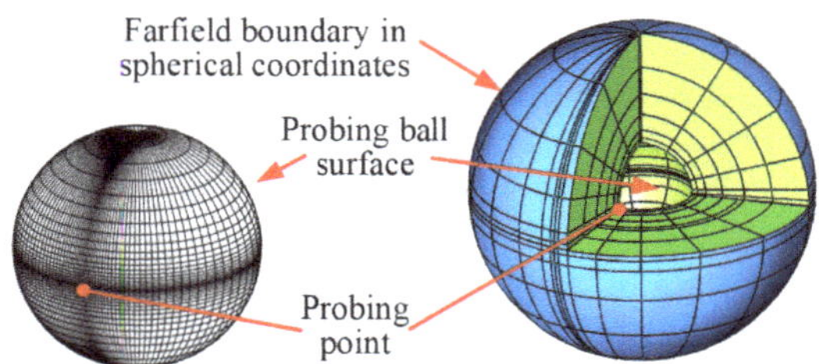

Figure 4. Schematic diagram of nonuniform gridding in spherical coordinates.

Electric field intensity distribution $\vec{E}$ can be obtained using Equation (4):

$$\vec{E} = -\nabla U \tag{4}$$

According to Gauss's law, charge density distribution σ on the surface of the probing ball can be expressed as Equation (5) since the probing ball is a conductive sphere,

$$\sigma = \varepsilon E, \tag{5}$$

where ε is the dielectric constant.

The distribution σ of the spherical surface charge on the probing ball can thus be obtained.

4. FEM Simulation Verification

In order to verify the effectiveness of the proposed method, FEM simulation is carried out by using COMSOL Multiphysics (COMSOL Inc, Burlington, MA, USA) to obtain the surface charge distribution σ over the spherical surface of the SSEP probing ball. The parameters used for the simulation are those commonly used in real situations. The diameter of the probing ball is ϕ1 mm, and the probing gap δ between the probing ball and the probed plane is 0.3 μm. The potential of the probing ball is set to be 1 V while the plane is grounded. The results for the surface charge distribution on the probing ball obtained with both methods are shown in Figure 5, in which the sphere stands for the probing ball of ϕ1 mm in diameter. An expanded view of surface charge distribution, in which the spherical surface is unfolded by polar angle θ and azimuthal angle ϕ, is shown in Figure 6 for clarity. The maximum relative differences in surface charge density on the probing point ($\theta = 90°$, $\phi = 180°$) of different probing balls between the modeling analysis and the simulation results are shown in Table 1, while the curves of the surface charge ratio as a function of the surface area ratio for different probing balls are shown in Figure 7. These curves are proposed to present the surface charge concentrating around the probing point, which is further discussed in Section 5.1. The surface charge ratio in this paper is referred to as the ratio between the surface charge on the area centering on the probing point and the total surface charge, while the surface area ratio is the ratio between the area centering on the probing point and the total surface area. It can be concluded that the results for the surface charge concentration obtained with the two methods perfectly match with each other, and the effectiveness of the method proposed can be verified through this comparison.

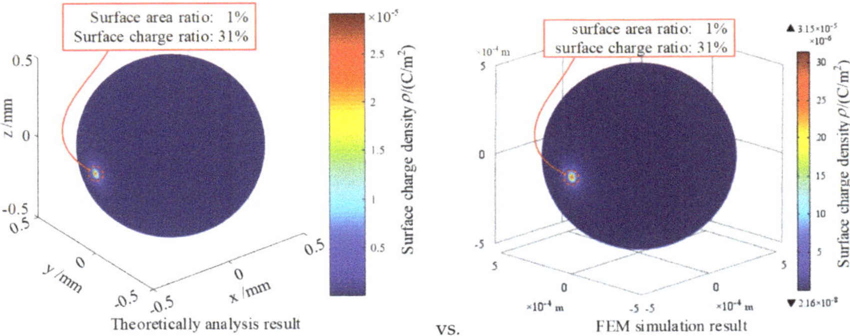

Figure 5. Modeling analysis and simulation results of surface charge distribution on the ϕ1 mm probing ball.

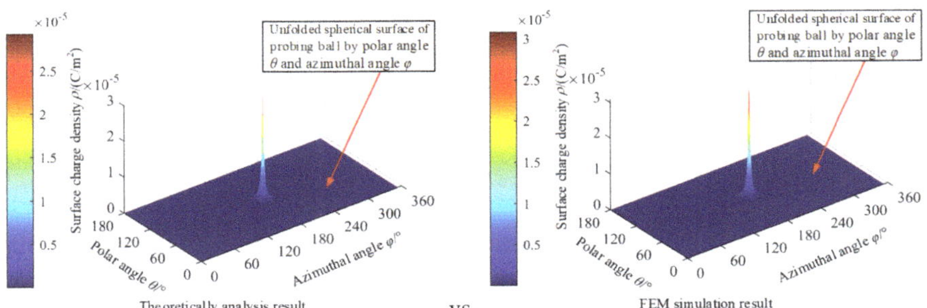

Figure 6. Modeling analysis and simulation results of surface charge distribution with the spherical surface unfolded.

Table 1. Maximum relative differences of surface charge density on the probing point of probing ball between modeling analysis and simulation results (δ = 0.3 µm).

Diameter of Probing Ball ϕ /mm	0.1	0.5	1	2
Maximum relative differences of surface charge density on probing ball	2.2%	5.9%	5.3%	4.2%

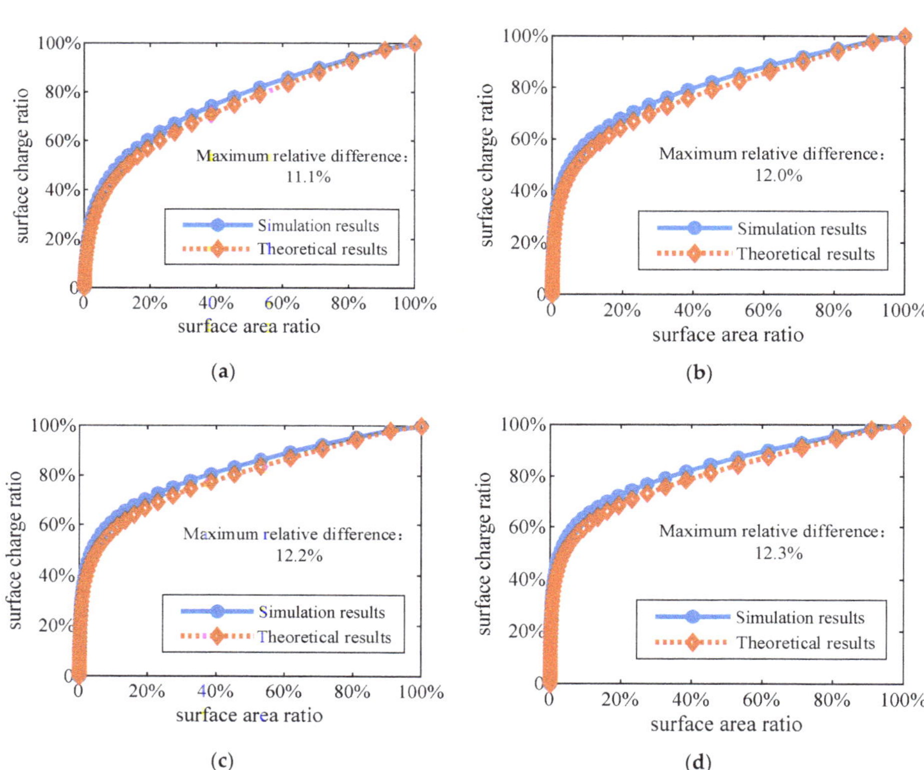

Figure 7. Surface charge ratio as a function of the surface area ratio for probing balls with different diameters. (**a**) ϕ0.1 mm; (**b**) ϕ0.5 mm; (**c**) ϕ1 mm; (**d**) ϕ2 mm.

5. Surface Charge Concentration Analysis

5.1. Surface Charge Concentration

To analyze the law of surface charge concentration, the surface charge density on the equatorial line (θ = 90°) of the probing ball is shown in Figure 8a, and the curve of the surface charge ratio as a function of the surface area ratio is shown in Figure 8b. For clarity, the small area centering on the probing point is referred to as the aiming area in this study. It can be seen that the surface charge density in the aiming area is much greater than that in other regions on the probing ball surface, with the diameter ϕ ranging from 0.1 mm to 2 mm. The analysis results in Figure 6 indicate that 31% of the total surface charge on the ϕ1 mm probing ball concentrates in the probing area, which accounts for 1% of the spherical surface area when a plane is probed by the gap of 0.3 µm.

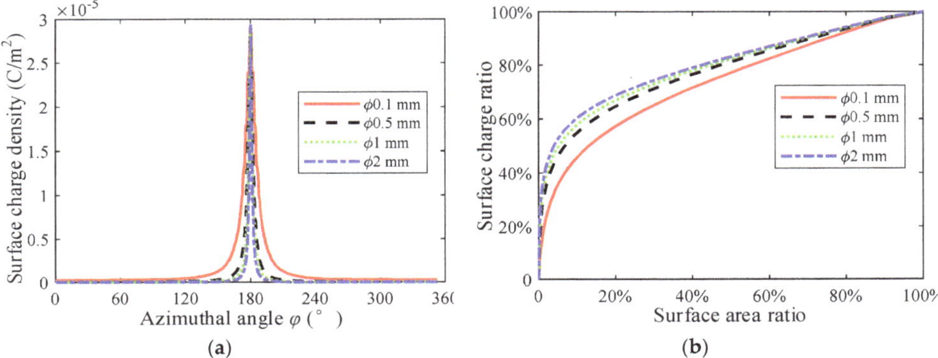

Figure 8. (**a**) Surface charge density on the equator of the probing ball; (**b**) surface charge ratio as a function of the surface area ratio.

5.2. Effect of the Probing Gap

The law of surface charge concentrations varying with probing gaps is shown in Figure 9 below. It can be seen that the relative charge amount on the aiming area increases as the probing gap decreases, and more than 42% of total surface charge stays in the small aiming area, which occupies only 1% of the total spherical surface area when the probing gap δ decreases to 0.01 μm, while the trend of surface charge concentration remains the same for different probing gaps.

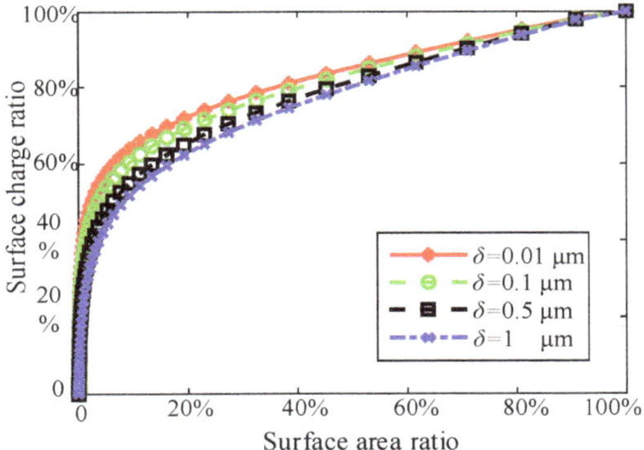

Figure 9. Surface charge ratio as a function of surface area ratio for different probing gaps.

5.3. Effect of Surface Geometry

In addition, to analyze the pattern of surface charge changes with the shape of the surface being measured, surface charge distributions on the probing ball are analyzed when the SSEP is used to probe the interior surfaces of cylindrical surfaces of ϕ6 mm and ϕ12 mm in diameter, and spherical surfaces of ϕ6 mm and ϕ12 mm in diameter. Other conditions are the same as the previous analyses, with ϕ1 mm probing ball and 0.3 μm probing gap δ. It can be seen from Figure 10 that the surface charge concentrations of the SSEP are all well in line when cylindrical surfaces, spherical surfaces, and planes are probed.

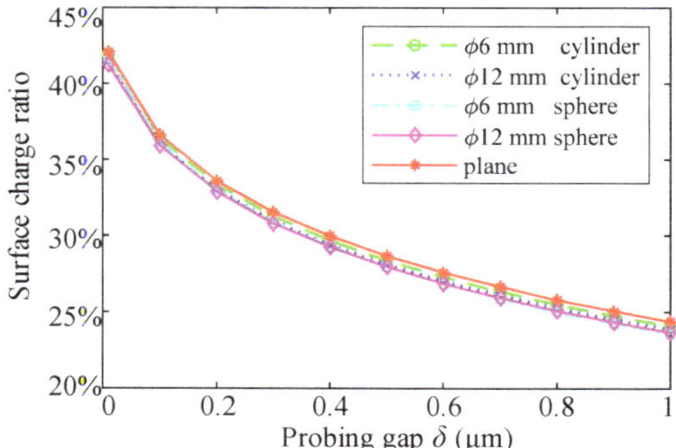

Figure 10. Surface charge ratio as a function of probing gaps on surfaces of different geometries (surface area ratio = 1%).

From the presentation in Figures 9 and 10, surface charge concentration on the probing ball is not only obvious but also similar when either the plane, the cylindrical surface, or the spherical surface is probed, and this concentration basically remains unchanged for different probing gaps in the sensing range. For common miniature internal structures with high aspect ratios, plane, cylindrical surface, and spherical surface are the main composition elements. It can therefore be concluded that the SSEP is of surface charge concentration characteristics when probing miniature internal structures of various free geometrical shapes.

5.4. Point Probing Characteristics

The variation of surface charge on the probing ball of the SSEP can be transformed into the variation of probing gap and even further displacement signal through detection of the charge signal. Similar to high surface charge density near the probing point, the electrical field intensity near the probing point is much stronger than that in any other region. Therefore the probing point and the tiny area centering on it have the strongest effect on the probing characteristics of the SSEP, and this unique characteristic proves the approximate point probing capacity of the SSEP. A virtual "needle" of electrical effect is formed, and thus the SSEP can pick up high-frequency spatial information. Together with its non-contact and 3D isotropy capability as well as nanometer resolution, the SSEP simultaneously possesses all the essentials for high-precision measurement of miniature structures with high aspect ratios, which no other single conventional measurement method possesses simultaneously.

As shown in Figure 11, the lack of measurement methods for internal structures with aspect ratios over 10:1, for which SEM is not suitable to be applied, could be filled with methods based on SSEP.

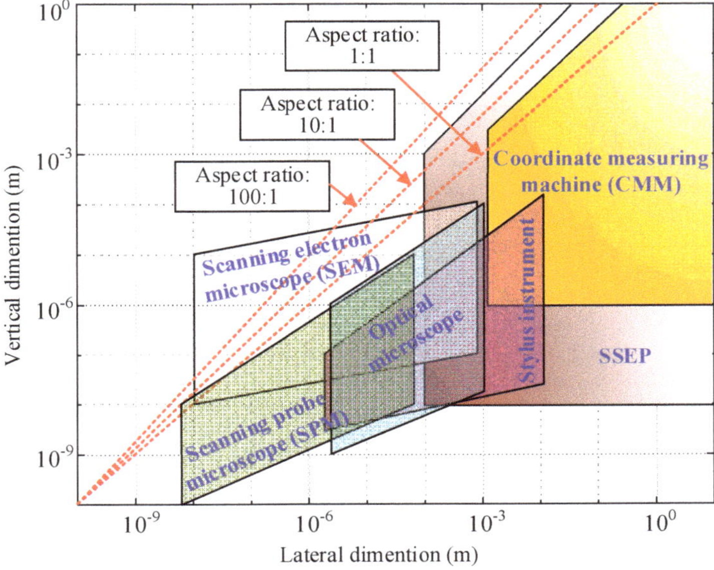

Figure 11. Measurement methods for dimensional micro- and nanometrology (inspired by Ref. [28], in this figure the "CMM" refers to a coordinate measuring machine (CMM) with traditional probes, and "SSEP" refers to SSEP used with CMM).

6. Conclusions

The spherical scattering electrical field of the SSEP formed during probing was modeled by a 3D seven-point finite difference method, and the characteristics and law of surface charge distribution on the probing ball of the SSEP were analyzed and verified with FEM simulation. Analysis results indicate that the surface charge on the probing ball of SSEP has the characteristic of concentrating towards the probing point. When a probing ball of ϕ1 mm in diameter probes a grounded plane with a microprobing gap of 0.3 μm, more than 31% of surface charge on the probing ball concentrates in the aiming area, which covers 1% of the total probing ball surface. The surface charge concentration remains the same when surfaces of different geometrical shapes are probed with different probing gaps in the sensing range, demonstrating the approximate point sensing capacity of the SSEP. Together with its non-contact and 3D isotropy sensing capability as well as nanometer resolution, the SSEP has all the necessary characteristics required for precision measurement of miniature internal structures with high aspect ratios, showing great potential to remedy the lack of measurement methods for internal structures with aspect ratios over 10:1. Our future work will focus on miniaturization of the SSEP.

Author Contributions: Conceptualization, J.C. and X.B.; Funding acquisition, J.C.; Methodology, X.B. and J.C.; Project administration, J.C. and J.T.; Software, X.B. and Y.L.; Validation, Y.Z. and Z.C.; Writing—original draft, X.B. All authors have read and agreed to the published version of the manuscript.

Funding: This research was funded by the National Natural Science Foundation of China, grant number 51675139, L1924059 and the Fundamental Research Funds for The Central Universities.

Conflicts of Interest: The authors declare no conflict of interest.

References

1. Tan, J.B.; Cui, J.N. Ultraprecision 3d probing system based on spherical capacitive plate. *Sens. Actuator A Phys.* **2010**, *159*, 1–6. [CrossRef]

2. Bian, X.; Cui, J.; Lu, Y.; Tan, J. Ultraprecision diameter measurement of small holes with large depth-to-diameter ratios based on spherical scattering electrical-field probing. *Appl. Sci.* **2019**, *9*, 242. [CrossRef]
3. Bos, E.J.C. Aspects of tactile probing on the micro scale. *Precis. Eng. J. Int. Soc. Precis. Eng. Nanotechnol.* **2011**, *35*, 228–240. [CrossRef]
4. Bauza, M.B.; Woody, S.C.; Woody, B.A.; Smith, S.T. Surface profilometry of high aspect ratio features. *Wear* **2011**, *271*, 519–522. [CrossRef]
5. Hansen, H.N.; Carneiro, K.; Haitjema, H.; De Chiffre, L. Dimensional micro and nano metrology. *CIRP Ann. Manuf. Technol.* **2006**, *55*, 721–743. [CrossRef]
6. Neugebauer, M.; Lüdicke, F.; Bastam, D.; Bosse, H.; Reimann, H.; Töpperwien, C. A new comparator for measurement of diameter and form of cylinders, spheres and cubes under clean-room conditions. *Meas. Sci. Technol.* **1997**, *8*, 849–856. [CrossRef]
7. Thalmann, R. A New High Precision Length Measuring Machine. In Proceedings of the 8th International Metrology Congress, Besançon, France, 20–23 October 1997; pp. 1–4.
8. Weckenmann, A.; Estler, T.; Peggs, G.; McMurtry, D. Probing systems in dimensional metrology. *CIRP Ann. Manuf. Technol.* **2004**, *53*, 657–684. [CrossRef]
9. Zhao, Y.; Li, P.; Wang, C.; Pu, Z. A novel fiber-optic sensor used for small internal curved surface measurement. *Sens. Actuator A Phys.* **2000**, *86*, 211–215. [CrossRef]
10. Weiqian, Z.; Jiubin, T.; Lirong, Q.; Limin, Z. A new laser heterodyne confocal probe for ultraprecision measurement of discontinuous contours. *Meas. Sci. Technol.* **2005**, *16*, 497.
11. Schmitt, R.; Mallmann, G. Evaluation of the usage of a fiber optic low-coherence interferometer for surface profile measurements using speckles analysis. In Proceedings of the Speckle 2010: Optical Metrology, Florianapolis, Brazil, 13–15 September 2010; p. 73870N.
12. Depiereux, F.; Konig, N.; Pfeifer, T.; Schmitt, R. Fiber-based white-light interferometer with improved sensor tip and stepped mirror. *IEEE Trans. Instrum. Meas.* **2007**, *56*, 2279–2283. [CrossRef]
13. Cui, J.; Li, L.; Li, J.; Tan, J.B. Fiber probe for micro-hole measurement based on detection of returning light energy. *Sens. Actuator A Phys.* **2013**, *190*, 13–18. [CrossRef]
14. Shiramatsu, T.; Kitano, K.; Kawada, M.; Mitsui, K. Development of measurement method for shape and dimension of micro-components. *JSME Int. J. Ser. C* **2004**, *47*, 369–376. [CrossRef]
15. McMurtry, D. The development of sensors for cmms. *Laser Metrol. Mach. Perform. VI* **2003**, *44*, 205–220.
16. Oiwa, T.; Nishitani, H. Three-dimensional touch probe using three fibre optic displacement sensors. *Meas. Sci. Technol.* **2004**, *15*, 84. [CrossRef]
17. Takamasu, K.; Chih-Che, K.; Suzuki, A.; Hiraki, M.; Furutani, R.; Ozono, S. Development of pneumatic ball probe for measuring small hole. In Proceedings of the International Conference on Precision Engineerin, Taipei, Taiwan, 20–22 November 1997.
18. Bowers, W.J.; Olson, D.A. A capacitive probe for measuring the clearance between the piston and the cylinder of a gas piston gauge. *Rev. Sci. Instrum.* **2010**, *81*, 035102. [CrossRef] [PubMed]
19. Dai, G.; Heidelmann, M.; Kübel, C.; Prang, R.; Fluegge, J.; Bosse, H. Reference nano-dimensional metrology by scanning transmission electron microscopy. *Meas. Sci. Technol.* **2013**, *24*, 085001. [CrossRef]
20. Peiner, E.; Doering, L. Nondestructive evaluation of diesel spray holes using piezoresistive sensors. *IEEE Sens. J.* **2013**, *13*, 701–708. [CrossRef]
21. Kang, H.; Lim, M.; Yun, K. Characterization of the electric field concentration between two adjacent spherical perfect conductors. *SIAM J. Appl. Math.* **2014**, *74*, 125–146. [CrossRef]
22. Morales, M.; Diaz, R.A.; Herrera, W.J. Solutions of laplace's equation with simple boundary conditions, and their applications for capacitors with multiple symmetries. *J. Electrost.* **2015**, *78*, 31–45. [CrossRef]
23. Johann, F.; Soergel, E. Quantitative measurement of the surface charge density. *Appl. Phys. Lett.* **2009**, *95*, 232906. [CrossRef]
24. Villeneuve-Faure, C.; Boudou, L.; Makasheva, K.; Teyssedre, G. Methodology for extraction of space charge density profiles at nanoscale from kelvin probe force microscopy measurements. *Nanotechnology* **2017**, *28*, 505701. [CrossRef] [PubMed]
25. Kumada, A.; Shimizu, Y.; Chiba, M.; Hidaka, K. Pockels surface potential probe and surface charge density measurement. *J. Electrost.* **2003**, *58*, 45–58. [CrossRef]
26. Deng, J.; Kumada, A.; Hidaka, K.; Zhang, G.; Mu, H. Residual charge density distribution measurement of surface leader with feedback electrostatic probe. *Appl. Phys. Lett.* **2012**, *100*, 192906. [CrossRef]

27. Okamoto, T.; Kitagawa, S.; Inoue, N.; Ando, A. Electric field concentration in the vicinity of the interface between anode and degraded BaTiO3-based ceramics in multilayer ceramic capacitor. *Appl. Phys. Lett.* **2011**, *98*, 072905. [CrossRef]
28. De Chiffre, L.; Kunzmann, H.; Peggs, G.N.; Lucca, D.A. Surfaces in precision engineering, microengineering and nanotechnology. *CIRP Ann.* **2003**, *52*, 561–577. [CrossRef]

Article

Quick Response Circulating Water Cooling of ±3 mK Using Dynamic Thermal Filtering

Yesheng Lu, Junning Cui *, Jiubin Tan and Xingyuan Bian

Center of Ultra-precision Optoelectronic Instrument, Harbin Institute of Technology, Harbin 150080, China; luyesheng@163.com (Y.L.); jbtan@hit.edu.cn (J.T.); bianxingyuan@hit.edu.cn (X.B.)

* Correspondence: cuijunning@hit.edu.cn; Tel.: +86-451-8640-2258

Received: 17 June 2020; Accepted: 6 August 2020; Published: 7 August 2020

Featured Application: The circulating cooling water machine proposed in this paper is conductive to the thermal management of equipment and facilities in the field of precision manufacturing and measurement. It has already been applied to our lithography development project.

Abstract: An enhanced circulating cooling water (CCW) machine is developed to simultaneously achieve high temperature stability and dynamic performance of CCW temperature control. Dynamic thermal filtering based on an auto-updatable thermal capacity medium is proposed to reduce the temperature fluctuation of the CCW. Agile thermal control is presented to realize a quick response and high resolution of temperature control, through thermal inertia minimization and bidirectional regulation of heating/cooling power. Experimental results indicate that a temperature stability of ±3 mK (peak to peak value) and a settling time of 128 s, corresponding to a 1 K step set value, are achieved. It can therefore be concluded that the developed machine can satisfy the challenging requirements of precision manufacturing.

Keywords: circulating cooling water; dynamic thermal filtering; precision manufacturing; quick response; temperature stability; thermal management

1. Introduction

Thermal management using circulating cooling water (CCW) is essential for precision manufacturing [1]. With precision manufacturing stepping into a nano-era, the challenging demand for CCW with mK level stability and a quick response is put forward by a variety of ultraprecision applications, such as ultraprecision optical machining with nanometer accuracy, and lithography which is stepping to a 7 nm technical node [2,3]. In an advanced lithography machine, CCW with a temperature stability of ± 0.01 K is required to cool down power devices like motors, power drivers, laser sources, and so on [4,5], especially to maintain a ±0.01 K temperature stability of objective lenses. Moreover, CCW is also used as a temperature reference medium to produce an 'air shower' of temperature stability, up to 2.3 mK/5 min, for air refractive index stabilization of laser interferometry. Meanwhile, a response time of CCW temperature control as short as dozens of seconds is demanded to suppress the cyclic thermal pollution of power devices in lithography. Similar requirements are also presented in ultraprecision optical machining. CCW with a temperature stability of ±0.1 K is needed to cool down electrical spindles [6,7], and the temperature of CCW needs to dynamically respond to the variation in rotation speed of spindles, and thus to realize 'servo' thermal pollution management of the spindles.

One effective technical route to achieve high temperature stability of CCW is to reduce the temperature fluctuation by increasing thermal inertia. Lawton et al., proposed a method to reach this goal through heat exchanging with an additionally introduced thermal capacity medium, such as water, and stainless steel balls [8,9]. Phase change material (PCM) has also been utilized, because of its large latent thermal capacity at the phase change temperature point [10–12]. However, the proposed methods only function in the very narrow range of the preset temperature point of the introduced thermal capacity medium, or the phase change temperature point of PCM. Therefore, they can only reduce the CCW temperature fluctuation in static water-cooling applications in which the output temperature of CCW is required to be constant.

New control schemes and algorithms have also been presented to improve the temperature stability and dynamic performance of CCW. For example, Madhavan et al., and Sarid et al., proposed temperature controllers achieving temperature stability at sub-millikelvin level [13–15]. However, the response time of CCW temperature control in these studies ranged from tens of minutes to hours, because the dynamic performance was not taken into consideration. Liu et al., proposed a 'servo' temperature control method for the cooling of spindles of precision machining tools [16]. Two CCW streams from two CCW machines with different temperature set values were used, then temperature of the mixed CCW was proportional to the mixing proportion of the two streams. Temperature stability at sub-kelvin level and a quick response of CCW temperature control were realized, at the cost of using two CCW machines and a complex flow rate control.

Therefore, it is difficult to simultaneously achieve high temperature stability and quick response of CCW temperature control. CCW with temperature stability at mK level and a response time at ~100 s level is still a challenging demand for precision manufacturing. Thus, this paper proposes an enhanced CCW machine using a dynamic thermal filtering method (DTFM) and bidirectional agile thermal control to solve this problem.

2. Principle

As shown in Figure 1, the proposed enhanced CCW machine mainly consists of a water circulating loop and a control sub-system. A frequency variable pump, a heating module, a cooling module, a DTFM attenuator, and a thermal load are cascaded as a water circulating loop. The pump drives the cooling water to circulate through the loop, and the temperature of CCW is regulated by the cooling and heating modules. Then, the DTFM attenuator is introduced to further improve the temperature stability of CCW. The chilled water supplier, which is normally a factory facility, serves as a cooling source to carry away dissipated thermal energy from the cooling module.

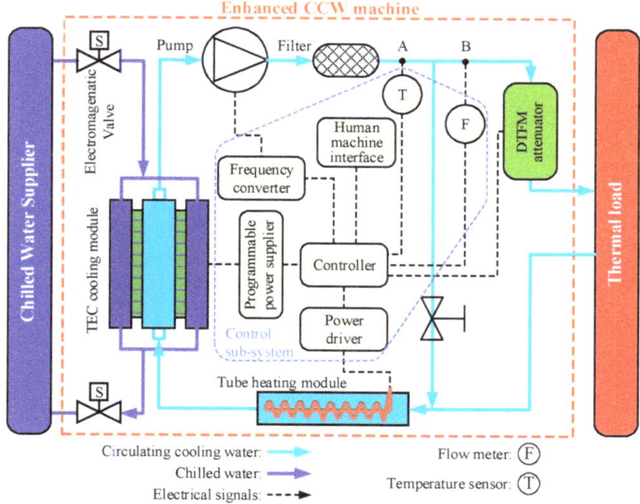

Figure 1. Scheme of an enhanced circulating cooling water (CCW) machine.

The control sub-system takes charge of the temperature feedback control, the flow rate feedback control, and the human-machine interface. The dynamic performance of CCW temperature control is improved by distributive thermal inertia minimization, and bidirectional agile control of the heating and cooling power.

3. Dynamic Thermal Filtering to Achieve High Temperature Stability

3.1. Principle

The scheme of the proposed DTFM is shown in Figure 2. The idea is to attenuate the CCW temperature fluctuation through heat exchanging with the thermal capacity medium. The temperature of the thermal capacity medium dynamically follows the set value of the CCW temperature, since the medium can automatically update. Thus, the temperature stability of CCW is further improved without any deterioration of the dynamic performance of CCW temperature control.

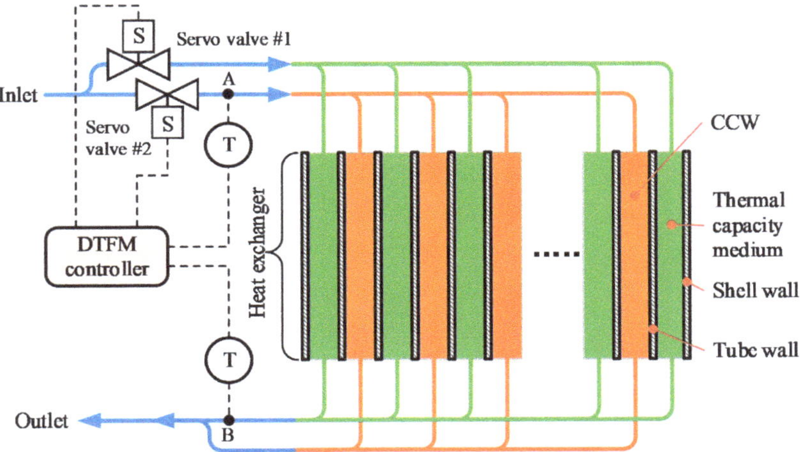

Figure 2. Scheme of an attenuator based on a dynamic thermal filtering method.

As shown in Figure 2, the DTFM attenuator mainly consists of a heat exchanger, two temperature sensors, two servo valves, and a controller. When the DTFM attenuator is working, some water is bypassed from CCW and blocked in one chamber of the heat exchanger, serving as a thermal capacity medium. Heat exchange between CCW and the thermal capacity medium, called 'thermal filtering effect', occurs when CCW flows through the other chamber of the heat exchanger. As a result, the temperature fluctuation of the CCW is significantly reduced, which means that the temperature stability of the CCW is effectively improved.

When the machine is applied in dynamic water-cooling applications in which the temperature of CCW varies with time, temperature sensor A detects the change of CCW temperature, then the attenuator starts the process of automatically updating the thermal capacity medium. The controller switches on servo valve #1 and switches off servo valve #2 to make the latest CCW flow into the thermal capacity medium chamber. Temperature sensor B monitors the temperature of outflow of the thermal capacity medium, and provides a trigger signal to the controller when the temperature at point B equals the temperature at point A. Then, the controller switches off servo valve #1 and switches on servo valve #2 to end the process of medium updating. Consequently, the change of the set value of the CCW temperature is dynamically followed and the dynamic performance of the CCW machine is maintained

3.2. Structure of DTFM Attenuator

A DTFM attenuator was designed using a custom-made tube-tank heat exchanger, as shown in Figure 3. A tube array is evenly arranged in a square stainless steel tank. CCW flows through the tubes, and the thermal capacity medium is blocked in the tank. A distributor is placed on the top of the exchanger and divides CCW into sub-streams flowing into tubes. An accumulator is placed at the bottom of the exchanger and makes all the sub-streams converge.

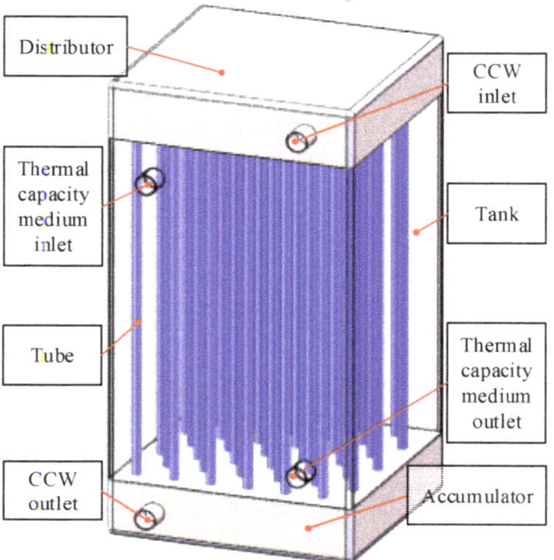

Figure 3. Three-dimensional structure of the tube-tank heat exchanger.

3.3. Modeling

The temperature fluctuation attenuation ratio of the DTFM attenuator was modeled using a transfer function method, based on the equations proposed by Lawton et al. [8]. Transfer functions of the distributor, tube-tank, and accumulator, expressed as r_{dis}, r_{att}, and r_{acc}, respectively, were defined as the temperature fluctuation ratios of the output CCW temperature to input CCW temperature of each part. In the distributor, convective heat transfer of CCW plays the key role in the temperature fluctuation attenuation. The temperature fluctuation attenuation of the tube-tank can be modeled through the double-pipe heat exchanger. The temperature fluctuation attenuation of accumulator resulted from CCW sub-streams mixing. Therefore, the transfer functions of the three parts of the DTFM attenuator can be derived as the following equations:

$$r_{\text{dis}}(f) = \exp\left[\frac{u_a - \sqrt{u_a^2 + 4(\lambda/\rho c_{\text{w}})2j\pi f C_{\text{a}}}}{2\lambda/\rho c_{\text{w}}}\right] \tag{1}$$

$$r_{\text{att}}(f) = \exp\left[-\frac{1}{f_{\text{m}} c_{\text{w}} R}\left(\frac{2j\pi f C_a}{2j\pi f C_a + 1}\right)\right] \tag{2}$$

$$r_{\text{acc}}(f) = \frac{1}{1 + (\rho V_{\text{acc}}/F_{\text{m}})2j\pi f} \tag{3}$$

where ρ is the density of water, V_{acc} is the volume of the accumulator, F_{m} is the mass flow rate of the CCW, f_{m} is the mass flow rate of each tube, u_{a} is the average water velocity in the accumulator, λ is the thermal conductivity of water, C_{a} is the thermal capacity of a single tube, L_{a} and L_{d} are height of the accumulator and distributor respectively, c_w is the specific thermal capacity of water, and R is the thermal resistance of the attenuator.

Then, the total temperature fluctuation attenuation ratio of the DTFM attenuator r is the product of transfer functions of the three parts.

$$r = r_{dis}(f) \times r_{acc}(f) \times r_{att}(f) \tag{4}$$

The main parameters of the designed DTFM attenuator are listed in Table 1.

Table 1. Parameters of the tube-tank heat exchanger and CCW.

Symbol	Parameter	Quantity and Unit
L	Length of tube	0.7 m
R_{tube}	Diameter of tube	8 mm
a	Center spacing between neighboring tubes	32 mm
N	Number of tubes	36
b	Width of tank	0.23 m
l	Height of tank	0.7 m
L_a	Height of accumulator	0.05 m
L_d	Length of distributor	0.05 m
F_m	Total flow rate	16 L/min

Before the DTFM attenuator is introduced into the CCW machine, the dynamic performance of the CCW temperature control is determined by the closed loop feedback temperature control of the CCW machine. After the DTFM attenuator is introduced, the response time of the CCW temperature control S_{sys} can be expressed as:

$$S_{sys} = \max(S_{mac}, S_{med}) \tag{5}$$

where S_{mac} is the response time of temperature control before the DTFM attenuator is introduced and S_{med} is the updating time of the thermal capacity medium of the DTFM attenuator, expressed as:

$$S_{med} = \frac{\left[b^2L - NL\pi(R_{tube}/2)^2\right]\rho}{F_m} \tag{6}$$

Therefore, $S_{mac} > S_{med}$ should be satisfied to guarantee the dynamic performance of the CCW temperature control.

In this research, the frequency band of the closed loop temperature feedback control of the CCW machine was ~0.005 Hz, so S_{mac} = 200 s. According to parameters listed in Table 1, S_{med} = 82 s. Therefore, $S_{mac} < S_{med}$ was well satisfied.

The theoretical modeling results of the temperature fluctuation attenuation ratio of the designed DTFM attenuator are shown in Figure 4. It can be seen from the curve that the attenuation has the characteristic of low-pass filtering. For the CCW temperature fluctuation at low frequency, from 0 to 0.005 Hz, the temperature fluctuation is within the closed loop temperature feedback control bandwidth of the CCW machine, and can be suppressed through closed loop control of the machine. For the CCW temperature fluctuation above 0.005 Hz, it can be reduced by the DTFM attenuator to tens of dB. As a result, the temperature stability of CCW is effectively improved without any deterioration of the dynamic performance of the CCW temperature control.

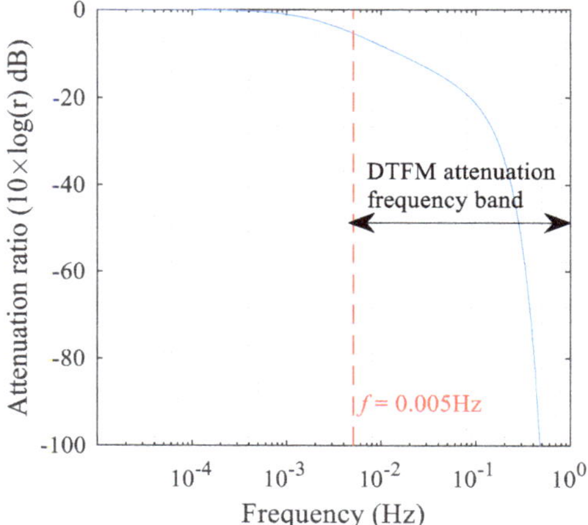

Figure 4. Modeling results of the temperature fluctuation ratio of the dynamic thermal filtering method (DTFM) attenuator.

To evaluate the performance of the developed DTFM attenuator, CCW with different temperature fluctuations was introduced as the input of the DTFM attenuator, and the temperature curves of the inlet and outlet of the DTFM attenuator were recorded as shown in Figure 5. The temperature fluctuation of CCW was significantly reduced by the developed DTFM attenuator.

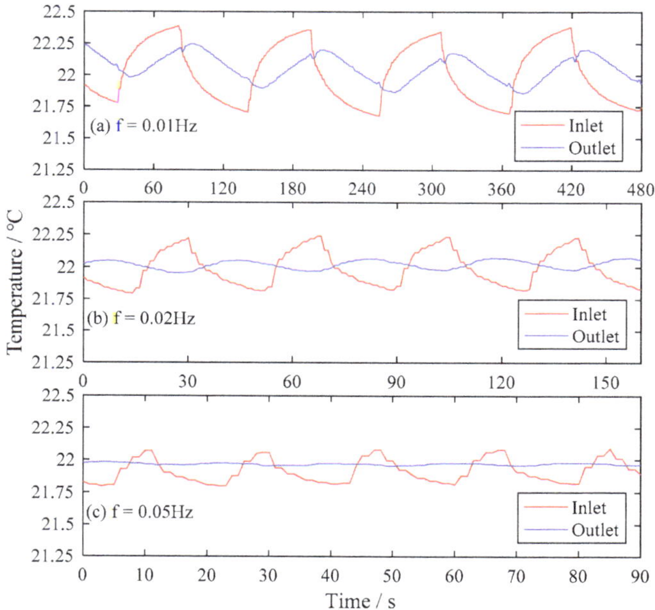

Figure 5. Modeling results of the temperature fluctuation ratio of the DTFM attenuator.

4. Agile Control to Achieve Quick Response and High Regulation Resolution

4.1. Principle

To achieve a quick response and high resolution of CCW temperature control, the agile control of CCW temperature, based on a thermal inertia minimization design and bidirectional regulation of heating/cooling power, was proposed. A thermal inertia minimization design of the cooling module was achieved by using an array of thermoelectric chips (TEC) as cooling elements and building a 'sandwich' structure with high heat exchange efficiency. The thermal inertia minimization design of the heating module was achieved based on a structure of 'solenoid-in-tube'. Bidirectional precision and quick temperature regulation of CCW were achieved through the cooperation of the heating and the cooling modules under the control of a Fuzzy Proportion Integration Differentiation (Fuzzy-PID) algorithm.

4.2. TEC Cooling Module

As shown in Figure 6, the cooling module based on TEC presents an integrated compact 'sandwich' structure, which was proposed to achieve high cooling efficiency.

A TEC can create a temperature difference between its two sides, when a DC current is applied, and then heating energy is transferred from the cold side to the hot side. Mechanical vibration and noise are avoided, since there are no moving objects in the TEC. Moreover, the cooling power of a TEC increases when the electrical power increases, so it is capable of linear and precision control of cooling power.

As shown in Figure 6, two layers of TEC arrays sandwiched the CCW chamber with the cold sides of TECs. The hot sides of the two TEC arrays were both covered with chilled water chambers. Each TEC array was placed in a frame, which was made of a thermal isolation material to avoid thermal conduction between the two sides of the TECs.

Since aluminum has a large thermal conductivity and low density, the CCW chamber and two chilled water chambers were both made of aluminum alloy, in which the CCW and chilled water traveled along the serpentine waterway to achieve a higher heat exchange efficiency.

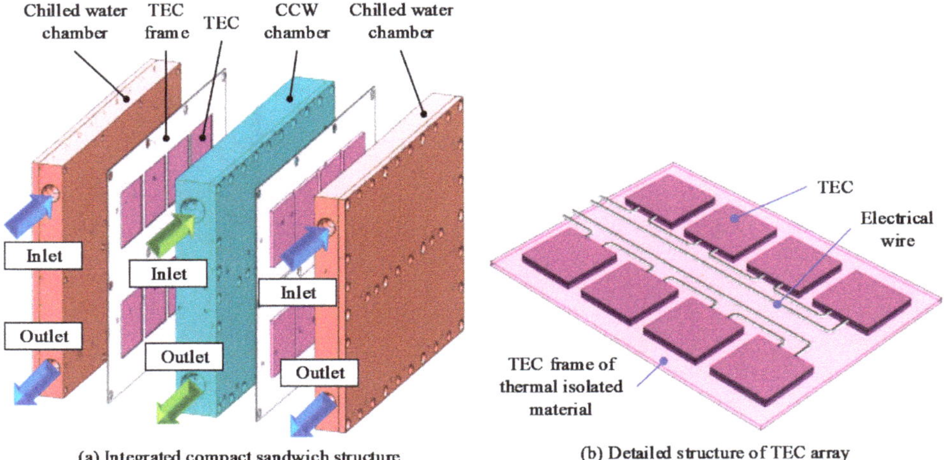

Figure 6. Scheme of the thermoelectric chip (TEC) cooling module.

Compared to the conventional cooling method, using compressors as cooling elements, the proposed TEC cooling module provided a large cooling power with a compact size, and featured sub-watt level precision and sub-second level response times of cooling power regulation [17,18]. Commercial TEC products (China KJLP, type TEC1-031504040, cooling power of 107 W) were employed as the cooling element of the cooling module. The developed TEC cooling module is 279 mm × 208 mm × 98 mm in size, and can provide a cooling power of 1.5 kW.

4.3. Tube Heating Module

As shown in Figure 7, the heating module based on a 'solenoid-in-tube' structure was proposed to achieve minimal thermal inertia. The electrical heating solenoid was assembled in a long thin tube that the CCW flows through. Since the heating solenoid features a sub-second response time, and its heating power is proportional to the electrical power applied, heating power can be controlled with sub-watt level resolution. Compared to conventional heating module design, in which the electrical solenoid is immersed in a water tank, the proposed heating module can achieve agile thermal control. The developed heating module is ϕ 30 mm × 200 mm (diameter × length) in size, and has a maximum heating power of 1.5 kW.

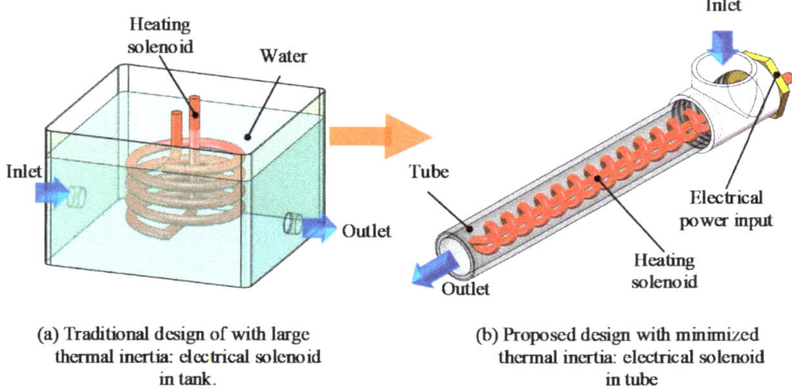

(a) Traditional design of with large thermal inertia: electrical solenoid in tank.

(b) Proposed design with minimized thermal inertia: electrical solenoid in tube

Figure 7. Scheme of the tube heating module.

4.4. Control Sub-System

As shown in Figure 8, the control sub-system of the enhanced CCW machine consists of the temperature feedback control loop and the flow rate feedback control loop. The temperature sensor at position A and the flowmeter at position B, as mentioned in Figure 1, provide temperature feedback of CCW, and flow rate feedback of CCW, to the controller, respectively. The heating module is driven by the power regulator, meanwhile, the cooling module is driven by the programmable power supplier.

A control algorithm based on a Fuzzy-PID was employed to obtain better performance of CCW temperature control, since the Fuzzy-PID can deal with nonlinearity, time variety, and large time delays of the CCW temperature control system [19], as shown in Figure 9.

Therefore, online PID parameter regulation was achieved by the fuzzy controller to gain better performance of CCW temperature control.

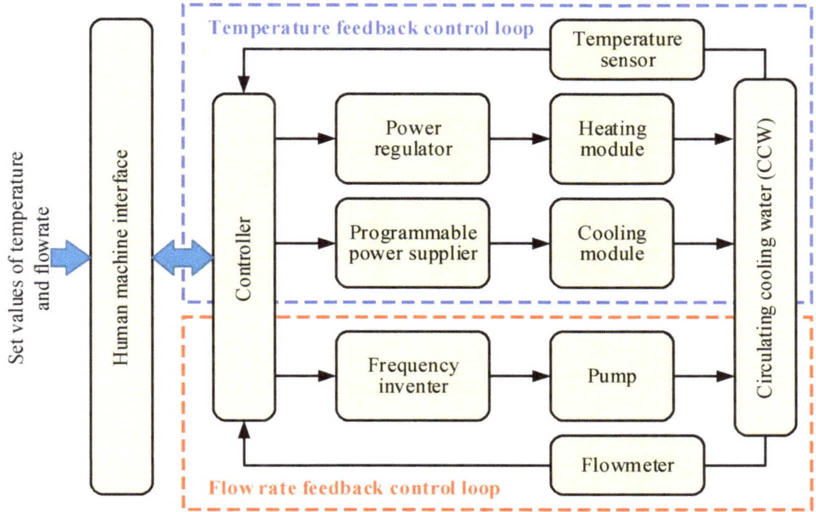

Figure 8. Schematic diagram of the control sub-system of the enhanced CCW machine.

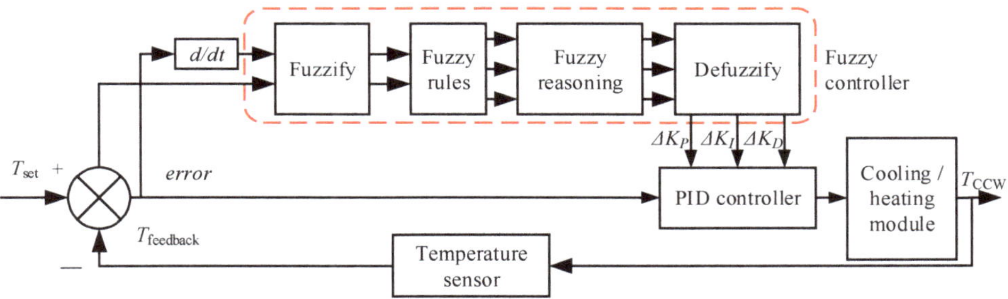

Figure 9. Schematic diagram of Fuzzy Proportion Integration Differentiation (Fuzzy-PID) introduced in the enhanced CCW machine.

5. Experiments and Discussion

5.1. Experimental Setup

The experimental setup is shown in Figure 10, the developed enhanced CCW machine can provide CCW of temperatures from 12 °C to 40 °C, with a maximum flow rate of 55 L/min. It has a maximum cooling power of 1.5 kW, and a maximum heating power of 1.5 kW. A commercial chilled water supplier was employed to supply chilled water of temperature of 12 °C serving as the cooling source for the system. A programmable heating device with maximum heating power of 1.5 kW served as the thermal load. A high-precision thermistor (Fluke, type 5641, accuracy of ±1 mK, resolution of 0.1 mK, with thermometer Fluke 1560) was employed to measure the temperature of CCW. The experimental setup was located in an air-conditioned machining workshop at 22 °C ± 3 °C.

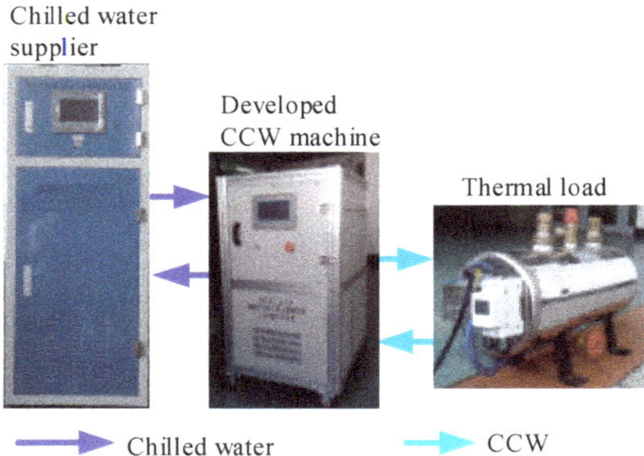

Figure 10. Experimental setup.

5.2. Temperature Stability Experiment

In this experiment, the set value of the CCW temperature was 22 °C, the heating power of the thermal load was 1 kW, and the flow rate of CCW was 15 L/min. The temperature of the CCW was recorded for 60 min with a sampling rate of 1 Hz. Experimental results are shown in Figure 11. It can be seen that, the peak-to-peak value of CCW temperature fluctuation for 60 min was ±3 mK, and its standard deviation was 1.2 mK.

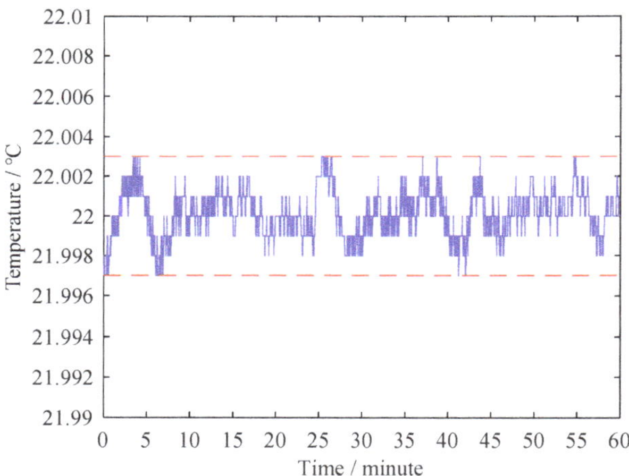

Figure 11. Experimental results of CCW temperature stability.

5.3. Dynamic Performance Experiments

The experiment of 1 K step response of the CCW temperature control was carried out to validate the dynamic performance of the developed CCW machine. The 1 K step of CCW temperature set value was from 22 °C to 23 °C, as shown in Figure 12. Experimental results show that the settling time was 128 s, and the overshoot was 0.03 K.

Then, the experiment of thermal impact response was carried out to validate the performance of thermal impact resistance of the developed CCW machine. A 500 W thermal load was introduced by the programmable heating device to make a step thermal impact. Experimental results indicated that the temperature of CCW recovered to a stable state in ~200 s after the thermal impact was introduced, and that the overshoot was about 0.14 K, as shown in Figure 13.

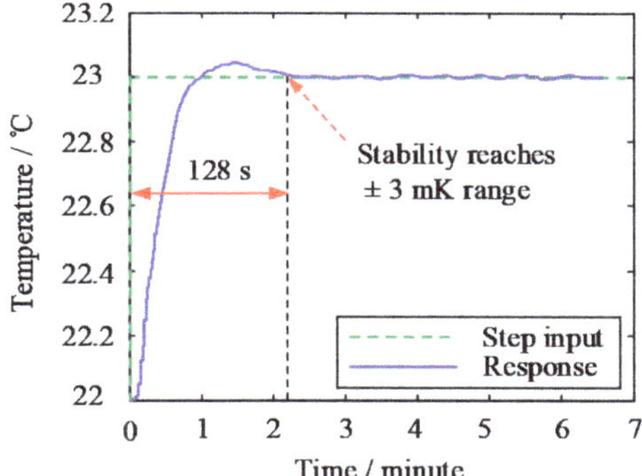

Figure 12. Experimental results of 1 K step response.

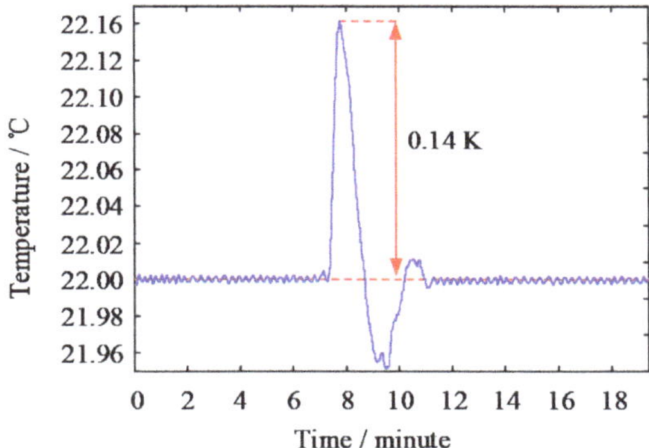

Figure 13. Experimental results of the 500 W thermal load impact response.

6. Conclusions

In this paper, an enhanced CCW machine, using a dynamic thermal filtering method and agile control, which can accomplish 'servo thermal management' for precision manufacturing, is proposed. It can be concluded that:

(1) A dynamic thermal filtering method is proposed. The temperature stability of CCW is significantly improved by the demonstrated temperature fluctuation attenuator, based on the proposed method, while the dynamic performance of the CCW temperature control is not degraded. According to the concept of agile control, a thermoelectric cooling module, with a compact multilayer sandwich structure, as well as a heating module with a 'solenoid-in-tube' structure, are proposed, and thus the thermal inertia of the modules is minimized. Furthermore, bidirectional regulation of thermal power is realized with the proposed cooling and heating modules, based on the control of Fuzzy-PID. Therefore, the excellent dynamic performance of CCW temperature control is achieved.

(2) Experiments were carried out to validate the performance of the enhanced CCW machine. The temperature stability was ±3 mK (peak-to-peak value), and its standard deviation was 1.2 mK. The settling time was 128 s, and the overshoot was 0.03 K for 1 K step of the set value of CCW temperature. The CCW temperature had a good performance against thermal impact.

Our following work will focus on the further improvement of the dynamic performance of the CCW machine, and the shortening of the response time of dozens of seconds.

Author Contributions: Conceptualization, Y.L., J.C. and J.T.; Funding acquisition, J.C. and J.T.; Investigation Y.L.; Methodology, J.C. and J.T.; Project administration, J.C. and J.T.; Resources, Y.L.; Visualization, X.B.; Writing—original draft, Y.L.; Writing—review & editing, Y.L., J.C. and J.T. All authors have read and agreed to the published version of the manuscript.

Funding: This research was funded by National Natural Science Foundation of China, grant number 51675139 and the Fundamental Research Funds for the Central Universities.

Conflicts of Interest: The authors declare no conflicts of interest.

References

1. Mayr, J.; Jedrzejewski, J.; Uhlmann, E.; Donmez, M.A.; Knapp, W.; Härtig, F.; Wendt, K.; Moriwaki, T.; Shore, P.; Schmitt, R.; et al. Thermal issues in machine tools. *CIRP Ann.* **2012**, *61*, 771–791. [CrossRef]
2. Zhao, Y.; Trumper, D.L.; Heilmann, R.K.; Schattenburg, M.L. Optimization and temperature mapping of an ultra-high thermal stability environmental enclosure. *Precis. Eng.* **2010**, *34*, 164–170. [CrossRef]
3. Manske, E.; Jäger, G.; Hausotte, T.; Füßl, R. Recent developments and challenges of nanopositioning and nanomeasuring technology. *Meas. Sci. Technol.* **2012**, *23*, 074001. [CrossRef]
4. Yamada, A. ArF immersion lithography for 45nm and beyond. In Proceedings of the SPIE—The International Society for Optical Engineering, Yukohama, Japan, 17–19 April 2007; Volume 6607, pp. 66071–66081.
5. Tay, A.; Chua, H.T.; Wang, Y.; Ngo, Y.S. Equipment design and control of advanced thermal-processing module in lithography. *IEEE Trans. Ind. Electron.* **2009**, *57*, 1112–1119. [CrossRef]
6. Chou, C.; DeBra, D.B. Liquid temperature control for precision tools. *CIRP Ann.* **1990**, *39*, 535–543. [CrossRef]
7. Cui, L.Y.; Zhang, D.W.; Gao, W.G.; Qi, X.Y.; Shen, Y. Thermal errors simulation and modeling of motorized spindle. *Adv. Mater. Res.* **2010**, *154*, 1305–1309. [CrossRef]
8. Lawton, K.M.; Patterson, S.R.; Keanini, R.G. Precision temperature control of high-throughput fluid flows: Theoretical and experimental analysis. *J. Heat Transf.* **2001**, *123*, 796–802. [CrossRef]
9. Lawton, K.M.; Patterson, S.R.; Keanini, R.G. Direct contact packed bed thermal gradient attenuators: Theoretical analysis and experimental observations. *Rev. Sci. Instruments* **2003**, *74*, 2886–2893. [CrossRef]
10. Oró, E.; De Gracia, A.; Castell, A.; Farid, M.; Cabeza, L.F. Review on phase change materials (PCMs) for cold thermal energy storage applications. *Appl. Energy* **2012**, *99*, 513–533. [CrossRef]
11. Charvat, P.; Klimes, L.; Stetina, J.; Ostry, M. Thermal storage as a way to attenuate fluid-temperature fluctuations: Sensible-heat versus latent-heat storage materials. *Mater. Tehnol.* **2014**, *48*, 423–427.
12. Alawadhi, E.M. Temperature regulator unit for fluid flow in a channel using phase change material. *Appl. Therm. Eng.* **2005**, *25*, 435–449. [CrossRef]
13. Unni, P.K.M.; Gunasekaran, M.K.; Kumar, A. ±30 μK temperature controller from 25 to 103 °C: Study and analysis. *Rev. Sci. Instrum.* **2003**, *74*, 231–242. [CrossRef]
14. Mann, G.; Hu, B.-G.; Gosine, R. Analysis of direct action fuzzy PID controller structures. *IEEE Trans. Syst. Man Cybern. Part B Cybern.* **1999**, *29*, 371–388. [CrossRef] [PubMed]
15. Lim, H.S.; Kang, Y.T. Optimization of influence factors for water cooling of high temperature plate by accelerated control cooling process. *Int. J. Heat Mass Transf.* **2019**, *128*, 526–535. [CrossRef]
16. Liu, T.; Gao, W.; Tian, Y.; Zhang, H.; Chang, W.; Mao, K.; Zhang, D. A differentiated multi-loops bath recirculation system for precision machine tools. *Appl. Therm. Eng.* **2015**, *76*, 54–63. [CrossRef]
17. Enescu, D.; Virjoghe, E.O. A review on thermoelectric cooling parameters and performance. *Renew. Sustain. Energy Rev.* **2014**, *38*, 903–916. [CrossRef]
18. Sun, K.; Qiu, Z.; Wu, H.; Xing, Y. Evaluation on high-efficiency thermoelectric generation systems based on differential power processing. *IEEE Trans. Ind. Electron.* **2018**, *65*, 699–708. [CrossRef]
19. Huang, C.-W.; Pan, S.-T.; Zhou, J.-T.; Chang, C.-Y. Enhanced temperature control method using ANFIS with FPGA. *Sci. World J.* **2014**, *2014*, 1–8. [CrossRef] [PubMed]

Article

An Innovative Dual-Axis Precision Level Based on Light Transmission and Refraction for Angle Measurement

Yubin Huang, Yuchao Fan, Zhifeng Lou, Kuang-Chao Fan * and Wei Sun

School of Mechanical Engineering, Dalian University of Technology, Dalian 116024, China; huangyubin@mail.dlut.edu.cn (Y.H.); wd_fyc@163.com (Y.F.); Louzf@dlut.edu.cn (Z.L.); sunwei@dlut.edu.cn (W.S.)

* Correspondence: fan@ntu.edu.tw

Received: 22 July 2020; Accepted: 26 August 2020; Published: 31 August 2020

Abstract: Currently, the widely used pendulum-type precision level cannot be miniaturized because reducing the size of the pendulum will reduce its displacement so as to decrease the measurement accuracy and resolution. Moreover, the commercial pendulum-type level can only sense one direction. In this paper, an innovative compact and high-accuracy dual-axis precision level is proposed. Based on the optical principle of light refraction and the reference of the invariant liquid level, the pendulum is no more needed. In addition, based on the light transmission design, there is no reflection signal to interfere with the true signal. Therefore, the level can achieve a high accuracy and small-sized design. The calibration result shows the error of the proposed precision level is better than ±0.6 arc-sec in the measurement range of ±100 arc-sec, and better than ±5 arc-sec in the full measurement range of ±800 arc-sec.

Keywords: dual-axis level; light refraction; light transmission; angle measurement

---

1. Introduction

In precision engineering, with the significant improvement in positioning sensor accuracy in the past few decades [1], the influence of angular error on the accuracy, repeatability, and stability of precision machinery has become increasingly important [2,3]. There are many types of angle measuring instrument that have been applied in precision engineering [4–6]. Some multi-degree-of-freedom measurement systems have been reported in recent years [7–9], and the roll angle measurement of a long stage is still affected by unstable laser beams [10,11]. For the advantage of measuring the roll error of the linear stage, and for the measurement of the inclination angle while installing the precision machinery, it is still necessary to know the precision level.

Traditionally, according to the different measurement principle, spirit levels can be divided into two types: the bubble type and the pendulum type. Due to the disadvantages of measurement resolution and the accuracy of the bubble-type sprit level, the bubble-type sprit level can only meet the requirement of rough adjustment of equipment horizontal installation. Currently, most commercial precision levels are designed based on a pendulum mechanism [4], and usually they are the one-axis pendulum type [4,5]. The pendulum-type precision level is also used as a sensor for monitoring the ground rotation of some large-scale scientific instruments, such as the Laser-Interferometer-Gravitational-wave-Observatory (LIGO) [12,13]. The resolution and accuracy of the mechanical pendulum-type level are dependent on the design of the pendulum and mechanism; for that, the size and weight of the pendulum-type precision level can hardly be reduced [14]. In addition, the design and manufacturing accuracy of the pendulum will directly affect the accuracy of the pendulum-type precision level, and the complex structure and requirement for high-precision manufacture processing mean that the cost of this type of level cannot be reduced. In order to improve the resolution and accuracy of the pendulum-type precision level, expensive precision sensors such as capacitive sensors further increase the cost of the level. These problems limit the further application of the precision level.

In order to miniaturize the size of the precision level, many novel measurement principles have been proposed. Using gravity as a reference, Ueda [15] proposed a high-precision micro capacitive inclination sensor, and Shimizu [16] proposed a roll error measurement system made by combining two micro capacitive inclination sensors to eliminate the influence of external disturbance. However, the accuracy and resolution of capacitive inclinometers are larger than 1 arc-sec. Based on the grating diffraction and reflection, Gao developed a three-axis autocollimator for detecting three angular error motions of a precision stage [17], and Shimizu [18] modified the system to make a liquid-surface-based three-axis inclination sensor for the measurement of stage tilt motions. However, the system was complex and the accuracy of the measured angles was still larger than 1 arc-sec. The dual-axis pendulum-type precision level was proposed by Fan [19], with the accuracy and resolution both less than 1 arc-sec. A brand-new design of a pendulum-free precision level was proposed by Torng [20] using the light refraction principle and the reference of the invariant liquid level. The dual-axis level could be small in size and light in weight. However, the first reflection beam from the liquid surface would interfere with the refracted beam and produce noise signals. Zhang [21] directly used the surface reflection beam to design a dual-axis level. However, it required high-energy laser power.

In this paper, a modified design of a refraction-type level from the author's previous system [20] is proposed. With the new principle of single-light refraction and transmission, an innovative dual-axis precision level is improved from the previous double refraction type. It can achieve a satisfactory accuracy and resolution in a compact and simple design. The measurement principle will be described in detail in Section 2, and the factors that affected the design of this type of precision level are discussed in Section 3. The calibration and application experiments of the prototype sensor will be presented in Section 4. At the end of this paper, some directions to improve the performance of the proposed precision level are discussed.

2. Measurement Principle

The optical configuration of the proposed dual-axis level is shown in Figure 1a. The reference laser emitted from the laser diode (LD) passes through the transparent liquid in a transparent container and is incident on an autocollimator unit composed of a focus lens (FL) and a quadrant photodetector (QPD). The mirrors M1 and M2 are used to reflect the reference laser to reduce the size. When the sensor is inclined, the liquid surface remains level. The reference laser emitted from the transparent liquid into the air is refracted. The angle change of the refracted reference laser is measured by the autocollimator unit. According to Snell's law, the relationship between the inclined angle and the angle change of the reference laser can be expressed as:

$$\begin{array}{c} n_1 \sin(\varepsilon_y) = n_2 \sin(\varepsilon_{ym}) \\ n_1 \sin(\varepsilon_x) = n_2 \sin(\varepsilon_{xm}) \end{array}, \tag{1}$$

where ε_y is the inclined angle around the Y axis, as shown in Figure 1b, and ε_x is the inclined angle around the X axis. ε_{ym} and ε_{xm} are the refraction angles in the X and Y directions, respectively, and n_1 and n_2 are the refractive index of the liquid and the air. When the inclined angle is small, the small angle is almost equal to its sine function value, ignoring the very small non-linear error. Equation (1) can be simplified to:

$$\begin{array}{c} n_1 \varepsilon_y = n_2 \varepsilon_{ym} \\ n_1 \varepsilon_x = n_2 \varepsilon_{xm} \end{array}. \tag{2}$$

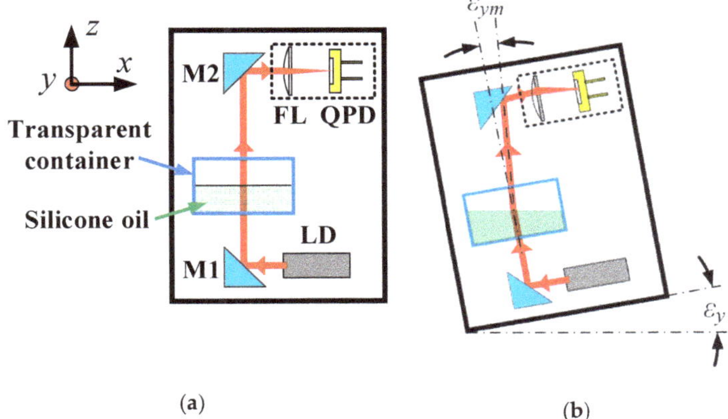

Figure 1. Schematic diagram of the dual-axis precision level: (**a**) optical configuration; (**b**) measurement principle.

According to the auto-collimation principle, the change in the reference laser's incident angle from ε_{ym} to ε_{xm} will cause the change in light spot's position from δ_{yQPD} to δ_{xQPD} on the sensitive surface of the QPD:

$$\varepsilon_{xm} = \frac{\delta_{xQPD}}{f}; \varepsilon_{ym} = \frac{\delta_{yQPD}}{f}, \tag{3}$$

where f is the focal length of FL, the change in the light spot's position from δ_{yQPD} to δ_{xQPD} would lead to a change in the light intensity received by each photodiode of the QPD sensor, resulting in a change in the photocurrent output by the QPD. The photocurrents of the QPD are converted to voltages by an I/V conversion circuit. Based on the working principle of QPD and Equations (2) and (3), the relationship between the output voltages of the QPD (V1, V2, V3, V4) and the inclined angle ε_y and ε_x can be expressed by:

$$\begin{array}{l} \varepsilon_x = \frac{n_2}{n_1}\varepsilon_{xm} = k_{\varepsilon x}\frac{(V_1+V_4)-(V_2+V_3)}{(V_1+V_2+V_3+V_4)} \\ \varepsilon_y = \frac{n_2}{n_1}\varepsilon_{ym} = k_{\varepsilon y}\frac{(V_1+V_2)-(V_3+V_4)}{(V_1+V_2+V_3+V_4)} \end{array}, \tag{4}$$

where $k_{\varepsilon x}$ and $k_{\varepsilon y}$ are the constants obtained by the calibration process, which will be mentioned in Section 4. The proposed measurement principle is modified from the author's previous work [20] to avoid the influence of the reflected light from the liquid surface on the QPD, which would cause a random error. In addition, compared with the measurement principle using the reflected light from the liquid surface [21], the intensity of the reference laser is lower and the signal of QPD is higher in our proposed measurement principle, leading to the better resolution performance of the proposed system.

It should be noticed that, although the bottom wall of the transparent container will produce the additional refraction of the laser beam, this refraction is always the same regardless of the angle of the level, because the LD, container, and autocollimator are tilted together. The principle of angle measurement by the autocollimator is to detect only the angle change in the incident laser beam, and this angle change is purely from the change in the refraction angle between the liquid surface and the laser beam. In addition, the autocollimator is not sensitive to the beam shift.

3. Structure Design

The main structure design of the sensor part of the proposed precision level is shown in Figure 2. The laser diode and micro autocollimator unit are mounted on the lower and upper sides of the container, respectively. A piece of optical window is fixed on the bottom of the container, allowing the reference laser to pass through while preventing the transparent liquid from leaking out. Above the container, a special design is used to prevent liquid from leaking when the sensor is tilted [20]. The structural design of the sensor is rigid and does not contain any moving parts to increase the stability of the sensor. In order to provide the damping ratio to the dynamic system of the sensor for the stabilization of the measurement signal, the transparent liquid uses silicone oil, which has viscosity and stable physical and chemical properties, which ensures the stable performance of the sensor. The total size of the sensor is determined by the volume of the container, and the design of the container is affected by the following factors.

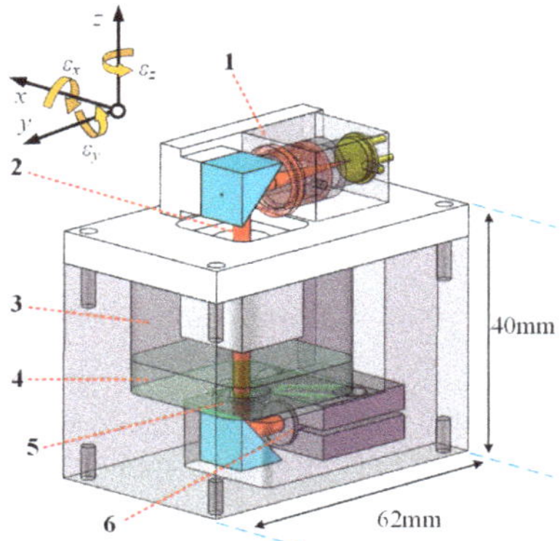

Figure 2. Structure design of the refractive precision level sensor: (1) micro autocollimator unit, (2) reference laser beam, (3) container, (4) transparent liquid, (5) optical window, (6) laser diode.

3.1. Surface Adhesion Effect

At the solid–liquid interface between the container and the liquid, the liquid surface will bend because of the influence of surface adhesion. As shown in Figure 3, if the refractive interface is concave, the linearity and stability of the sensor signal will be affected. Therefore, the container must have sufficient size and the liquid must have sufficient depth to ensure that the liquid surface which the reference laser passes through is flat. Through a simple analysis, the size of the container's cross-section was designed to be 30×30 mm^2, and the depth of the silicone oil was set to 7 mm.

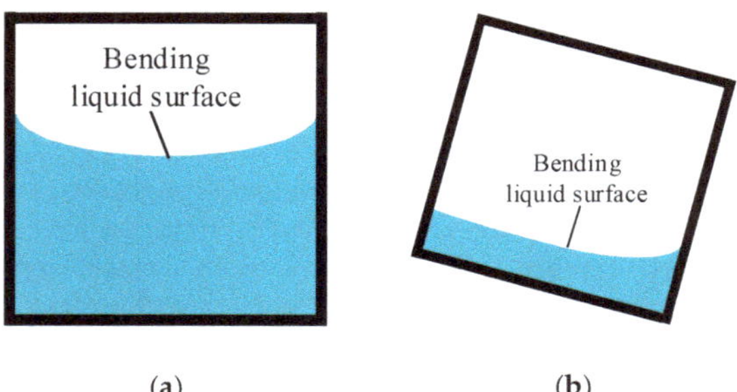

Figure 3. The influence of the surface adhesion effect. (**a**) Too small container size; (**b**) too shallow liquid depth.

3.2. Leakage Prevention Design

When the level is not in use, it will be put in a carrying case. Hence, the container needs to be sealed to prevent liquid leakage. However, if we use the optical window to seal the container on the top, the liquid will inevitably remain on the optical window after the sensor returns from the rest pose, especially for silicone oils with a higher viscosity. This is the natural phenomenon of surface wettability. That will significantly reduce the measurement accuracy and the stability of the precision level. In order to avoid this problem, the container has been specially designed to ensure that the liquid will not leak out in any posture without using an optical window to completely seal the container [20]. A tube with a certain length is mounted on the top of the container. The hollow tube allows the reference laser to pass through. When the level is tilted, the tube wall can prevent the liquid in the container from leaking out, as shown in Figure 4.

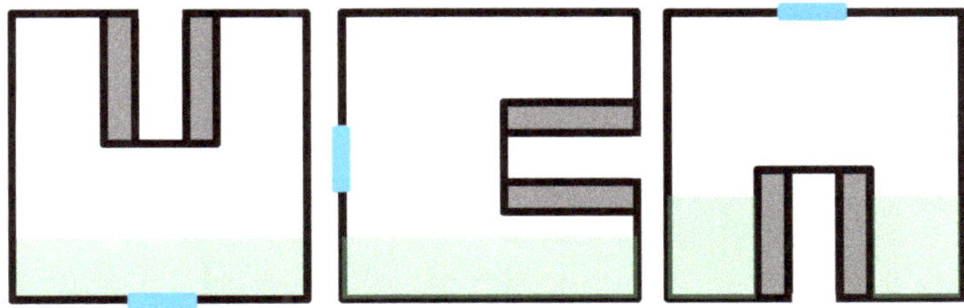

Figure 4. The leakage prevention design: liquid in the container in different tilted states.

Considering the influence of the above factors on the structure design, the final designed container size is $30 \times 30 \times 25$ mm^3, as shown in Figure 4, and the size of the prototype sensor is $60 \times 37 \times 55$ mm^3.

4. Calibration and Application Experiments

The prototype of the manufactured dual-axis precision level based on the measurement principle proposed in Section 2 is shown in Figure 5. The level body and the electronic box are separate so that they can be mounted together or separately according to the installation space. The laser diode has a wavelength of 635 nm (model DI635, Huanic, Xian, China,). The micro-autocollimator set, consisting of a high precision QPD (QP5.8-6-TO5, First Sensor, Berlin, Germany) and a focus lens (FL1, φ10, Tokyo, Hitachi, Japan), was constructed to detect 2D angle changes. The signal acquisition electronic device is to acquire and process the signal of the QPD using an analog-to-digital converter (ADC 7606, Analog Devices, Norwood. USA) and a micro-control unit (MCU, ARM SAMD21 Cortex-M0+, Microchip Technology). The measurement data is wirelessly transmitted to the computer via Bluetooth to avoid the influence of pulling the data cable when moving the level.

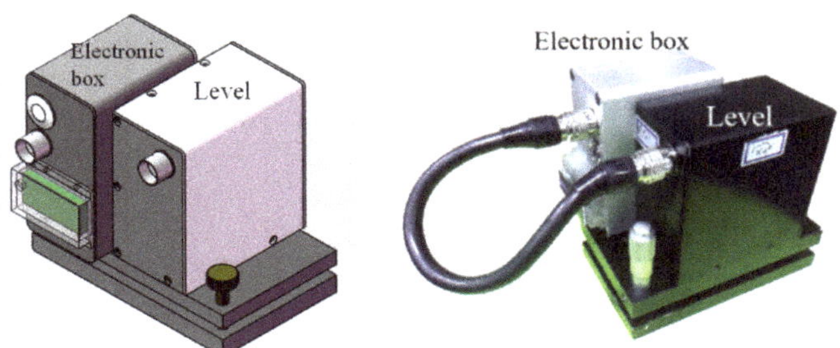

Figure 5. The prototype of the dual-axis precision level: 3D design and photo.

4.1. Calibration Results

The proposed precision level was calibrated by a commercial autocollimator (AutoMAT Co. model 5000U; resolution: 0.01 arc-sec; accuracy: ±0.1 arc-sec; repeatability: 0.05 arc-sec; uncertainty: 0.119 arc-sec). It is known that the output signal of QPD will show obvious nonlinearity in a large range [22]. In order to obtain the most suitable linear sensitivity coefficient on different measurement ranges, the two inclined directions of the dual-axis precision level were calibrated for the different measurement ranges of ±100 arc-sec, ±400 arc-sec, and ±800 arc-sec, separately. The results of the averaged linearity curve and residual curve of five times calibration are shown in Figure 6. It can be seen that the residual error increases with the increase in the measurement range, but the peak-to-valley value of residual errors never exceeds 1% of the measurement range. For the measurement range of ±100 arc-sec, the peak-to-valley value of the residual was less than 0.5% of the measurement range. The calibration results show that an accuracy of ±0.6 are-sec in the range of ±100 arc-sec was obtained, and in the full measurement range of ±800 arc-sec, the accuracy was ±4 are-sec. The resolution was 0.1 arc-sec. The overall performance is very satisfying.

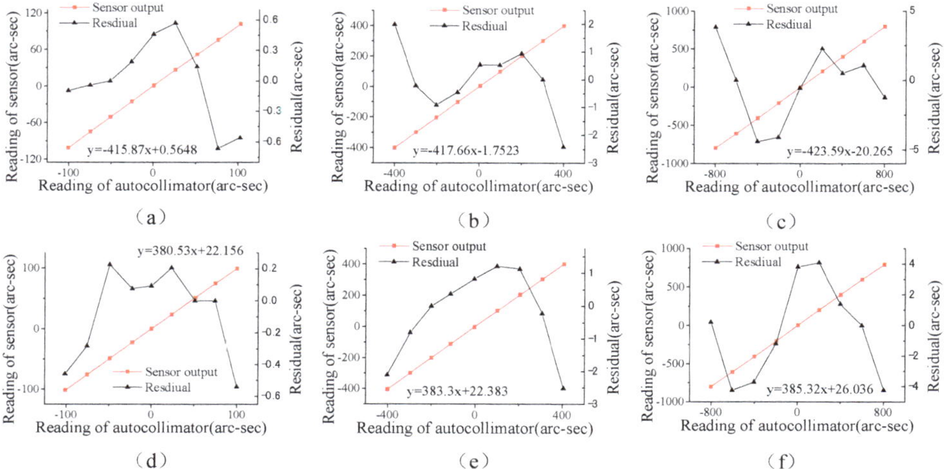

Figure 6. Calibration results of the dual-axis precision level: **(a)** ε_x in ±100 arc-sec; **(b)** ε_x in ±400 arc-sec; **(c)** ε_x in ±800 arc-sec; **(d)** ε_y in ±100 arc-sec; **(e)** ε_y in ±400 arc-sec; **(f)** ε_y in ±800 arc-sec.

4.2. Comparison Measurement Experiment

To verify the measurement accuracy of the proposed precision level, a comparison measurement experiment was designed. The set-up of the comparison experiment is shown in Figure 7a. The precision level was mounted on a linear stage, and the pitch error of the linear stage was simultaneously measured by the proposed precision level and the autocollimator. After the performance of one direction was tested, the proposed precision level was turned to 90° to compare the other direction. The pitch error of the tested linear stage was adjusted to be close to 100 arc-sec in order to compare the accuracy of the proposed precision level in a large measurement range. The comparison results are shown in Figure 7b. It can be seen that the peak-to-valley value of the residual error is less than 5% of the total pitch error. The experiment results proved that the accuracy of the proposed precision level is acceptable.

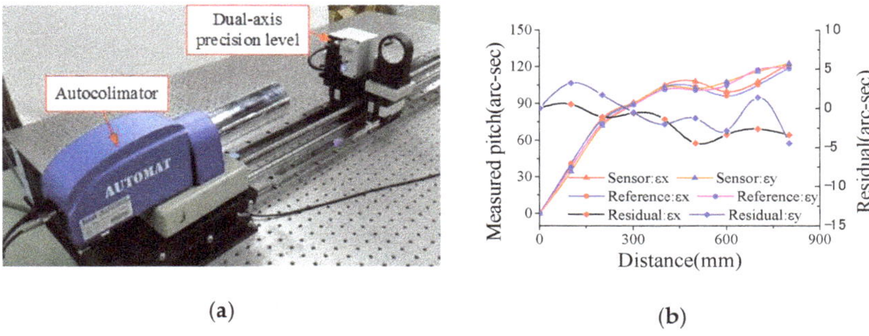

(a) (b)

Figure 7. Comparison experiment: (**a**) experiment set up; (**b**) results.

4.3. Measurement Repeatability Tests

The measurement repeatability of the proposed precision level was obtained by measuring 5 times at 9 positions on a class 00 granite table, as shown in Figure 8. The experiment results are shown in Figure 9. The measurement uncertainty expressed by $\pm 3\sigma$ of the dual-axis precision level in the X and Y directions are ± 1.1 arc-sec and ± 0.92 arc-sec, respectively. This result proves that the measurement repeatability of the proposed precision level is satisfactory.

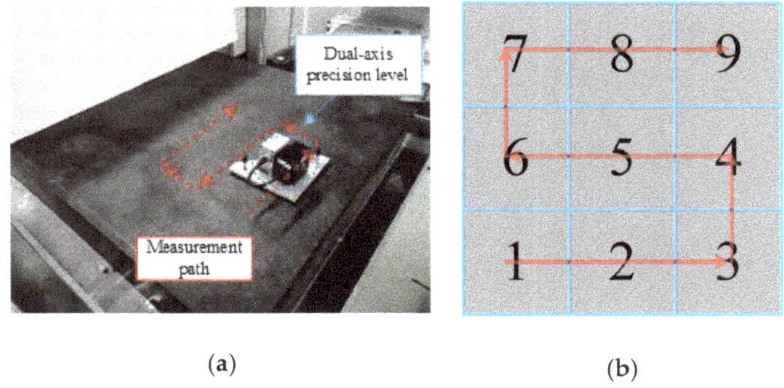

(a) (b)

Figure 8. Measurement repeatability test: (**a**) photo of the experiment set-up; (**b**) measured positions.

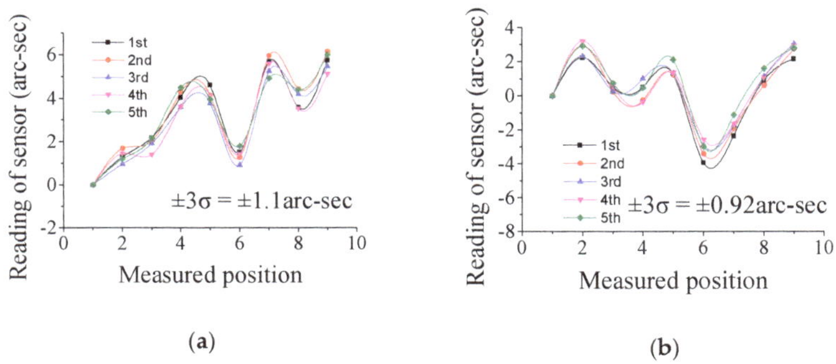

(a) (b)

Figure 9. Measurement repeatability: (**a**) ε_x; (**b**) ε_y.

4.4. Response Time Test

According to the International Standard Organization (ISO) technical report [11], the response time of a precision level is an important performance factor that determines whether the precision level is practical or not. To verify the response time of the proposed dual-axis precision level, the precision level was fixed on a linear stage of a CNC (computer numerical control) lathe. When the linear stage moved 100 mm at a speed of 60 mm/s, the raw data of the proposed dual-axis precision level from the start to the end of the movement were saved by the software with the sampling rate of 1000 samples per second. When the jitter of the reading of the precision of the dual-axis level was less than 0.1 arc-sec, the reading of the level was considered to be stable. The response data are shown in Figure 10. It can be seen that the settling time to steady-state condition was less than 1 s. This performance is much better than our previous system that needs a 3 s settling time [21].

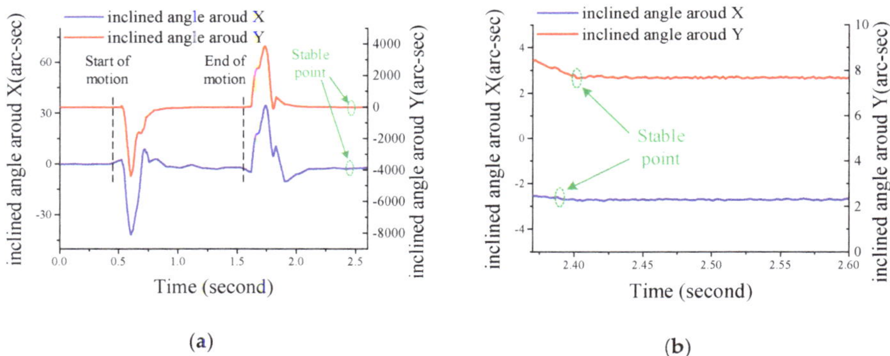

Figure 10. The response time of the proposed precision level: (**a**) all the response raw data; (**b**) partial view.

It should be noted that the response time of the precision level designed according to the measurement principle proposed in Section 2 is affected by the viscosity of the liquid in the container. The filtering algorithm used in the signal processing system would also affect the response time. The liquid used in the prototype of the precision dual-axis level is a synthetic silicone oil with a viscosity of about 200 centistokes. The filtering algorithm used in the signal processing system was a moving average filter, and the length of the sampling window of the moving average filter is 10.

4.5. Stability Tests

The stability of the proposed precision level was tested in a non-temperature-controlled environment. While measuring the stability of the precision level, it is directly mounted on an optical bench. The one-hour stability data of the developed level sensor are shown in Figure 11. The drift of the reading is less than 1 arc-sec within one hour. The stability of the proposed dual-axis precision level is satisfactory. It can also be seen from the stability data that the short-term reading jitter is only about 0.1 arc-sec, which can also represent the resolution of the proposed sensor.

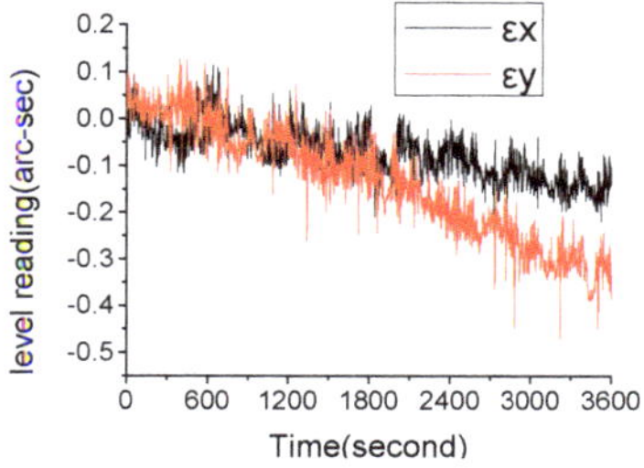

Figure 11. Stability data of the proposed precision level.

5. Measurement Uncertainty Analysis

Benefiting from the simple measurement principle and the compact structure design of the inclination sensor, the measurement uncertainty of the proposed dual-axis precision level is mainly affected by the performance of the QPD and the electronic device. Thus, it is possible to evaluate the measurement uncertainty of the proposed precision dual-axis level based on the data obtained by the experiment proposed in Section 4. The combined standard measurement uncertainty of the inclination angle around the X axis $u_{\varepsilon x}$ could be obtained by:

$$u_{\varepsilon x} = \sqrt{u_{cal\_x}^2 + u_{res\_x}^2 + u_{drf\_x}^2 + u_{rep\_x}^2} \tag{5}$$

The first influential factor, $u_{cal\_\varepsilon x}$, is the standard uncertainty of the system calibration, which can be obtained from Figure 6. Taking the P-V data of the ±100 arc-sec range and using a rectangular distribution, the standard uncertainty of this part was evaluated to be 0.652 arc-sec. The second factor, $u_{res\_\varepsilon x}$, is the standard uncertainty of the resolution, which can be obtained from the steady-state signal variation in 60 min. From the experiment, the P-V data was 0.09 arc-sec in a rectangular distribution, so that its uncertainty was 0.057 arc-sec. The third factor, $u_{drf\_\varepsilon x}$, is the standard uncertainty of the drift due to environmental effect. According to Figure 11, the maximum drift in the X-direction was 0.22 arc-sec in one hour and was in a rectangular distribution. Its standard uncertainty was 0.127 arc-sec. The fourth factor, $u_{rep\_\varepsilon x}$, is the standard uncertainty of the measurement repeatability, and its probability distribution is Gaussian. From Figure 9a, it was evaluated to be 0.351 arc-sec. Table 1 summarizes the evaluated results of the combined uncertainty from these four sources.

Table 1. Uncertainty of the inclination angle around the X-axis of the proposed precision level.

Sources of Uncertainty	Symbol	Type	Probability Distribution	Standard Uncertainty
Calibration of the sensor (including the contribution of systematic error and the uncertainty of the reference autocollimator)	$u_{cal\_\varepsilon x}$	B	Rectangular	0.663 arc-sec
Resolution of sensor	$u_{res\_\varepsilon x}$	B	Rectangular	0.057 arc-sec
Drift	$u_{drf\_\varepsilon x}$	B	Rectangular	0.127 arc-sec
Repeatability	$u_{rep\_\varepsilon x}$	A	Gaussian	0.351 arc-sec
Standard uncertainty of ε_x	$u_{\varepsilon x}$			0.762 arc-sec

It should be noticed that the uncertainty of calibration $u_{cal\_\varepsilon x}$ has already taken the systematic error sources into consideration, such as the systematic error of the autocollimator measurement method (including the manufacturing error of the focusing lens and the defocus error of the QPD installation), the nonlinearity caused by the initial refraction angle of the reference laser, and the nonlinearity error of the QPD sensor in the large-range error measurement. Therefore, the used optical components must meet a certain manufacturing accuracy, and the standard operating procedure (SOP) should be followed to ensure that the systematic error caused by the assembly errors can be suppressed. The combined standard measurement uncertainty of the inclination angle around the Y axis $u_{\varepsilon y}$ could be obtained in the same way; the evaluated combined standard uncertainty $u_{\varepsilon y}$ and its sources are listed in Table 2.

Table 2. Uncertainty of the inclination angle around the Y-axis of the proposed precision level.

Sources of Uncertainty	Symbol	Type	Probability Distribution	Standard Uncertainty
Calibration of the sensor (Including the contribution of systematic error and the uncertainty of reference autocollimator)	$u_{cal\_\varepsilon y}$	B	Rectangular	0.454 arc-sec
Resolution of sensor	$u_{res\_\varepsilon y}$	B	Rectangular	0.057 arc-sec
Drift	$u_{drf\_\varepsilon y}$	B	Rectangular	0.242 arc-sec
Repeatability	$u_{rep\_\varepsilon y}$	A	Gaussian	0.301 arc-sec
Standard uncertainty of ε_y	$u_{\varepsilon y}$			0.599 arc-sec

As the final result, the combined uncertainties of the proposed precision dual-axis level were evaluated to be 0.762 arc-sec and 0.599 arc-sec in the ε_x and ε_y direction, respectively. The expanded uncertainties were thus evaluated to be 1.506 arc-sec and 1.174 arc-sec in the ε_x and ε_y direction (coverage factor $k = 2$, 95% confidence), respectively.

6. Discussion

Compared with the traditional precision level, the inclination measurement principle based on the light refraction proposed in this report has some significant advantages.

(1) The proposed inclination measurement principle does not require any moving parts, so the systematic error caused by the improper design and manufacturing error of the pendulum or hinge can be avoided.
(2) Compared with the commercial precision level of other inclination angle measurement principle based on the light reflection or refraction of the liquid level surface, the measurement principle proposed in this report if simpler, uses fewer optical components, and effectively avoids the influence of stray light on the measurement accuracy.
(3) The cost of the sensor and light source used is lower, and there are fewer components that require a high manufacturing accuracy; therefore, while the measurement accuracy is ensured, the cost of the proposed precision level can be kept very low, which provides convenience for the application.

On the basis of this research, the final goal is to design a precision level as a sensor to monitor the inclination angle of a precision instrument or machine. To achieve this goal, the following issues still need to be further studied.

(1) A mathematical model of the surface adhesion at the liquid–solid interface should be built and verified. Thus, the design of the sensor can be optimized through the mathematical model, improving the dynamic performance and reducing the size of the inclination sensor.
(2) The range of the inclination angle measurement should be increased by compensating the systematic error of QPD in a large-range measurement or using PSD as the detector in the precision level, under the premise of ensuring the accuracy and resolution of it.
(3) In order to reduce the size and improve the reliability of the proposed precision level, a further integrated optimization design is still necessary.

7. Conclusions

An innovative dual-axis precision level is proposed in this paper. Different from the traditional precision level based on the pendulum measurement principle, the proposed dual-axis precision level is based on the light refractive principle. Compared with the traditional measurement principle of the precision level, this novel inclination angle measurement principle can achieve a high resolution and accuracy at a lower cost and compact design. The proposed light refractive and transmission measurement principle can significantly reduce the size of the precision level while maintaining the measurement precision and accuracy. A prototype precision level sensor was made to verify the feasibility of the instrument. The results of the verification experiments are satisfactory. The resolution of the proposed precision level is 0.1 arc-sec, and the measurement uncertainty is less than ±1.2 arc-sec. The calibration results show that, for the measurement range of ±100 arc-sec, the accuracy of the proposed level is within ±0.6 arc-sec; for the full range of ±800 arc-sec, the accuracy of the proposed level is within ±4 arc-sec. Benefitting from the simplicity of the measurement principle, the accuracy and uncertainty of the proposed precision level are almost determined by the photoelectric sensor and its signal processing electronic device, making it easy to improve its performance. In addition, with a proper size design and response test, the quick settling time of about 1 s exhibits an advantage in practical applications. Future works will concern the angular error measurement of precision machines. At the same time, rigorous research on the surface adhesion effect will help us understand how this phenomenon affects the error source and dynamic characteristics of this type of inclination angle sensor. The establishment of a mathematical model could help to optimize the design of the proposed precision level, which will also be the focus of future research.

Author Contributions: Conceptualization, K.-C.F. and Y.H.; methodology, K.-C.F. and Y.H.; software, Y.F; validation, Y.H., Y.F.; resources, Z.L.; data curation, Y.H; writing—original draft preparation, Y.H.; writing—review and editing, K.-C.F.; supervision, W.S.; funding acquisition, K.-C.F. and Z.L. All authors have read and agreed to the published version of the manuscript.

Funding: This research was funded by the fund of The National Key Research and Development Program of China (2017YFF0204800) and the Liaoning Provincial Fund (No. 2020JH6/10500017).

Conflicts of Interest: The authors declare no conflict of interest.

References

1. Gao, W.; Kim, S.-W.; Bosse, H.; Haitjema, H.; Chen, Y.; Lu, X.; Knapp, W.; Weckenmann, A.; Estler, W.; Kunzmann, H. Measurement technologies for precision positioning. *CIRP Ann.* **2015**, *64*, 773–796. [CrossRef]
2. Gao, W. *Precision Nanometrology: Sensors and Measuring Systems for Nanomanufacturing*; Springer Science & Business Media: Berlin/Heidelberg, Germany, 2010.
3. *ISO 230-1. Test Code for Machine Tools—Part 1: Geometric Accuracy of Machine Operating under No-Load or Quasi-Static Conditions*; International Organization for Standardization: Switzerland, Geneva, 2012.
4. *ISO/TR 230-11. Test Code for Machine Tools—Part 11: Measuring Instruments and its Application to the Machine Tool Geometry Tests*; International Organization for Standardization: Switzerland, Geneva, 2018.
5. Schwenke, H.; Knapp, W.; Haitjema, H.; Weckenmann, A.; Schmitt, R.; Delbressine, F. Geometric error measurement and compensation of machines—An update. *CIRP Ann.* **2008**, *57*, 660–675. [CrossRef]
6. Mutilba, U.; Gomez-Acedo, E.; Kortaberria, G.; Olarra, A.; Yagüe-Fabra, J.A. Traceability of on-machine tool measurement: A review. *Sensors* **2017**, *17*, 1605. [CrossRef] [PubMed]
7. Liu, C.S.; Pu, Y.F.; Chen, Y.T.; Luo, Y.T. Design of a measurement system for simultaneously measuring six-degree-of-freedom geometric errors of a long linear stage. *Sensors* **2018**, *18*, 3875. [CrossRef] [PubMed]
8. Fan, K.C.; Wang, H.Y.; Yang, H.W.; Chen, L.M. Techniques of multi-degree-of-freedom measurement on the linear motion errors of precision machines. *Adv. Opt. Technol.* **2014**, *3*, 375–386. [CrossRef]
9. Yu, X.; Gillmer, S.R.; Woody, S.C.; Ellis, J.D. Development of a compact, fiber-coupled, six degree-of-freedom measurement system for precision linear stage metrology. *Rev. Sci. Instrum.* **2016**, *87*, 065109. [CrossRef] [PubMed]
10. Cai, Y.; Sang, Q.; Lou, Z.F.; Fan, K.C. Error Analysis and Compensation of a Laser Measurement System for Simultaneously Measuring Five-Degree-of-Freedom Error Motions of Linear Stages. *Sensors* **2019**, *19*, 3833. [CrossRef] [PubMed]
11. Cai, Y.; Yang, B.H.; Fan, K.C. A robust roll angular error measurement method for precision machines. *Opt. Express* **2019**, *27*, 8027–8036. [CrossRef] [PubMed]
12. Lantz, B.; Schofield, R.; O'Reilly, B.; Clark, D.E.; DeBra, D. Requirements for a ground rotation sensor to improve advanced LIGO. *Bull. Seismol. Soc. Am.* **2009**, *99*, 980–989. [CrossRef]
13. Venkateswara, K.; Hagedorn, C.A.; Turner, M.D.; Arp, T.; Gundlach, J.H. A high-precision mechanical absolute-rotation sensor. *Rev. Sci. Instrum.* **2014**, *85*, 015005. [CrossRef] [PubMed]
14. Ishikawa, T.; Tan, S.L.; Shimizu, Y.; Gao, W.; Hata, R.; Nakagawa, S.; Yoshioka, K. Evaluation of an electronic clinometer for precision measurement of inclination. In Proceedings of the International Conference on Leading Edge Manufacturing in 21st century: LEM21, Miyagi, Japan, 7–8 November 2013; Japan Society of Mechanical Engineers: Tokyo, Japan, 2013.
15. Ueda, H.; Ueno, H.; Itoigawa, K.; Hattori, T. Development of micro capacitive inclination sensor. *IEEJ Trans. Sens. Micromachines* **2006**, *126*, 637–642. [CrossRef]
16. Shimizu, Y.; Satoshi, K.; Gao, W. High resolution clinometers for measurement of roll error motion of a precision linear slide. *Chin. J. Mech. Eng.* **2018**, *31*, 92. [CrossRef]
17. Gao, W.; Saito, Y.; Muto, H.; Arai, Y.; Shimizu, Y. A three-axis autocollimator for detection of angular error motions of a precision stage. *CIRP Ann.* **2011**, *60*, 515–518. [CrossRef]
18. Shimizu, Y.; Kataoka, S.; Ishikawa, T.; Chen, Y.L.; Chen, X.; Matsukuma, H.; Gao, W. A liquid-surface-based three-axis inclination sensor for measurement of stage tilt motions. *Sensors* **2018**, *18*, 398. [CrossRef] [PubMed]

19. Fan, K.C.; Wang, T.H.; Lin, S.Y.; Liu, Y.C. Design of a dual-axis optoelectronic level for precision angle measurements. *Meas. Sci. Technol.* **2011**, *22*, 055302. [CrossRef]
20. Torng, J.S.; Wang, C.H.; Huang, Z.N.; Fan, K.C. A novel dual-axis optoelectronic level with refraction principle. *Meas. Sci. Technol.* **2013**, *24*, 035902. [CrossRef]
21. Zhang, C.; Duan, F.; Fu, X.; Liu, C.; Liu, W.; Su, Y. Dual-axis optoelectronic level based on laser auto-collimation and liquid surface reflection. *Opt. Laser Technol.* **2019**, *113*, 357–364. [CrossRef]
22. Vo, Q.; Zhang, X.; Fang, F.Z. Extended the linear measurement range of four-quadrant detector by using modified polynomial fitting algorithm in micro-displacement measuring system. *Opt. Laser Technol.* **2019**, *112*, 332–338. [CrossRef]

Article

Investigation on the Differential Quadrature Fabry–Pérot Interferometer with Variable Measurement Mirrors

Yi-Chieh Shih [1,*], Pi-Cheng Tung [1], Wen-Yuh Jywe [2], Chung-Ping Chang [3], Lih-Horng Shyu [4] and Tung-Hsien Hsieh [2]

1 Department of Mechanical Engineering, National Central University, Taoyuan 320, Taiwan; t331166@ncu.edu.tw
2 Smart Machinery and Intelligent Manufacturing Research Center, National Formosa University, Yunlin 632, Taiwan; jywe@nfu.edu.tw (W.-Y.J.); p98951078@nfu.edu.tw (T.-H.H.)
3 Department of Mechanical and Energy Engineering, National Chiayi University, Chiayi 600, Taiwan; cpchang@mail.ncyu.edu.tw
4 Department of Electro-Optical Engineering, National Formosa University, Yunlin 632, Taiwan; lhshyu@nfu.edu.tw
* Correspondence: 105383003@cc.ncu.edu.tw

Received: 7 July 2020; Accepted: 4 September 2020; Published: 6 September 2020

Abstract: Due to the common path structure being insensitive to the environmental disturbances, relevant Fabry–Pérot interferometers have been presented for displacement measurement. However, the discontinuous signal distribution exists in the conventional Fabry–Pérot interferometer. Although a polarized Fabry–Pérot interferometer with low finesse was subsequently proposed, the signal processing is complicated, and the nonlinearity error of sub-micrometer order occurs in this signal. Therefore, a differential quadrature Fabry–Pérot interferometer has been proposed for the first time. In this measurement system, the nonlinearity error can be improved effectively, and the DC offset during the measurement procedure can be eliminated. Furthermore, the proposed system also features rapid and convenient replacing the measurement mirrors to meet the inspection requirement in various measuring ranges. In the comparison result between the commercial and self-developed Fabry–Pérot interferometer, it reveals that the maximum standard deviation is less than 0.120 μm in the whole measuring range of 600 mm. According to these results, the developed differential Fabry–Pérot interferometer is feasible for precise displacement measurement.

Keywords: differential Fabry–Pérot interferometer; homodyne interferometer; nonlinearity error; linear displacement

1. Introduction

From the comparison between the resolution and measuring range of different measurement devices, high resolution and contactless measurement can be realized by the laser interferometer in various measuring ranges. Hence, the industrial application, including the calibration of the linear axis for the machine tool, the positioning of the microelectromechanical equipment, and the wafer dicing, must rely on it to ensure the quality of production and the high-precision inspection requirement [1–4].

Currently, the Michelson interferometer is the primary tool for displacement measurement in a large dynamic range. In the optical structure, the laser beam is divided into a reference beam and a measurement beam by a non-polarizing beam splitter (BS), and corresponding mirrors reflect each beam and then form the interferometric signal. The phase of this signal depends on the optical path difference between the reference beam and the measurement beam. Because the reference path is separate from the measurement path, this kind of interferometer is susceptible to relative environmental changes.

In contrast, the Fabry–Pérot interferometer is based on the common path structure. For this reason, laser beams propagate in the same optical path, and this interferometer possesses high resistance of environmental disturbances that contain thermal expansion, micro-flow gradient, and the tiny vibration [5–7].

The optical arrangement of the conventional Fabry–Pérot interferometer, which contains a laser light source, a detector, and an optical cavity composed of two plane mirrors (PMs) had been proposed by Charles Fabry and Alfred Pérot in 1897 (illustrated in Figure 1a) [8]. The laser light source is incident into the optical cavity in a nearly vertical direction, and the laser beam is divided into numerous transmitted beams from the cavity reciprocally. Then the interferometric signal can be acquired by the detector. Because the optical cavity of the conventional Fabry–Pérot interferometer is composed of two PMs with a high reflectance, which results in the discontinuous signal distribution with high finesse. Furthermore, displacement or vibration can only be realized by the fringe counting method, and it is rarely utilized in large-scale dynamic displacement measurements.

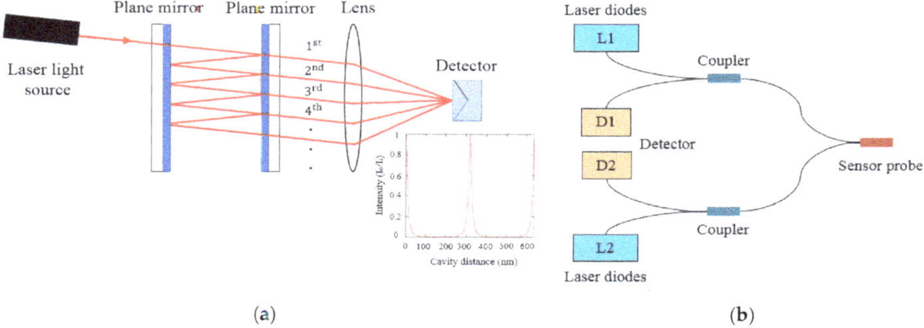

Figure 1. Optical arrangement of Fabry–Pérot interferometer: (**a**) Conventional structure, and (**b**) quadrature phase-shifted fiber-optic structure.

According to the development of the optical encoder and the interferometer system, to improve the measurement accuracy, correlated technologies, including orthogonal signal processing and signal subdivision, have been proposed. Therefore, the quadrature phase-shifted fiber-optic Fabry–Pérot interferometer demonstrated in Figure 1b had been presented by Kent A. Murphy et al. in 1999 [9]. This structure is based on the spatial phase-shifted method to form the orthogonal signal, which will be obtained by two detectors (D_1, D_2). The phase shift is mainly decided by the position and angle of the probe and the measured surface. For this reason, the measuring range is limited by the performance of the optical-mechanical alignment.

In the recent study, polarized Fabry–Pérot interferometer with low finesse, shown in Figure 2, had been proposed by Chang et al. in 2013 [10,11]. The octadic-wave plate is placed in the optical cavity, and its polarization axis must be the same as that of the polarizing beam splitter (PBS) to introduce the orthogonal phase shift between two interferometric signals. By reducing the reflectance of the plane mirror, continuous signal distribution can be acquired, and the interferometric signals (I_s, I_p) can be detected by the photodiodes (PD_1, PD_2). The orthogonal interferometric signal, shown in Figure 3a, can be expressed by Equations (1) and (2) [12], where A_0 is the amplitude of the incident beam, R and T are the reflectance and transmittance of the coated plane mirror, T′ is the transmittance of the optical cavity, and δ is the phase difference. In the simulation, optical parameters of R, T, and T′ are 0.25, 0.75, and 0.86, respectively. The analysis method of the mean normalization is adopted to evaluate the amplitude and DC offset of the orthogonal signal in this study. Theoretically, if the signal amplitude is uniform change and there is no DC offset existing, the center (offset) of the orthogonal signal will be at the zero points, and it will become a circle shape after the mean normalization processing.

Interferometric signals (I_s, I_p) can be processed by the mean normalization ($I_{s\text{-nom}}$, $I_{p\text{-nom}}$) individually represented in Equations (3) and (4). Then, the normalized orthogonal signal illustrated in Figure 3b can be obtained, where $I_{\_max}$, $I_{\_min}$, and μ are the maximum, minimum, and average intensity.

$$I_s = \frac{\frac{1}{2}{A_0}^2 \times T^2 \times T\prime}{1 + R^2 \times T'^2 - 2 \times T\prime \times R \times \cos\left(\delta - \frac{\pi}{4}\right)} \tag{1}$$

$$I_p = \frac{\frac{1}{2}{A_0}^2 \times T^2 \times T\prime}{1 + R^2 \times T'^2 - 2 \times T\prime \times R \times \cos\left(\delta + \frac{\pi}{4}\right)} \tag{2}$$

$$I_{s-nom} = \frac{I_s - \mu}{I_{s\_max} - I_{s\_min}} \tag{3}$$

$$I_{p-nom} = \frac{I_p - \mu}{I_{p\_max} - I_{p\_min}} \tag{4}$$

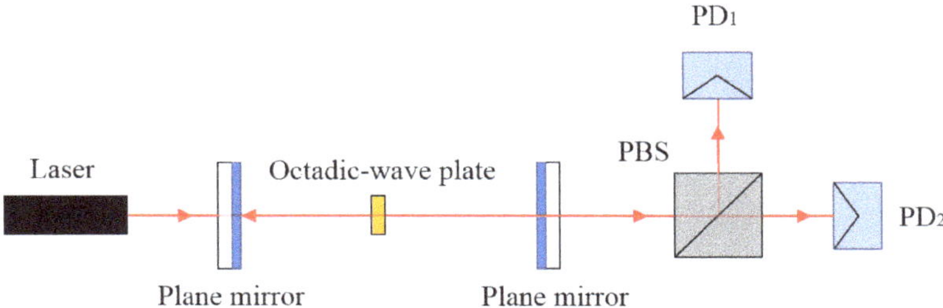

Figure 2. The optical arrangement of polarized Fabry–Pérot interferometer. PBS, polarizing beam splitter; PD, photodiode.

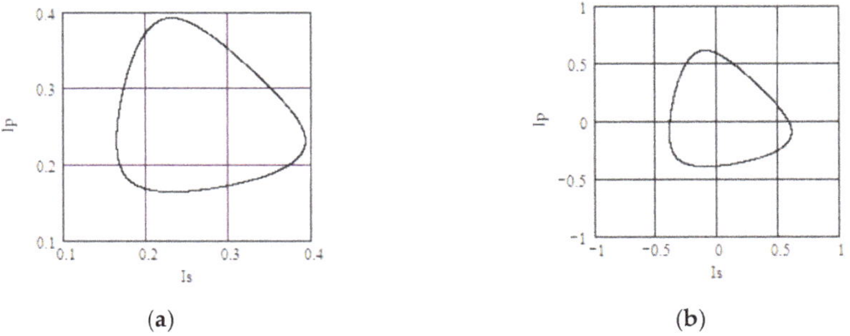

Figure 3. Orthogonal signal of polarized Fabry–Pérot interferometer. (**a**) Original signal; (**b**) signal with mean normalization processing.

In this structure, there are two critical issues, including the existing large nonlinearity error in the signal and involving the complex signal processing for the elimination of the direct current (DC) offset. The nonlinearity error is caused by the DC offset, unequal alternating current (AC) amplitudes, and quadrature phase errors that occur in the orthogonal signal. In the simulation results in Figure 3, the orthogonal signal after the mean normalization processing reveals that the signal possesses the DC offset and unequal AC amplitudes, which will lead to the nonlinearity error. Therefore, the measurement accuracy of this measurement structure will be affected by the nonlinearity error during the measurement. The formula for the analysis of the nonlinearity error is revealed in

Equation (5), where I_x and I_y are the interferometric signals, ψ represents the ideal phase, and m is a constant [13–15].

$$\text{Nonlinearity error} = \arctan\,(I_y/I_x) - \psi + m\pi \tag{5}$$

In accordance with the measurement structure of the polarized Fabry–Pérot interferometer, the analysis of the nonlinearity error is based on the simulated orthogonal signal, shown in Figure 3. The simulation result (Figure 4) reveals that the nonlinearity error is ranging from −48 to 131 nm in the polarized Fabry–Pérot interferometer.

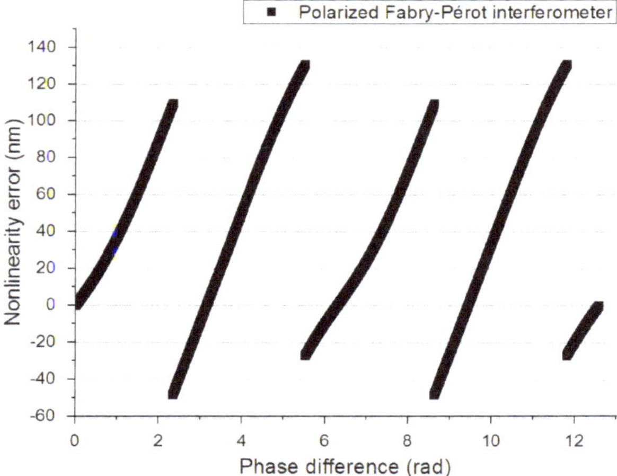

Figure 4. Nonlinearity error of polarized Fabry–Pérot interferometer.

For subsequent signal subdivision, significant interpolation error will occur. Furthermore, the conventional correction method of nonlinearity error is also difficult to implement in this signal [16,17]. Subsequently, a look-up table method was provided to carry out the signal subdivision with the resolution of 0.1 nm in the polarized Fabry–Pérot interferometer [18]. Still, this method can only be processed for a single Lissajous signal under specific measurement conditions, e.g., measuring speed and range, etc. If the measurement signal changes, this method will not be able to carry out. In the correlated study, the Lissajous signal of the start-point and the endpoint in the whole measuring range of 500 mm is different, which means that the signal phase and amplitude are changing [11]. If this signal processing method is utilized in this situation, the inherent measurement performance with the resolution of 0.1 nm will not be achieved in the actual experiment. Furthermore, to avoid the signal drift during the measurement procedure, the DC offset compensation module is also introduced in the polarized Fabry–Pérot interferometer. The conversion of analog to digital and digital to analog is performed in real-time to acquire the signal center and then is to eliminate the DC offset. Since signal processing involves more complex feedback control, if the noise occurs during the measurement, the measurement accuracy will be affected severely. Therefore, this kind of interferometer cannot be performed in high precision and high-speed industrial applications.

By reviewing the development of the Fabry–Pérot interferometer employed for the displacement measurement, the major problems in critical structures are summarized in Table 1. Therefore, how to provide a measurement system based on the Fabry–Pérot interferometer with high measurement performance for employing in the large range is an essential issue. From the above-mentioned description, a differential quadrature Fabry–Pérot interferometer with variable measurement mirrors employing in different measuring ranges is proposed in this study. By the integration of the differential optical structure and the Fabry–Pérot interferometric technique, the nonlinearity error can be improved effectively, and the DC offset can also be eliminated during the measurement process to realize high precision measurement performance conveniently and flexibly.

Table 1. Comparison between the previous Fabry–Pérot interferometers.

Fabry–Pérot Interferometer \ Item		Signal Distribution	Signal Processing Method	Problems
Conventional structure	Reference 8	Discontinuous (high finesse)	Fringe counting	Not suitable for dynamic measurement in large range
Polarized structure	Reference 11	Continuous (low finesse)	Quadrature-based sensing with 2 photodiodes	1. With large nonlinearity error 2. Involved with the complex signal processing

2. Proposed Differential Quadrature Fabry–Pérot Interferometer

Based on the pending patent (application number: 108142443), the optical structure of the proposed differential quadrature Fabry–Pérot interferometer is extended and realized experimentally. It enables the arrangement with two measurement types simultaneously for determining the linear displacement. This structure contains a laser source, an optical cavity, and an optical element and signal sensing module (Figure 5a). The laser beam is incident to the optical cavity module from the laser source module and the optical element and signal sensing module. Then the laser beam is divided into multiple beams and output from the cavity, and PDs placed in the optical element and signal sensing module will inspect the interferometric signals. The laser source module is composed of the high stabilized He-Ne laser and the optical isolator. The purpose of placing the optical isolator is to avoid the laser beam to reflect back from other optical components. The optical cavity module contains two measurement types depending on the selection of different measurement mirrors (Figure 5b). One is the PM type whose optical cavity is composed of two PMs, and the other is the corner cube retro-reflector (CCR) type whose optical cavity consists of a PM and a CCR. In the CCR type, due to its folded measurement structure, the optical resolution can be improved by a factor of two compared to the type of the PM [19]. In addition, owing to CCR bearing large tilt angles of the measurement mirror, this type is more suitable for the displacement measurement within the large range. Due to the multi-beam interferometric measurement structure, the incident beam travels forward and backward in the cavity and is divided into numerous reflected beams (order number of beam: 1st, 2nd, 3rd … Nth). The Nth beam will pass through the quarter-wave plate twice more than the N-1th beam, so the polarization direction of each laser beam will be orthogonally converted. Then they will be transmitted back to the optical element and signal sensing module. In the optical element and signal sensing module, it contains a quarter-wave plate, two BS, two PBS, and four PDs. According to the proposed optical structure, the interferometric signal can be acquired by PDs for further displacement measurement.

The linearly polarized laser beam with a polarization direction angle of $-\pi/4$ with respect to the y-axis will be acquired after passing through the optical isolator, and it is converted to the right-handed circularly polarized beam by the first quarter-wave plate, whose fast axis is along the y-axis. Then the laser beam will be incident to the BS_1 and the optical cavity. In this cavity, the beam is transmitted to the second quarter-wave plate whose fast axis is at a zero angle with respect to the y-axis multiple times. Numerous reflected laser beams output by the optical cavity are sequentially converted into the orthogonally polarized beams (Figure 5c). The backward laser beam is split into two beams by BS_2, and then each beam is transmitted to the PBS_1 and the PBS_2, respectively. The optical axis of the PBS_2 is rotated around the incident beam by $\pi/4$ relative to that of PBS_1. In the optical arrangement, two pairs of the interferometric signal with a phase difference of π can be obtained. And by subtracting each pair signal, the orthogonal signal for displacement measurement will be determined.

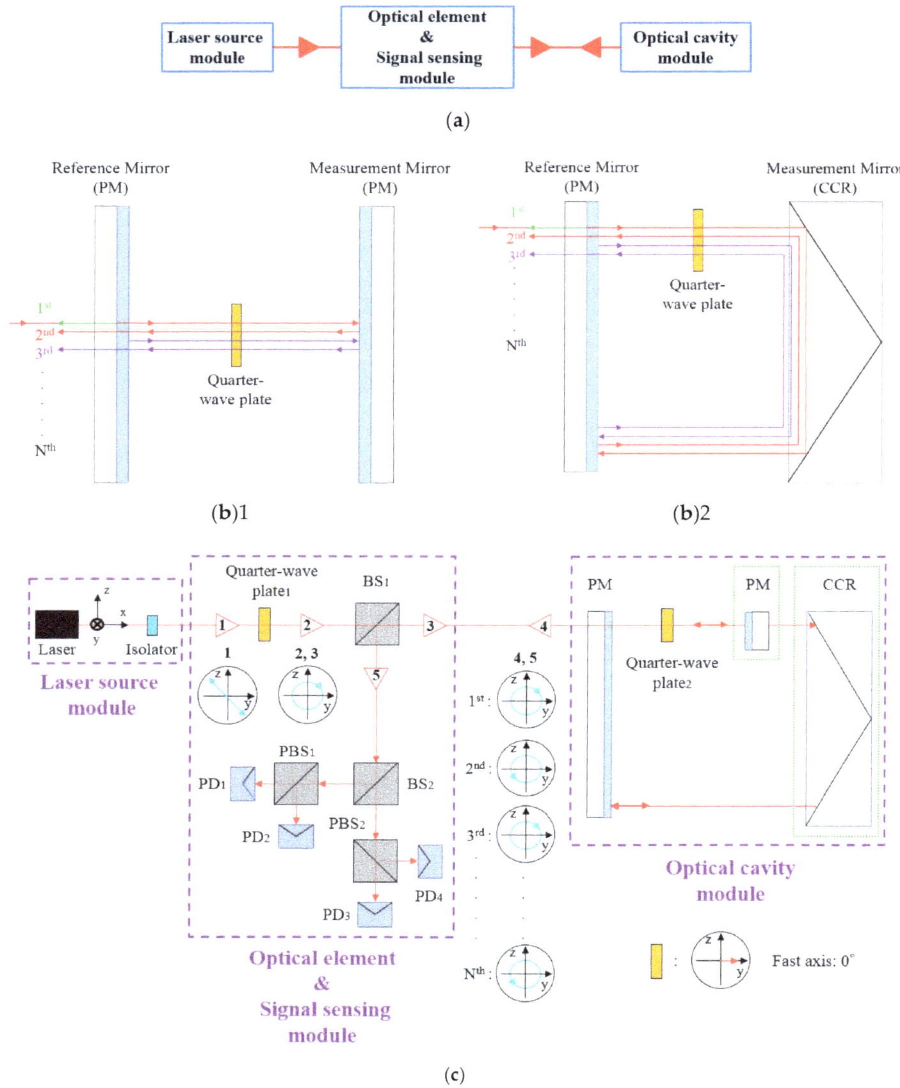

Figure 5. Proposed differential quadrature Fabry–Pérot interferometer: (**a**) Structure composition; (**b**) transmission of laser beams in two measurement types (**b**)1 plane mirrors (PM) type, (**b**)2 cube retro-reflector (CCR) type, (**c**) optical arrangement and the polarization direction of laser beams. PD, photodiode, PBS, polarizing beam splitter; BS, non-polarizing beam splitter.

The Jones matrix of the incident beam and the relevant optical elements in this structure is as Equation (6) to Equation (11).

$$E_{(\Phi)} = \begin{pmatrix} \cos\Phi \\ \sin\Phi \end{pmatrix} \tag{6}$$

$$PF = e^{i\cdot\delta} = e^{i\cdot\frac{4\pi nd}{\lambda}} \tag{7}$$

$$\mathrm{BS} = \frac{1}{\sqrt{2}} \cdot \begin{pmatrix} 1 & 0 \\ 0 & 1 \end{pmatrix} \tag{8}$$

$$\mathrm{QWP}(\theta) = \begin{pmatrix} \cos(\theta) & -\sin(\theta) \\ \sin(\theta) & \cos(\theta) \end{pmatrix} \cdot \begin{bmatrix} e^{-i\cdot(\frac{\pi}{4})} & 0 \\ 0 & e^{i\cdot(\frac{\pi}{4})} \end{bmatrix} \cdot \begin{pmatrix} \cos(\theta) & \sin(\theta) \\ -\sin(\theta) & \cos(\theta) \end{pmatrix} \tag{9}$$

$$\mathrm{PBS_o}(\alpha) = \begin{pmatrix} \sin(\alpha)^2 & -\sin(\alpha)\cdot\cos(\alpha) \\ -\sin(\alpha)\cdot\cos(\alpha) & \cos(\alpha)^2 \end{pmatrix} \tag{10}$$

$$\mathrm{PBS_e}(\alpha) = \begin{pmatrix} \cos(\alpha)^2 & \sin(\alpha)\cdot\cos(\alpha) \\ \sin(\alpha)\cdot\cos(\alpha) & \sin(\alpha)^2 \end{pmatrix} \tag{11}$$

Where E (Φ) is the electric field of the linearly polarized laser beam after traveling the optical isolator, the angle of the polarization direction with respect to the y-axis is Φ. PF is defined as the phase factor ($e^{i\cdot\delta}$) accounts for the optical phase difference (δ) caused by the changing of the cavity distance (d). δ equals $4\pi nd/\lambda$, where λ is the laser wavelength, and n is the refractive index. For the type of the plane mirror and the corner cube retro-reflector, the optical phase difference equals to $4\pi nd/\lambda$ (δ) and $8\pi nd/\lambda$ (2δ), respectively. BS is the matrix of the non-polarizing beam splitter, and QWP (θ) represents a matrix that corresponds to a quarter-wave plate whose fast axis is at an angle θ with respect to the y-axis. $\mathrm{PBS_o}$ (α) and $\mathrm{PBS_e}$ (α) represent two separated beams rotated by an angle α relative to the x-axis.

The sum of the electric field can be express as Equation (12), where R and T are the reflectance and transmittance of the coated plane mirror, respectively, Q is the transmittance of the optical cavity, and N is the order number of the backward reflected beam.

$$E_{sum} = \left[\mathrm{BS}^3\cdot\sqrt{R} + \mathrm{BS}^3\cdot\left(\sqrt{T}\right)^2\cdot\sum_{N=2}^{\infty}\left(\sqrt{R}\right)^{2N-3}\cdot Q^{N-1}\cdot \mathrm{PF}^{N-1}\cdot \mathrm{QWP}(\theta_2)^{2\cdot N-2}\right]\cdot \mathrm{QWP}(\theta_1)\cdot E(\Phi) \tag{12}$$

For PD_1 to PD_4, the corresponding electric fields of each incident beam are shown from Equation (13) to Equation (16).

$$E_1 = \mathrm{PBS_o}(\alpha_1)\cdot E_{sum} \tag{13}$$

$$E_2 = \mathrm{PBS_e}(\alpha_1)\cdot E_{sum} \tag{14}$$

$$E_3 = \mathrm{PBS_o}(\alpha_2)\cdot E_{sum} \tag{15}$$

$$E_4 = \mathrm{PBS_e}(\alpha_2)\cdot E_{sum} \tag{16}$$

The light intensity can be expressed as Equation. (17) and two orthogonal signals can be determined by Equation (18) and Equation (19).

$$I = E\cdot E^* \tag{17}$$

$$I_x = I_1 - I_2 \tag{18}$$

$$I_y = I_3 - I_4 \tag{19}$$

For the proposed differential measurement structure, the mean normalization processing is similar to the analysis method in the polarized Fabry–Pérot interferometer, shown in Section 1. By the mean normalization ($I_{x\text{-nom}}$, $I_{y\text{-nom}}$) for each interferometric signal (I_x, I_y) indicated in Equations (20) and (21), the orthogonal signal with normalization processing can be acquired. In order to illustrate this method, an assumed ellipse is utilized to perform the mean normalization. Its center is in the coordinate (0, 0), the length of the semi-major axis, and the semi-minor axis equal 2 and 1, respectively. The original graph, shown in Figure 6a, can be converted into a circle (Figure 6b) by this normalization. In summary, the reflectance of the coated plane mirror and transmittance of the optical cavity are 0.25 and 0.86.

And the angle of Φ, θ, α_1 and α_2 are $\pi/4$, 0, 0, and $\pi/4$, respectively. The Lissajous signal of the proposed differential Fabry–Pérot interferometer is demonstrated in Figure 7a,b.

$$I_{x-nom} = \frac{I_x - \mu}{I_{x\_max} - I_{x\_min}} \tag{20}$$

$$I_{y-nom} = \frac{I_y - \mu}{I_{y\_max} - I_{y\_min}} \tag{21}$$

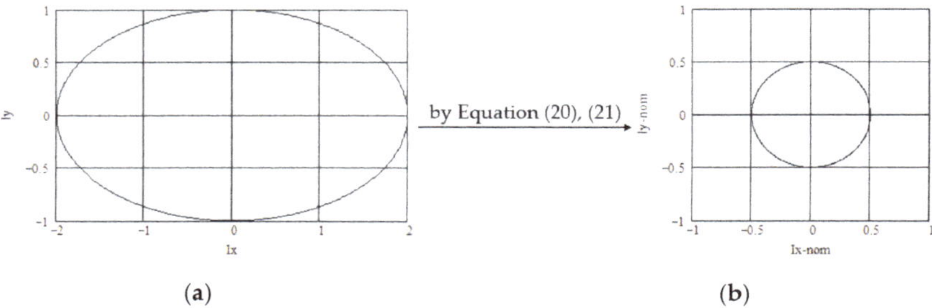

Figure 6. Assumed ellipse. (**a**) Original graph; (**b**) Graph with mean normalization processing.

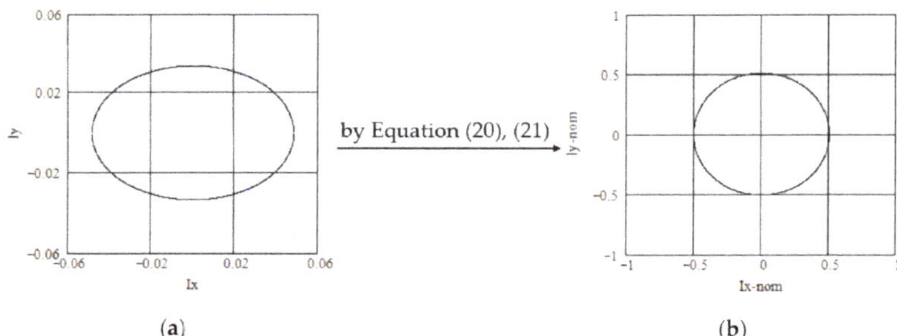

Figure 7. Orthogonal signal of proposed differential quadrature Fabry–Pérot interferometer. (**a**) Original signal; (**b**) Signal with mean normalization processing.

According to the proposed differential Fabry–Pérot interferometer, the simulated results illustrated in Figure 7 reveal that it is an ellipse orthogonal signal without the DC offset. It means that by subtracting two pairs of the signal with a phase difference of π, the DC offset can be eliminated. Then, the equal AC amplitudes can be realized by the hardware circuit, and that minimized the nonlinearity error. Therefore, by the integration of the differential optical structure, and common path interferometric technique, the DC offset can be eliminated, and the nonlinearity error can also be improved. The analysis of the nonlinearity error is similar to the polarized Fabry–Pérot interferometer. Compared to the polarized structure, the result reveals that the nonlinearity error is significantly reduced to less than one nanometer, which equals to two magnitude orders (Figure 8).

In summary, by the proposed differential measurement structure, the DC offset occurs during the measurement can be eliminated without a complex signal processing module. The nonlinearity error can also be improved significantly to realize the precise industrial and scientific applications.

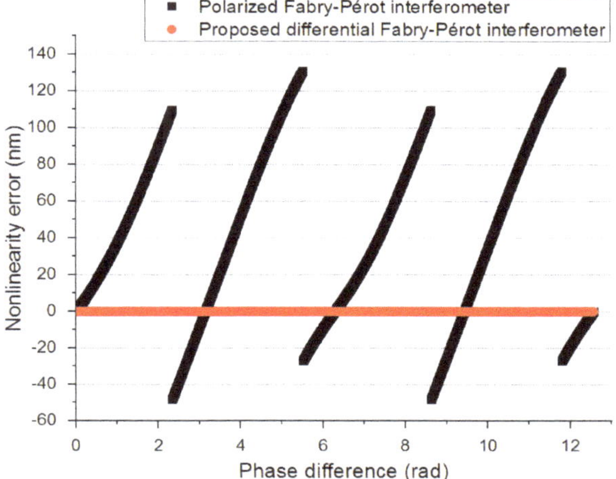

Figure 8. Comparison of nonlinearity error between polarized and proposed differential Fabry–Pérot interferometer.

3. Experimental Results

To verify the measurement performance of the proposed differential quadrature Fabry–Pérot interferometer, a commercial interferometer with the resolution of 1 nm is employed as a reference standard to carry out the comparison experiment simultaneously. After the experiment, the deviation can be obtained by the difference between the measured values of two interferometers. For determining the measurement repeatability, the experimental standard deviation is calculated by 10 deviations (10 repeated cycles) at each position, and then the linearity can be confirmed through dividing three times the maximum standard deviation by the whole measuring range [20]. The experimental arrangement is demonstrated in Figure 9. In this experiment, the measurement structure of PM type in the small measuring range and that of CCR type in the large range have been utilized. Results can be gained in both structures. The Fabry–Pérot interferometer and the commercial interferometer are installed on the left and right sides of the linear stage, and the corresponding measurement mirror is fixed on it to carry out the measurement procedure.

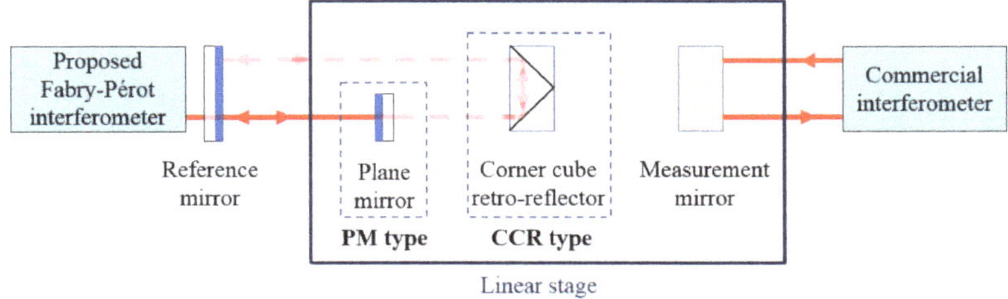

Figure 9. The optical arrangement of the comparison experiment between the commercial and proposed interferometer.

In the comparison experiment between the Fabry–Pérot interferometer of the PM type with the resolution of $\lambda/8$ and commercial interferometer, the forward displacement is regulated from 0 mm to 150 mm with a step interval of 15 mm, and each cycle is repeated 10 times. According to the measurement result (Figure 10), the maximum deviation between the two interferometers is 0.219 μm, the maximum standard deviation is 0.076 μm, and the linearity is 1.52×10^{-6} F.S. in the whole range.

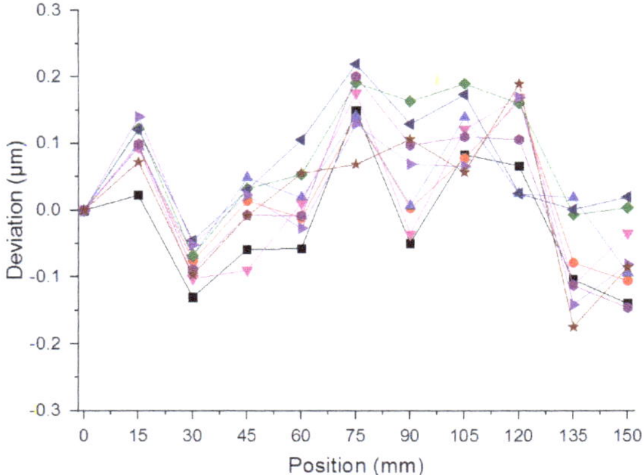

Figure 10. Comparison result between the commercial interferometer and proposed Fabry–Pérot interferometer of PM type.

The comparison experiment between the Fabry–Pérot interferometer of the CCR type with the resolution of λ/16 and commercial interferometer is similar to the above-mentioned procedure. The measuring range is extended to 600 mm, which is conducted with a step interval of 60 mm. The measurement results, shown in Figure 11, demonstrates that the maximum deviation, the maximum standard deviation, and the linearity are 0.427 μm, 0.120 μm, and 6×10^{-7} F.S., respectively. From these comparison results between two measurement types demonstrated in Table 2, the linearity of CCR type is better than the PM type. Because of CCR bearing large tilt angles of the measurement mirror, the linearity of CCR type would be better than that of the PM type within the same measuring range.

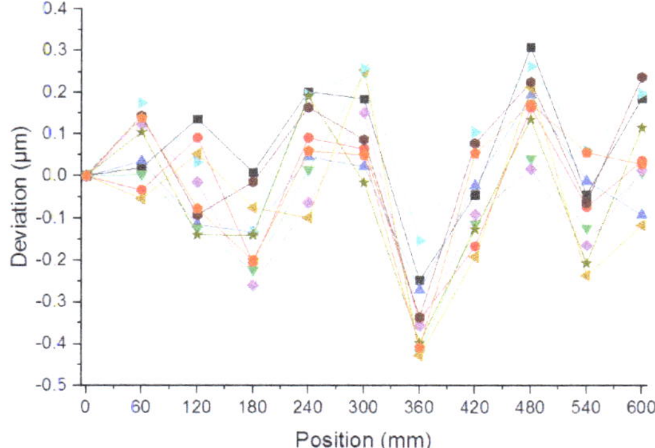

Figure 11. The comparison result between the commercial interferometer and proposed Fabry–Pérot interferometer of CCR type.

In order to evaluate the improvement of the measurement performance in the large measuring range, comparison results between the published polarized Fabry–Pérot interferometer in reference 11 and the proposed differential Fabry–Pérot interferometer of CCR type are illustrated in Table 3. It reveals that the linearity in the published structure within the range of 500 mm is 8.76×10^{-7} F.S., and the linearity in the proposed structure within the extended range of 600 mm is 6×10^{-7} F.S. In summary, the maximum deviation and linearity can be improved by 48% and 32%, respectively by the proposed measurement system. This measurement performance can be utilized for the linear displacement measurement to meet the high-precision industrial demand.

Table 2. The comparison result of the proposed measurement structure between the PM type and the CCR type.

Measurement Performance \ Type			PM	CCR
Range		mm	150	160
Maximum	standard deviation (σ)	μm	0.076	0.120
	standard deviation (3σ)		0.228	0.360
Linearity ($\frac{3\sigma}{\text{Range}}$)		F.S.	1.52×10^{-6}	6×10^{-7}

Table 3. The comparison result between the published and the proposed measurement structure.

Measurement Performance \ Item			Reference 11	Proposed Structure	Improvement (%)
Range		mm	500	600	
Maximum	deviation	μm	0.824	0.427	48
	standard deviation (σ)		0.146	0.120	
	standard deviation (3σ)		0.438	0.360	
Linearity ($\frac{3\sigma}{\text{Range}}$)		F.S.	8.76×10^{-7}	6×10^{-7}	32

4. Conclusions

In this study, the differential quadrature Fabry–Pérot interferometer with variable measurement mirrors for employing in different measuring ranges has been proposed. By the integration of the differential optical structure and common path interferometric technique, the nonlinearity error can be effectively reduced, which equals two magnitude orders, and the DC offset can also be eliminated during the measurement procedure. The measurement performance of the developed Fabry–Pérot interferometer has been verified. From the experimental result, the measurement repeatability is less than 0.120 μm, and the linearity is 6×10^{-7} F.S. within the whole range of 600 mm. By the comparison of the measurement performance between the published polarized Fabry–Pérot interferometer and the proposed differential Fabry–Pérot interferometer, the maximum deviation and linearity can be improved by 48% and 32%, respectively. The capability of the proposed system is conducive to industrial applications to realize the high precision displacement measurement.

Author Contributions: Conceptualization, Y.-C.S., P.-C.T., W.-Y.J., C.-P.C., L.-H.S. and T.-H.H.; Investigation, Y.-C.S., L.-H.S., P.-C.T., W.-Y.J., C.-P.C., T.-H.H.; Writing—original draft, Y.-C.S. and P.-C.T.; Writing—review and editing, P.-C.T., Y.-C.S., W.-Y.J., C.-P.C., L.-H.S. and T.-H.H. All authors have read and agreed to the published version of the manuscript.

Funding: This research was funded by the Ministry of Science and Technology under the Grant number MOST 108-2221-E-224-044, and also by the Ministry of Economic Affairs under the Grant number108-EC-17-A-05-S5-001.

Acknowledgments: For the technical suggestions and the experimental analyses, authors would like to be grateful to Prof. Yung-Cheng Wang and Mr. Teng-Chi Wu in National Yunlin University of Science and Technology. We acknowledge the financial support by the Ministry of Science and Technology under the Grant number MOST 108-2221-E-224-044, and also by the Ministry of Economic Affairs under the Grant number108-EC-17-A-05-S5-001.

Conflicts of Interest: The authors declare no conflict of interest.

References

1. John, L.; Ernest, K. Michelson interferometry with 10 pm accuracy. *Rev. Sci. Instrum.* **2000**, *71*, 2669.

2. Gerd, J. Three-Dimensional Nanopositioning, Nanomeasuring Machine with a Resolution of 0.1 nm. *Optoelectron. Instrum. Data Proc.* **2010**, *46*, 318–323.
3. Walter, D.P. Inference of cosmological parameters from gravitational waves: Applications to second generation interferometers. *Phys. Rev. D* **2011**, *86*, 3011.
4. Gargi, R. Atomic Force Microscopy as a Nanometrology Tool: Some Issues and Future Targets. *Mapan* **2013**, *28*, 311–319.
5. Pietraszewski, K.A.R.B. Recent Developments in Fabry-Pérot Interferometer. *ASP Conf. Ser.* **2000**, *195*, 591–596.
6. John, R.L. Fabry-Pérot metrology for displacements up to 50 mm. *J. Opt. Soc. Am. A Opt. Image. Sci. Vis.* **2005**, *22*, 2786–2798.
7. Wang, Y.C.; Shyu, L.H.; Chang, C.P. The Comparison of Environmental Effects on Michelson and Fabry-Perot Interferometers Utilized for the Displacement Measurement. *Sensors* **2010**, *10*, 2577–2586. [CrossRef] [PubMed]
8. Vaughan, J.M. *Fabry–Perot Interferometer History Theory Practice and Applications*; Adam Hilger: Bristol, UK; Boca Raton, FL, USA, 1989; pp. 1–43.
9. Murphy, K.A.; Gunther, M.F.; Vengsarkar, A.M.; Claus, R.O. Quadrature phase-shifted, extrinsic Fabry–Perot optical fiber sensors. *Opt. Lett.* **1991**, *16*, 273–275. [CrossRef] [PubMed]
10. Chang, C.P.; Tung, P.C.; Shyu, L.H.; Wang, Y.C.; Manske, E. Fabry–Perot displacement interferometer for the measuring range up to 100 mm. *Measurements* **2013**, *46*, 4094–4099. [CrossRef]
11. Chang, C.P.; Tung, P.C.; Shyu, L.H.; Wang, Y.C.; Manske, E. Modified Fabry-Perot interferometer for displacement measurement in ultra large measuring range. *Rev. Sci. Instr.* **2013**, *84*, 053105. [CrossRef] [PubMed]
12. Shyu, L.H.; Chang, C.P.; Wang, Y.C. Influence of intensity loss in the cavity of a folded Fabry-Perot interferometer on interferometric signals. *Rev. Sci. Instr.* **2011**, *82*, 063103. [CrossRef] [PubMed]
13. Cui, J.; He, Z.; Jiu, Y.; Tan, J.; Sun, T. Homodyne laser interferometer involving minimal quadrature phase error to obtain subnanometer nonlinearity. *Appl. Opt.* **2016**, *55*, 55. [CrossRef] [PubMed]
14. Fu, H.; Ji, R.; Hu, P.; Wang, Y.; Wu, G.; Tan, J. Measurement Method for Nonlinearity in Heterodyne Laser Interferometers Based on Double-Channel Quadrature Demodulation. *Sensors* **2018**, *18*, 2768. [CrossRef] [PubMed]
15. Hu, P.; Zhu, J.; Zhai, X.; Tan, J. DC-offset-free homodyne interferometer and its nonlinearity compensation. *Opt. Express* **2015**, *23*, 8399–8408. [CrossRef] [PubMed]
16. Heydemann, P.L.M. Determination and correction of quadrature fringe measurement errors in interferometers. *Appl. Opt.* **1981**, *20*, 3382–3384. [CrossRef] [PubMed]
17. Eom, T.; Kim, J.; Jeong, K. The dynamic compensation of nonlinearity in a homodyne laser interferometer. *Meas. Sci. Technol.* **2001**, *12*, 1734–1738. [CrossRef]
18. Shyu, L.H.; Wang, Y.C.; Chang, C.P.; Shih, Y.C.; Manske, E. A signal interpolation method for Fabry–Perot interferometer utilized in mechanical vibration measurement. *Measurements* **2016**, *92*, 83–88. [CrossRef]
19. Rabinowitz, P.; Jacobs, S.F.; Shultz, T.; Gould, G. Cube-Corner Fabry-Perot Interferometer. *J. Opt. Soc. Am.* **1962**, *52*, 452. [CrossRef]
20. Enderton, H.B. Vladimir A. Uspensky. Godel's incompleteness theorem. A reprint of LV 889 with minor corrections. Theoretical computer science, vol. 130 (1994), pp. 239–319. *J. Symb. Log.* **1995**, *60*, 1320. [CrossRef]

Article

An Off-Axis Differential Method for Improvement of a Femtosecond Laser Differential Chromatic Confocal Probe

Chong Chen, Yuki Shimizu *, Ryo Sato, Hiraku Matsukuma and Wei Gao

Precision Nanometrology Laboratory, Department of Finemechanics, Tohoku University, Sendai 980-8579, Japan; chongchen@nano.mech.tohoku.ac.jp (C.C.); satoryo@nano.mech.tohoku.ac.jp (R.S.); hiraku.matsukuma@nano.mech.tohoku.ac.jp (H.M.); gaowei@cc.mech.tohoku.ac.jp (W.G.)

* Correspondence: yuki.shimizu@nano.mech.tohoku.ac.jp; Tel.: +81-22-795-6950

Received: 8 September 2020; Accepted: 15 October 2020; Published: 16 October 2020

Abstract: This paper presents an off-axis differential method for the improvement of a femtosecond laser differential chromatic confocal probe having a dual-detector configuration. In the proposed off-axis differential method employing a pair of single-mode fiber detectors, a major modification is made to the conventional differential setup in such a way that the fiber detector in the reference detector is located at the focal plane of a collecting lens but with a certain amount of off-axis detector shift, while the fiber detector in the measurement detector is located on the rear focal plane without the off-axis detector shift; this setup is different from the conventional one where the difference between the two confocal detectors is provided by giving a defocus to one of the fiber detectors. The newly proposed off-axis differential method enables the differential chromatic confocal setup to obtain the normalized chromatic confocal output with a better signal-to-noise ratio and approaches a Z-directional measurement range of approximately 46 μm, as well as a measurement resolution of 20 nm, while simplifying the optical alignments in the differential chromatic confocal setup, as well as the signal processing through eliminating the complicated arithmetic operations in the determination of the peak wavelength. Numerical calculations based on a theoretical equation and experiments are carried out to verify the feasibility of the proposed off-axis differential method for the differential chromatic confocal probe with a mode-locked femtosecond laser source.

Keywords: chromatic confocal probe; femtosecond laser; off-axis differential method; tracking local minimum method

1. Introduction

A confocal probe is an optical displacement sensor often employed in microscopy for form measurement of three-dimensional microstructures for the quality control of ultra-precision machined surfaces [1–6]. Due to a high lateral resolution and a good depth discrimination performance by the Z-directional depth-sectioning effect, a confocal probe can also be employed in the surface profilometry of various materials and medical applications in the observation of living cells [7–10]. Chromatic confocal microscopy [11] has been proposed for the elimination of the axial-scanning of the conventional confocal microscopy. To obtain a better lateral resolution and an in-depth discrimination performance while eliminating the influence of nonuniform surface reflectivity of a target object in the three-dimensional surface topography measurement at nanometric or sub-nanometric level, differential methods and the corresponding signal processing algorithms for a confocal probe [12–26] have been proposed. Most of the differential methods for a confocal probe employ two confocal outputs captured by two detectors with different apertures or focusing conditions.

The differential methods can be categorized into two major types. The first type [25] employs two photodetectors having apertures with different sizes. Each of the apertures is placed at the focal plane of the corresponding collecting lens in the confocal setup, and the height information of an object surface under inspection can be decoded from the linear relationship between the axial displacement of the target surface and the normalized intensity ratio of the two confocal outputs. Some of the differential probes of this type can carry out measurement without the scanning along the optical axis. However, the in-depth measurement range is relatively small [26].

The second type employs two detectors having identical confocal apertures [17–24]. One confocal aperture is located in front of the focal plane of a corresponding collecting lens, while the other is placed behind the focal plane of another corresponding collecting lens to obtain two different confocal outputs. In the case of employing two slits as the apertures for the confocal setups, the in-depth resolution could be degraded [17]. However, this can be overcome by employing two identical pinholes as the apertures for the confocal setup [18–24]. For this type of differential method, two corresponding signal processing algorithms have been designed. The first signal processing algorithm [21–24] is designed in such a way that the normalized output is defined as the ratio of intensity subtraction of the two confocal outputs. Meanwhile, this algorithm cannot remove the influence of the nonuniform surface reflectivity of an object under inspection, as well as the influence of the light source. To address these issues, a second signal processing algorithm [18–20] has been designed. In the algorithm, the ratio of intensity difference and intensity sum (or maximum intensity) of the two confocal outputs are utilized as a confocal output. It should be noted that these two types can be combined with each other [26].

In the previous work by the authors of [27–31], the dual-detector confocal setup based on the second type, as well as the signal processing algorithm, has been modified for the femtosecond laser differential chromatic confocal probe, where a mode-locked femtosecond laser is employed as the laser source. In the modified differential chromatic confocal setup, the two identical fiber detectors are arranged in such a way that the fiber detector in the measurement detector unit is placed at the focal plane of the collecting lens of the detector unit, as shown in Figure 1a, while the fiber detector in the reference detector unit is placed at a position with a displacement (defocus) d along with the axial position, as shown in Figure 1b. For acquiring the in-depth information, two confocal outputs with these detectors in different confocal setups are utilized. For the signal processing, the algorithm referred to as the tracking local minimum (TLM) method has been proposed [27]. In the TLM method, the normalized confocal output is defined as the ratio of the measurement detector output to the reference detector output. This algorithm can theoretically eliminate the influences of the nonuniform spectrum of a mode-locked femtosecond laser source. Meanwhile, the intensity around the first local minimum of the reference detector output spectrum is relatively small, and hence the normalized output can easily be influenced by out-of-focus noise [25,32–34].

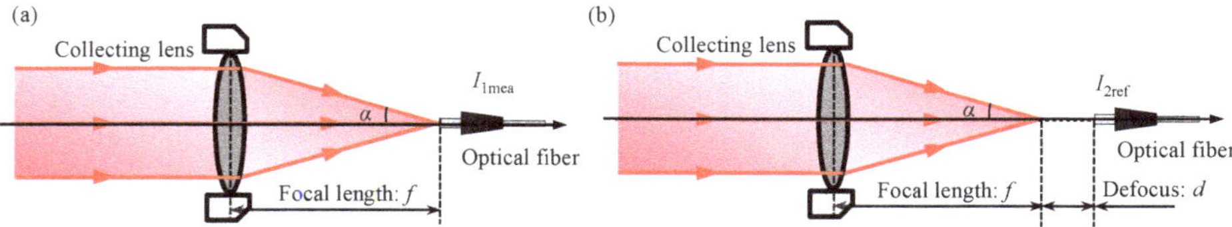

Figure 1. Schematics of the relative position between the fiber core and the collecting lens in measurement and reference detectors in the femtosecond confocal probe. (**a**) A setup for the measurement detector. (**b**) A setup for the reference detector with a defocus d.

To address the issue, in this paper, a major modification is made to the conventional differential chromatic confocal setup in the mode-locked femtosecond laser differential chromatic confocal probe. Figure 2 shows a schematic of the confocal setup of the reference detector unit employed in the newly proposed method. As can be seen in the figure, the fiber detector is placed at the focal plane of the collecting lens while an off-axis detector shift d_m is given to the fiber detector, instead of giving a defocus d; this modification improves the signal-to-noise ratio of the normalized chromatic confocal output to be obtained in the differential confocal setup. This optical configuration is also expected to make the optical alignment of the differential confocal setup easier than the conventional setup shown in Figure 1. Theoretical analysis and experimental examination are implemented to verify the feasibility of the proposed off-axis differential method for the differential chromatic confocal probe with a mode-locked femtosecond laser source. It should be noted that the work described in this paper is mainly intended to focus on the proposal of the preliminary idea of an off-axis differential method for the improvement of the signal-to-noise ratio (SNR), the measurement range and the measurement resolution of the femtosecond laser chromatic confocal probe. The subsequent application of this new method is needed in the imaging of three-dimensional (3D) microstructures in our future work. Regarding the spectral range (from 1480 to 1640 nm) and the chromatic objective lens employed in the experiments, a measurement range of 46 μm is expected to be achieved through the above improvements by the proposed off-axis differential method.

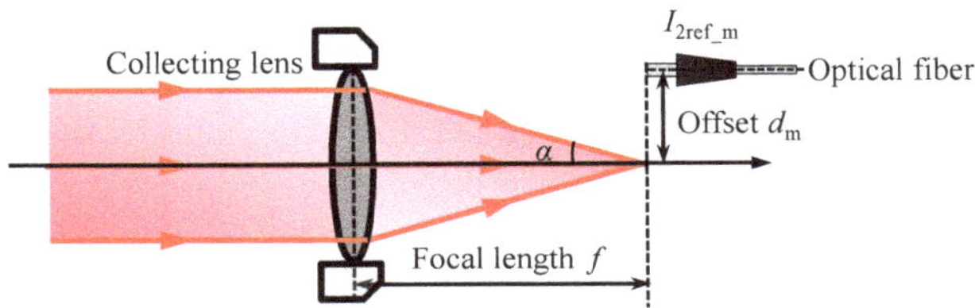

Figure 2. Reference detector unit with an off-axis detector shift d_m in the newly proposed method.

2. Measurement Principle

2.1. Image Formation Principle

Figure 3 shows a schematic diagram of the differential chromatic confocal configuration [27–31]. A mode-locked femtosecond laser beam from the laser source is made to focus on a target surface by a chromatic objective lens. The reflected laser beam from the target surface is then divided into two sub-beams by a beam splitter, and the sub-beams are then captured by two fiber detector units composed of two identical fibers and two identical collecting lenses, L_1 and L_2. One of the sub-beams is captured by one fiber detector unit (referred to as the measurement detector unit) set at the focal plane of L_1, while the other is captured by another fiber detector unit (referred to as the reference detector unit) with the collecting lens L_2.

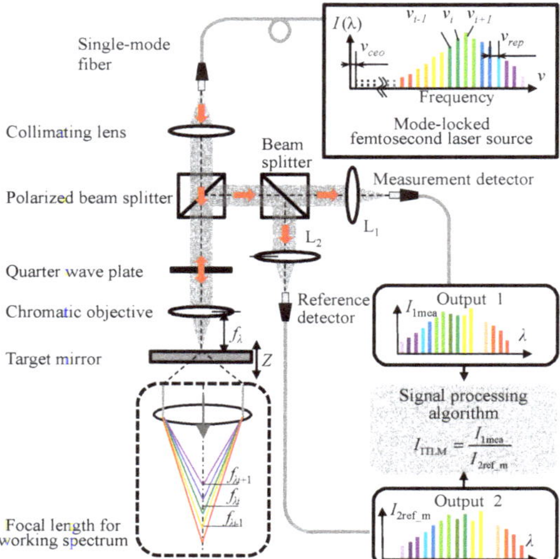

Figure 3. A schematic diagram of the differential chromatic confocal configuration with a mode-locked femtosecond laser source.

The confocal output $I_{1mea}(u)$ to be captured by the measurement detector unit, whose optical configuration is shown in Figure 1a, can be expressed by the following equation [27,35–37].

$$I_{1mea}(u) = \left| \int_0^1 \exp(ju\rho^2) P_{1eff}(\rho) P_{2eff}(\rho) \rho d\rho \right|^2 \quad (1)$$

where ρ is the normalized radius of the imaging lens, j is the imaginary unit and P_{1eff} and P_{2eff} stand for the effective pupil functions of the imaging lens (L1 or L2) and illuminating lens (collimating lens), respectively. The parameters v and u represent the optical coordinates related to the actual radial and axial coordinates r and z, respectively, that can be given as follows [27,35–37]. It should be noted that only the axial (Z-) response is considered and measured in the experiments. Therefore, the variable v in Equation (1) is treated to be zero [27] and is omitted in the equation.

$$v = \frac{2\pi}{\lambda} r \sin\beta \quad (2)$$

$$u = \frac{2\pi}{\lambda} z \sin^2\beta \quad (3)$$

where sinβ and λ are the numerical aperture of the collecting lens and the light wavelength, respectively.

Figure 4 shows schematic diagrams of the reference fiber detector unit by the proposed off-axis differential method and the conventional differential method for the chromatic confocal probe system. As can be seen in Figure 4a, in the reference fiber detector unit based on the conventional differential method [27], the single-mode fiber detector is placed with a defocus d with respect to the back focal plane of the collecting lens L_2. In this case, the confocal output $I_{2ref}(u;u_d)$ can be represented as follows:

$$I_{2ref}(u; u_d) = \left| \int_0^1 \exp[j(u + u_d/2)\rho^2] P_{1eff}(\rho) P_{2eff}(\rho) \rho d\rho \right|^2 \quad (4)$$

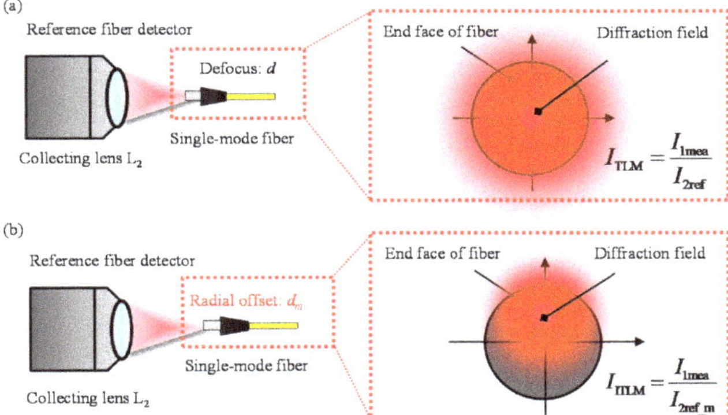

Figure 4. Schematic of the reference fiber detector settings by the proposed off-axis differential method and conventional differential method. (**a**) Schematic diagram of the reference fiber detector setting by the conventional differential method; (**b**) schematic diagram of the reference fiber detector setting by the proposed off-axis differential method.

By using these two confocal outputs, the height of a surface under measurement can be detected through the arithmetic operation based on the algorithm referred to as the tracking local minimum (TLM) method [30]. Meanwhile, in the newly proposed off-axis differential method, a major modification is made to the reference detector unit: the single-mode fiber detector is located on the rear focal plane of the collecting lens L_2 by the optical axis but with an off-axis detector shift d_m, as shown in Figure 4b. In this case, the confocal output $I_{2ref\_m}(u;d_m)$ can be represented as follows [38]:

$$I_{2ref\_m}(u, d_m) = \left| \int_0^1 J_0\left(\frac{d_m}{r_f}\rho v_f\right) \exp\left(ju\rho^2\right) P_{1eff} P_{2eff} \rho d\rho \right|^2 \tag{5}$$

where r_f and v_f are the radius and the normalized radius of the single-mode fiber, respectively. As can be seen in Equations (4) and (5), the confocal output becomes different in the modified setup. According to Equation (5), the effect of employing the proposed off-axis differential method is similar to reducing the core diameter of a fiber detector acting as a pinhole aperture in the confocal setup; namely, the rejection of more scattered light and out-of-focus noise in the reference fiber detector can be realized. This effect is expected to improve the signal-to-noise (S/N) ratio of the confocal output, as well as the resolution, in the differential chromatic confocal probe.

2.2. Signal Processing Algorithm

In the tracking local minimum (TLM) algorithm [30], the normalized chromatic output I_{TLM} is defined as the intensity ratio of the two fiber detector outputs [27–30]. Denoting the chromatic output to be captured by the off-axis reference detector unit as I_{2ref_m}, the normalized chromatic output in the newly proposed off-axis differential method can be expressed as follows:

$$I_{ITLM}(u, d_m) = \frac{I_{1mea}(u)}{I_{2ref\_m}(u, d_m)} \tag{6}$$

Figure 5a shows a graphical diagram of the TLM algorithm [30]. The wavelength at a specific peak position of the normalized chromatic output I_{ITLM}, which corresponds to the wavelength at the first local minimum of the confocal output I_{2ref_m}, changes linearly with respect to the Z-directional displacement of the target surface. A displacement of the target mirror can thus be acquired by tracking the wavelength at the first local minimum of the confocal output through detecting the peak wavelength in the normalized chromatic output I_{ITLM}. Figure 5b shows a schematic of the wavelength-to-displacement encoding by the signal processing algorithm. In the tracking local minimum algorithm [27], the asymmetric peak in the normalized chromatic output I_{ITLM} could make it difficult under practical conditions to precisely determine the peak wavelength due to the out-of-focus noise [30]. To address the issue, a centroid wavelength of the peak is employed as the peak wavelength.

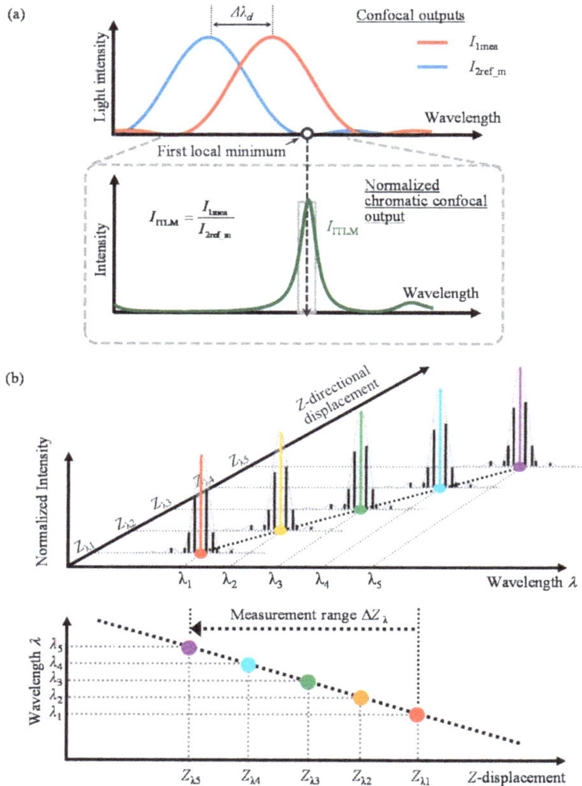

Figure 5. A schematic of the tracking local minimum (TLM) algorithm. (**a**) The normalized output obtained from the reference output and the measurement output [30]; (**b**) a schematic of the wavelength-to-displacement encoding in the TLM method.

3. Experiments and Discussions

3.1. Experimental Confocal Configuration

To verify the feasibility of the proposed off-axis differential method, experiments were carried out by using the developed differential chromatic confocal system, a schematic diagram of which is shown in Figure 6. For a fair comparison, the optical components in the configuration are the same as those adopted in the previous work by the authors of [30] based on the defocus setup, except a manual stage for the adjustment of the off-axis detector shift d_{m} of the fiber detector in the reference fiber detector unit. In this paper, d_m in the positive Y-direction was treated to be positive.

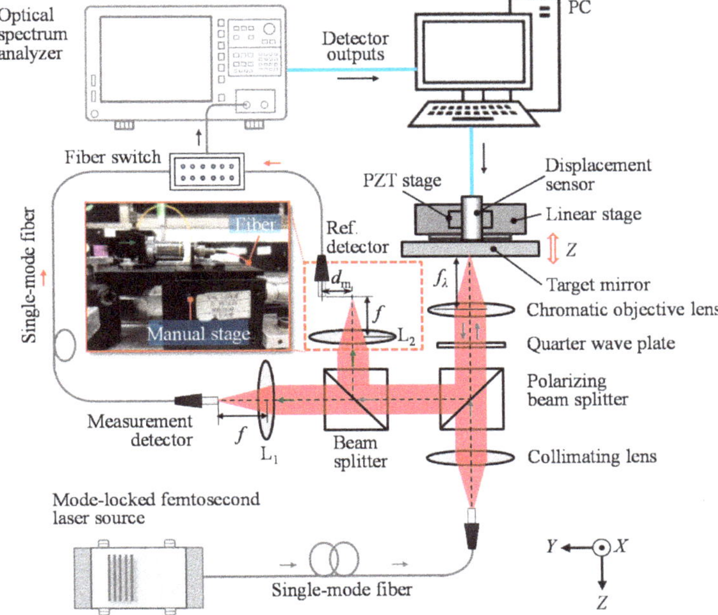

Figure 6. A schematic diagram of the proposed off-axis differential chromatic confocal probe configuration with a mode-locked femtosecond laser source [30].

At first, the coarse adjustment of the off-axis fiber in the reference detector was carried out, since too much offset d_m could result in the local minimum in the reference detector output out of the wavelength range of the optical spectrum analyzer employed in the setup. The amount of d_m given to the fiber detector in the reference detector unit was adjusted so that the first local minimum wavelength of the confocal output I_{2ref_m} could be captured at each Z-directional position of a target surface. It should be noted that the single-mode fiber (SMF-28e, Corning Inc.) had a core diameter of 8.2 μm and an NA of 0.14. The detector was composed of a single-mode fiber and an optical spectrum analyzer. At each Z-position, it took about 8 s to capture the spectrum by the measurement and reference detectors. A bandwidth of the optical spectrum analyzer was set to be 0.5 nm with a sampling interval of 0.1 nm. Figure 7 shows the noise reduction performances in the comparison between the conventional defocus differential method and the proposed off-axis differential method. As can be seen in the figures, the out-of-focus noises at the ends of the measurement range (positions A1 and B1) could be found in the conventional defocus differential method. Meanwhile, the out-of-focus noises at the ends of the measurement range (positions A2 and B2) were found to be reduced in the newly proposed off-axis differential method; this contributes to achieving a wider measurement range. These results mean that a better signal-to-noise ratio (SNR) of the normalized chromatic output I_{ITLM} can be expected by the newly proposed off-axis differential method. It should be noted that this can also contribute to carrying out stable detection of the peak wavelength to be employed for the detection of the displacement of a surface under inspection through the wavelength-to-displacement encoding.

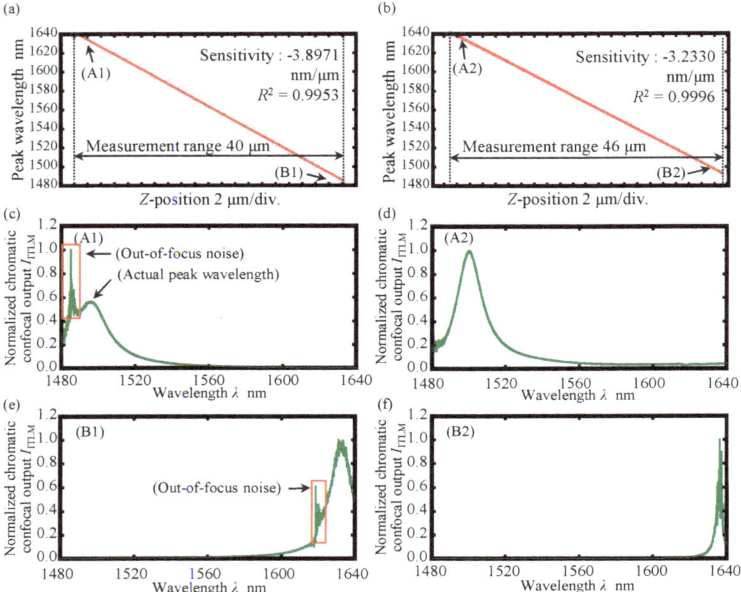

Figure 7. Noise reduction performances in the two methods. (**a**) Measurement range of the conventional defocus differential method; (**b**) measurement range of the proposed off-axis differential method; (**c**) normalized chromatic confocal output I_{ITLM} in (A1) position; (**d**) normalized chromatic confocal output I_{ITLM} in (A2) position; (**e**) normalized chromatic confocal output I_{ITLM} in (B1) position; (**f**) normalized chromatic confocal output I_{ITLM} in (B2) position.

Figure 8a shows the spectrum of the mode-locked femtosecond laser employed in the experiments, and Figure 8b shows the typical spectra of the two confocal outputs, I_{1mea} and I_{2ref_m}, captured by the dual-fiber detector units. As can be seen in Figure 8b, due to the nonuniform spectrum of the utilized mode-locked femtosecond laser source shown in Figure 8a, it is not so easy to figure out peak wavelengths in the obtained spectra. Figure 8c shows the normalized chromatic confocal output I_{ITLM} obtained from the spectra shown in Figure 8a,b. As can be seen in Figure 8c, it is much easier to figure out the peak in the spectrum with a much smaller full width at half maximum (FWHM) value.

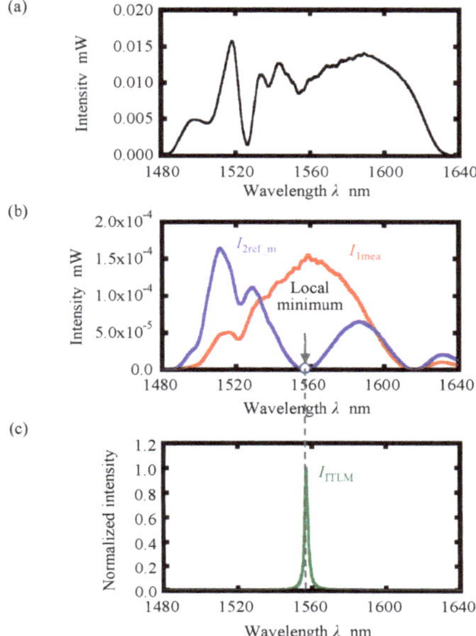

Figure 8. Schematic of the wavelength-to-displacement encoding in the proposed off-axis differential chromatic confocal probe system. (**a**) A spectrum of the mode-locked femtosecond laser employed in this paper; (**b**) chromatic confocal outputs obtained by the detector units; (**c**) the normalized chromatic confocal output obtained through the arithmetic operation with I_{1mea} and I_{2ref_m}.

3.2. Optimum of the Misalignment Value

The measurement resolution of the femtosecond laser differential chromatic confocal probe based on the TLM algorithm could be affected by the sharpness of the peak in the normalized chromatic confocal output [30]. In this paper, the appropriate off-axis detector shift d_m was investigated through experiments by using the full width at half maximum (FWHM) of the peak in the normalized chromatic confocal output I_{ITLM}. At first, numerical calculations [28] were carried out based on Equation (5) to estimate the variation of the FWHM due to the change in d_m. Figure 9 shows the result. As can be seen in the figure, the FWHM was estimated to reduce as there was an increase in the off-axis displacement d_m. It should be noted that the simulation based on Equation (5) is valid under the ideal condition of a simplified optical configuration without any aberrations.

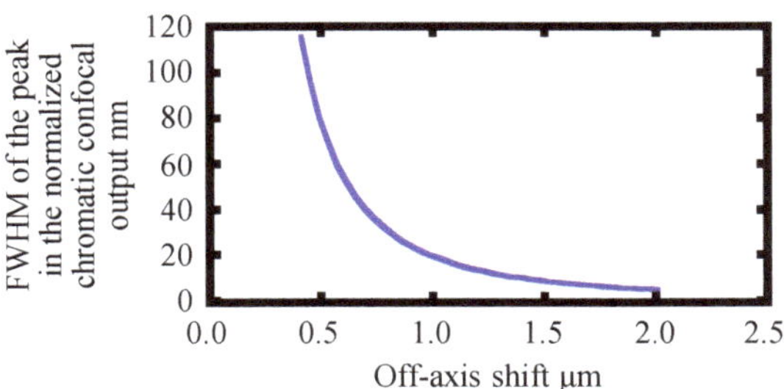

Figure 9. Variation of the full width at half maximum (FWHM) of the peak in the normalized chromatic confocal output estimated by the numerical calculations.

Following the numerical calculations, experiments were carried out. In the experiments, the central wavelength λ_0 was set to be approximately 1560 nm. The off-axis displacement d_m of the fiber detector in the reference detector was changed in a step of 1 µm, while obtaining the normalized chromatic confocal output at each d_m. Figure 10 shows the results. As can be seen in the figure, the FHWM of the peak in I_{ITLM} was found to deviate with the change in d_m. Figure 11 summarizes the variation of the FWHM observed in Figure 10. As can be seen in the figures, FWHM values can clearly be figured out when the mismatch displacement (off-axis shift) is smaller than 3 µm. Due to the effects of the numerical aperture (mismatch) on the normalized chromatic confocal output I_{ITLM}, it becomes difficult to find out the FWHM value for the case having large d_m. The FWHM value was found to be minimized around d_m =1 µm. The root cause of the difference observed between the results of numerical calculations and the experiments is not clear, and further investigation will be carried out in future work. Regarding the experimental results, d_m was set to be 1 µm in the following experiments. It should be pointed out that the determination of the appropriate off-axis detector shift d_m is a less time-consuming task compared with the determination of the appropriate defocus in the conventional setup, where the defocus d needs to be changed in a wide range up to hundreds of micrometers depending on the confocal setup; this is another advantage of the newly proposed off-axis differential method compared with the conventional defocus method.

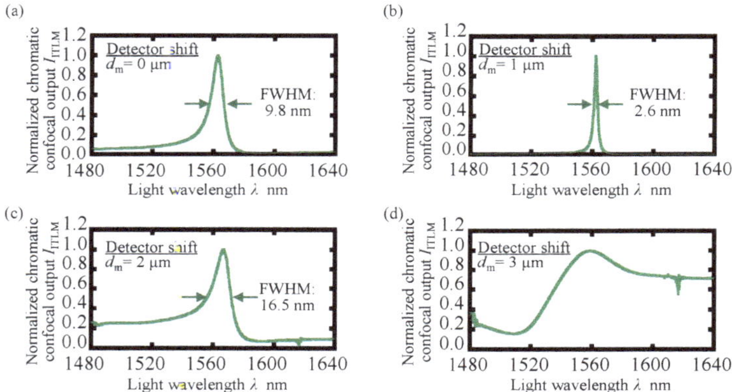

Figure 10. Normalized chromatic confocal output at each off-axis displacement d_m of the fiber detector in the reference detector. (**a**) d_m = 0 µm; (**b**) d_m = 1.0 µm; (**c**) d_m = 2.0 µm; (**d**) d_m = 3.0 µm.

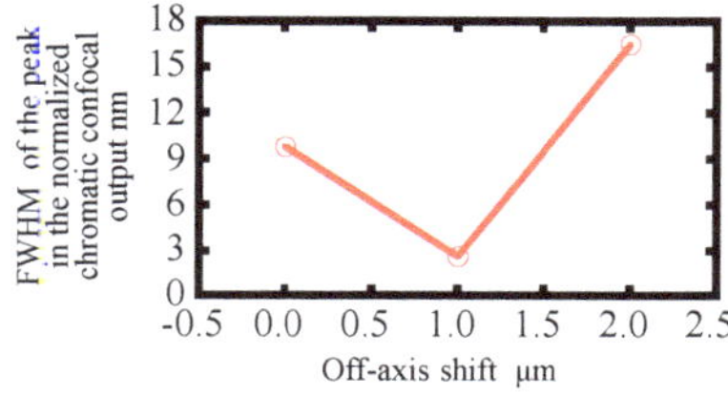

Figure 11. Variation of the FWHM of the peak in the normalized chromatic confocal output due to the change in the off-axis displacement d_m of the fiber detector in the reference detector.

3.3. Evaluation of the Measurement Range and Resolution to Be Achieved by the Developed Setup

Experiments were extended to evaluate the measurement range of the femtosecond laser differential chromatic confocal probe based on the off-axis differential method. In experiments, a target mirror was moved along the Z-direction in a step of 2 µm by a piezoelectric (PZT) stage equipped with a capacitance displacement sensor (P-621.2CL, Physik Instrumente, Karlsruhe, Germany). Two confocal outputs were obtained at each Z-position of the target mirror. Figure 12a shows the normalized chromatic confocal output I_{ITLM} obtained through arithmetic operations based on Equation (6) in a wavelength range of 1550–1610 nm. Figure 12b shows the variation of the peak wavelength detected from the obtained I_{ITLM} at each Z-position of the target mirror. The measurement sensitivity, defined as the ratio of the change in the peak wavelength to the Z-displacement of the target mirror, was evaluated to be −3.233 nm/μm with good linearity of R^2 = 0.9996 over a Z-directional range of 46 μm. It should be noted that the above results contained the uncertainty that came from the PZT stage employed in the experiments. Meanwhile, the PZT stage had a positioning resolution of 0.2 nm, which was much smaller than that obtained by the femtosecond laser chromatic confocal probe and was enough for the experiments. The verification experiments in this manuscript have been done as the first step of the investigation of the proposed off-axis differential method. Analyses on the measurement uncertainty of the femtosecond laser chromatic confocal probe with the proposed off-axis method will be carried out in future work while considering the positioning uncertainty of the PZT stage.

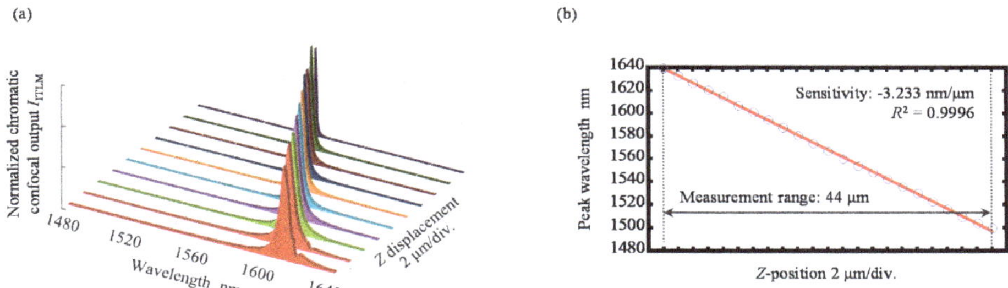

Figure 12. Experimental results of the wavelength-to-displacement encoding in the proposed off-axis differential chromatic confocal probe system. (**a**) Normalized chromatic confocal outputs in the range of 1550–1610 nm obtained at each Z-position; (**b**) variation of the peak wavelength over a range of 46 μm.

Figure 13 shows the centroid wavelength $\lambda_{\text{ITLM\_C}}$ [30] and peak wavelength λ_{ITLM} of I_{ITLM} encoded from the normalized chromatic confocal output obtained at each Z-position of the target mirror. As can be seen in the figure, the deviations between the peak wavelengths λ_{ITLM} and centroid wavelengths $\lambda_{\text{ITLM\_C}}$ of I_{ITLM} were found to be within ± 1.0 nm, which is negligibly small compared with the actual measurement resolution, and is much smaller than that observed in the conventional defocus method (maximum > 40 nm) [30]. These results mean that the peak wavelength λ_{ITLM} detected in the normalized chromatic confocal output in the newly proposed off-axis differential method can directly be adopted in the wavelength-to-displacement encoding, without the complex and time-consuming arithmetic operations for obtaining the centroid wavelength $\lambda_{\text{ITLM\_C}}$.

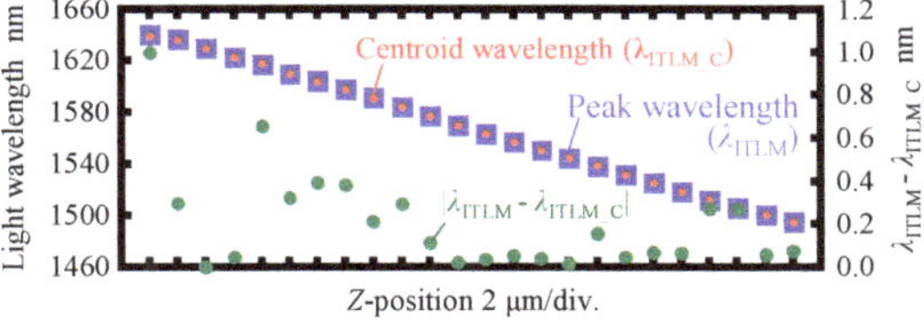

Figure 13. Peak wavelength λ_{ITLM} and centroid wavelength $\lambda_{\text{ITLM\_C}}$ of I_{ITLM} encoded from the normalized chromatic confocal output obtained at each Z-position of the target mirror by the chromatic confocal probe with the proposed off-axis differential method.

Experiments were then carried out to investigate the measurement resolution. In the following experiments, the target mirror was made to move along the Z-direction in a small step by the PZT stage. Figure 14a–c show the variation of the detected peak wavelength in the normalized chromatic output with a Z-directional step of 40, 30 and 20 nm, respectively. For each moving interval, five repetitive measurements were made, and the mean value from the repeated measurements at each period is plotted in Figure 14. As can be seen in the figures, the change in the peak wavelength with respect to the step Z-displacement given to the target mirror was clearly distinguished in each of the cases. Figure 14d shows a standard deviation of the five repetitive measurements at each step observed in the case of Figure 14c. The standard deviation was evaluated to be smaller than 0.14 nm. It should be noted that the above results contained the uncertainty that came from the PZT stage employed in the experiments. Meanwhile, the PZT stage had a positioning resolution of 0.4 nm and a linearity error of 0.02% (corresponding to 1.2 nm regarding a travel range of 6 μm in Figure 14c), which were much smaller than those obtained by the femtosecond laser chromatic confocal probe and were enough for the experiments. The verification experiments in this manuscript have been done as the first step of the investigation of the proposed off-axis differential method. Analyses on the measurement uncertainty of the femtosecond laser chromatic confocal probe with the proposed off-axis method will be carried out in future work while considering the positioning uncertainty of the PZT stage.

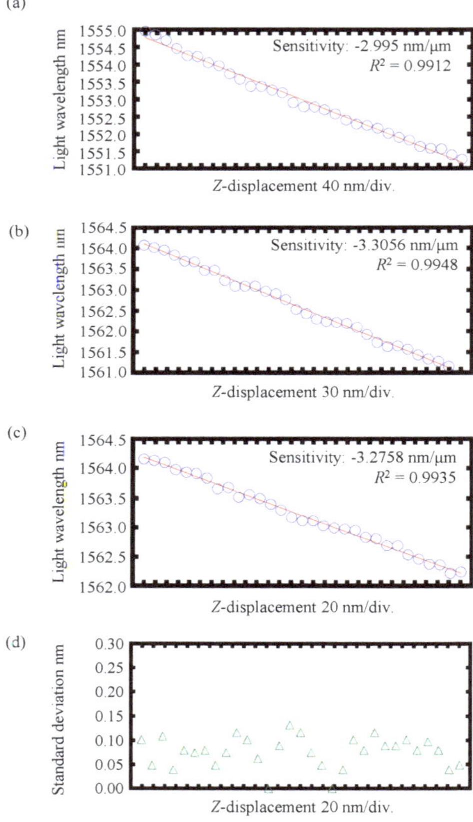

Figure 14. Evaluation of the measurement resolution of the mode-locked femtosecond laser differential chromatic confocal probe with the off-axis differential method. (**a**) Measured variation of the stage position moved in a step of 40 nm; (**b**) measured variation of the stage position moved in a step of 30 nm; (**c**) measured variation of the stage position moved in a step of 20 nm; (**d**) standard deviation of the repeated measurements in (**c**).

4. Conclusions

In this paper, an off-axis differential method for a femtosecond laser differential chromatic confocal probe has been proposed to improve the measurement performance, as well as to simplify the optical alignments and the data processing in the peak wavelength-Z position encoding. A major modification has been made to the conventional differential chromatic confocal setup in such a way that the fiber detector in the reference detector unit has been placed to have an offset with respect to the optical axis of the collecting lens in the detector unit, rather than giving a defocus to the fiber detector. This modification enables the newly proposed off-axis setup to obtain a narrower peak having a profile in the normalized chromatic confocal output, as well as to simplify the optical alignment of the differential confocal setup. With the enhancement of the tracking local minimum method, an improved Z-directional measurement range of 46 μm has been achieved by the newly proposed off-axis differential method. The possibility of achieving a resolution of 20 nm has also been verified.

In this paper, attention has been paid to the verification of the feasibility of the newly proposed off-axis method in the femtosecond chromatic confocal probe compared with the conventional setup with the defocus method [27–30]. Further work contains the investigation on the Z-directional measurement resolution with another signal processing algorithm such as the tracking intersection method [30]. It should also be noted that this paper has focused on the proposal of a preliminary idea of an off-axis differential method for the improvement of the signal-to-noise ratio (SNR), the measurement range and measurement resolution of the chromatic confocal probe. A more detailed investigation of the novelty of the displaced probe, as well as the corresponding application in three-dimensional microstructure imaging, will be carried out in future work. Further optimization and simplification of the established differential chromatic confocal probe system for a higher measurement resolution and investigation on the difference between the results of numerical calculations and experimental results described in this paper, as well as the extension of the measurement range by the optimization of the chromatic objective lens, will also be carried out. The application of the developed mode-locked femtosecond chromatic differential confocal probe to the three-dimensional surface profile measurement also remains as a challenge to be addressed and will be carried out in future work.

Author Contributions: Conceptualization, W.G.; methodology, C.C. and R.S.; software, C.C. and R.S.; validation, C.C., H.M. and R.S.; formal analysis, C.C., R.S. and Y.S.; investigation, C.C. and Y.S.; resources, W.G. and Y.S.; Data curation, C.C.; writing—original draft preparation, C.C.; writing—review and editing, Y.S., C.C.; visualization, W.G. and Y.S.; supervision, W.G. and Y.S.; administration, W.G.; funding acquisition, W.G., Y.S. and H.M. All authors have read and agreed to the published version of the manuscript.

Funding: This research was funded by Japan Society for the Promotion of Science (JSPS) 18H01345.

Conflicts of Interest: The authors declare no conflict of interest. The funders had no role in the design of the study; in the collection, analyses, or interpretation of data; in the writing of the manuscript, and in the decision to publish the results.

References

1. Gao, W.; Kim, S.W.; Bosse, H.; Haitjema, H.; Chen, Y.L.; Lu, X.D.; Knapp, W.; Weckenmann, A.; Estler, W.T.; Kunzmann, H. Measurement technologies for precision positioning. *CIRP Ann.-Manuf. Technol.* **2015**, *64*, 773–796. [CrossRef]
2. Gao, W.; Haitjema, H.; Fang, F.Z.; Leach, R.K.; Cheung, C.F.; Savio, E.; Linares, J.M. On-machine and in-process surface metrology for precision manufacturing. *CIRP Ann.* **2019**, *68*, 843–866. [CrossRef]
3. Fan, K.C.; Chu, C.L.; Mou, J.I. Development of a low-cost autofocusing probe for profile measurement. *Meas. Sci. Technol.* **2001**, *12*, 2137. [CrossRef]
4. Chen, L.C.; Chang, Y.W. Innovative simultaneous confocal full-field 3D surface profilometry for in situ automatic optical inspection (AOI). *Meas. Sci. Technol.* **2013**, *21*, 1–12.
5. Chen, L.C.; Chang, Y.W. High accuracy confocal full-field 3-D surface profilometry for micro lenses using a digital fringe projection strategy. *Key Eng. Mater.* **2008**, *364*, 113–116. [CrossRef]

6. Cai, Y.; Yang, B.; Fan, K.C. Robust roll angular error measurement system for precision machines. *Opt. Express* **2019**, *27*, 8027–8036. [CrossRef]
7. Stephens, D.J.; Allan, V.J. Light Microscopy Techniques for Live Cell Imaging. *Science* **2003**, *300*, 82–86. [CrossRef] [PubMed]
8. Yoneyama, T.; Watanabe, T.; Tamai, S.; Miyashita, K.; Nakada, M. Bright spot analysis for photodynamic diagnosis of brain tumors using confocal microscopy. *Photodiagnosis Photodyn. Ther.* **2019**, *25*, 463–471. [CrossRef]
9. Bohn, S.; Sperlich, K.; Allgeier, S.; Bartschat, A.; Prakasam, R.; Reichert, K.M.; Stolz, H.; Guthoff, R.; Mikut, R.; Kohler, B.; et al. Cellular in vivo 3D imaging of the cornea by confocal laser scanning microscopy. *Biomed. Opt. Express* **2018**, *9*, 2511–2525. [CrossRef] [PubMed]
10. Hickey, P.C.; Swift, S.R.; Roca, M.G.; Read, N.D. Live-cell imaging of filamentous fungi using vital fluorescent dyes and confocal microscopy. *Methods Microbiol.* **2004**, *34*, 63–87.
11. Blateyron, F. Chromatic Confocal Microscopy. In *Optical Measurement of Surface Topography*; Springer: Berlin/Heidelberg, Germany, 2012; pp. 71–106.
12. Zhao, W.; Jiang, Q.; Qiu, L.; Liu, D. Dual-axes differential confocal microscopy with high axial resolution and long working distance. *Opt. Commun.* **2011**, *284*, 15–19. [CrossRef]
13. Lee, C.H.; Wang, J. Noninterferometric differential confocal microscopy with 2-nm depth resolution. *Opt. Commun.* **1997**, *135*, 233–237. [CrossRef]
14. Tan, J.; Liu, J.; Wang, Y. Differential confocal microscopy with a wide measuring range based on polychromatic illumination. *Meas. Sci. Technol.* **2010**, *21*, 054013. [CrossRef]
15. Sánchez-Ortiga, E.; Sheppard, C.J.; Saavedra, G.; Martínez-Corral, M.; Doblas, A.; Calatayud, A. Subtractive imaging in confocal scanning microscopy using a CCD camera as a detector. *Opt. Lett.* **2012**, *37*, 1280–1282. [CrossRef]
16. Qiu, L.; Zhao, W.; Feng, Z.; Ding, X. A lateral super-resolution differential confocal technology with phase-only pupil filter. *Optik* **2007**, *118*, 67–73. [CrossRef]
17. Kobayashi, K.; Akiyama, K.; Suzuki, T.; Yoshizawa, I.; Asakura, T. Laser-scanning imaging system for real-time measurements of 3-D object profiles. *Opt. Commun.* **1989**, *74*, 165–170. [CrossRef]
18. Butler, D.J.; Horsfall, A.; Hrynevych, M.; Kearney, P.D.; Nugent, K.A. Confocal profilometer with nanometric vertical resolution. *Opt. Commun.* **1993**, *100*, 87–92. [CrossRef]
19. Liu, J.; Tan, J.; Bin, H.; Wang, Y. Improved differential confocal microscopy with ultrahigh signal-to-noise ratio and reflectance disturbance resistibility. *Appl. Opt.* **2009**, *48*, 6195–6201. [CrossRef]
20. Qiu, L.; Liu, D.; Zhao, W.; Cui, H.; Sheng, Z. Real-time laser differential confocal microscopy without sample reflectivity effects. *Opt. Express* **2014**, *22*, 21626–21640. [CrossRef]
21. Zhao, W.; Tan, J.; Qiu, L. Bipolar absolute differential confocal approach to higher spatial resolution. *Opt. Express* **2004**, *12*, 5013–5021. [CrossRef]
22. Tan, J.; Wang, F. Theoretical analysis and property study of optical focus detection based on differential confocal microscopy. *Meas. Sci. Technol.* **2002**, *13*, 1289. [CrossRef]
23. Zhao, W.; Tan, J.; Qiu, L.; Zou, L. A new laser heterodyne confocal probe for ultraprecision measurement of discontinuous contours. *Meas. Sci. Technol.* **2005**, *16*, 497. [CrossRef]
24. Zhao, W.; Tan, J.; Qiu, L. Tri-heterodyne confocal microscope with axial superresolution and higher SNR. *Opt. Express* **2004**, *12*, 5191–5197. [CrossRef]
25. Lee, D.R.; Kim, Y.D.; Gweon, D.G.; Yoo, H. Dual-detection confocal fluorescence microscopy: Fluorescence axial imaging without axial scanning. *Opt. Express* **2013**, *21*, 17839–17848. [CrossRef] [PubMed]
26. Chen, L.C.; Nguyen, D.T.; Chang, Y.W. Precise optical surface profilometry using innovative chromatic differential confocal microscopy. *Opt. Lett.* **2016**, *41*, 5660–5663. [CrossRef] [PubMed]
27. Chen, X.G.; Nakamura, T.; Shimizu, Y.; Chen, C.; Chen, Y.L.; Matsukuma, H.; Gao, W. A chromatic confocal probe with a mode-locked femtosecond laser source. *Opt. Laser Tech.* **2018**, *103*, 359–366. [CrossRef]
28. Chen, C.; Sato, R.; Shimizu, Y.; Nakamura, T.; Matsukuma, H.; Gao, W. A Method for Expansion of Z-Directional Measurement Range in a Mode-Locked Femtosecond Laser Chromatic Confocal Probe. *Appl. Sci.* **2019**, *9*, 454. [CrossRef]
29. Sato, R.; Shimizu, Y.; Chen, C.; Matsukuma, H.; Gao, W. Investigation and Improvement of Thermal Stability of a Chromatic Confocal Probe with a Mode-Locked Femtosecond Laser Source. *Appl. Sci.* **2019**, *9*, 4084. [CrossRef]

30. Sato, R.; Chen, C.; Matsukuma, H.; Shimizu, Y.; Gao, W. A new signal processing method for a differential chromatic confocal probe with a mode-locked femtosecond laser. *Meas. Sci. Technol.* **2020**, *31*, 094004. [CrossRef]
31. Chen, C.; Matsukuma, H.; Sato, R.; Chen, X.; Shimizu, Y.; Gao, W. Theoretical investigation on measurement range of a femtosecond laser chromatic confocal probe by utilizing side-lobe of axial response. In Proceedings of the 2018 IEEE International Conference on Advanced Manufacturing (ICAM), Yunlin, Taiwan, 16–18 November 2018; pp. 362–364.
32. Gu, M. *Principles of Three Dimensional Imaging in Confocal Microscopes*; World Scientific: Singapore, 1996; pp. 117–220.
33. Cox, I.J.; Sheppard, C.J.; Wilson, T. Improvement in resolution by nearly confocal microscopy. *Appl. Opt.* **1982**, *21*, 778–781. [CrossRef]
34. Pawley, J. (Ed.) *Handbook of Biological Confocal Microscopy*; Springer Science & Business Media: Berlin, Germany, 2006; Volume 236, pp. 17–20.
35. Wilson, T.; Sheppard, C. *Theory and Practice of Scanning Optical Microscopy*; Academic Press: London, UK, 1984; pp. 17–20.
36. Wilson, T.; Carlini, A.R. Size of the detector in confocal imaging systems. *Opt. Lett.* **1987**, *12*, 227–229. [CrossRef] [PubMed]
37. Kimura, S.; Wilson, T. Confocal scanning optical microscope using single-mode fiber for signal detection. *Appl. Opt.* **1991**, *30*, 2143–2149. [CrossRef] [PubMed]
38. Barrell, K.F.; Pask, C. Optical fibre excitation by lenses. *Opt. Acta Int. J. Opt.* **1979**, *26*, 91–108. [CrossRef]

Article

A Fast Laser Adjustment-Based Laser Triangulation Displacement Sensor for Dynamic Measurement of a Dispensing Robot

Zhuojiang Nan, Wei Tao *, Hui Zhao and Na Lv

Department of Instrument Science and Engineering, Shanghai Jiao Tong University, Shanghai 200240, China; nanzhuojiang@sjtu.edu.cn (Z.N.); zhaohui@sjtu.edu.cn (H.Z.); nana414526@sjtu.edu.cn (N.L.)
* Correspondence: taowei@sjtu.edu.cn

Received: 23 September 2020; Accepted: 20 October 2020; Published: 22 October 2020

Abstract: Height measurement and location by a laser sensor is a key technology to ensure accurate and stable operation of a dispensing robot. In addition, alternation of dynamic and static working modes of a robot, as well as variation of surface and height of a workpiece put forward strict requirements for both repeatability and respond speed of the location system. On the basis of the principle of laser triangulation, a displacement sensor applied to a dispensing robot was developed, and a fast laser adjustment algorithm was proposed according to the characteristics of static and dynamic actual laser imaging waveforms on different objects. First, the relationship between the centroid position of static waveform and peak intensity for different measured objects was fitted by least square method, and the intersection point of each curve was solved to confirm the ideal peak intensity, and therefore reduce the interference of different measured objects. Secondly, according to the dynamic centroid difference threshold of two adjacent imaging waveforms, the static and dynamic working modes of the sensor were distinguished, and the peak intensity was adjusted to different intervals by linear iteration. Finally, a Z direction reciprocating test, color adaptability test, and step response test were carried out on the dispensing robot platform; the experiments showed that the repeatability accuracy of the sensor was 2.7 um and the dynamic step response delay was 0.5 ms.

Keywords: laser triangulation displacement sensor (LTDS); dispensing robot; location system; actual laser imaging waveform; centroid difference; repeatability accuracy; dynamic response speed

1. Introduction

With the development of modern science and technology, miniaturization of intelligent equipment has become a trend, which puts forward more stringent requirements for the processing technology of the equipment. In order to ensure the reliability of equipment processing, protective measures need to be taken. Precision dispensing technology effectively solves this problem and is widely used in many fields such as bottom filling, precision coating, chip encapsulation, component fixation, and so on [1–3].

Traditional precision dispensing depends on manual work under a microscope [4], which has the disadvantages of low production efficiency, high work intensity, and inevitable human error. With the continuous development of automation and robot technology, a series of dispensing robots have been developed to improve dispensing efficiency. All the dispensing robots developed by the Nordson Corporation (USA), the Datron Corporation (Germany), and the Soonchunhyang University (Korea) [5,6] have realized automatic high-speed precision dispensing, but the price has been expensive and they have been costly to maintain. Therefore, development of a high-precision and low-cost dispensing robot has potential application prospects. The location technology of the dispensing robot is the key to ensure measurement accuracy. Generally, height information of the actuator relative to the workpiece is accurately fed back to a robot through the photoelectric detection system, in order to carry out the operation [7,8].

A laser triangulation displacement sensor (LTDS) is a kind of ranging sensor which combines high directivity and brightness of laser with the principle of triangular measurement. It has the advantages of wide application, large measuring range, high precision, and fast response [9–12], and can be applied for precisely measuring the height information of a dispensing robot. Considering the working environment of a dispensing robot, the technical difficulties to be overcome when using LTDS to detect dispensing height information are as follows:

1. The problem of the adaptability of different objects, i.e., the different textures and colors of an object on the dispensing substrate affect the measurement, and the sensor must have good adaptability to different objects;
2. The problem of dynamic response speed, i.e., dispensing is a high-speed dynamic process, because the height of the workpiece on the dispensing substrate is not uniform, the sensor should have fast dynamic response speed during the measurement process.

Concerning the adaptability of different measured objects, for several years, researchers have studied the accuracy optimization methods of LTDS and have obtained a number of achievements. Klemen [13] introduced a dynamic symmetric projection and hyperbolic fitting compensation algorithm to solve robust alignment problems for different material objects, but the installation and measurement steps of this method were relatively complex. In contrast, it was relatively convenient to control the intensity of the laser imaging spot using an algorithm. Jung [14] improved the adaptability of the sensor to different colors by controlling the beam intensity of the laser diode so that the power received on the photoelectric device was constant when the object color changed. Similarly, Keyence [15] proposed the ABLE algorithm to adapt to the problem of light intensity inconsistency caused by different object colors. Song [16] established a mathematical model for the relationship between the imaging point position and the surface characteristic parameters of the measured object and modified the imaging point position according to the surface parameters to adapt to different measured objects. In addition, Li Sansi [17] analyzed the measurement error caused by different surface colors based on the Phong reflector model and established a color error function library to compensate for the error caused by the color change of the measured object. Although the above researches effectively improved the measurement accuracy of the LTDS for different objects, they did not involve the study of the dynamic response performance of the LTDS.

For the dynamic response speed problem, Zhang [18] developed a new type of LTDS which could measure the surface of a train with a speed of 64 km/h, but the accuracy was low. Scichi [19] proposed a dual-view triangulation structure to reduce the measurement error caused by the change of laser speckle when the measured object moved laterally. However, the structure was too large to be integrated. Thus, the LK-G series LTDS developed by Keyence [20,21] could achieve high-speed and high-precision measurement synchronously (50 KHz sampling frequency and accuracy of 0.05% F.S) through a highly integrated design. However, the LK-G series sensor took 4096 times, on average, to improve the measurement repeatability, and it relied on the sampling speed of customized imaging devices, with the disadvantage of complex steps that resulted in higher production costs.

In addition, imaging characteristics influence the accuracy of an optical measurement system. With the development of computer pattern recognition, researchers have carried out recognition and feature extraction for various optical images. Jeong [22] realized the detection of a dynamic laser spot by auto-associative multilayer perceptron (AAMLP) but did not extract the detailed information of the spot. Huffman [23] adopted a supervised classification algorithm based on a *t*-test filter for spectral feature classification in a complex data environment. Wang [24] proposed a method combining deep learning with a physical model to recover the information of the measured object from a single diffraction intensity diagram, but the imaging time of this method was too long. Manzo [25] developed an object recognition framework based on a fast convolution neural network, which could further extract more detailed features of the image and improved the speed of the model. Although the above studies had forward-looking significance, they had high requirements for the performance of the hardware system, which was not conducive to integrated development and promotion of LTDS. Therefore, we need a simpler laser imaging recognition method that requires less hardware. Moreover, in the particular application of a dispensing robot, it is necessary to quickly distinguish the dynamic and static imaging and to adjust the imaging to ensure measurement accuracy.

In this paper, an LTDS was proposed for a dispensing robot. By using the adjustment algorithm of laser imaging waveform, a method of "replacing hardware with software" was adopted to save hardware cost, and high-precision measurements in a dynamic scene were realized to meet the application requirements of a dispensing robot. First, the principle of dispensing robot system and the laser spot location method of the LTDS were introduced. Secondly, from the static and dynamic point of view, we studied the characteristics of an actual laser imaging waveform, in which the measured objects were a white diffuse reflector and a black frosted reflector, respectively. The optimal peak intensity of the imaging waveform was determined by least square fitting. Then, considering the actual working scene of a dispensing robot, a fast waveform adjustment algorithm was proposed. According to the dynamic centroid position shift of the imaging waveform, the static and dynamic working modes of the sensor were distinguished, and the peak intensity of the waveform was adjusted to different intervals by linear iteration. Finally, based on the dispensing robot experimental platform, we verified the repeatability accuracy and the step dynamic response characteristics of the sensor.

2. Dispensing Robot System and Laser Spot Location Principle

2.1. Principe of the Dispensing Robot System

The location and height measurement system of the dispensing robot is shown in Figure 1. The LTDS was fixed on the end of the manipulator to measure the workpiece in real time. It was designed based on direct-injection laser triangulation imaging [26]. The laser light emitted by a laser diode (LD) was focused by a collimated lens and projected vertically onto the object to form reflected light. The reflected light propagated in space by way of diffuse reflection, and then the reflected light was focused by the receiver lens in the direction of the observation angle ε and formed a laser imaging spot on the photosensitive imaging device (CMOS). The position of the imaging spot on the CMOS was obtained by the spot location algorithm. When the object moves s, the position of the imaging spot moves Δx on the CMOS accordingly, and the displacement s can be calculated according to Equation (1), where l and f_{po} are the object distance and image distance of the receiver lens group, and β is the imaging angle:

$$s = \frac{\Delta x \cdot l \sin\beta}{f_{po} \cdot \sin\varepsilon - \Delta x \cdot l \sin(\alpha + \beta)} \tag{1}$$

During the high-speed movement when the dispensing robot was working, the surface and height of the workpiece could suddenly change, which inevitably led to distortion of the laser imaging spot and reduced the measurement accuracy. The different surfaces and heights of the workpiece led to a change in the shape and position shift of the imaging waveform, respectively. According to the characteristics of an actual imaging waveform, considering the static and dynamic working mode of the robot, we proposed a fast laser intensity iterative adjustment method to obtain an ideal waveform for laser spot location, in order to satisfy measurement accuracy and response speed in both.

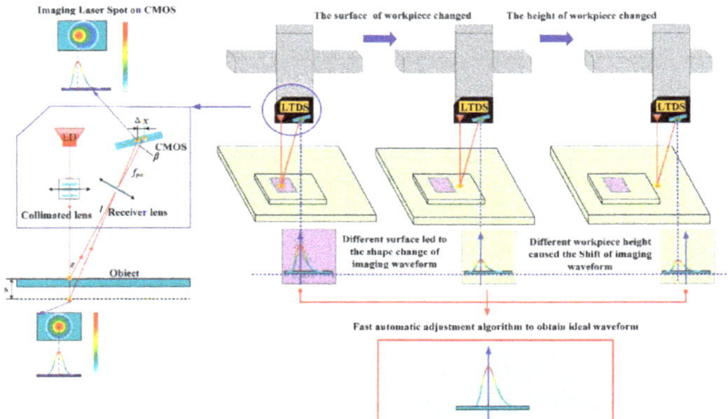

Figure 1. Dispensing robot location system and its measurement method.

2.2. Laser Imaging Spot Location

The laser spot location algorithm is the most important key technology to improve measurement accuracy. The gray centroid method [27,28] and function fitting method [29–32] are the mainstream methods. Under ideal conditions, the laser spot energy presents a Gaussian distribution. The function fitting method fits the Gaussian distribution of the spot energy by a simplified mathematical function, which can effectively and quickly obtain the position of the imaging point on the CMOS. In practice, however, due to optical distortion and the structure of the optical path, the imaging waveform was tilted and stretched, which led to a skewed distribution. Therefore, the function method causes some calculation deviation. In contrast, the gray centroid method can calculate the centroid of distorted waveforms to achieve the spot subpixel location.

To reduce the influence of random noise and increase the weight value near the peak, in this paper, the square weighted gray centroid method is adopted to calculate position x of the laser spot on the CMOS, as shown in Equation (2), where x_i is the position on the CMOS and y_i is the waveform intensity. In addition, we convert the position x from the pixel coordinates to the world coordinates by the calibration method of linear piecewise interpolation, and finally obtain the measured value.

$$x = \frac{\sum_i^N x_i \cdot y_i^2}{\sum_i^N y_i^2} \tag{2}$$

3. Characteristics of the Actual Laser Imaging Waveform

The quality of imaging waveform affects the result of the centroid calculation, thus the sensor developed, in this paper, adjusted the peak intensity of the imaging waveform by controlling the CMOS exposure time. In order to improve the efficiency of the adjustment algorithm and the adaptability to different objects, it was necessary to further analyze the characteristics of the imaging waveform.

3.1. Experiment Platform for Analyzing Imaging Waveform Characteristics

Although many practical factors, such as air flow, humidity, and temperature, have been considered in recent research on laser triangulation imaging [33,34], most of these researches have focused on simulation analysis and lacked complete analysis of the actual imaging waveform. In addition, there was a lack of research on the dynamic imaging waveform, so we comprehensively analyzed the actual laser imaging waveform from the static and dynamic perspectives.

An experimental platform for analysis of laser imaging waveform was established, as shown in Figure 2a. The LTDS developed, in this paper, was placed at a certain distance from the mobile platform.

It was highly integrated with a field programmable gate array (FPGA) and STM32 microprocessor for signal processing. The key components in the sensor included a laser diode (Rohm Corporation, wavelength 650 nm, power 2 mw), a linear CMOS (Hamamatsu Corporation), and an optical lens, which were purchased without special customization, therefore, improving productivity and reducing costs. The measured object was fixed on the mobile platform (resolution 1 um and precision 2 um). When its movement was controlled by the motion controller, the imaging waveforms on the CMOS were collected synchronously by the host computer. The black frosted and white diffuse reflectors were selected as the measured objects to carry out the actual waveform analysis experiment, as shown in Figure 2b.

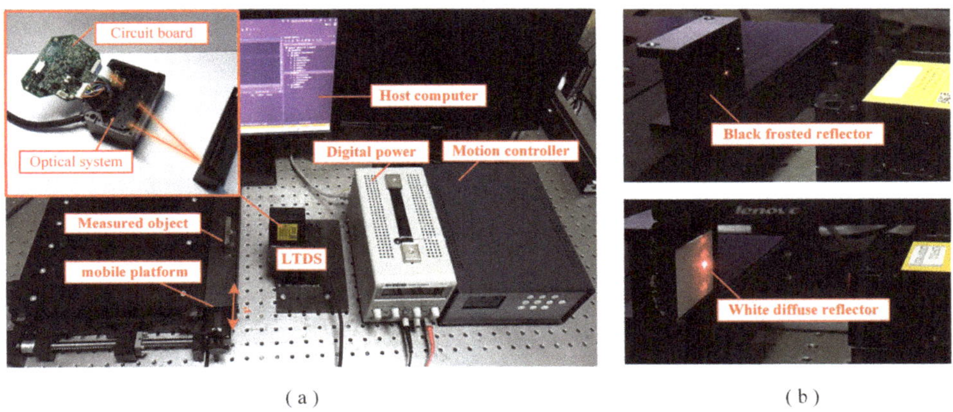

Figure 2. Imaging waveform experiment platform. (**a**) Hardware platform; (**b**) Measured object.

3.2. Characteristics of Static Laser Imaging Waveforms

By controlling the exposure time of the CMOS to obtain a gray-scale image, the characteristics of static imaging waveforms of two different measured object were studied. First, 10 groups of waveforms were obtained for two measured objects, and the change rule of waveform shape under different exposure times was analyzed. Then, the centroids of the waveforms were calculated, and the relationships among the centroid positions of the imaging waveform and the peak intensity were studied.

The imaging waveforms corresponding to different exposure times are shown in Figure 3. Although the imaging waveforms of the two different objects had obvious morphological differences (Figure 3a,b), the waveforms had a consistent trend when the peak intensity was in a certain range (350–800). The relationship between the exposure time and the peak intensity was further analyzed, as shown in Figure 3c. For different objects, the relationship between the peak intensity and the CMOS exposure time was different; the CMOS exposure time required to reach the target peak intensity could not be simply calculated from a function. Establishing multiple functions to calculate the CMOS exposure time for different measured objects was costly in terms of both software and hardware. Therefore, we intended to develop an iterative algorithm to automatically adjust the CMOS exposure time for different objects.

To confirm the ideal peak intensity, we considered the relationship between the peak intensity and the centroid position. As shown in Figure 4a, there was a certain difference in the centroid positions calculated from the waveforms with different peaks. To reduce this difference and to eliminate the interference caused by different colors, we fit the "centroid position-peak intensity" curve equation of black and white objects, respectively, as shown in Equation (3):

$$X(p) = a_k p^k + a_{k-1} p^{k-1} + a_{k-2} p^{k-2} + \cdots\cdots + a_1 p + a_0 \tag{3}$$

where a_k (k = 0, 1, 2, ...) is the coefficient term, p is the peak intensity, and X (p) is the centroid position when the peak intensity was p.

The least square method is used to confirm the coefficient term, and the sum of squares of error $I(a_0, a_1, \ldots, a_k)$ is constructed, where C_i is the actual centroid position:

$$I(a_0, a_1, \cdots, a_k) = \sum_{i=1}^{n} [C_i - x(p_i)]^2 = \sum_{i=1}^{n} \left(y_i - a_0 - a_1 p_1 - \cdots - a_k p_i^k\right) \tag{4}$$

Solving the corresponding coefficient $a_0{}^*, a_1{}^*, \ldots, a_k{}^*$ such that:

$$I(a_0{}^*, a_1{}^*, \ldots, a_k{}^*) = \min_{a_0{}^*, a_1{}^*, \ldots, a_k{}^*} I(a_0, a_1, \ldots, a_k) \tag{5}$$

According to the least square approximation method described above, the centroid position-peak intensity equation can be written as:

$$\begin{aligned} X(p) &= \begin{bmatrix} X_w \\ X_b \end{bmatrix} = \begin{bmatrix} 1.216e-19 \\ 5.878e-19 \end{bmatrix} p^7 + \begin{bmatrix} -3.39e-16 \\ -1.856e-15 \end{bmatrix} p^6 + \begin{bmatrix} 3.916e-13 \\ 2.406e-12 \end{bmatrix} p^5 \\ &+ \begin{bmatrix} -2.574e-10 \\ -1.649e-9 \end{bmatrix} p^4 + \begin{bmatrix} 1.178e-7 \\ 6.383e-7 \end{bmatrix} p^3 + \begin{bmatrix} -4.255e-5 \\ -0.0001399 \end{bmatrix} p^2 + \begin{bmatrix} 0.01124 \\ 0.01872 \end{bmatrix} p + \begin{bmatrix} 568 \\ 567.3 \end{bmatrix} \end{aligned} \tag{6}$$

where X_w and X_b are the centroid positions when the measured object is white diffuse reflector and black frosted reflector, and the intersection point of the equation curves is the ideal peak intensity (520 in this paper), as shown in Figure 4b.

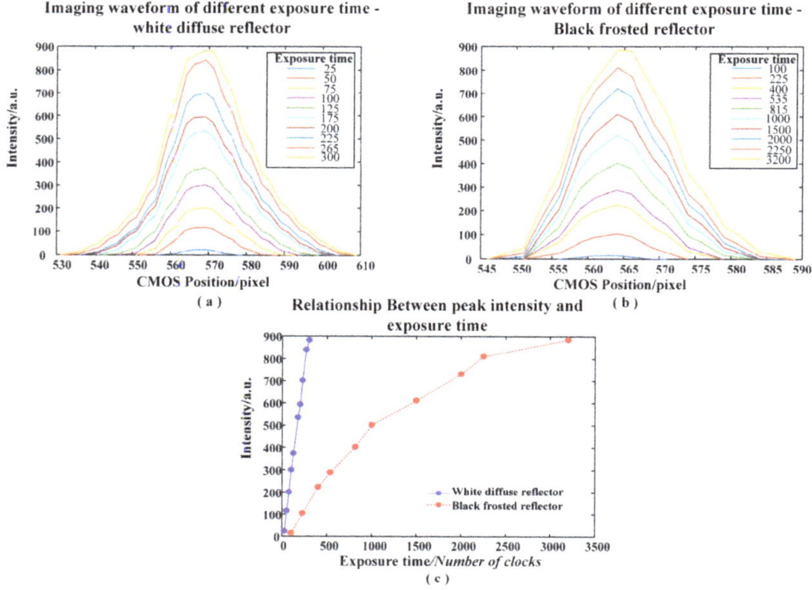

Figure 3. Imaging waveforms corresponding to different photosensitive imaging device (CMOS) exposure times. (**a**) Imaging waveforms of white diffuse reflector; (**b**) Imaging waveforms of black frosted reflector; (**c**) Relationship between peak intensity and exposure time of imaging waveform for different objects.

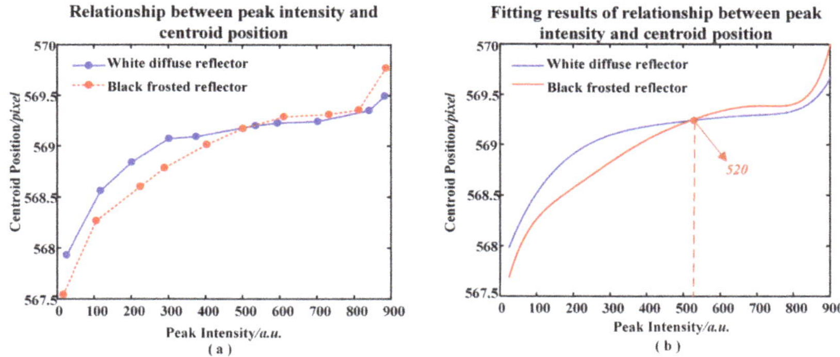

Figure 4. Analysis of the relationship between peak intensity and centroid position of imaging waveform. (**a**) Relationship between centroid position and peak intensity of imaging waveform for different objects; (**b**) Fitting results of relationship between centroid position and peak Intensity.

3.3. Characteristics of Dynamic Laser Imaging Waveform

In order to study the change law of imaging waveform when the measured object moved, the mobile platform was driven by the host computer to enable the object to move 0.1 mm horizontally at a speed of 20 mm/s, which was a tiny displacement to avoid the interference caused by the texture change of the object surface, and five groups real-time imaging waveforms were collected (similar to static experiments, two different objects were selected). On the basis of the result of the static waveform experiment, we controlled the peak intensity of the imaging waveform to be as close as possible to the ideal value (520) by giving a constant CMOS exposure time.

The dynamic waveform change is shown in Figure 5. Although we tried to obtain a stable waveform by giving a constant exposure time, the imaging waveform was still distorted in the dynamic measurement process. Moreover, when the measured object was a black frosted reflector, the distortion of the imaging waveform changed more obviously than that of a white diffuse reflector, and the peak intensity oscillated violently. We considered that it was inevitable that the laser energy distribution on the CMOS was not uniform because of the superposition of light and the motion vector in the process of laser transmission. Therefore, it was necessary to develop a dynamic waveform adjustment strategy for dynamic waveform distortion.

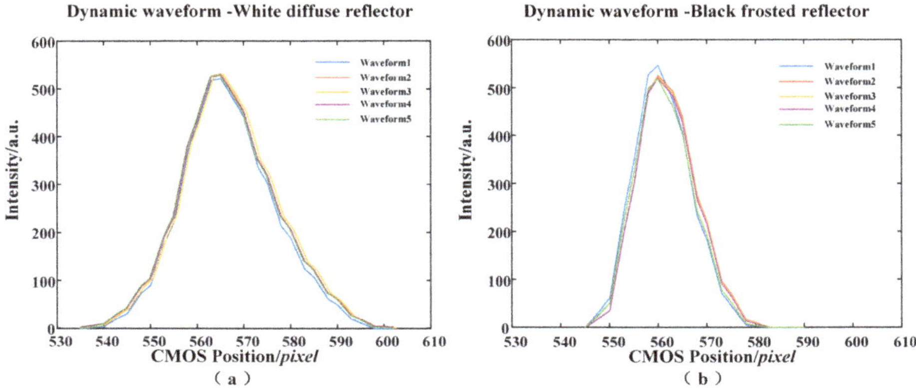

Figure 5. Dynamic imaging waveform. (**a**) Imaging waveform of white diffuse reflector; (**b**) Imaging waveform of black frosted reflector.

Before developing the dynamic waveform adjustment algorithm, we quantitatively compared the static and dynamic waveforms, as shown in Table 1. The peak intensity fluctuation of dynamic waveforms was two to seven times that of static waveforms, and the maximum centroid position shift was 24 times that of static waveforms. The peak intensity and centroid position drift of dynamic waveform were taken as the important indexes of dynamic waveform adjustment.

Table 1. Change of centroid position and peak intensity of static and dynamic imaging waveforms.

	Centroid Shift (Pixels)		Intensity Difference (a.u.)	
	Static	Dynamic	Static	Dynamic
White diffuse reflector	0.0259	0.5312	4	9
Black frosted reflector	0.0352	0.8641	4	28

In summary, based on the analysis of actual imaging waveforms of laser on the CMOS, the characteristics of static and dynamic imaging waveforms were significantly different. Therefore, we needed to fully consider the difference between the two measurement scenarios to develop the control algorithm.

4. Fast Automatic Adjustment Algorithm of Imaging Waveform

On the basis of the characteristics of the static and dynamic actual imaging waveforms, and according to the centroid position shift of two adjacent frames of imaging waveforms, an automatic fast waveform adjustment algorithm was proposed. This method could satisfy the dynamic response of the dispensing robot during its high-speed motion, and could also ensure its static measurement accuracy. The algorithm flow is shown in Figure 6.

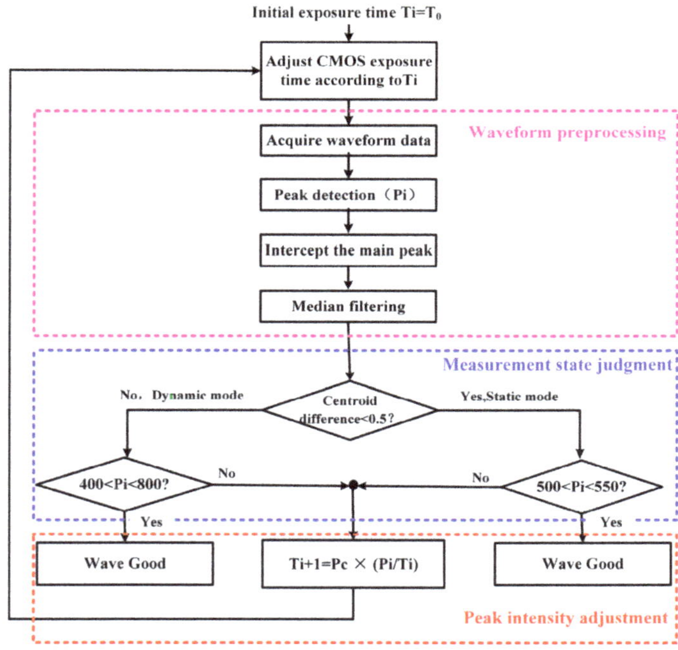

Figure 6. Flow chart of fast waveform adjustment algorithm based on centroid difference.

The proposed algorithm mainly included three main parts, i.e., waveform preprocessing, measurement mode judgment, and peak intensity adjustment which are described as follows: (1) The waveform preprocessing filtered out the noise interference and obtained the effective waveform information. (2) The measurement mode judgment determined whether the measurement mode was static or dynamic by setting the centroid difference threshold of two waveforms, and different peak intensity adjustment intervals were selected for these two measurement modes. (3) During the peak intensity adjustment part, a linear approximation prediction method was presented to quickly adjust the imaging waveform.

4.1. Imaging Waveform Preprocessing

Before entering the algorithm loop, we gave the initial CMOS exposure time T_0 to obtain the first frame image, and based on it, each subsequent waveform image was preprocessed as follows:

Step 1 Read the raw data of CMOS imaging waveform, identify the peak (P_i), and record its position x_p.

Step 2 To filter out the edge noise, select 15% of the peak intensity as the threshold, and the waveform whose energy is higher than the threshold is intercepted as the main peak.

Step 3 Apply a median smoothing filter to the main peak obtained by Step 2, taking into account both the filtering effect and speed.

4.2. Automatic Judgment of Measurement Mode

The actual working state of the dispensing robot alternated between dynamic and static. First, the dispensing robot moved to the target position at a high speed in the dynamic state, and then measured and dispensed in the static state.

Therefore, we set the dynamic centroid difference threshold between two adjacent frames of imaging waveform to distinguish the dynamic and static states and adopted different algorithm strategies. When it was the static working mode, the peak intensity was strictly controlled in an interval close to the ideal value (520), and therefore reduced the interference caused by different measured objects. Then, 64 times of smooth filter was applied to improve the measuring stability. In the dynamic mode, considering that the sharp fluctuation of the peak intensity of the dynamic waveform decreased the dynamic response speed of the sensor, we enlarged the peak intensity control interval. Different from the static mode, we did not filter the measured value in dynamic mode.

In the above algorithm, the dynamic centroid difference threshold and the peak intensity control interval of different working modes were identified as two important parameters.

4.2.1. Dynamic Centroid Difference Threshold

Table 1 shows the comparison of centroid shift of imaging waveforms between two adjacent frames, the centroid position shift caused by object motion is obvious, and the state of the measured object can be judged by the centroid difference. When the object moved, the centroid of the imaging waveform shifted at least 0.5312 pixel, therefore, we set the dynamic centroid difference threshold as 0.5 pixel. When the dynamic centroid difference was higher than 0.5 pixel, the sensor entered the dynamic mode. Otherwise, the sensor entered the static mode.

4.2.2. Peak Intensity Adjustment Interval

It was difficult to adjust the peak intensity to the ideal value at a time. Therefore, according to the static and dynamic working modes, different adjustment convergence intervals were determined to improve the peak intensity adjustment efficiency.

On the basis of the result of Equation (6), the "centroid position-peak intensity" fitting curve of different objects are subtracted to obtain the static centroid difference $Xc(p)$ caused by the change of the peak intensity of different measured objects, as shown in Figure 7a, which was different from the dynamic centroid difference of two adjacent waveforms. We define the absolute deviation evaluation function $\phi(p)$, as shown in Equation (7), where φ is the deviation threshold. The peak intensity adjustment intervals under different working modes are determined according to φ:

$$\Phi(p) = |X_c(p)| = |X_b(p) - X_w(p)| \leq \varphi \tag{7}$$

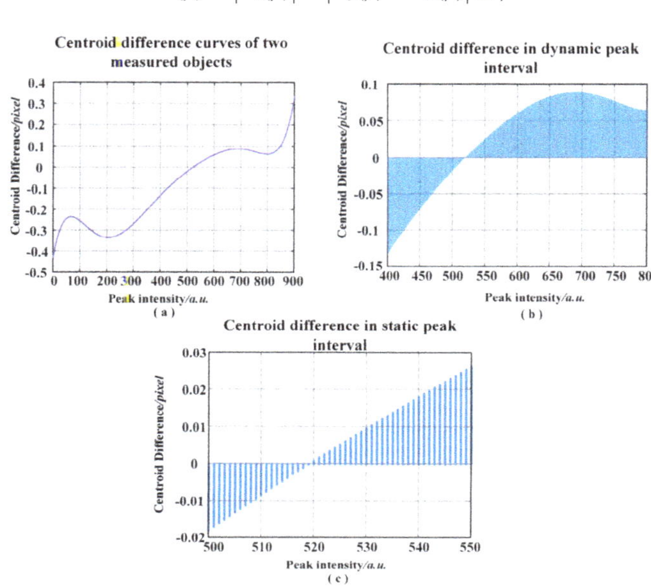

Figure 7. Analysis of centroid shift caused by different peak intensity. (**a**) Centroid difference curve; (**b**) Centroid difference in dynamic peak intervals; (**c**) Centroid difference in static peak intervals.

When $\varphi = 0.125$ pixel, the corresponding dynamic peak intensity interval is [400,800], as shown in Figure 7b. When $\varphi = 0.025$ pixel, the corresponding static peak intensity interval is [500,550], as shown in Figure 7c. The pixel coordinates are converted to the world coordinates, the theoretical errors of dynamic and static measurement caused by φ are 2.7 um and 0.5 um respectively. Therefore, the selection of the above peak intervals satisfied the needs of high-precision measurement.

4.3. Peak Intensity Linear Iterative Adjustment

In order to further improve the convergence speed of peak intensity adjustment iteration, we proposed a linear iterative method, which simplified the exposure time and the peak intensity as a linear relationship, as shown in Figure 8. The relationship between the current waveform peak intensity and the ideal peak intensity was established, and the ideal peak intensity was approximated iteratively in the next frame.

First, we judge whether the waveform peak intensity Pi of the current frame is in the ideal interval. If it is in the ideal interval, no adjustment is needed, otherwise, the peak intensity adjustment is required. According to the coordinate point (T_i, P_i) consisting of the CMOS exposure time, T_i, and the peak intensity, P_i, a linear relationship is established, and the exposure time, T_{i+1}, required for the next frame is calculated from Equation (8) for waveform adjustment. This cycle is iterated until the peak intensity is adjusted to the ideal interval.

$$T_{i+1} = \frac{P_{i+1}}{P_i} T_i \tag{8}$$

In order to approach the ideal peak intensity, P_{i+1}, Equation (8) is replaced by the ideal peak intensity, P_c, as shown in Equation (9):

$$T_{i+1} = \frac{P_c}{P_i} T_i \tag{9}$$

where P_c is a constant, determined to be 520 by the analysis of the actual imaging waveform. On the basis of the above peak intensity linear iterative approximation method, the peak could be quickly adjusted to an ideal interval, and therefore improve the response performance of the sensor.

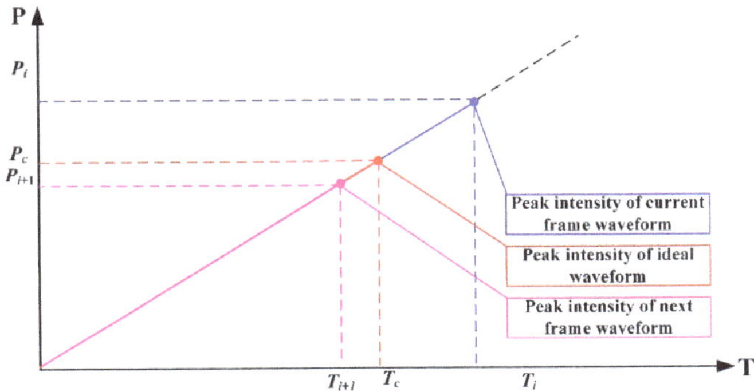

Figure 8. Schematic diagram of peak intensity linear prediction.

5. Dispensing Robot Experiment

5.1. Experimental Platform and Method

We verified the performance of the sensor developed in this paper on the dispensing robot platform by experiments. The DS-200C dispensing robot developed by the Mingseal Robot company (in China) was taken as the experimental platform, as shown in Figure 9a. The robot could move in three dimensions, i.e., X, Y, and Z. The maximum moving speeds in the X-Y direction and Z direction were 800 and 400 mm/s, respectively, and its repeat positioning accuracy was 10 um.

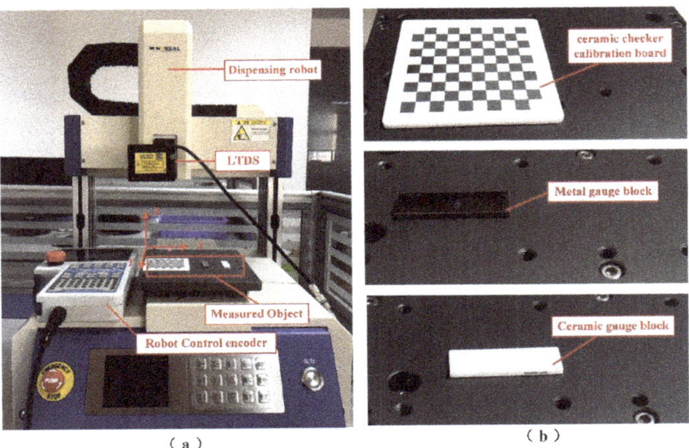

Figure 9. Experiment platform of dispensing robot. (**a**) Dispensing robot; (**b**) Different measured objects.

The sensor was fixed to the mechanical arm in the Z direction of the robot and different measured objects were placed on the X-Y platform (Figure 9b) for experiments, including the Z direction reciprocating test, color mutation adaptability test, and step response test. It should be noted that the robot stayed at each measured position for 30 ms to avoid interference caused by shaking of the manipulator. Considering the processing errors of the measured objects and the working table of the dispensing robot, the measurement performance of the sensor was evaluated for repeatability accuracy.

5.2. Z Direction Reciprocating Test Results and Analysis

To test the stability of the sensor in the Z direction of the dispensing robot, we fixed the sensor on the robot arm and moved back and forth at a speed of 400 mm/s along the Z direction, reaching five preset positions and staying for 30 ms, where the distance between each two-adjacent positions was 10 mm, as shown in Figure 10.

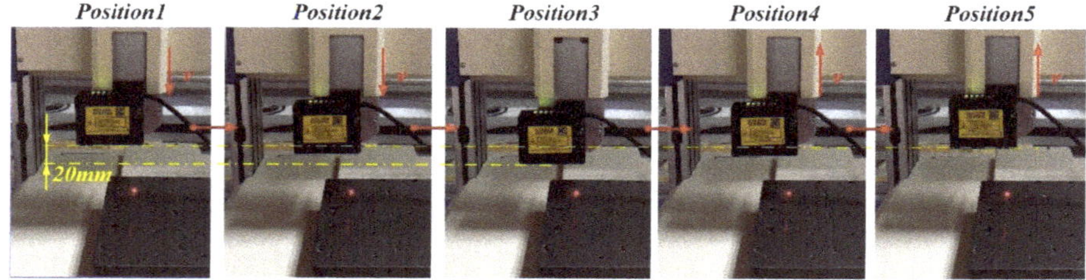

Figure 10. Z direction reciprocating test.

The displacement change curve of the sensor was recorded, as shown in Figure 11. When the sensor moved in the Z direction, its displacement curve was smooth and there was no obvious noise point, therefore, the sensor had good stability. Among them, the repeatability accuracy of five pause positions are shown in Table 2, with an average repeatability accuracy of 2 um.

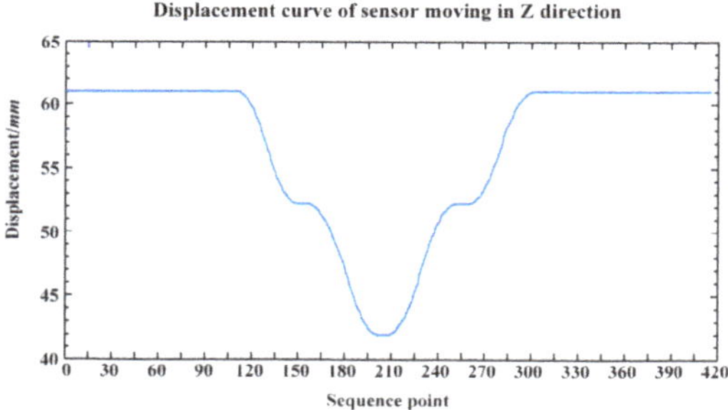

Figure 11. Displacement curve of the sensor when it moved in the Z direction.

Table 2. Repeatability accuracy of pause positions in the Z direction reciprocating test.

Position	Max Displacement (mm)	Min Displacement (mm)	Repeatability Accuracy (um)
1	61.0265	61.0251	1.4
2	52.2737	52.2719	1.8
3	42.0042	42.0034	0.8
4	52.2767	52.2746	2.1
5	61.0265	61.0249	1.6
Average	/	/	2.0

5.3. Color Adaptability Test Results and Analysis

We placed a ceramic checkered calibration board (its total size was 50 × 50 mm, each square size was 5 × 5 mm, and surface flatness was 5 um) on the working platform of the dispensing robot. The black and white squares were used to simulate the scene of color mutation, and the adaptability of the sensor to the color of the measured object was tested.

As shown in Figure 12, we selected 11 alternate measured positions on the checkerboard for static and dynamic tests. First, the dispensing robot moved slowly to the measured positions. When the system was stable, the repeatability accuracy of each position was recorded, sequentially, in the static state. Then, the dynamic test was carried out, the dispensing robot moved at a speed of 800 mm/s, and it stopped for 30 ms at each measured position. The repeatability accuracy of each position during the short-time stay were also recorded, as shown in Table 3.

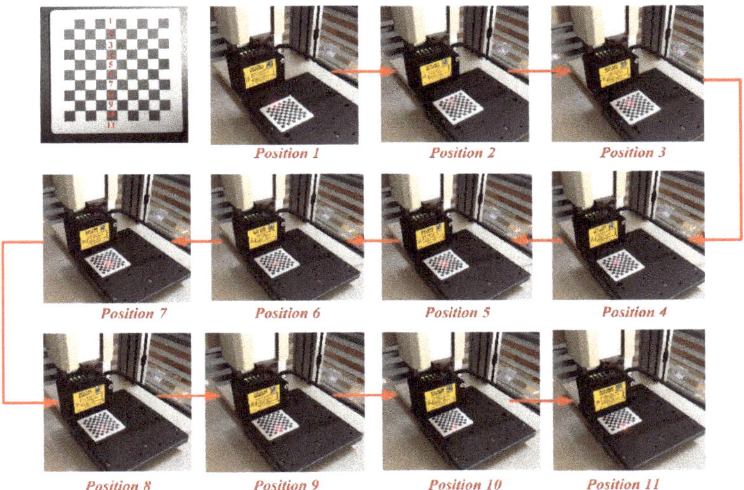

Figure 12. Adaptability test when the color of the measured object changed suddenly.

Table 3. Repeatability accuracy test results when measured object color changed.

NO.	1	2	3	4	5	6	7	8	9	10	11	Average
Static (um)	0.9	0.5	1.5	0.3	1.6	0.3	1.4	5.4	0.8	1	1.9	1.42
Dynamic (um)	1.6	1	1	1.5	1.5	2.6	1.2	1.2	2.7	2.6	0.5	1.58

According to the results of Table 3, the dynamic measurement repeatability (1.58 um) of the sensor was very close to the static repeatability (1.42 um). This indicated that the sensor could quickly adjust to a stable working state within 30 ms under dynamic test conditions and verified that the sensor had good adaptability to the color dynamic change of the measured object.

5.4. Step Response Test Results and Analysis

Two different steps were constructed for the step response test as follows: (1) A metal gauge block (2 mm in height, 10 mm in width, and 75 mm in length) was placed on the working table of the dispensing robot to construct a step without color change. (2) Similarly, a step with color change was constructed by a white ceramic gauge block (the same size as the metal gauge block). As shown in Figure 13, based on the steps constructed, the reciprocating dynamic experiments were conducted five times. The X-Y worktable of the robot moved at a speed of 800 mm/s, and stayed for 30 ms at the starting position (Position 1) and the measured position (Position 2) on the gauge block. During this process, the sensor recorded the data at a sampling frequency of 2 KHz.

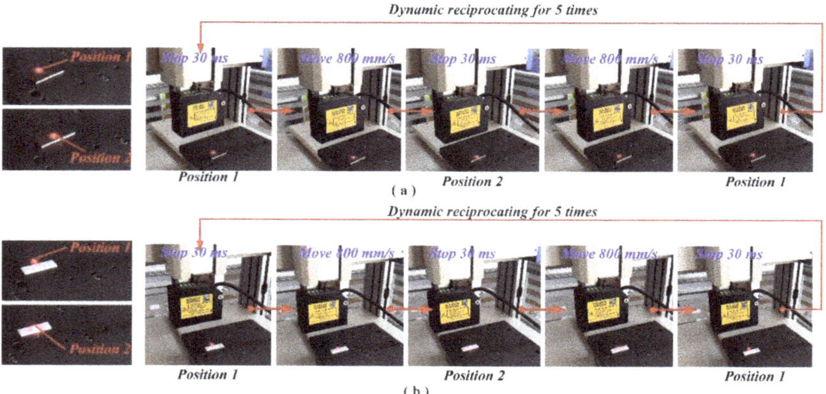

Figure 13. Step reciprocating dynamic test. (**a**) Metal gauge step dynamic test; (**b**) Ceramic gauge step dynamic test.

First, we investigated the step response characteristics of the sensor in the dynamic test. As shown in Figure 14, the dynamic displacement curve of the sensor could completely express the changes of the two different steps' height information, during the five reciprocating tests. However, there were obvious noise points at the edge of the step, which resulted in a delay of 0.5 ms for the measurement, therefore, the dynamic step response delay of the sensor we developed was 0.5 ms. We considered that there were two main reasons for the delay, one was the random tremble of the dispensing robot, the other was the instantaneous distortion of the laser spot when it moved from the worktable to the gauge block.

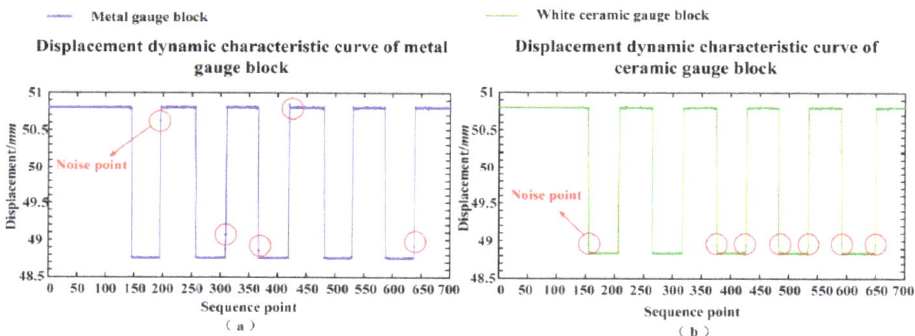

Figure 14. Displacement curve of the step dynamic reciprocating test. (**a**) Metal block step displacement dynamic characteristic curve; (**b**) White ceramic block step displacement dynamic characteristic curve.

Furthermore, we investigated the repeatability accuracy of the sensor when it reached the measured position at a quick speed and passed through the step, as shown in Table 4. The repeatability accuracies of the sensor for the metal gauge block and the ceramic gauge block were 2.7 um and 1.48 um, respectively, within 30 ms after the mechanical arm stopped, which met the needs of high-precision measurement.

Table 4. Repeatability of measured position on the gauge block in the step dynamic reciprocating test.

No.	Metal Gauge Step			Ceramic Gauge Step		
	Max (um)	Min (um)	R (um)	Max (um)	Min (um)	R (um)
1	48.8753	48.8706	4.7	48.9579	48.9569	1
2	48.8727	48.8707	2	48.9579	48.9565	1.4
3	48.8721	48.8706	1.5	48.9575	48.9559	1.6
4	48.8726	48.8703	2.3	48.9577	48.9558	1.9
5	48.8735	48.8704	3.1	48.9586	48.9571	1.5
Average	48.8732	48.8705	2.7	48.9579	48.9564	1.48

To sum up, we verified the measurement repeatability and dynamic respond speed of the LTDS we proposed in various working scenarios of the dispensing robot through Z direction motion, color dynamic mutation. and step response experiments. The dynamic respond delay was 0.5 ms and the maximum repeatability error among the test scenarios was 2.7 um.

5.5. Comparison and Discussion

We added three different laser adjustment strategies for comparison, as shown in Figure 15. The laser adjustment method that only considered dynamic mode, the laser adjustment method that only considered static working mode, and the Proportional Integral Derivative (PID) control were selected, and these algorithms were implemented on a FPGA by program.

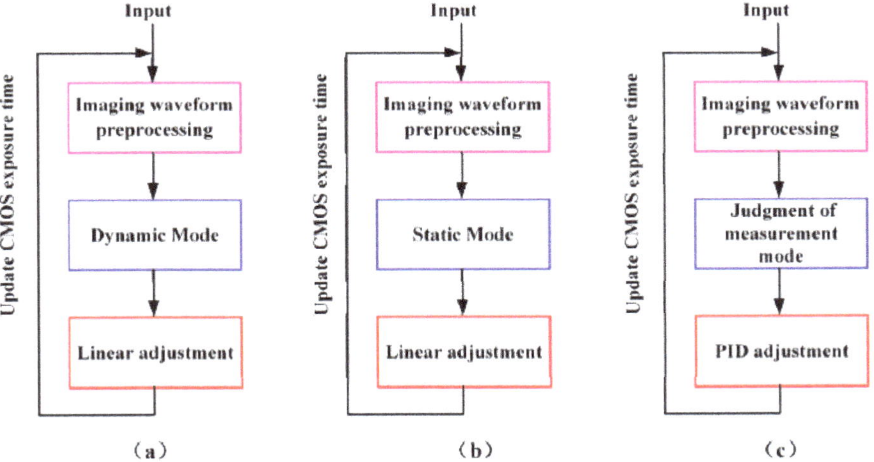

Figure 15. Different laser adjustment methods. (**a**) The method only considered dynamic mode; (**b**) The method only considered static mode; (**c**) PID laser adjustment method.

For the above different methods, we took the z direction stability, color adaptability, and step response tests, successively, under the same conditions (the measured object moved at the same speed). The measurement repeatability and dynamic response speed of each method were calculated. In addition, we defined the concept of positional noise density, σ, to comprehensively evaluate the accuracy and response speed of measurement, as shown in Equation (10).

$$\sigma = \sqrt{\frac{\sum_{i=1}^{n} (u_i - u)^2}{w}} \tag{10}$$

where u is the average measured displacement, u_i is the ith measured displacement, n is the total measured data, and w is the dynamic response speed. The measurement results were compared with the method we proposed, as shown in Table 5.

Table 5. Comparison of different laser adjustment methods.

Method	Repeatability	Dynamic Delay	σ
Only dynamic mode	16.2 um	0.5 ms	0.65 um/√Hz
Only static mode	2.9 um	10 ms	0.38 um/√Hz
PID adjustment	10.9 um	2 ms	0.85 um/√Hz
The method proposed	2.7 um	0.5 ms	0.08 um/√Hz

First, we discuss the necessity of working mode judgement. It was obvious that high measurement repeatability and fast dynamic response speed could not be achieved at the same time in the previous method. The dynamic mode only lead to low measurement accuracy, and the dynamic response speed in high precision static mode was very slow. Our new method combined the advantages of both static and dynamic mode by auto mode judgment. Its measurement accuracy was similar to the static mode, and it had the same response speed as the dynamic mode.

In addition, compared with the relatively complex PID control algorithm, the linear adjustment method had better repeatability accuracy and faster response speed, which we considered to be caused by the operation consumption of PID algorithm on hardware.

Finally, we discuss the positional noise density, σ, for different methods. The method proposed in this paper had the lowest σ, which also indicated that it had much better performance in both accuracy and speed.

Therefore, the method proposed greatly improved the dynamic response speed while ensuring the measurement accuracy and it had good adaptability to the alternating static and dynamic working scenarios.

6. Conclusions

In this paper, a laser triangulation displacement sensor for a dispensing robot location system was developed based on the comprehensive characteristics of the actual image waveform of the laser. The "centroid position-peak intensity" relationship of static imaging waveforms of different objects was fitted by the least square method, and the optimal peak intensity was solved to eliminate the interference of different measured objects. On the basis of the centroid difference of two dynamic imaging waveforms, the static and dynamic working modes were distinguished, and the peak intensity was adjusted to the ideal interval by the linear iterative method. According to the dynamic test experiment of the dispensing robot, the step response speed of the sensor was 0.5 ms under the dynamic measurement scene, and the repeatability accuracy was 2.7 um for different measured objects. However, the repeatability of the sensor still had room for improvement. In future work, we would increase the sampling frequency to collect more data for an average filter, and research a method to quickly obtain more feature information on the imaging laser spot, in order to improve the repeatability accuracy. Moreover, the LTDS proposed in this paper can also be applied to other high-precision dynamic measurement scenarios, in addition to the dispensing robot.

Author Contributions: Conceptualization, Z.N. and W.T.; data curation, N.L.; methodology, Z.N.; software, H.Z.; writing—original draft, Z.N. All authors have read and agreed to the published version of the manuscript.

Funding: This work was supported by the National Natural Science Foundation of China under grant no. 51975374.

Conflicts of Interest: The authors declare no conflict of interest.

References

1. Deng, G.; Cui, W.; Zhou, C.; Li, J. A piezoelectric jetting dispenser with a pin joint. *Optik* **2018**, *175*, 163–171. [CrossRef]
2. Tran, M.-S.; Hwang, S.-J. Design and Experiment of a Moving Magnet Actuator Based Jetting Dispenser. *Appl. Sci.* **2019**, *9*, 2911. [CrossRef]
3. Lu, S.; Yao, Y.; Liu, Y.; Zhao, Y. Design and experiment of a needle-type piezostack-driven jetting dispenser based on lumped parameter method. *J. Adhes. Sci. Technol.* **2015**, *29*, 716–730. [CrossRef]
4. Das, A.N.; Sin, J.; Popa, D.O.; Stephanou, H.E. On the precision alignment and hybrid assembly aspects in manufacturing of a microspectrometer. In Proceedings of the IEEE International Conference on Automation Science and Engineering, Arlington, VA, USA, 23–26 August 2008; pp. 959–966.
5. Yanwei, L.; Guiling, D. The Influence of Fluid Viscosity of Fluid Jetting Dispensing. In Proceedings of the 2007 International Symposium on High Density packaging and Microsystem Integration, Shanghai, China, 26–28 June 2017; pp. 1–4.
6. Kwon, K.S. Speed measurement of ink droplet by using edge detection techniques. *Measurement* **2009**, *42*, 44–50. [CrossRef]
7. Wang, N.; Liu, J.; Wei, S.; Zhang, X. A Vision Location System Design of Glue Dispensing Robot. *Lect. Notes Comput. Sci.* **2015**, *9246*, 536–551.
8. Ting, Y.; Chen, C.; Feng, H.; Chen, S. Apply Computer Vision and Neural Network to Glue Dispenser Route Inspection. In Proceedings of the 2007 International Conference on Mechatronics and Automation, Harbin, China, 5–8 August 2007; pp. 3882–3887.
9. Donges, A.; Noll, R. Laser Triangulation. In *Laser Measurement Technology: Fundamentals and Applications*; Springer: Berlin/Heidelberg, Germany, 2015; pp. 247–278. [CrossRef]
10. Wu, C.; Chen, B.; Ye, C. Detecting defects on corrugated plate surfaces using a differential laser triangulation method. *Opt. Lasers Eng.* **2020**, *129*, 106064. [CrossRef]
11. Peng, T.; Zhang, Z.; Chen, F.; Zeng, D. Dimension Measurement and Key Point Detection of Boxes through Laser-Triangulation and Deep Learning-Based Techniques. *Appl. Sci.* **2020**, *10*, 26. [CrossRef]
12. Xing-Qiang, L.; Zhong, W.; Lu-Hua, F. A Fast and in-Situ Measuring Method Using Laser Triangulation Sensors for the Parameters of the Connecting Rod. *Sensors* **2016**, *16*, 1679. [CrossRef]
13. Žbontar, K.; Mihelj, M.; Podobnik, B.; Povše, F.; Munih, M. Dynamic symmetrical pattern projection based laser triangulation sensor for precise surface position measurement of various material types. *Appl. Opt.* **2013**, *52*, 2750–2760. [CrossRef]
14. Jung, J.K.; Kang, S.G.; Nam, J.S.; Park, K.H. Intensity Control of Triangulation Based PSD Sensor Independent of Object Color Variation. *IEEE Sens. J.* **2011**, *11*, 3311–3315. [CrossRef]
15. Amann, M.C.; Bosch, T.; Lescure, M.; Myllyla, R.; Rioux, M. Laser Ranging: A Critical Review of Unusual Techniques for Distance Measurement. *Opt. Eng.* **2001**, *40*, 10–19.
16. Song, D.; Xia, D.; Wang, Z. Impact of Standard Deviation and Reflectance of the Measured Surface on Laser Diode-Position Sensitive Detector System. In *MATEC Web of Conferences*; EDP Sciences: Ulis, France, 2016; Volume 61, p. 06001. [CrossRef]
17. Li, S.; Jia, X.; Chen, M.; Yang, Y. Error analysis and correction for color in laser triangulation measurement. *Optik* **2018**, *168*, 165–173. [CrossRef]
18. Zhang, Z.; Feng, Q.; Gao, Z.; Kuang, C.; Fei, C.; Li, Z.; Ding, J. A new laser displacement sensor based on triangulation for gauge real-time measurement. *Opt. Laser Technol.* **2008**, *40*, 252–255. [CrossRef]
19. Ibaraki, S.; Kitagawa, Y.; Kimura, Y.; Nishikawa, S. On the limitation of dual-view triangulation in reducing the measurement error induced by the speckle noise in scanning operations. *Int. J. Adv. Manuf. Technol.* **2017**, *88*, 731–737. [CrossRef]
20. Da Silva Maciel, L.; Spelt, J.K. Influence of process parameters on average particle speeds in a vibratory finisher. *Granul. Matter* **2018**, *20*, 65. [CrossRef]
21. Xu, Y.Y.; Jiang, D.H.; Ma, G.T.; Deng, Z.G.; Zheng, J.; Zhang, W.H. Dynamic Response Characteristics of a High-Temperature Superconducting Maglev Vehicle under Laterally Unbalanced Load Conditions. *J. Supercond. Nov. Magn.* **2014**, *27*, 35–39. [CrossRef]

22. Jeong, S.; Jung, C.; Kim, C.S.; Shim, J.H.; Lee, M. Laser spot detection-based computer interface system using autoassociative multilayer perceptron with input-to-output mapping-sensitive error back propagation learning algorithm. *Opt. Eng.* **2011**, *50*, 084302. [CrossRef]
23. Huffman, C.; Sobral, H.; Terán-Hinojosa, E. Laser-induced breakdown spectroscopy spectral feature selection to enhance classification capabilities: A *t*-test filter approach. *Spectrochim. Acta Part B Spectrosc.* **2019**, *162*, 105721. [CrossRef]
24. Wang, F.; Bian, Y.; Wang, H.; Lyu, M.; Pedrini, G.; Osten, W.; Barbastathis, G.; Situ, G. Phase imaging with an untrained neural network. *Light Sci. Appl.* **2020**, *9*, 77. [CrossRef]
25. Manzo, M.; Pellino, S. FastGCN+ARSRGemb: A novel framework for object recognition. *arXiv* **2020**, arXiv:2002.08629.
26. Nan, Z.; Feng, Y.; Zhao, H.; Tao, W. Research on laser source drift with temperature of laser triangular displacement sensor. In Proceedings of the Ninth International Symposium on Precision Mechanical Measurements, Chongqing, China, 18–21 October 2019; p. 1134326. [CrossRef]
27. Assen, H.C.v.; Egmont-Petersen, M.; Reiber, J.H.C. Accurate object localization in gray level images using the center of gravity measure: Accuracy versus precision. *IEEE Trans. Image Process.* **2002**, *11*, 1379–1384. [CrossRef]
28. Xian, W.; Qinwei, M.; Shaopeng, M.; Hongtao, W. A Marker Locating Method Based on Gray Centroid Algorithm and its Application to Displacement and Strain Measurement. In Proceedings of the 2011 Fourth International Conference on Intelligent Computation Technology and Automation, Shenzhen, China, 28–29 March 2011; pp. 932–935.
29. Otepka, J.; Fraser, C. Accuracy Enhancement of Vision Metrology through Automatic Target Plane Determination. *Int. Arch. Photogramm. Remote Sens. Spat. Inf. Sci.* **2004**, *35*, 873–879.
30. Xin, L.; Xu, L.; Cao, Z. Laser spot center location by using the gradient-based and least square algorithms. In Proceedings of the 2013 IEEE International Instrumentation and Measurement Technology Conference (I2MTC), Minneapolis, MN, USA, 6–9 May 2013; pp. 1242–1245.
31. Wang, X.; Zhao, Q.; Ling, Q. Robust image processing method of laser spot center location in complex industrial environment. In Proceedings of the 2017 International Symposium on Intelligent Signal Processing and Communication Systems (ISPACS), Xiamen, China, 6–9 November 2017; pp. 651–656. [CrossRef]
32. Liu, X.; Lu, Z.; Wang, X.; Ba, D.; Zhu, C. Micrometer accuracy method for small-scale laser focal spot centroid measurement. *Opt. Laser Technol.* **2015**, *66*, 58–62. [CrossRef]
33. Beermann, R.; Quentin, L.; Stein, G.; Reithmeier, E.; Kaestner, M. Full simulation model for laser triangulation measurement in an inhomogeneous refractive index field. *Opt. Eng.* **2018**, *57*, 114107. [CrossRef]
34. Ciddor, P.E. Refractive index of air: 3. The roles of CO2, H2O, and refractivity virials: Erratum. *Appl. Opt.* **2002**, *41*, 7036. [CrossRef]

Publisher's Note: MDPI stays neutral with regard to jurisdictional claims in published maps and institutional affiliations.

Article

Photonic Microwave Distance Interferometry Using a Mode-Locked Laser with Systematic Error Correction

Wooram Kim [1], Haijin Fu [1,2], Keunwoo Lee [1,3], Seongheum Han [1,4], Yoon-Soo Jang [1,5,*] and Seung-Woo Kim [1,*]

1 Department of Mechanical Engineering, Korea Advanced Institute of Science and Technology (KAIST), 291 Daehak-ro, Yuseong-gu, Daejeon 34141, Korea; kwr0704@kaist.ac.kr (W.K.); haijinfu@hit.edu.cn (H.F.); kwl@nanosystemz.com (K.L.); sh_han@kimm.re.kr (S.H.)
2 Ultra-Precision Optoelectronic Instrument Engineering Center, Harbin Institute of Technology, School of Electrical Engineering and Automation, Harbin 150001, China
3 NanoSystem Co. Ltd., 90 Techno 2-ro, Yuseong-gu, Daejeon 34014, Korea
4 Department of Ultra-Precision Machines and Systems, Korea Institute of Machinery & Materials (KIMM), Daejeon 34103, Korea
5 Length Standard Group, Physical Metrology Division, Korea Research Institute of Standards and Science (KRISS), 267 Gajeong-ro, Yuseong-gu, Daejeon 34113, Korea
* Correspondence: ysj@kriss.re.kr (Y.-S.J.); swk@kaist.ac.kr (S.-W.K.)

Received: 29 September 2020; Accepted: 28 October 2020; Published: 29 October 2020

Abstract: We report an absolute interferometer configured with a 1 GHz microwave source photonically synthesized from a fiber mode-locked laser of a 100 MHz pulse repetition rate. Special attention is paid to the identification of the repeatable systematic error with its subsequent suppression by means of passive compensation as well as active correction. Experimental results show that passive compensation permits the measurement error to be less than 7.8 μm (1 σ) over a 2 m range, which further reduces to 3.5 μm (1 σ) by active correction as it is limited ultimately by the phase-resolving power of the phasemeter employed in this study. With precise absolute distance ranging capability, the proposed scheme of the photonic microwave interferometer is expected to replace conventional incremental-type interferometers in diverse long-distance measurement applications, particularly for large machine axis control, precision geodetic surveying and inter-satellite ranging in space.

Keywords: absolute distance measurement; system error correction; femtosecond laser

1. Introduction

Distance interferometry employing lasers has long been used for precision ranging and positioning in diverse industrial applications [1,2]. Continuous-wave monochromatic sources such as He–Ne gas lasers or semiconductor diode lasers are preferably used to attain sub-wavelength resolutions by performing phase-measuring interferometry based on homodyne or heterodyne principles [3,4]. Such continuous-wave laser interferometry basically leads to incremental displacement measurement, whereas absolute positioning interferometry requires incorporating additional source functionality such as intensity modulation [5], frequency sweeping [6], and multiple wavelengths [7]. In general, most industrial applications favor absolute positioning interferometry, but its measurement accuracy when attained with continuous-wave lasers is limited by the practical difficulties encountered in extending the source functionalities. As a result, incremental displacement interferometry using continuous-wave lasers is still widely used, with absolute positioning being achieved by the digital accumulation of instantaneous distance increments. At the same time, the zero-datum of absolute positioning has to be set up arbitrary by installing a fiducial reference in advance.

With the aim of providing a new light source for absolute positioning interferometry, mode-locked pulse lasers are being investigated to provide more efficient functionalities beyond continuous-wave lasers [8–10]. As a result, by making the most of the unique spectral and temporal characteristics of mode-locked lasers, various novel methods for absolute positioning have so far been demonstrated; they are conveniently referred to as synthetic wavelength interferometry [11,12], spectral-resolved interferometry [13,14], dual-comb interferometry [15,16], optical cross-correlation [17,18] and multi-wavelength interferometry [19,20]. Each method has its own capability of utilizing mode-locked lasers as an alternative source for absolute positioning, and particularly the method of synthetic wavelength interferometry draws attention as it utilizes microwaves that are photonically synthesized directly from the pulse repetition rate of a mode-locked laser. Compared with conventional microwaves that are electrically generated by means of intensity or frequency modulation, the photonic microwaves offer superior stability in frequency and intensity so as to implement long-distance ranging with high precision. Accordingly, efforts are being made to apply photonic microwaves to large-scale industrial metrology [21] and space missions such as multiple satellite formation flying [22,23].

In this investigation, we describe a 1 GHz microwave distance interferometer based on the 10th inter-mode harmonic of a 100 MHz pulse repetition rate of a fiber mode-locked laser. A comprehensive analysis is made for correction of the systematic error through pre-calibration with respect to a commercial incremental He–Ne laser interferometer. The photonic microwave created in this study is able to offer an accurate length ruler with a 0.3 m wavelength stabilized to the Rb clock, with a fractional instability of 10^{-11} at 1 s averaging. The repeatable systematic measurement error attributable to the optical and radio frequency (RF) electrical disturbance embedded in the interferometer system is identified and corrected by passive and active suppression schemes. Experimental results reveal that absolute positioning can be achieved within a residual error of 3.5 μm (1 σ) over a 2.0 m distance, which corresponds to a phase resolution of ~0.0084°, which is in fact comparable to the phase resolving power of the phasemeter configured in this study using a lock-in amplifier.

2. Photonic Microwave Distance Interferometer

2.1. Interferometer Configuration

Figure 1 describes the distance interferometer devised in this investigation. The light source is an Er-doped fiber mode-locked laser emitting 120 fs short pulses at a repetition rate (f_r) of 100 MHz. When the laser pulses are observed using a photodetector over a time period, the resulting electrical signal exhibits an RF comb structure in the frequency domain. As illustrated in Figure 1a, the resulting RF comb is in fact a down-converted version of the optical comb of the mode-locked source. Each RF comb mode element represents an integer-multiple harmonic frequency of the pulse repetition rate f_r, which can be isolated from other modes through an electrical bandpass filter. An N-th harmonic frequency has the wavelength (Λ_N), expressed as $\Lambda_N = \frac{c}{Nf_r}$, with c being the speed of light in a vacuum. Lower harmonic frequencies, i.e., longer wavelengths, are preferable as the interferometer source to enlarge the non-ambiguity range of distance measurement, whereas higher harmonic frequencies offer finer measurement resolutions.

The source beam is launched into free space and divided into the reference beam (green) and measurement beam (blue) with a 50:50 beam splitter as shown in Figure 1b. Both the reference beam and measurement beam are converted into electrical signals and mixed with a local oscillator signal (1 GHz–100 kHz) stabilized to the Rb atomic clock by phase-locked loop (PLL) control as shown in Figure 1c. The reference beam directly goes to a reference photodetector, while the measurement beam reaches a measurement photodetector after reflecting from a target mirror installed on a motorized stage with a 2.5 m travel range. In this study, the 10th harmonic of 1 GHz is selected to be used as the microwave source. The phase delay (θ_N) of the 1 GHz source microwave signal is measured by down-shifting to a 100 kHz signal using an electrical local oscillator set to operate near 1 GHz. The target distance (L) is then determined as:

$$L = \frac{\Lambda_N}{2n_{air}}(m_N + \theta_N) \tag{1}$$

where n_{air} denotes the refractive index of air and m_N is an integer. The phase delay θ_N between the measurement and measurement paths is measured with respect to the particular photonic wavelength Λ_N. Finally, the super-heterodyned electrical signals of 100 kHz are processed through a lock-in amplifier that is used as a phase meter to determine the interferometric phase delay θ_N. The phase measurement resolution is found to be 0.01° with an accuracy of 0.02°, from which the measurement precision is estimated to be 4.2 μm.

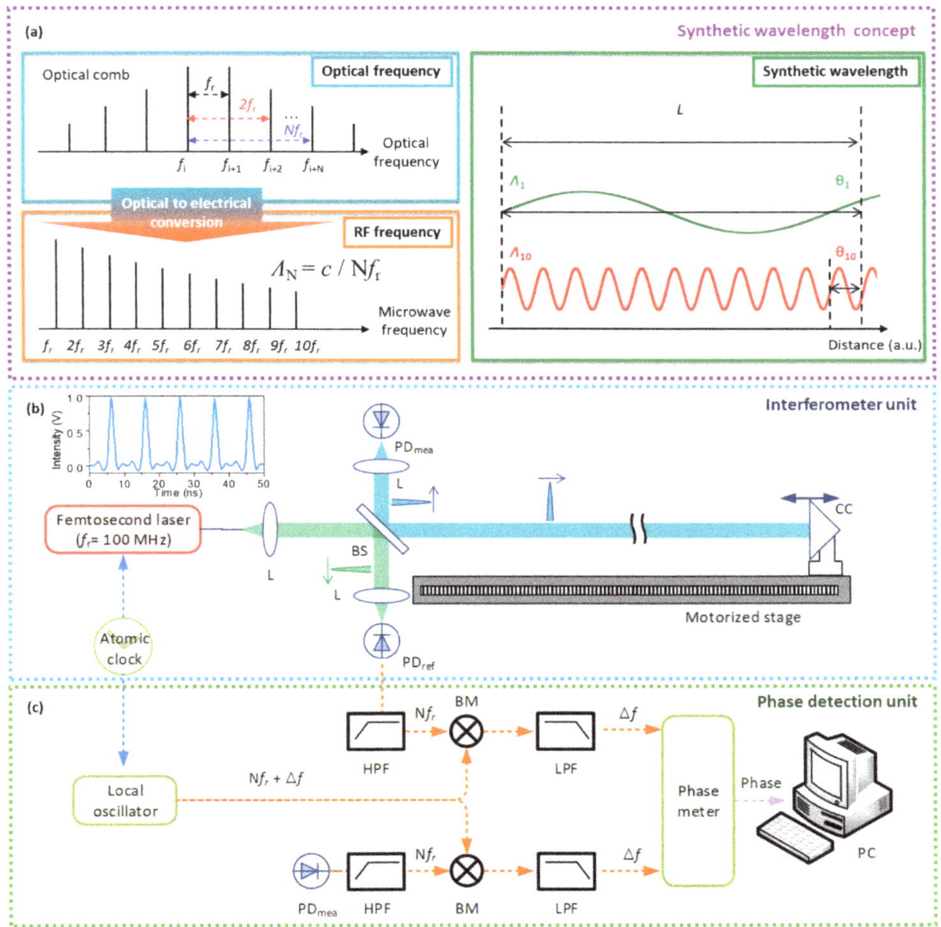

Figure 1. Photonic microwave distance interferometer. (**a**) Optical spectrum of a mode-locked laser and its radio frequency (RF) synthetic wavelengths. (**b**) Optical configuration for absolute distance measurement. (**c**) RF electrical circuit for interferometric phase detection. BM: balanced mixer, BS: beam splitter, c: speed of light in air, CC: corner cube, HPF: high pass filter, L: lens, LPF: low pass filter, PC: personal computer, PD: photodetector, *ref*: reference, *mea*: measurement, Δf: beat frequency, f_r: pulse repetition rate, Nf_r: N-th harmonics of f_r, Λ: synthetic wavelength, θ: interferometric phase.

2.2. Measurement Repeatability

The measurement repeatability was evaluated at a fixed target distance of about 1 m. Figure 2a shows a typical time-trace of distance measurement with an update rate of 10 Hz (blue) and 0.1 Hz (yellow), respectively. Over a time period of 1000 s, the measured distance fluctuates within a few tens of micrometers as clearly seen in the measurement histogram, appearing to be a Gaussian shape. For quantitative evaluation of the measurement repeatability, the Allan deviation of the time-trace measurement was calculated as shown in Figure 2b, from which the measurement repeatability is estimated to be 9.5 and 2.5 μm at an averaging of 0.1 s and 10 s, respectively. It is important to note that the best measurement repeatability achievable near 10 s averaging is practically limited by the detection resolution of the used phasemeter of lock-in amplifier type, not by the photonic microwave signal stability of the mode-locked laser source.

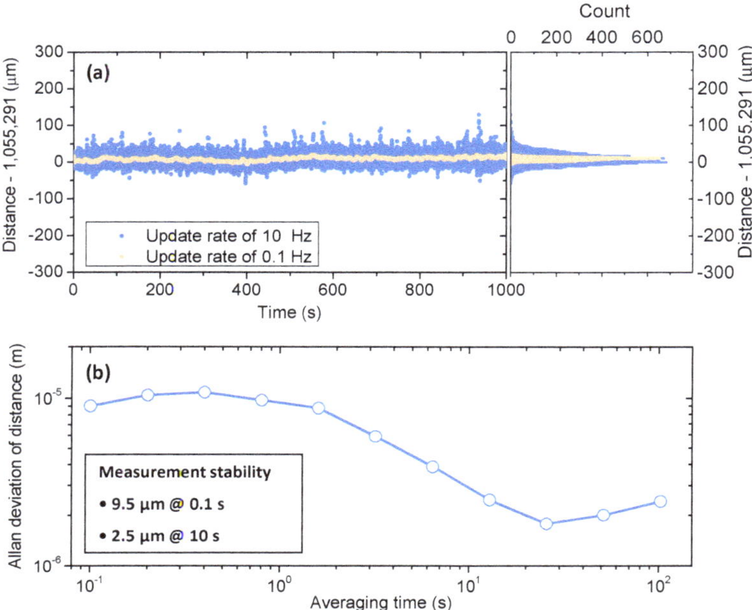

Figure 2. Measurement repeatability. (**a**) Measurement fluctuation of a 1-m fixed distance traced over 1000 s along with its histogram. (**b**) Allan deviation from 0.1 to 100 s averaging.

3. Systematic Error Correction

3.1. Systematic Error Analysis

We performed distance measurements repeatedly over a range of 0.5 to 2.5 m back and forth with 10 mm steps as illustrated in Figure 3. Each position value is represented by an average of 60 consecutive measurements taken over a period of 6 s. The averaged distance was compared with its counterpart value obtained simultaneously with a commercial incremental-type He–Ne laser interferometer, of which the discrepancy was plotted in Figure 3a. At each distance, with respect to the He–Ne laser interferometry, the measurement error of our interferometer was evaluated by processing eight individual data sets. The measurement error appears to be a sum of two separable terms; one is the non-repeatable random error and the other is the repeatable systematic error. In principle, the random error determines the measurement precision ultimately achievable, whereas the systematic error can be eliminated through correction at each position. Further, as depicted in Figure 3b, it is also important to note that the repeatability of the systematic error can be either periodic (orange line) or non-periodic (purple line).

In general, it is known that the periodic systematic error is caused by the signal leakage occurring in the interferometer optics or electronics, with the period appearing to be identical to the wavelength of the microwave signal used in distance measurement [24]. On the other hand, the non-periodic systematic error arises from the amplitude-to-phase error conversion due to the light intensity fluctuation as well as the imperfect RF electronics of nonlinear behaviors [25,26]. For the 1 GHz microwave wavelength employed in this study, the periodic error is found to have an amplitude of 22.6 μm with an actual period of 0.15 m. The periodic error is attributable to the cyclic cross talk of the measurement signal with unwanted signals. Yielding no position-dependency, the periodic error is generated by the RF circuit noise, as well as the partially-reflected laser beam from the interferometer optics for beam splitting and recombination. The effect of cross talk on the measurement error can be modelled as [24]:

$$\cos\left(\frac{2\pi f}{c}2L\right) + \varepsilon_{\text{crosstalk}}\cos(\theta_{\text{crosstalk}}) = \cos\left(\frac{2\pi f}{c}2L + \theta_{\text{periodic error}}(L)\right) \quad (2)$$

where f is the carrier frequency (here 1 GHz), c is the speed of light, L is the target distance, $\varepsilon_{\text{crosstalk}}$ is the cross talk ratio to the measurement signal, $\theta_{\text{crosstalk}}$ is the constant phase of the cross talk components and $\theta_{\text{periodic error}}$ (L) is the periodic error induced by cross talk as a function of the target distance L. In our case, the cross talk ratio $\varepsilon_{\text{crosstalk}}$ is found to be 0.1%, as illustrated in Figure 3c.

The non-periodic systematic error shown in Figure 3a is found to be ±300 μm, with a position dependency with a strong correlation with the optical power actually received by the photodetector, as shown in Figure 3d. This confirms that the non-periodic error is attributable to the amplitude-to-phase conversion caused by the photodiode and RF components. However, the microwave phase is not reciprocal for increasing and decreasing the direction of the received optical power.

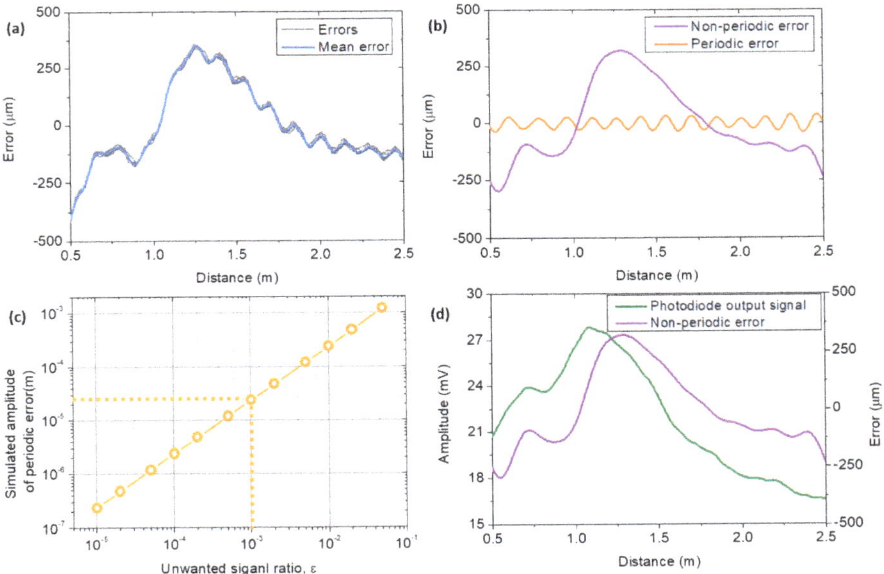

Figure 3. Systematic measurement error. (**a**) Experimental data of measurement errors over a range from 0.5 to 2.5 m. (**b**) Periodic error vs. and non-periodic error. (**c**) Cyclic error amplitude vs. unwanted signal ratio (**d**) Photodiode output amplitude (proportional to the received optical power) in comparison to the non-cyclic error.

3.2. Passive Correction by Post-Processing

Figure 4 shows the passive correction results of the systematic error, of which the mean line before correction is shown in Figure 3a. The residual error after correction is found to be ±7.8 μm (1 σ) over a 2 m range. The accuracy of the phasemeter used in the calibration is 0.02°, which corresponds to 8.4 μm. This implies that our passive correction almost reached the ultimate limitation imposed by the phasemeter. For more comprehensive analysis, the residual error after correction was Fourier transformed as shown in Figure 4b. The result revealed that the periodic error of an amplitude of 22.6 μm is completely suppressed; whereas the residual error after correction is mainly dominated by the non-periodic error.

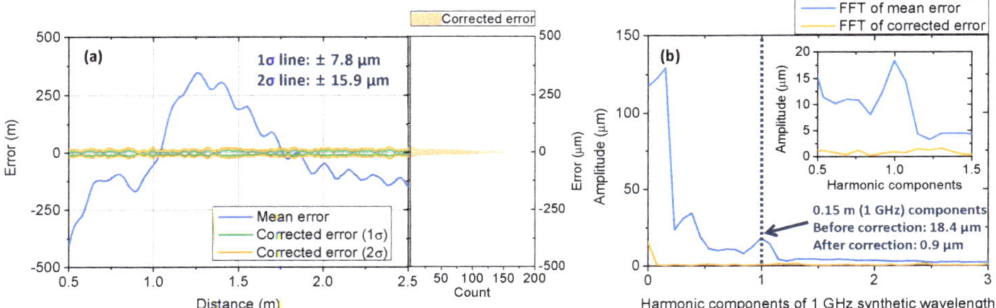

Figure 4. Passive systematic error correction. (**a**) Mean systematic error (blue) and residual error after passive correction—1 σ line (green) and 2 σ line (orange). (**b**) Fourier transformed result of the mean error (blue) and corrected error (orange). FFT means a Fourier transformed result.

3.3. Active Correction by Compensating Optical Power

Figure 5 shows the result of active correction of the systematic error. The optical power received by the interferometer photodetectors in the measurement and reference arms varies while the target is translated on the motorized stage. The optical power fluctuation is detected within the lock-in amplifier as illustrated Figure 5a, and stabilized by controlling the gain of the RF amplifier installed in the measurement arm circuit. As a consequence, the amplitude of the measurement signal into the lock-in amplifier is kept constant with a maximum deviation of 0.014% as depicted in Figure 5b. As discussed in the previous section, because the non-periodic error is dominantly influenced by the amplitude-to-phase conversion in RF components, the optical power stabilization is able to actively eliminate the non-periodic error. As a result, the systematic error is observed only in the form of cyclic error without the non-periodic error over a 2-m travel range, as shown in Figure 5c. The residual cyclic error can be corrected by post-processing so that the total measurement error remains within the level of ±3.5 μm (1 σ).

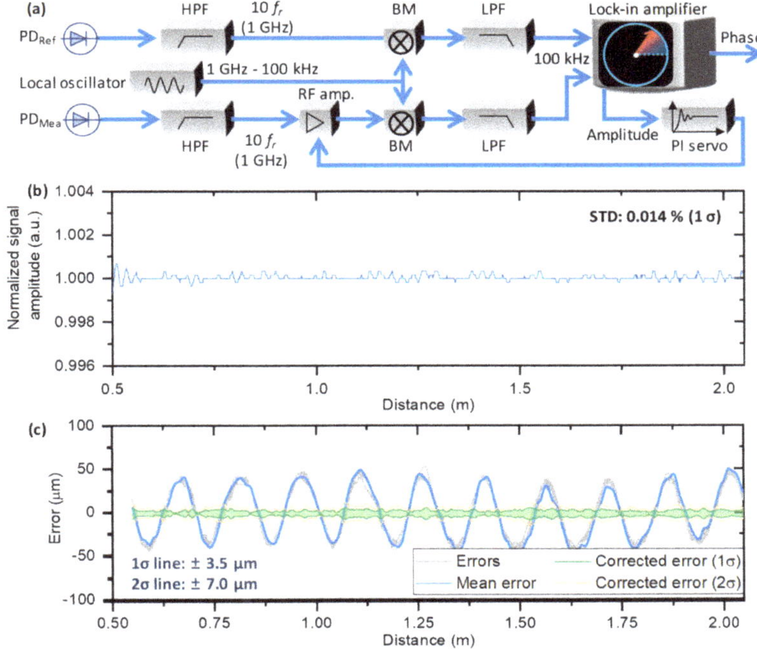

Figure 5. Active systematic error correction. (**a**) Phase measurement circuit with feedback control for signal power compensation. (**b**) The result of optical power stabilization. (**c**) Position error after power compensation (blue) and its corrected error (green). BM: balanced mixer, HPF: high pass filter, LPF: low pass filter, PD: photodetector, RF: radio frequency, *ref*: reference, *mea*: measurement, f_r: pulse repetition rate, PI: proportional integral, STD: standard deviation.

4. Conclusions

We have demonstrated a photonic scheme of an absolute distance interferometer using a 1 GHz microwave synthesized by the 10th harmonic of the pulse repetition rate of a fiber mode-locked laser. The photonic microwave offers an accurate length ruler with a 0.3 m wavelength stabilized to the Rb clock with a fractional instability of 10^{-11} at 1 s averaging. However, the measured interferometric phase is disturbed by optical and RF electrical components comprising the interferometer system, which were identified by measurements and subsequently corrected by passive and active schemes of systematic error suppression. Experimental results revealed that the systematic error can be corrected to a residual level of 7.8 μm in terms of the standard deviation (1 σ), which is a phase measurement error of ~0.0084°, comparable to the resolution of the lock-in amplifier used in phase measurement. It is expected that the measurement error can be enhanced to a sub-micron level by selecting a higher microwave frequency, along with high frequency RF components that will be available in the near future. It is also important to note that the scheme of photonic microwave interferometry can be a promising candidate for practical use of the frequency comb-based distance measurement with traceability to the time/frequency standard, which can be applied to high-precision machine axis control as well as long distance measurement space missions such as geodetic survey and inter-satellite ranging.

Author Contributions: Conceptualization, Y.-S.J. and S.-W.K.; methodology, W.K.; software, H.F.; validation, W.K., H.F. and Y.-S.J.; formal analysis, W.K.; investigation, H.F.; resources, K.L.; data curation, S.H.; writing—original draft preparation, Y.-S.J.; writing—review and editing, S.-W.K.; visualization, W.K.; supervision, Y.-S.J.; project administration, S.-W.K.; funding acquisition, S.-W.K. All authors have read and agreed to the published version of the manuscript.

Funding: This work was supported by the National Research Foundation of the Republic of Korea (NRF-2012R1A3A1050386).

Acknowledgments: Y.-S. Jang acknowledges support from the KRISS (20011030) and S. Han appreciates support from the KIMM (NK224A).

Conflicts of Interest: The authors declare no conflict of interest.

References

1. Gao, W.; Kim, S.-W.; Bosse, H.; Haitjema, H.; Chen, Y.L.; Lu, X.D.; Knapp, W.; Weckenmann, A.; Estler, W.T.; Kunzmann, H. Measurement technologies for precision engineering. *CIRP Ann. Manuf. Technol.* **2015**, *64*, 773–796.
2. Berkovic, G.; Shafir, E. Optical methods for distance and displacement measurements. *Adv. Opt. Photon.* **2012**, *4*, 441–473. [CrossRef]
3. Bobroff, N. Recent advances in displacement measuring interferometry. *Meas. Sci. Technol.* **1993**, *4*, 907–926. [CrossRef]
4. Joo, K.-N.; Ellis, J.D.; Buice, E.S.; Spronck, J.W.; Schmidt, R.H.M. High resolution heterodyne interferometer without detectable periodic nonlinearity. *Opt. Express* **2010**, *18*, 1159–1165. [CrossRef] [PubMed]
5. Fujima, I.; Iwasaki, S.; Seta, K. High-resolution distance meter using optical intensity modulation at 28 GHz. *Meas. Sci. Technol.* **1998**, *9*, 1049–1052. [CrossRef]
6. Dale, J.; Hughes, B.; Lancaster, A.J.; Lewis, A.J.; Reichold, A.J.H.; Warden, M.S. Multi-channel absolute distance measurement system with sub ppm-accuracy and 20 m range using frequency scanning interferometry and gas absorption cells. *Opt. Express* **2014**, *22*, 24869–24893. [CrossRef]
7. Dandliker, R.; Thalmann, R.; Prongue, D. Two-wavelength laser interferometry using superheterodyne detection. *Opt. Lett.* **1988**, *13*, 339–341. [CrossRef] [PubMed]
8. Kim, S.-W. Metrology: Combs rule. *Nat. Photon.* **2009**, *3*, 313–314. [CrossRef]
9. Jin, J. Dimensional metrology using the optical comb of a mode-locked laser. *Meas. Sci. Technol.* **2016**, *27*, 022001. [CrossRef]
10. Jang, Y.-S.; Kim, S.-W. Distance measurements using mode-locked laser: A review. *Nanomanuf. Metrol.* **2018**, *1*, 131–147. [CrossRef]
11. Minoshima, K.; Matsumoto, H. High-accuracy measurement of 240-m distance in an optical tunnel by use of a compact femtosecond laser. *Appl. Opt.* **2000**, *39*, 5512–5517. [CrossRef] [PubMed]
12. Jang, Y.-S.; Kim, W.; Jang, H.; Kim, S.-W. Absolute distance meter operating on a free-running mode-locked laser for space mission. *Int. J. Precis. Eng. Manuf.* **2018**, *19*, 975–981. [CrossRef]
13. Joo, K.-N.; Kim, S.-W. Absolute distance measurement by dispersive interferometry using a femtosecond pulse laser. *Opt. Express* **2006**, *14*, 5954–5960. [CrossRef] [PubMed]
14. Van Den Berg, S.A.; Persijn, S.T.; Kok, G.J.P.; Zeitouny, M.G.; Bhattacharya, N. Many-wavelength interferometry with thousands of lasers for absolute distance measurement. *Phys. Rev. Lett.* **2012**, *108*, 183901. [CrossRef]
15. Coddington, I.; Swann, W.C.; Nenadovic, L.; Newbury, N.R. Rapid and precise absolute distance measurements at long range. *Nat. Photon.* **2009**, *3*, 351–356. [CrossRef]
16. Zhang, H.; Wei, H.; Wu, X.; Yang, H.; Li, Y. Absolute distance measurement by dual-comb nonlinear asynchronous optical sampling. *Opt. Express* **2014**, *22*, 6597–6604. [CrossRef]
17. Lee, J.; Kim, Y.-J.; Lee, K.; Lee, S.; Kim, S.-W. Time-of-flight measurement using femtosecond light pulses. *Nat. Photonics* **2010**, *4*, 716–720. [CrossRef]
18. Han, S.; Kim, Y.-J.; Kim, S.-W. Parallel determination of absolute distances to multiple targets by time-of-flight measurement using femtosecond light pulses. *Opt. Express* **2015**, *23*, 25874–25882. [CrossRef]
19. Jin, J.; Kim, Y.-J.; Kim, Y.; Kim, S.-W.; Kang, C.-S. Absolute length calibration of gauge blocks using optical comb of a femtosecond pulse laser. *Opt. Express* **2006**, *14*, 5968–5974. [CrossRef]
20. Hyun, S.; Kim, Y.-J.; Kim, Y.; Jin, J.; Kim, S.-W. Absolute length measurement with the frequency comb of a femtosecond laser. *Meas. Sci. Technol.* **2009**, *20*, 095302. [CrossRef]
21. Jang, Y.-S.; Wang, G.; Hyun, S.; Chun, B.J.; Kang, H.J.; Kim, Y.-J.; Kim, S.-W. Comb.-referenced laser distance interferometer for industrial nanotechnology. *Sci. Rep.* **2016**, *6*, 31770. [CrossRef]

22. Lee, J.; Lee, K.; Jang, Y.-S.; Jang, H.; Han, H.; Lee, S.-H.; Kang, K.-I.; Lim, C.-W.; Kim, Y.-J.; Kim, S.-W. Testing of a femtosecond pulse laser in outer space. *Sci. Rep.* **2014**, *4*, 5134.
23. Lezius, M.; Wilken, T.; Deutsch, C.; Giunta, M.; Mandel, O.; Thaller, A.; Schkolnik, V.; Schiemangk, M.; Dinkelaker, A.; Kohfeldt, A.; et al. Space-borne frequency comb metrology. *Optica* **2016**, *3*, 1381–1387. [CrossRef]
24. Kikuta, H.; Iwata, K.; Nagata, R. Absolute distance measurement by wavelength shift interferometry with a laser diode: Some systematic error sources. *Appl. Opt.* **1987**, *26*, 1654–1660. [CrossRef]
25. Phung, D.-H.; Merzougui, M.; Alexandre, C.; Lintz, M. Phase Measurement of a Microwave Optical Modulation: Characterisation and Reduction of Amplitude-to-Phase Conversion in 1.5 µm High Bandwidth Photodiodes. *J. Lightwave Technol.* **2014**, *32*, 3759–3767. [CrossRef]
26. Guillory, J.; Gardia-Marques, J.; Alexandre, C.; Truong, D.; Wallerand, J.-P. Characterization and reduction of the amplitude-to-phase conversion effects in telemetry. *Meas. Sci. Technol.* **2015**, *26*, 084006. [CrossRef]

Article

Surface Texture Measurement on Complex Geometry Using Dual-Scan Positioning Strategy

Fang Cheng [1], Shaowei Fu [2,3,*] and Ziran Chen [1]

1 Engineering Research Center of Mechanical Testing Technology and Equipment (Ministry of Education), Chongqing Key Laboratory of Time Grating Sensing and Advanced Testing Technology, Chongqing University of Technology, Chongqing 400054, China; chf19chf19@hotmail.com (F.C.); czr@cqut.edu.cn (Z.C.)

2 School of Mechanical and Aerospace Engineering, Nanyang Technological University, Singapore 639798, Singapore

3 JM VisTec System Pte Ltd., Singapore 417942, Singapore

* Correspondence: FUSH0009@e.ntu.edu.sg

Received: 3 November 2020; Accepted: 24 November 2020; Published: 26 November 2020

Featured Application: This work is potentially applicable to in-line surface measurement. Integrated with robots or other manipulation devices, the proposed technology will be able to capture surface finishing parameters, including roughness and surface texture uniformity.

Abstract: In this paper, a surface measurement method based on dual-scan positioning strategy is presented to address the challenges of irregular surface patterns and complex geometries. A confocal sensor with an internal scanning mechanism was used in this study. By synchronizing the local scan, enabled by the internal actuator in the confocal sensor, and the global scans, enabled by external positioners, the developed system was able to perform noncontact line scan and area scan. Thus, this system was able to measure both surface roughness and surface uniformity. Unlike laboratory surface measurement equipment, the proposed system is reconfigurable for in situ measurement and able to scan free-form surfaces with a proper stand-off distance and approaching angle. For long-travel line scan, which is needed for rough surfaces, a surface form tracing algorithm was developed to ensure that the data were always captured within the sensing range of the confocal sensor. It was experimentally verified that in a scanning length of 100 mm, where the surface fluctuation in vertical direction is around 10 mm, the system was able to perform accurate surface measurement. For area scan, XY coordinates provided by the lateral positioning system and the Z coordinate captured by the confocal sensor were plotted into one coordinate system for 3D reconstruction. A coherence scanning interferometer and a confocal microscope were employed as the reference measurement systems to verify the performance of the proposed system in a scanning area of 1 mm by 1 mm. Experimental data showed that the proposed system was able to achieve comparable accuracy with laboratory systems. The measurement deviation was within 0.1 μm. Because line scan mechanisms are widely used in sensor design, the presented work can be generalized to expand the applications of line scan sensors.

Keywords: surface texture measurement; confocal sensing; surface form tracing; 3D reconstruction; roughness

---

1. Introduction

An important tool for product quality assessment, surface texture measurements are traditionally performed based on laboratory systems, due to the high degree of measurement resolution required [1,2]. With the rapid development of advanced manufacturing technologies, such as five-axis machining and additive manufacturing (AM), new surface measurement solutions are needed to address the challenges of complex geometries and irregular surface patterns [3,4]. Stylus profilometry is widely used for conventionally machined samples [5]. Equipped with a coordinate measuring machine (CMM), the measurement space of stylus profilometry can be enlarged to accommodate samples of large volume and complicated geometry [6]. However, stylus profilometry is not an ideal solution for measuring surfaces with irregular patterns due to its contact measurement mechanism and flanking errors [7,8]. Another limitation of stylus profilometry is its low efficiency for areal surface texture measurement. Therefore, based on ISO 4287 [9], stylus profilometry is mainly used in line scan mode for conventionally machined surfaces with known directional surface patterns. 3D microscopy [10], such as confocal microscopy [11–13], coherence scanning interferometry [14,15] and focus variation microscopy [16,17], are suitable for area scans. 3D microscopic systems are able to determine surface roughness and surface isotropy based on ISO 25178-2 [18,19]. Using objectives with high numerical aperture (NA), 3D microscopic systems are able to achieve much higher lateral resolution compared with stylus profilometry. However, the vertical measurement range relies on mechanical scanning, which significantly limits measurement efficiency. In order to avoid mechanical movement, the concept of chromatic confocal microscopy (CCM) [20,21] has been developed in the past decade. However, for high-accuracy measurement, CCM's sensing range is within a few hundreds of microns, which limits its application to free-form surface measurement. Furthermore, when measuring high-roughness surfaces, long scans are required to differentiate surface roughness from surface waviness and surface form [22]. Therefore, a large vertical measurement range is needed to cover surface fluctuations over a long scan range.

Based on the above literature review and analysis, the motivation of this presented work is to develop an optical surface texture measurement system able to perform both line and area scans for challenging surfaces with complex geometry and irregular surface patterns. Unlike a laboratory solution, the proposed system has the potential to be utilized for in situ measurement [23,24].

2. Principle of Confocal Sensing with Internal Scanning

A laser confocal sensor, Keyence LT9011 (KEYENCE, Osaka, Japan), was utilized for surface height measurement in this study. Figure 1a shows the general principle of a single-point confocal sensor. Only when the measurement point is in focus will the detector receive the peak intensity of the reflected laser. Figure 1b shows the working principle of the Keyence LT9011. The optics are driven by a tuning fork to generate high-frequency vibration, enabling vertical scanning. In the meantime, the focus position can be detected and height can be measured accordingly. The optics are also driven by an oscillating unit for lateral scanning, with a frame rate of 2.8 fps covering a scan length of 1.1 mm. This product was originally developed for height or thickness measurements.

The vertical resolution of the LT9011 is 0.1 μm. The laser beam spot size is 2 μm with a scanning spacing of 2 μm, which complies with the requirements for surface roughness measurements based on ISO 4287 [9]. In this study, this sensor was further developed for surface texture measurement.

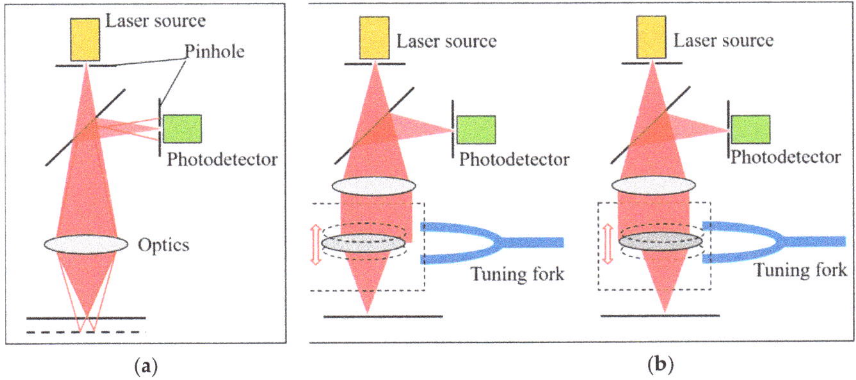

Figure 1. Principle of confocal sensing with internal scanning: (**a**) principle of confocal sensor; (**b**) internal scanning mechanism.

With the internal scanning mechanism, this sensor is able to measure surface roughness subject to two conditions:

(1) the surface is relatively flat; the surface fluctuation does not exceed the sensing range of 0.6 mm;
(2) the surface is relatively smooth; the measurement of roughness parameters does not require a roughness evaluation length longer than 1.1 mm.

Figure 2 shows the measured surface profile of a calibrated roughness master (178–602, Mitutoyo Corporation, Kawasaki-shi, Japan) with known *Ra* and *Rt* values of 2.97 μm and 9.4 μm, respectively.

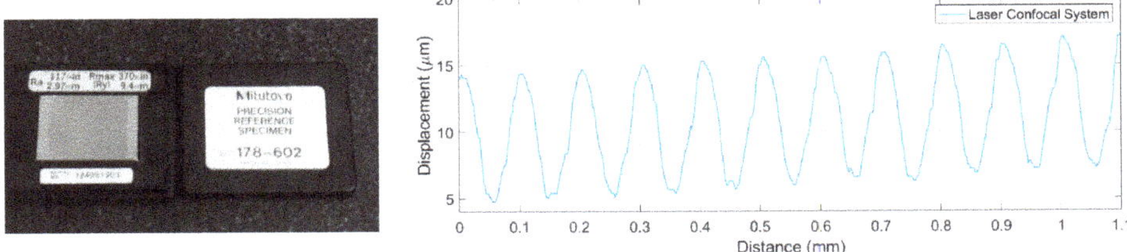

Figure 2. Line scan for roughness measurement using confocal sensing.

The arithmetical mean roughness (*Ra*) can be calculated using Equation (1), where N is the number of data points acquired in a measurement and Z_i is the ith data point in height direction of the roughness profile. The total height of roughness profile (*Rt*) is defined as the difference between the largest profile peak height and the largest profile valley depth within the evaluation length [9].

$$R_a = \frac{1}{N}\sum_{i=1}^{N}|Z_i|. \quad (1)$$

Therefore, with Z_i values captured by the confocal sensor LT9011, the surface roughness parameter *Ra* over a scan length of 1.1 mm can be calculated. To verify the measurement capability of the laser confocal sensor LT9011, measurement results of the roughness master are shown in Table 1. The nominal *Ra* and *Rt* values of the roughness master are 2.97 μm and 9.40 μm, respectively. The mean error and standard deviation are 0.06 μm and 0.04 μm for *Ra* measurements and 0.10 μm and 0.08 μm for *Rt* measurements. The measurement error is partially due to short sampling length.

Table 1. Roughness master measurement result (unit: μm).

	1	2	3	4	5	Mean	Std. Dev.
Ra (μm)	3.02	3.02	3.10	3.00	3.01	3.03	0.04
Rt (μm)	9.51	9.43	9.67	9.46	9.48	9.50	0.08

3. 3D Surface Topography Measurement with Dual-Scan Scanning

Integrated with an external motorized stage (PT1/M-Z8, Thorlabs Inc., Newton, NJ, USA), a 3D surface topography measurement system was configured, as shown in Figure 3a. By synchronizing the internal and external scan, the 3D surface topography of the measured area was reconstructed, as shown in Figure 3b. The scanning area was 1.1 mm by 1 mm, with point spacing and line spacing of 2 μm. The total measurement time was 250 s.

The motivation for developing this system was to address the challenge of measuring geometrically complicated samples. Instead of accommodating the sample into the measurement space of a laboratory system, the measurement system was designed to approach the samples for adaptive measurement.

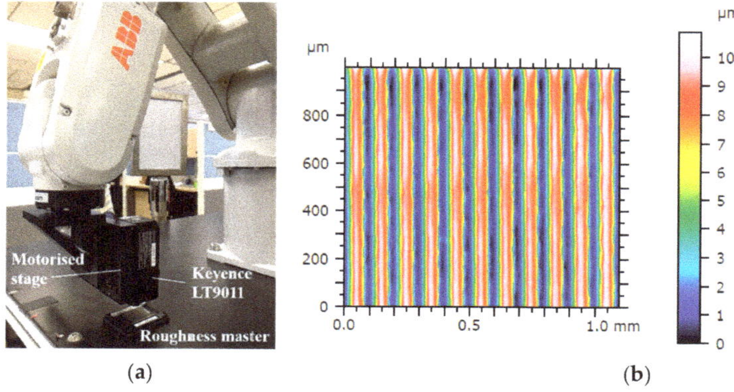

Figure 3. 3D surface topography measurement based on confocal sensing: (**a**) system configuration; (**b**) 3D surface topography.

Based on the captured 3D surface topography, surface texture parameters were calculated as follows:

(1) height parameters Sa and Sq, for general understanding of the surface roughness;
(2) statistic parameters Ssk and Sku, to evaluate the dominant feature of the surface, peak dominant or valley dominant, as reference for further surface finishing process;
(3) spatial parameters Str and Std, to characterize the uniformity and analyze the directional patterns of the surface texture, if any.

Section 5.1 will provide the experimental data to verify the measurement performance.

4. Surface Tracing Strategy for Long Scan Length

In order to expand both the lateral and vertical measurement ranges of the line scan sensor, a dual-scan surface tracing algorithm was developed in this study. For rough surfaces, a long scanning length is needed to separate the information of roughness, waviness and form. Based on ISO 4288, for example, if the *Ra* value is greater than 10 μm, the cut-off length is recommended to be 8 mm, and five cut-offs are recommended for averaging computations. The entire scanning length would be 40 mm. For a free-form surface, the profile over such a scan length is very likely to exceed the vertical sensing range of a laser confocal sensor. In this study, therefore, a surface tracing control algorithm was developed to extend the vertical measurement range.

4.1. Surface Tracing Algorithm

For high-resolution surface measurement, the confocal sensing range in the vertical direction is typically less than 1 mm. The Keyence LT9011 laser confocal sensor used in this study has a vertical sensing range of 0.6 mm. When measuring rough surfaces with significant curvature, the measurement is very likely to fail due to insufficient sensing range, as shown in Figure 4a. In order to ensure surface information was always captured within the sensing range, an adaptive surface tracing algorithm was developed.

The methodology is illustrated in Figure 4b. The sensor moved in piece-wise mode. Each unit travel length was 1 mm. Since the oscillating unit inside the senor was able to generate a local scan length of 1.1 mm, the scanning range of the neighboring scans had an overlapping section of 0.1 mm. This overlapping section was used for profile reconstruction based on the best fitting algorithm [25].

In this system, a vertical positioner (LS-110, Physik Instrumente GmbH & Co. KG, Karlsruhe, Germany) was used to position the sensor to ensure that at each lateral scanning location, the data capture was within the sensing range. The positioner LS-110 was able to achieve a maximum speed of 90 mm/s and a positioning resolution of 50 nm in a range of 26 mm. The actual speed needed in this study depends on the lateral movement speed and the height variation of the measured surface. Since the lateral scanning speed generated by the internal positioner was 2 mm/s, this vertical positioner LS-110 had sufficient capability for most use cases.

At each location, the sensor LT9011 collected height information over a scan length of 1.1 mm. With the lateral position information determined by the sampling spacing and sequence number, a second-order polynomial $f(x)$ was obtained based on least-squares fitting. Then, the height to position the sensor for the next scan was determined as

$$h_{j+1} = \frac{1}{2}f'(x_n) + f(x_n). \quad (2)$$

where $f'(x)$ is the first derivate of $f(x)$, l is the lateral positioning increment and n is the number of data points acquired in l. In this study, l and n were set as 1.1 mm and 551, respectively.

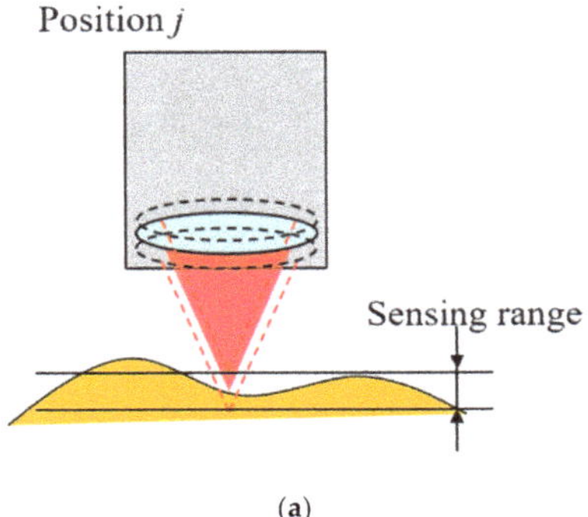

(a)

Figure 4. *Cont.*

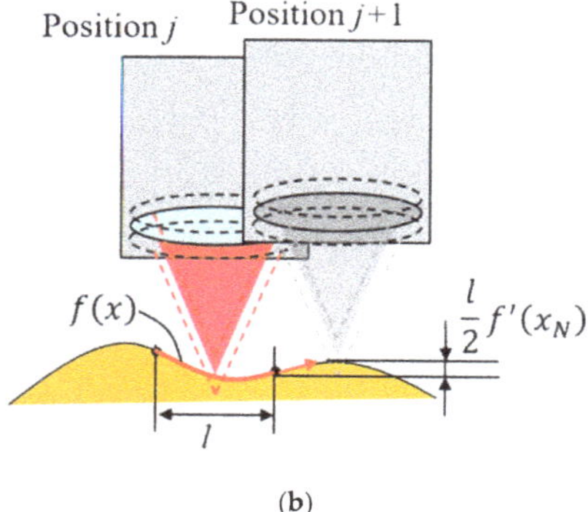

(**b**)

Figure 4. Surface tracing strategy to extend the measurement range: (**a**) measurement out of sensing range; (**b**) surface tracing.

The vertical positioner has an embedded optical encoder to record the position. Therefore, the height $z_{i,j}$ measured by the system included two portions: the height measured by the senor LT9011 and the height read from the optical encoder.

Figure 5a shows the setup for measuring a free-form surface. The sample was carried by a linear stage (TSA100-B, Zolix Instruments Co., Ltd., Beijing, China), the displacement of which was measured by an in-house developed image grating system [26,27]. Using this integrated system, the scanning range can be expanded to 100 mm. In this study, at each location, time taken for positioning and local scan was 1 s.

Figure 5b shows the scanning result. The vertical measurement range can be expanded using the proposed surface tracing algorithm.

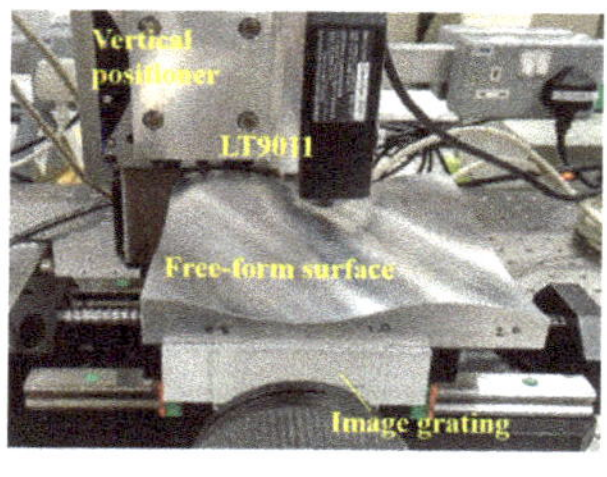

(**a**)

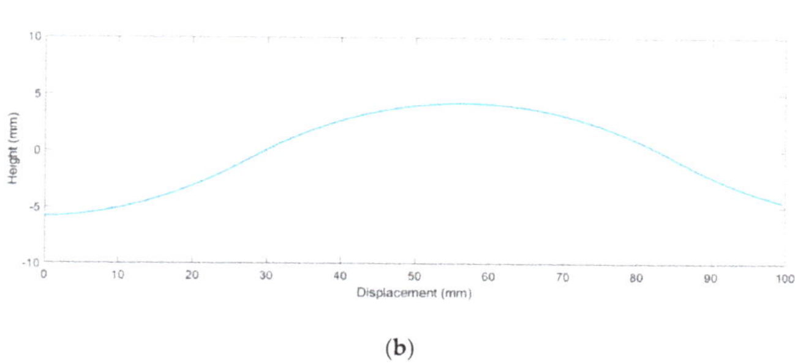

(**b**)

Figure 5. Free-form surface measurement: (**a**) system configuration; (**b**) scan data.

4.2. Correction of Misalignment

The misalignment of the internal and external scan needed to be calibrated. Instead of physically parallelizing the two moving axes, a pre-test on an optical flat (Figure 6a) was conducted to determine the misalignment. With the data collected from all sections plotted into one coordinate system, the measured profile is illustrated in Figure 6b.

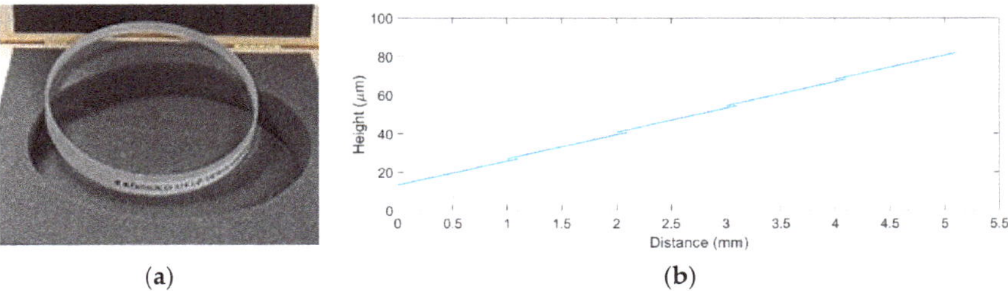

Figure 6. Profile data misalignment illustration: (**a**) optical flat; (**b**) measured surface profile.

Discontinuity between neighboring sections can be corrected by introducing a vertical shift Δd to compensate for the misalignment. The corrected height values can be expressed as

$$\begin{cases} z'_{i,0} = z_{i,0} + 0; \\ z'_{i,1} = z_{i,1} + \Delta d; \\ z'_{i,2} = z_{i,2} + 2\Delta d; \\ \ldots \\ z'_{i,m-1} = z_{i,m-1} + m\Delta d. \end{cases} \tag{3}$$

where m is the number of data sections. These equations can be generalized as

$$z'_{i,j} = z_{i,j} + j{\cdot}\Delta d. \tag{4}$$

where $z_{i,j}$ is the height value captured by the LT9011 at the ith position in the jth section, and $z'_{i,j}$ is the value after misalignment correction. When the section length was set as 1.1 mm with a sampling spacing of 2 μm, and the external positioning interval was set as 1 mm, there was an overlapping of 0.1 mm with 50 sampling points between neighboring sections.

The least-squares principle can be applied to determine the value of Δd. The optimal Δd should minimize the residual sum of squares (RSS) of the deviation between the neighboring overlaps:

$$\begin{aligned} RSS &= \sum_{j=0}^{m-1} \sum_{i=0}^{n} (z'_{i+(n-N),j} - z'_{i,j+1})^2; \\ &= \sum_{j=0}^{m-1} \sum_{i=0}^{n} [z_{i+(n-N),j} + j{\cdot}\Delta d - z_{i,j+1} - (j+1)\Delta d]^2; \\ &= \sum_{j=0}^{m-1} \sum_{i=0}^{n} [z_{i+(n-N),j} - z_{i,j+1} - \Delta d]^2. \end{aligned} \tag{5}$$

where m, n and N represent the number of sections, the number of data points in each section and the number of the data points in each overlap. In order to obtain the minimum value of RSS, Δd should comply with the following condition:

$$\frac{\partial RSS}{\partial \Delta d} = 0. \tag{6}$$

With Equation (5) taken into Equation (6):

$$-2\sum_{j=0}^{m-1}\sum_{i=0}^{N}(z_{i+(n-N),j} - z_{i,j+1} - \Delta d) = 0. \tag{7}$$

$$\Delta d = [\sum_{j=0}^{m-1}\sum_{i=0}^{N}(z_{i+(n-N),j} - z_{i,j+1})]/[N/(m-1)]. \tag{8}$$

4.3. Profile Restoration

Using Equations (2) and (6), linear misalignment can be well compensated. However, misalignment could be eliminated only when the linear guide had perfect straightness and repeatability. In practical tests, further compensation on the residual mismatch was still needed. Figure 7 shows the concept of compensating the residual mismatch, demonstrated by simulation data. The residual mismatches were compensated by adjusting every alternate data section. Assuming k is an odd number, the mismatches between the kth, $(k-1)$th and $(k \pm 1)$th sections are

$$\begin{cases} \Delta d_k = \frac{1}{N}\sum_{i=0}^{N} z'_{i+(n-N),k-1} - z'_{i,k}; \\ \Delta d_{k+1} = \frac{1}{N}\sum_{i=0}^{N} z'_{i,k+1} - z'_{i+(n-N),k}. \end{cases} \tag{9}$$

Then, the further corrected height values are

$$z''_k = z'_k + \Delta d_k + \frac{i}{N}(\Delta d_{k+1} - \Delta d_k). \tag{10}$$

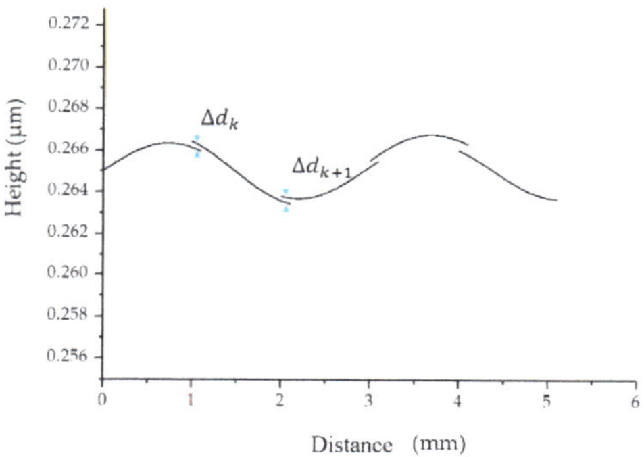

Figure 7. Compensation of residual mismatch.

Considering the redundant data in the overlapping area, in each adjusted section $\{z''_k\}$, the first N data points and the last N data points were removed to connect with the neighboring sections.

4.4. Dual-Scan Positioning Control for Surface Profiling

In the above sections, development of enabling technologies for surface profiling was discussed. As a summary, in this section the algorithm of dual-scan positioning control and surface reconstruction is presented.

In practical measurements, if the surface is relatively flat, the system will scan the surface without vertical positioning control. Misalignment between the internal and external positioning was corrected, as discussed in Section 4.2, and residual mismatch was eliminated as discussed in Section 4.3.

For free-form surface measurements, the system followed the process shown in Figure 8.

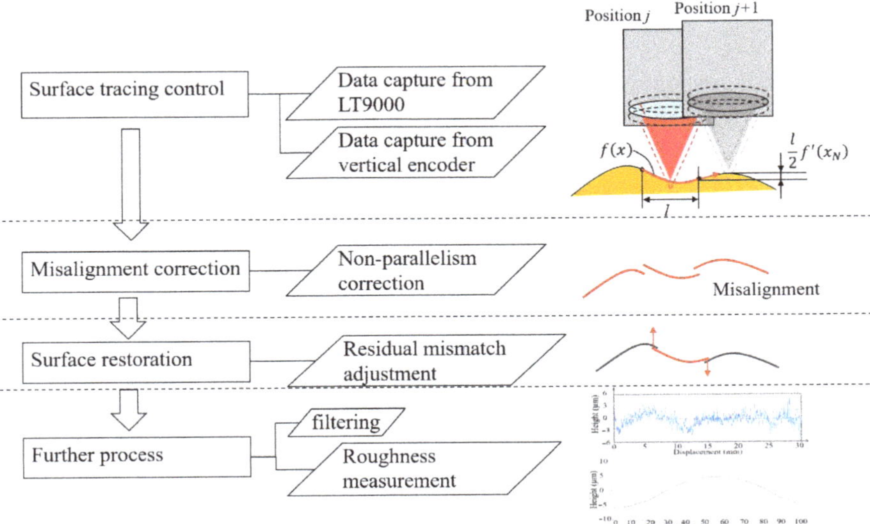

Figure 8. Surface measurement workflow.

With the surface tracing algorithm discussed in Section 4.1, the system was able to capture the surface data by combining the height information captured by LT9011 and the vertical positioner. However, since the data captured in each section were from the internal scan of the LT9011, misalignment between the internal scan and external lateral positioning was not considered.

With the misalignment correction algorithm discussed in Section 4.2, the non-parallelism between the moving axes of internal scan and external lateral positioning can be compensated. After this misalignment correction, the profile may still show mismatches between neighboring sections. This mainly is due to the vertical positioning error and straightness of the lateral movement.

With the profile restoration algorithm discussed in Section 4.3, the mismatches between neighboring sections can be compensated by adjusting data sections with even sequence numbers.

5. Experimental Verification

Experiments were conducted to verify the measurement performance of the proposed dual-scan scanning method. Because temperature-induced deformation of the measurement modules may affect measurement accuracy [28], the measurement tests were performed in an air-conditioned workshop, with temperature controlled at 20 ± 2 °C. The reference systems were also working in a laboratory with the temperature controlled at 20 ± 1 °C. In order to minimize the influence of temperature variation, the data capture processes were completed within 3 min.

An area of 1 mm × 1 mm on an additive manufacturing sample was scanned to verify the areal measurement method as discussed in Section 3. A line scan of 25 mm was conducted to verify the long-travel surface tracing strategy discussed in Section 4.

5.1. Robotic Vibration Test

Since the proposed systems were portable and reconfigurable, the proposed methodology has the potential to be utilized for in situ measurement [29]. Figure 9 shows the robotic application. Vibration tests were conducted on a collaborative robot (UR5, Universal Robots, Odense, Denmark) to address the concern of measurement stability. The measurements were conducted at five different timings and repeated fifty times on the same spot at each timing. The test results (mean ± st. dev) are presented in Table 2. It can be concluded that robotic vibration in standby mode has minimum effects for roughness measurement on the micron to sub-micron level.

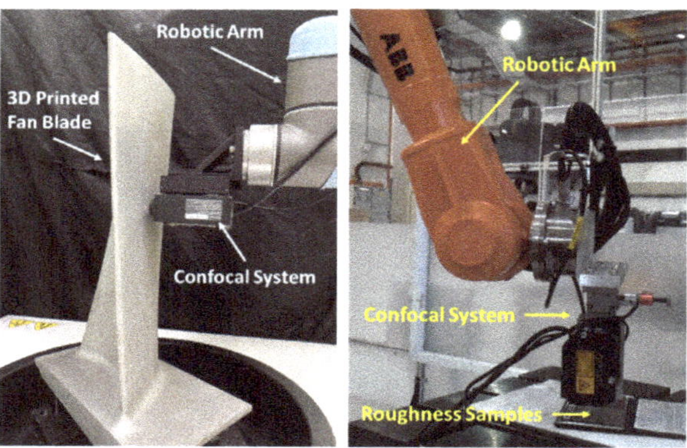

Figure 9. Robotic setup for in situ measurement [25].

Table 2. Measurement results of robotic system vibration.

Timing No.	1	2	3	4	5
Test Result (μm)	63.38 ± 0.02	63.33 ± 0.02	63.30 ± 0.01	63.13 ± 0.02	62.62 ± 0.03

5.2. Area Scan

An external positioner was installed with the moving axis perpendicular to the internal scan of the laser confocal sensor LT9011. The system was introduced in Section 3. As a representative surface with irregular patterns, an additive manufacturing (AM) sample was selected for measurement. A coherence scanning microscope (Talysurf CCI HD, AMETEK Inc., Berwyn, PA, USA) and a spinning-disk confocal microscope (Smartproof 5, Carl Zeiss AG, Oberkochen, Germany) were used as the reference systems for the comparison study. The surface topography of the AM sample measured by different instruments is shown in Figure 10. It can be observed that the measurement results showed consistent surface patterns.

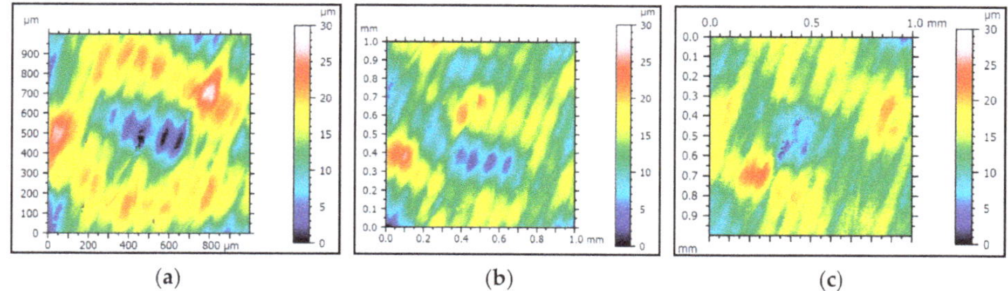

Figure 10. Surface topography measurement of an additive manufacturing (AM) sample: (**a**) by CCI; (**b**) by Smartproof 5; (c) by the proposed system.

For quantitative evaluation, height parameters Sa, Sq and S10z, statistical parameters Ssk and Sku, and surface isotropy parameters Str and Std were employed to evaluate the measurement performance. Comparison results are shown in Table 3, with the standard deviation of five repeats for each measurement.

Table 3. Measurement results of area scans.

	Taylor Hobson CCI	Carl Zeiss Smartproof 5	Proposed System
Sa (μm)	2.50 ± 0.10	2.42 ± 0.13	2.41 ± 0.12
Sq (μm)	3.26 ± 0.16	3.09 ± 0.19	3.04 ± 0.17
S10z (μm)	24.95 ± 0.16	23.02 ± 0.27	22.09 ± 0.17
Ssk	−0.18 ± 0.01	−0.14 ± 0.01	−0.17 ± 0.02
Sku	3.40 ± 0.04	3.58 ± 0.04	3.37 ± 0.08
Str	0.19 ± 0.01	0.21 ± 0.00	0.18 ± 0.02
Std (°)	69.97 ± 0.14	71.04 ± 0.10	70.58 ± 0.13

5.3. Long-Travel Line Scan

A free-form sample was fabricated to verify the long-travel measurement capability of the proposed dual-scan surface tracing system. The system setup is shown in Figure 5a. Areas with different roughness were intentionally made on the sample surface. The surface roughness measurement results are shown in Table 4, with a standard deviation of five repeats for each measurement. A stylus profilometer (Talysurf PGI 800, AMETEK Inc., Berwyn, PA, USA) was used as the reference system.

Table 4. Long-travel roughness measurement results.

		PGI 800	Proposed System	Absolute Error	Relative Error
Zone 1	*Ra* (μm)	0.56 ± 0.01	0.58 ± 0.02	0.02	3.6%
	Rq (μm)	0.69 ± 0.01	0.72 ± 0.03	0.03	4.3%
Zone 2	*Ra* (μm)	0.93 ± 0.02	0.97 ± 0.05	0.04	4.3%
	Rq (μm)	1.14 ± 0.02	1.19 ± 0.07	0.05	4.4%
Zone 3	*Ra* (μm)	1.36 ± 0.01	1.42 ± 0.04	0.06	4.4%
	Rq (μm)	1.70 ± 0.01	1.78 ± 0.06	0.08	4.7%

6. Conclusions and Future Work

In this paper, surface texture measurement methodology based on a dual-scan scanning mechanism has been proposed for surface texture measurement. The motivation for this study was to address the challenges of measuring surfaces with geometric complexity and irregular surface pattern.

By setting the internal and external scanning in perpendicular directions, the system was able to perform area measurements. Areal parameters to evaluate the surface quality were obtained, including height parameters *Sa*, *Sq* and *S10z*, statistic parameters *Ssk* and *Sku*, and surface isotropy parameters *Str* and *Std*.

By setting the internal and external scanning in parallel directions, the system was able to perform line measurements over long travels. A misalignment compensation algorithm was developed to reduce the measurement error due to non-parallelism and straightness errors.

Commercial 3D microscopes (Talysurf CCI and Zeiss Smartproof 5) and a tactile stylus profilometer (Talysurf PGI 800) were employed to evaluate the measurement performance of the proposed methodology. For area measurement in a range of 1 mm by 1 mm, and for line measurement in a scan length up to 100 mm, experimental results showed that the proposed systems were able to achieve comparable accuracy with the commercial systems.

Future work will focus on further analyzing the effects of robotic vibration and environmental temperature variation and on developing advanced noise filtering and error compensation algorithms. In addition, the developed in situ measuring system will be fully integrated into a robotic polishing cell to conduct in situ measurements as a tool of quality verification.

Author Contributions: Conceptualization, F.C. and S.F.; methodology, F.C. and S.F.; software, S.F.; validation, S.F.; formal analysis, S.F.; investigation, S.F.; resources, F.C. and Z.C.; data curation, S.F.; writing—original draft preparation, F.C.; writing—review and editing, F.C. and S.F.; visualization, F.C. and S.F.; supervision, F.C.; project administration, F.C. and S.F.; funding acquisition, Z.C. All authors have read and agreed to the published version of the manuscript.

Funding: This work was funded by the National Natural Science Foundation of China (Grant No. 51775075, 51935004).

Conflicts of Interest: The authors declare no conflict of interest.

References

1. Davim, J.P. *Surface Integrity in Machining*; Springer: London, UK, 2010.
2. Whitehouse, D. *Surfaces and Their Measurement*; Butterworth-Heinemann: London, UK, 2004.
3. Fay, M.F.; Badami, V.G.; De Lega, X.C. Characterizing additive manufacturing parts using coherence scanning interferometry. In Proceedings of the ASPE Spring Topical Meeting on Dimensional Accuracy and Surface Finish in Additive Manufacturing, Berkely, CA, USA, 13–16 April 2014; pp. 135–138.
4. Triantaphyllou, A.; Giusca, C.L.; Macaulay, G.D.; Leach, R.K.; Milne, K.A. Surface texture measurement for additive manufacture. *Surf. Topogr. Metrol. Prop.* **2015**, *3*, 024002. [CrossRef]
5. Whitehouse, D.J. Surface metrology. *Meas. Sci. Technol.* **1997**, *8*, 955–972. [CrossRef]
6. Bradley, C. Automated Surface Roughness Measurement. *Int. J. Adv. Manuf. Technol.* **2000**, *16*, 668–674. [CrossRef]
7. Whitehouse, D. A new look at surface metrology. *Wear* **2009**, *266*, 560–565. [CrossRef]
8. Cheng, F.; Fu, S.W.; Leong, Y.S. Research on optical measurement for additive manufacturing surfaces. In Proceedings of the SPIE International Conference on Optical and Photonics Engineering, Chengdu, China, 26–30 September 2016.
9. British Standards Institution. *BS EN ISO 4287 Profile Method—Terms, Definitions and Surface Texture Parameters*; BSI: London, UK, 1998.
10. Sandoz, P.; Tribillon, G.; Gharbi, T.; Devillers, R. Roughness measurement by confocal microscopy for brightness characterization and surface waviness visibility evaluation. *Wear* **1996**, *201*, 186–192. [CrossRef]
11. Paddock, S.W. *Confocal Microscopy*; Humana Press: Totowa, NJ, USA, 1998.
12. Rishikesan, V.; Samuel, G.L. Evaluation of Surface Profile Parameters of a Machined Surface Using Confocal Displacement Sensor. *Procedia Mater. Sci.* **2014**, *5*, 1385–1391. [CrossRef]
13. Yoshizawa, T. *Handbook of Optical Metrology: Principles and Applications*; CRC Press: Boca Raton, FL, USA, 2009.
14. Hocken, R.J.; Chakraborty, N.; Brown, C. Optical Metrology of Surfaces. *CIRP Ann. Manuf. Technol.* **2005**, *54*, 169–183. [CrossRef]
15. Viotti, M.R.; Albertazzi, A.; Fantin, A.V.; Pont, A.D. Comparison between a white-light interferometer and a tactile formtester for the measurement of long inner cylindrical surfaces. *Opt. Lasers Eng.* **2008**, *46*, 396–403. [CrossRef]
16. Danzl, R.; Helmli, F.; Scherer, S. Focus variation—A robust technology for high resolution optical 3D surface metrology. *J. Mech. Eng.* **2011**, *57*, 245–256. [CrossRef]
17. Newton, L.; Senin, N.; Gomez, C.; Danzl, R.; Helmli, F.; Blunt, L.; Leach, R. Areal topography measurement of metal additive surfaces using focus variation microscopy. *Addit. Manuf.* **2019**, *25*, 365–389. [CrossRef]
18. British Standards Institution. *BS EN ISO 25178-2 Terms, Definitions and Surface Texture Parameters*; BSI: London, UK, 2012.
19. Leach, R. *Characterisation of Areal Surface Texture*; Springer: Berlin/Heidelberg, Germany, 2013; p. 17.
20. Chen, L.C. Full-field chromatic confocal surface profilometry employing digital micromirror device correspondence for minimizing lateral cross talks. *Opt. Eng.* **2012**, *51*, 081507. [CrossRef]
21. Blateyron, F. *Chromatic Confocal Microscopy. Optical Measurement of Surface Topography*; Springer: Berlin/Heidelberg, Germany, 2011; pp. 71–106.
22. British Standards Institution. *BS EN ISO 16610-21 Linear Profile Filters: Gaussian Filters*; BSI: London, UK, 2012.
23. Fu, S.; Cheng, F.; Tjahjowidodo, T. Surface Topography Measurement of Mirror-Finished Surfaces Using Fringe-Patterned Illumination. *Metals* **2020**, *10*, 69. [CrossRef]
24. Fuh, Y.-K.; Fan, J.R.; Huang, C.Y.; Jang, S.C. In-process surface roughness measurement of bulk metallic glass using an adaptive optics system for aberration correction. *Measurement* **2013**, *46*, 4200–4205. [CrossRef]

25. Fu, S.; Cheng, F.; Tjahjowidodo, T.; Zhou, Y.; Butler, D. A Non-Contact Measuring System for In-Situ Surface Characterization Based on Laser Confocal Microscopy. *Sensors* **2018**, *18*, 2657. [CrossRef] [PubMed]
26. Fu, S.; Cheng, F.; Tjahjowidodo, T.; Liu, M. Development of an Image Grating Sensor for Position Measurement. *Sensors* **2019**, *19*, 4986. [CrossRef] [PubMed]
27. Fu, S.; Cheng, F.; Tjahjowidodo, T.; Liu, M. Image grating: A novel technology for position measurement. In Proceedings of the Seventh International Conference on Optical and Photonic Engineering, Phuket, Thailand, 16–20 July 2019.
28. Grochalski, K.; Wieczorowski, M.; Pawlus, P.; H'Roura, J. Thermal Sources of Errors in Surface Texture Imaging. *Materials* **2020**, *13*, 2337. [CrossRef] [PubMed]
29. Fu, S.; Kor, W.S.; Cheng, F.; Seah, L.K. In-situ measurement of surface roughness using chromatic confocal sensor. *Procedia CIRP* **2020**, *94*, 780–784. [CrossRef]

Article

From Light to Displacement: A Design Framework for Optimising Spectral-Domain Low-Coherence Interferometric Sensors for In Situ Measurement

Tom Hovell *, Jon Petzing, Laura Justham and Peter Kinnell

Wolfson School of Mechanical, Electrical and Manufacturing Engineering, Loughborough University, Loughborough, LE11 3TU, UK; J.Petzing@lboro.ac.uk (J.P.); L.Justham@lboro.ac.uk (L.J.); P.Kinnell@lboro.ac.uk (P.K.)

* Correspondence: T.Hovell@Lboro.ac.uk

Received: 23 October 2020; Accepted: 26 November 2020; Published: 30 November 2020

Abstract: Growing requirements for in situ metrology during manufacturing have led to an increased interest in optical coherence tomography (OCT) configurations of low coherence interferometry (LCI) for industrial domains. This paper investigates the optimisation of spectral domain OCT hardware and signal processing for such implementations. A collation of the underlying theory of OCT configured LCI systems from disparate sources linking the journey of the light reflected from the object surface to the definition of the measurand is presented. This is portrayed in an applicable, comprehensible design framework through its application to profilometry measurements for optimising system performance.

Keywords: in-process; metrology for machining; optical coherence tomography

1. Introduction

In situ metrology is a growing requirement for many high value manufacturing processes, ensuring traceability of parts with high tolerances and complex geometric forms, in addition to process control [1,2]. Current in situ forms of metrology commonly involve the use of sensors collecting information on parameters such as temperature, pressure, vibration and torque [3,4] to quantify part quality. Verification of quality becomes increasingly difficult via these methods with adaptive production lines creating made to measure custom parts such as for biomedical implants. Current systems used for geometric measurements on-machine are often tactile touch trigger probes, due to the measurement speed of tactile probes only sparse datasets can be practically obtained in production line scenarios, leading to the ability to only capture simple geometric forms requiring a low datapoint density to accurately describe the structure [2,5,6]. Such methods of metrology do not allow for comprehensive analysis of component manufacture progress in-process nor do they allow for a complete verification of the parts adherence to required tolerances. Such a system would enable quick adaptation to change in component design, ensure traceability of each stage of the manufacturing process and identification of issues at the point of occurrence, then corrective actions can be taken without needing to remove and re-fixture the workpiece.

In order to achieve this goal direct measurement of the component is required through the migration of complex metrology systems from the measurement lab and into the domain of manufacturing, allowing for geometric measurement data to be collected while a manufacturing process is underway, or soon after it is completed. However, due to spatial constraints and impacts from the environment, direct measurement of geometric form is not easily achieved using conventional, currently developed metrology tools.

Low-coherence interferometry (LCI) is a well-established optical method used for obtaining absolute geometric measurements present in many metrology laboratories around the world. However, typical LCI instruments are large and not suited to direct implementation into machining processes, often utilising white light scanning configurations prone to error from external environmental factors and requiring dedicated workstations. Optical coherence tomography (OCT) is a branch of applied LCI developed for operating in non-ideal environments within biomedical science [7]. OCT has been developed for over three decades and demonstrates high data integrity with high spatial resolution even whilst operating within the hostile environment of the human body [8]. OCT configurations have demonstrated straightforward coupling with fibre optics, allowing sensors to be located in situ and at a distance from the readout sensor and electronics, meaning only the chemically inert, extremely compact and robust optical fibre needs to be integrated within the manufacturing process environment. This allows for potential sensor integration into hostile areas in the presence of issues such as high temperatures, electromagnetic interference, corrosive fluids and radioactive activity, whereas current electronic sensors would be prevented from working properly or require extensive shielding. Additionally, the sensors require relatively inexpensive optical components that are readily available making this technology highly accessible. These characteristics make OCT configurations of LCI particularly interesting for in situ measurement where integration using current sensor technology is currently not possible for profilometry.

Although several review papers have been generated from literature on the developments in OCT systems [8–11], these focus on its application in the biomedical domain. As a result, areas of the underlying theoretical principles of parameters affecting the captured signal and methods of its processing for such configurations remain largely dispersed and isolated amongst the literature. An in-depth end-to-end demonstration of the theory of operation, from light source to spatial measurement in non-ideal, realistic scenarios with the additional focus for manufacturing regimes, has not yet been completed. As this technology transitions over to greater use in different engineering sectors and applications, a complete practical methodology for its design and application into such environments is required.

The purpose of this article is to create an optical and data processing design framework for system hardware characteristic selection and signal processing to optimise desired system performance. This is achieved though collation of the underlying theory of spectral domain OCT systems from disparate sources with an application to industrial metrology, linking the journey from the light reflected from the object surface to the definition of the measurand, portraying it in an applicable, comprehensible format through its application to profilometry.

Sensor Applications in Industry

It can be seen from Table 1 that a diverse range of applications have already had initial investigation into the use of LCI sensors for monitoring processes. Recent work demonstrating the use of LCI in the domain of a liquid environment has shown that the system is able to provide stable, accurate data during profilometry measurements across various step-heights [12]. An extension of this work demonstrates an embedded LCI sensor for absolute distance measurement of samples operating within a jet of water has also shown significant data stability [13], demonstrating the potential usefulness of the LCI sensor for post-processing verification of parts, utilising a liquid jet for flushing of debris and coolant during measurement. This growing interest in the use of LCI sensors in industrial applications indicated a requirement of a text focused on optimising systems for such environments, along with expected outcomes and potential issues commonly found.

Although there are several implementations of OCT configurations, due to the sensitivity advantages of Fourier-domain approaches over time-domain ones [14] and the prohibitive cost of operating with laser-swept light sources, this paper focuses on the use of spectrometer or spectral-domain based implementations.

Table 1. Publication breakdown by application domain, table ordered by first appearance of application in literature.

First Seen in Literature	Application Domain	Publications
1984	Absolute ranging for detecting faults in fibre optic cables	[15–18]
2004	Art and Archaeology inspection	[10,19,20]
2005	Measurement during laser machining	[21–25]
2005	Material investigation	[26–28]
2007	Investigation of paper structure	[29,30]
2008	Electronic board and micro-electronics	[31,32]
2008	Profiling measurements	[33–38]
2009	Nuclear research reactor	[39]
2009	Surface roughness measurements	[40,41]
2011	Measurement of tablet coatings offline and online	[11,42]
2013	Measurement of optical components such as GRIN material	[43]
2013	Tracking film growth	[44]
2020	Measurement in liquid jets	[13]

2. Fundamental Principles

Spectral domain LCI investigates the optical path length and magnitude of backreflected light from a sample to determine the samples' distance offset. This involves analysing the achieved interferometric signal frequency modulation which is dependent on the optical path difference between a stable reference signal and the signal returned from the measured sample; this is commonly achieved through conversion of the signal into the Fourier domain via a Fast Fourier Transform (FFT). This behaviour is well outlined in literature especially for measurement of biological systems with OCT [45,46]. This section will clarify the step by step underlying theory of how to achieve a spatial distance measurement from a simple LCI system.

A Michelson interferometer configuration is commonly used for the interferometric portion of the LCI sensor which is outlined in Figure 1a, the mathematical description of the light signal is also shown at each stage of progress. Figure 1b,c shows optical components can also be added at the exit of the fibre, allowing for the control of specific beam parameters (shape, depth of focus, focal length, and intensity distribution). Additionally, scanning methods can be added to perform profilometry as shown in Figure 1b. However, when applying such a sensor to operate within the spatially constrained non-ideal environment present in many manufacturing procedures scanning systems may not allow for the required robustness and spatial tolerances imposed within the measurement area, thus scanning procedures such as shown in Figure 1c may be more appropriate.

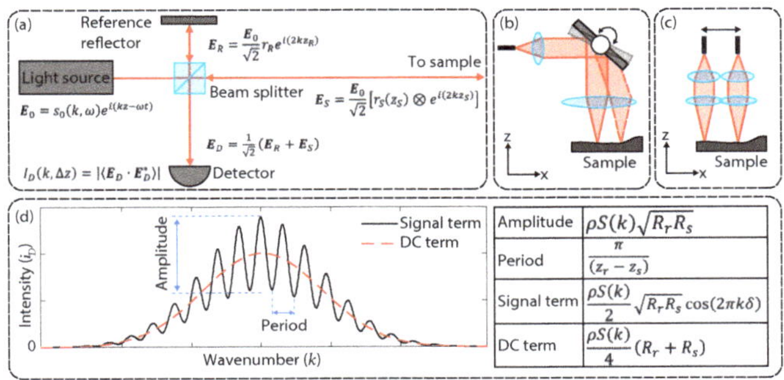

Figure 1. (**a**) schematic of Michelson interferometer for LCI sensor, (**b**) scanning by galvanometer mirror, (**c**) scanning via moving probe or moving sample, (**d**) signal term for a singular sample reflector and corresponding equations relating to signal features. For multiple reflectors, the cross-correlation component is a superposition of sinusoids.

2.1. Mathematical Formulation

The progress of light through a Michelson configured LCI is shown in Figure 2; offshoots at each stage in the process show key parameters which will be affected by the system components. By understanding the impacting factors on signal quality at each stage of this progression, optimisation of system design and signal processing can be undertaken to maximise achieved signal quality.

The theoretical principles of LCI have been detailed in literature for an object with M discrete back-scattering reflectors [45,46]. This section collates that information in an end-to-end theoretical breakdown from light source to spatial measurement.

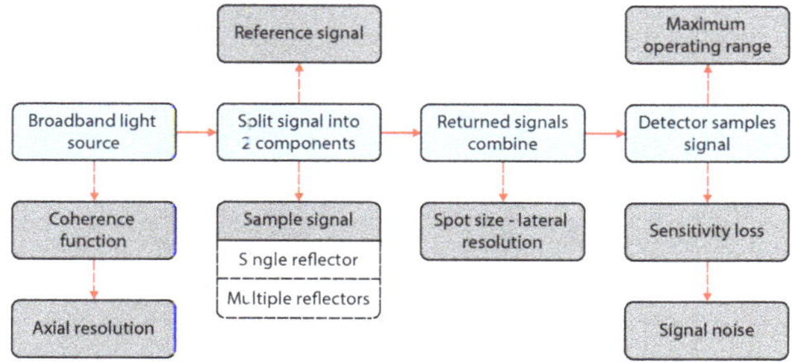

Figure 2. Flowchart of the signal progress throughout a Michelson interferometer lit by a broadband light source. The grey boxes indicate effects resulting from each stage.

The interferometric signal that is generated by the superposition of reference and sample signals is shown in Figure 1d and Equations (2) and (3), respectively; these are composed of the same original signal from the light source represented by Equation (1) but with a phase delay due to the optical path difference between them.

$$\mathbf{E}_0 = s_0\left(k,\omega\right)e^{i(kz-\omega t)}, \tag{1}$$

$$\mathbf{E}_R = \mathbf{A}_R e^{i(2kz_R)}; \mathbf{A}_R = r_R\frac{1}{\sqrt{2}}s_0\left(k,\omega\right), \tag{2}$$

$$\mathbf{E}_S = \mathbf{A}_S * e^{i(2kz_S)}; \mathbf{A}_R = r_S\frac{1}{\sqrt{2}}s_0\left(k,\omega\right), \tag{3}$$

$$\mathbf{E}_D = \frac{1}{\sqrt{2}}\left(\mathbf{E}_S+\mathbf{E}_R\right) = \frac{1}{\sqrt{2}}\left[\mathbf{A}_R e^{i(2kz_R)}+\mathbf{A}_S e^{i(2kz_S)}\right], \tag{4}$$

where s_0 represents the light source spectrum as a function of wavenumber k also known as spatial frequency and angular frequency ω. The phase offset (kz) can be set to zero for the initial signal in Equation (1). The reference signal and sample signal are shown by Equations (2) and (3), respectively, having their corresponding reflectivity represented by r_R and $r_S(z_S)$, where z_S denotes the path length variable in the sample arm from the beam splitter, the factor $1/\sqrt{2}$ representing a 50:50 distribution of the optical power of the source signal from the beam splitter. The interferogram signal is composed of the superposition of the returned reference and signal electric fields as shown by Equation (4), the factor $1/\sqrt{2}$ represents that only half of the returned signal power will be passed towards the detector as shown in Figure 1a; this signal takes the form shown in Figure 1d.

To digitise the interferometric signal, an optical detector is used; this comes in the form of a charge-coupled device (CCD) or complementary metal-oxide-semiconductor (CMOS). This means that the recorded intensity I is proportional to the time averaged electric field multiplied by its complex conjugate. The signal obtained by the spectrometer is shown in Equation (5), where the photocurrent is generated by a square-law detector expressed as a function of spatial frequency. Using the knowledge that the modulus of a complex number has the relationship: $|\tilde{z}| \equiv \sqrt{\tilde{z}\tilde{z}^*}$ the following equation for intensity can be determined:

$$\begin{aligned} I_D\,(k,\Delta z) &= |\langle \mathbf{E}_D \cdot \mathbf{E}_D^* \rangle|, \\ &= \langle \mathbf{E}_S \cdot \mathbf{E}_S^* \rangle + \langle \mathbf{E}_R \cdot \mathbf{E}_R^* \rangle + 2\mathcal{R}\left\{\langle \mathbf{E}_S \cdot \mathbf{E}_R^* \rangle\right\}, \end{aligned} \tag{5}$$

where $\mathcal{R}$ represents the real component and the angled brackets denote a time-average which can be displayed as:

$$I_D\,(k,\Delta z) = \lim_{T\to\infty} \frac{1}{2T} \int_{-T}^{T} \mathbf{E}_D \cdot \mathbf{E}_D^* dt, \tag{6}$$

where I_D is the photocurrent generated by the spectrometer. Substituting Equations (2) and (3) for the reference, E_R and signal, E_S components in Equation (5) will yield Equation (7). Here, the light source intensity spectrum is denoted as $S(k) = |s(k)|^2$. This represents a modulating signal term combined with a carrier signal term "DC term" to produce the modulated signal shown in Figure 1d.

$$\begin{aligned} I_D(k) &= \frac{\rho}{4} S\,(k) \left[R_R + \sum_{n=1}^{N} R_{Sn} \right] \\ &+ \frac{\rho}{2} \left[S\,(k) \sum_{n=1}^{N} \sqrt{R_R R_{Sn}} (cos\,[2kn\,(z_R - z_{Sn})]) \right] \\ &+ \frac{\rho}{4} \left[S\,(k) \sum_{n\neq m=1}^{N} \sqrt{R_{Sn} R_{Sm}} (cos\,[2kn\,(z_{Sn} - z_{Sm})]) \right], \end{aligned} \tag{7}$$

where ρ is the responsivity of the line scan camera, R_R is the reflectance of the reference, R_S is the reflectance of the sample reflector, $2(z_R - z_{Sn})$ is the round trip optical path difference (OPD) between the reference and n^{th}; and m^{th}. sample reflectors. It can be seen from Equation (7) that the intensity of the signal is composed from the sum of three terms:

The 1st term is a constant "DC" signal, independent of changes in the OPD between the reference and sample signals. This term dominates the measured signal with an amplitude proportional to the power reflectivity of the reference and sample, scaled by the light source optical intensity as a function of wavenumber. This dominance can be reduced through changing the beam splitter ratio such that a greater percentage of the signal is directed towards the sample arm of the interferometer. Additionally, as the reference signal optical power is typically far greater than the samples returned signal optical power, this term will limit the signal strength before the spectrometer detector becomes saturated; by reducing the reference signal, the signal strength can be increased without saturation.

The 2nd term represents the reflected signals from the measured sample. The signal magnitude contribution from this term is typically far less than the DC signal contribution, due to the optical intensity being proportional to the square root of the reference and sample reflectivity.

The 3rd term represents the interference between different sample reflectors for a system where the reference and sample paths are separate as in Figure 1a. This term will appear as additional peak artefacts in the Fourier domain of the measured signal; reduction can be achieved through increasing the reference signal power through changing the beam splitter ratio directing light along each path. As shown in Equation (7), the 2nd term signal amplitude is dependent on the reflectivity of the reference and sample whilst the 3rd term is dependent on only the sample signal reflectivity. However, for a common-path configuration, this represents the reflected signals from the measured sample.

Depth information is obtained by performing a Fourier transform on Equation (7) to form Equation (8), allowing for determination of the modulating frequencies corresponding to sample reflector locations:

$$\begin{aligned} i_D(z) &= \frac{\rho}{8}\gamma(z)[R_R + R_{S1} + R_{S2} + ...] + \frac{\rho}{4}\gamma(z) * \sum_{n=1}^{N} \sqrt{R_R R_{Sn}}(\delta[z \pm 2(z_R - z_{Sn})]) \\ &+ \frac{\rho}{8}\gamma(z) * \sum_{n \neq m=1}^{N} \sqrt{R_{Sn} R_{Sm}}(\delta[z \pm 2(z_{Sn} - z_{Sm})]), \end{aligned} \tag{8}$$

where $i_D(z)$ is intensity recorded as a function of OPD and $\gamma(z) = FT\{S(k)\}$ is the coherence function manifesting as a convolution of the Dirac delta function which represents frequency elements of the Fourier transformed signal at set bins. Hence, this term determines the point spread function (PSF) over which the spread of interference fringes occur for a set sample reflector.

Signal Analysis for Singular Reflecting Objects

When taking measurements of opaque, singular reflecting objects such as metallic surfaces, as is often the case in manufacturing scenarios ($N = 1$); this allows for a reduction of Equation (7) to the form shown in Equation (9). This removes the interference noise, which occurs between reflectors at different OPDs for $N > 1$:

$$I_D(k) = \frac{\rho}{4}S(k)[R_R + R_{S1}] + \frac{\rho}{2}S(k)\left[\sqrt{R_R R_{S1}}(cos[2kn(z_R - z_{S1})])\right]. \tag{9}$$

Depth information can be obtained by performing an inverse Fourier transform on Equation (9) to form Equation (10):

$$i_D(z) = \frac{\rho}{8}\gamma(z)[R_R + R_{S1}] + \frac{\rho}{4}\gamma(z) * \left[\sqrt{R_R R_{S1}}(\delta[(z \pm 2n(z_R - z_{S1})])\right]. \tag{10}$$

2.2. Extracting Spatial Depth from Spectral Interferogram

From Equations (8) and (10), the spatial distance measurement which is of interest is a sinusoidal frequency dependent on $cos(2n(z_R - z_S))$; hence, the period of the wave is $\pi/n(z_R - z_S)$.

The Fourier transformed result is composed from a series of discrete steps dependent on the sampling resolution δk of the spectrometer with an array of N_s pixels. The spectral sampling resolution can be determined by $\delta\lambda = \Delta\lambda/N_s$ or $\delta k = \delta\lambda/\lambda_0^2$, hence the sampling interval in the z-domain is given by $\delta\hat{z} = 2\pi/2\delta_s k$, where $\hat{z} = 2z$ to account for the apparent depth-doubling factor shown in the modulation frequency term $cos(2n(z_R - z_S))$. Hence, a peak located at pixel N_i on the spectrometer reading will correspond to a depth shown in Equation (11):

$$z_i = \left(\frac{\pi}{\delta k}\right)\left(\frac{N_i}{N_s}\right). \tag{11}$$

However, systems are not perfectly representative of their theoretical counterparts, errors in various parts of the system such as pixel spacing and diffraction grating ruling errors will bleed through into the measurement data if such a linear fit approach is taken. As such, further calibration of the system can be undertaken to improve the accuracy of this transfer function. Previous work discovered for one setup that, during measurements in air, this would lead to errors of up to approximately 50 nm and whilst operating in water approximately 100 nm [12]. Hence, depending on the measurement accuracy requirements, further action may be required to compensate for nonlinearities in the system.

To calibrate the LCI system, sensor readings were obtained from a mirror sample at a range of offsets from the fibre tip. Simultaneous reference measurements were taken using a reference interferometer (Renishaw XL-80, United Kingdom), allowing accurate determination of δk and Δk across the useable range of operation.

2.3. Signal Noise

The noise in a detected signal is a result of noise introduced from the measurement hardware, and as result of non-ideal measurement conditions. Noise sources resulting from the measurement hardware can be grouped generally as: readout noise, photon noise, relative intensity noise (RIN) and dark noise [47]. These can be further classified as temporal or spatial noise sources. Temporal sources such as photon and dark noise vary with time and thus can be reduced through averaging of frames. Spatial noise arises from non-uniform outputs in dark current non-uniformity and photo response.

Photon noise also known as shot noise is a statistical variation in the arrival rate of photons on the camera detector governed by Poisson statistics and as such the magnitude is linked to the square root of the signal optical intensity as shown in Equation (12). Dark noise results from a charge accumulation in each pixel due to thermal fluctuations in the silicon which can release electrons. Readout noise arises from converting the pixel charge carriers into a voltage signal and processing from analog-to-digital conversion with the largest contributor being the amplifier electronics. RIN describes the instability in the power level of the light source and is typically independent of laser power.

Spectrometer CCD's commonly have cooling which results in contributions to noise from dark current being insignificant. Noise contributions from photon variance reduces as photon flux increases due to the square root relationship with signal. When the measured signal has a low optical intensity, the read noise will be greater than the photon noise and the signal is termed read-noise limited. Increasing the optical power input into the spectrometer or increasing the integration time will allow the pixels to collect more photons, which in turn increases the SNR, there will then come a crossover point where photon noise exceeds read and dark noise, and the signal is termed photon or shot-noise limited:

$$\sigma_{photon} = \sqrt{\frac{\rho\eta\tau}{hv_0}\frac{P_0}{N}(\gamma_S R_S + \gamma_R R_R)}, \tag{12}$$

where η is the quantum efficiency of the detector, τ is the exposure time , h is the Planck constant, v_0 is the centre frequency of the light source spectrum, P_0 is the optical power, and γ_S and γ_R denote the quantity of input power that will exit the interferometer from each arm, assuming $R_S = R_R = 1$.

2.4. Measurement Range

The useable measurement range of an LCI system is dependent on: the sampling rate of the signal due to Nyquist's theory, the sensitivity fall-off and the returned signal strength from measured regions such that the returned signal can be distinguished from the systems noise. This section looks at factors impacting on these areas.

2.4.1. Maximum Range—Nyquist Theory

Since the analogue signal is being digitised through its sampling via the optical detector pixels the maximum sample distance from the fibre that the sensor can resolve is limited by the sampling rate of the signal by the detector. The signal is digitised through the sampling of its wavelength components intensities across an optical detectors pixel array and thus is victim to Nyquist theory of requiring a sampling rate of at least twice that of the measured signal frequency. This limitation due to hardware can be calculated through the use of Equation (13):

$$z_{max} = \left(\frac{\lambda_0^2}{4}\right)\left(\frac{N_s}{\Delta\lambda}\right). \quad (13)$$

2.4.2. Sensitivity Fall-Off

When using spectrometer based LCI techniques, there is an apparent fall-off in signal strength following a Gaussian rule with increasing OPD between the reference and sample signal leg, even if the reflected signal amplitude is constant. This is caused by the following two factors: finite pixel width and finite spectrometer resolution.

Finite Pixel Width

The diffraction grating within the spectrometer spectrally separates the interferometric signal in terms of wavelength, directing spectral components of the signal onto discrete pixel detectors across the pixel array. The optical intensity of each of these spectrally dispersed spots is sampled across each pixel. The optical spot intensity is integrated over the pixel area which in effect imposes a rectangular function as a convolution of the Gaussian beam spot. This signal decay factor as a function of the finite pixel width (R_{PW}) manifests as a non-normalized Sinc function as shown in Equation (14) along with the Rect function [47]:

$$R_{PW} = sinc^2\left(\frac{\pi z}{2 \cdot z_{max}}\right), Rect\left(\frac{k}{\delta k}\right) = \begin{cases} 0 & \text{for } |k| > \frac{\delta k}{2} \\ \frac{1}{2} & \text{for } |k| = \frac{\delta k}{2}. \\ 0 & \text{for } |k| < \frac{\delta k}{2} \end{cases} \quad (14)$$

The result of this is a 4 dB drop in the signal-to-noise ratio (SNR) near the Nyquist limit [47]. However, the finite pixel width does not introduce a decay of the noise level due to the statistical independence of the shot noise between neighbouring pixels of the array [46].

Finite Spectrometer Resolution

The finite spectrometer resolution determined by the Gaussian beam profile of the spot size incident on the optical detector pixel array also introduces a sensitivity decay (R_{SR}) as shown in Equation (15):

$$R_{SR} = exp\left[-\frac{\omega^2}{2ln2}\left(\frac{\pi z}{2 \cdot z_{max}}\right)^2\right], \quad (15)$$

where $\omega = \frac{\delta k'}{\Delta k}$ and $\delta k'$ is the spectrometer's spectral resolution dependent on the size of the beam spot in the focal plane of the detector pixel. Thus, the total signal decay in the system as a function of depth comes in the form of a convolution of the finite pixel size with the Gaussian spectral resolution. After applying a Fourier transform, this leads to a multiplication of both of the decay functions which gives the expression shown in Equation (16) for sensitivity reduction R, as a function of imaging depth z [48]:

$$R(z) = sinc^2\left(\frac{\pi z}{2 \cdot z_{max}}\right) exp\left[-\frac{\omega^2}{2ln2}\left(\frac{\pi z}{2 \cdot z_{max}}\right)^2\right]. \quad (16)$$

The depth-dependent sensitivity of the system presented in this work is shown in Figure 3a, the SNR sharply declines with depth and measurements near the Nyquist limit become difficult to reliably analyse as shown in Figure 3b.

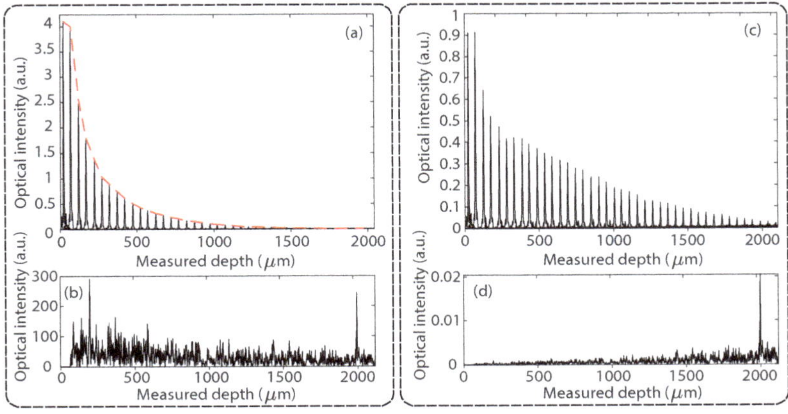

Figure 3. Depth-dependent loss in signal sensitivity from sample reflector. (**a**) overlaid signals for sample measurements up to 2.17 mm, (**b**) an example signal at a depth of 2 mm, (**c**) dataset (**a**) after normalisation, (**d**) the same signal in (**b**) after the normalisation relationship.

A normalisation procedure can be applied in the Fourier domain to compensate for fall-off by multiplying the frequency components of the signal by the inverse of the signal drop-off envelope, indicated by the red dashed line in Figure 3a. This improvement can be seen in Figure 3c across the range of operation, Figure 3d shows the same signal as in Figure 3b with the normalisation curve applied to it. However, care should be taken to optimise this normalisation curve as it will also amplify high frequency noise and dampen low frequency signal components.

Signal Strength

An additional factor limiting a systems ability to measure over large distances is signal strength resulting from the amount of light backreflected or backscattered into the interferometer. The surface which is being measured will also have a large impact on the signal magnitude and quality along with the orientation of the surface in relation to the axis of measurement. Environmental impacts may also degrade the signal, from attenuation due to operating medium, or the effects of vibration and particulates in solution causing scattering of the light source.

Some improvement can be achieved using optics, focussing the light on a constrained small region to improve re-coupling of back-reflected light into the system from the region of interest. However, for in situ measurements or for integration into many manufacturing regions, tight geometric constraints along with the influence of non-ideal environmental factors may limit the ability to add such optical elements into the system.

3. LCI System Design Framework

The prospect of designing an LCI system from first principles can be somewhat daunting. For practical implementation, the system will be built with a design specification of requirements, thus a robust methodology and knowledge of the various parameters which affect the outcome of critical operating qualifiers such as system resolution, operating range and measurement speed is required. Furthermore, many parameters are inter-linked making the design process somewhat iterative and, in many cases, trade-offs or compromises will need to be made to achieve a system that can perform in the desired manner under the required circumstances of operation. Figure 4 shows a system design framework based around metrology design operating criteria depicted by the blue highlighted elements with the offshoots in grey representing parameters impacting them. Some of the parameters are broken down into smaller elements with the positive and negative symbols representing if the parameter should be increased or decreased to maximise each design criteria based on our exemplar; other designs may vary based on prioritisation of certain requirements.

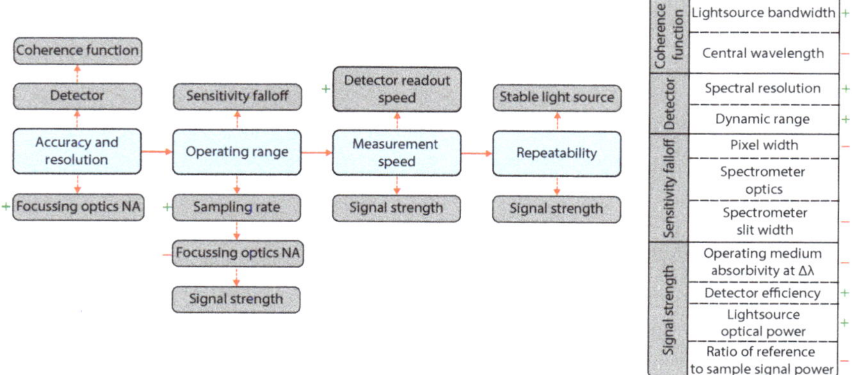

Figure 4. LCI sensor design framework, optimising hardware choices for key operating criteria.

The operating domain should also be taken into account during the design phase, such as operating medium, environmental conditions and sample surface characteristics and their impact on the mentioned design parameters outlined in Figure 4. The following sections consider each step of the flowchart in Figure 4 with more in depth discussion on the various parameters and the cross-linking that occurs.

3.1. Axial Resolution

The axial resolution in an LCI system is determined by the smallest change in OPD that the system can detect, determined by the spectral resolution of the spectrometer. Previous work utilising the LCI system described in Section 3.2.1 has demonstrated submicrometric accuracy across step heights ranging from 5 μm to 8 μm with a precision of ±56 nm [12]. However, if transparent samples are measured with multiple reflective layers, the smallest discernible distance between layers will be limited by the coherence length of the light source and the axial resolution will be as shown in Equation (17). The larger the bandwidth of the light that is sent into the interferometer, the narrower the coherence function/PSF upon performing an FFT. Increasing the bandwidth of the light source for a detector of set pixel number will result in a reduction in the maximum achievable measurement range as shown in Equation (13); thus, a balance must be found:

$$\delta_z = \frac{2ln2}{\pi} \times \frac{\lambda_0^2}{\Delta\lambda}. \tag{17}$$

When performing tomographic measurements, the resolution of the instrument is limited by the coherence function due to potential reflectors interfering with one another. It has been shown for measurements of singular reflecting surfaces the accuracy of measurement is not limited by the peak PSF with step-height measurements with submicrometric accuracy [12].

3.2. Lateral Resolution

The lateral resolution of the system is controlled by the optics of the sensing arm of the interferometer. These fixed optics will allow for an axial field of view (FOV_{axial}) as shown in Equation (18) in which the lateral resolution (δx) will be approximately equivalent to the focused spot size shown from Equation (19). Outside of this field of view, the lateral resolution will deteriorate:

$$FOV_{axial} = \frac{0.221\lambda}{sin^2\left[\frac{sin^{-1}(NA)}{2}\right]}. \tag{18}$$

$$\delta x = 0.37\frac{\lambda_0}{NA}. \tag{19}$$

Hence, from Equations (18) and (19), it can be seen that, as lateral resolution is increased due to a large NA focusing optic, the axial field of view over which the spot size is constant decreases.

The refractive index of the operating medium may also impact the lateral resolution [34]. Additionally, for in situ measurements during manufacturing processes, tight spatial constraints may be in place along with environmental factors such as vibration and non-ideal operating regions containing dust and liquid contaminants which may affect optical components. In such a situation, the use of un-lensed optical fibres is a desirable solution due to the compact size and robustness nature. However, as highlighted above, the lateral resolution severely degrades as a result of the light diffraction upon exiting the fibre.

3.2.1. Operating outside the Region of Focus

To demonstrate degradation in lateral resolution outside of a focused region, profile measurements were taken at two offsets of 235 µm and 1020 µm from a British five pence coin using the setup described in Figure 5. The system was configured as a fully fibre-enclosed implementation of an LCI with a common path for both the reference and sample signal leg, taking advantage of reduced sensitivity to vibrations, thermal fluctuations, and humidity, and removing the requirement for dispersion compensation between signals [49]. The system was un-lensed with the bare cleaved fibre having a numerical aperture of 0.13. This provides a micrometre-level footprint and form factor of the sensor. The system in this configuration has a confocal parameter of 83.5 µm and a $1/e^2$ transverse resolution of 5 µm, hence both measured regions will be outside of the focused beam region.

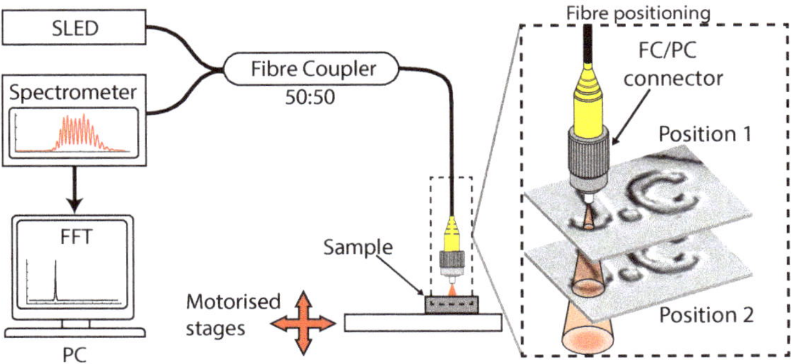

Figure 5. Experimental setup of a common path low coherence interferometer.

The system components consist of a broadband light source provided by a super luminescent diode (EXS210068-01, Beratron, 850 nm) with a 3-dB bandwidth of 58 nm and an emitting power of 5.14 mW at 160 mA. A single-mode fibre coupler with a splitting ratio of 50:50 for beam splitting and coupling was used, with only one branch of signal output being used as the common path for the signal and reference. A spectrometer (MayaPro2000, Ocean Optics) operating at 125 Hz with a 2048×64 pixel array, starting wavelength of 756 nm and spectral range of 174 nm with a resolution of 0.21 nm was used, giving a theoretical axial operating range of 2.15 mm before aliasing as defined from Equation (13).

A region of the coin was focused on which contains the artisan's initials "J.C" to highlight the variation in lateral resolution with the distance of the coin from the fibre. Unilateral profile measurements were taken with the coin being translated at a constant speed of 100 μm/s on the x-axis over 1100 μm with a measurement integration time of 8 ms and a single spectrum captured per datapoint. This translated to a resolution of approximately 5 μm per datapoint. Steps of 5 μm were taken on the y-axis over a range of 800 μm.

Figure 6 shows the results from the two measurements, the red circles indicate a predicted spot size calculated from the NA of the fibre for an offset of 235 μm and 1020 μm. A clear reduction in the lateral resolution is present between the two offsets which manifests in a blurring of the measured feature. This degradation is a result of the increased spot on the surface of the sample reflecting light back from a larger area of focus. This results in an integration over the illuminated area which can shift the focus to other regions as the primary returned signal. From the measurement data, an increase of 1784% in the theoretical spot size area relating to an increase of 88% in the measured area covered by the "J.C" was observed; it is worth noting that an uncertainty metric for this measurement has not been completed. However, the values reflect the reduction in lateral resolution seen in Figure 6, where the achievable resolution is not equal to the spot size; this is due to the optical intensity distribution across the spot and that, with large incident angles to the sample surface, it is less likely for the light component to be recoupled into the system upon back reflection. Although the comparison may appear dramatic for a change in offset of approximately 900 μm, the lateral scale of measurement should be taken into account which is approximately 700 μm; thus, for many applications, this introduced blur will not impact measurement data integrity. However, for some small features such as shown in Figure 6, or surface roughness, this may be an unacceptable compromise and thus measurement offset must be taken into account or additional steps such as focussing optics to improve results must be used.

Several methods of improving the lateral resolution of an OCT profilometry measurement have been shown in literature [50–52]. A promising technique is the use of synthetic-aperture radar (SAR) [53] which has been used for improving the lateral resolution of radar systems for decades; due to LCI operating characteristics being analogous to radar, the same principles can be applied whilst performing line scan measurements to drastically improve the measurement resolution. This may circumvent the issue of lateral resolution whilst operating with an un-lensed system or outside of a systems lateral field of view. Application of SAR to LCI has been demonstrated with some success in literature with a technique named 'Interferometric Synthetic Aperture Microscopy' (ISAM) [51].

3.3. Motion Artefacts and Fringe Washout

For in situ profilometry, data measurement on the fly is necessary. However, the effect of sample motion can negatively impact on data integrity due to the signal being integrated over a period of time. This results in a reduction in SNR and image degradation [54]. The magnitude of motion artefacts and SNR reduction is controlled by the total axial or transverse displacement during a singular signal acquisition time. Hence, for a given sample motion, if the capture rate is increased or axial and transverse displacements are decreased, then the impact of motion artefacts will reduce.

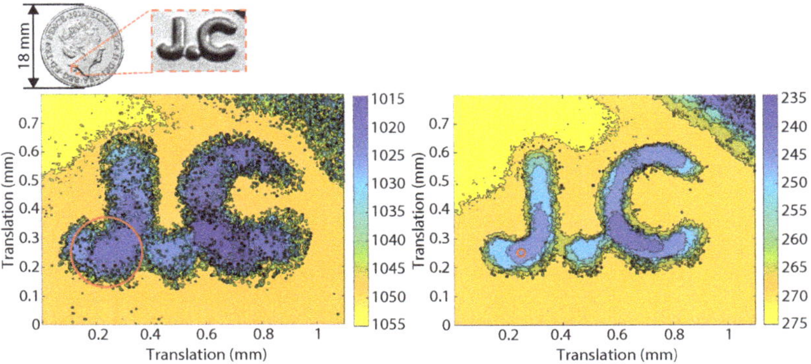

Figure 6. Effect of offset distance on artefact measurement.

3.3.1. Axial Motion Effects

Motion in the axial plane may occur due to vibrations of the measured sample or sensor leading to SNR reduction. This manifests as fringe washout due to the continuous phase changes of fringes over the integration period of the measurement. Equation (20) represents the detected number of electrons per wavenumber, $N(k)$:

$$N(k) = N_0 sinc(k\Delta z), \tag{20}$$

where N_0 is the number of electrons obtained when the sample and sensor are stationary and $\Delta z = |v_z T|$ is the axial displacement of the sample during the integration time (T) with axial velocity (v_z). Performing a Fourier transform on Equation (20) also allows for the approximation of the factor $sinc(k\Delta z)$ to be seen as $sinc(k_0\Delta z)$ for the case $|\Delta z| << |z - z_b|$, where z_b is the zero path length location. Thus, the axial motion of the sample or sensor gives rise to an SNR penalty of $sinc^2(k_0\Delta z)$. For $k\Delta z >> 1$, broadening of the axial resolution occurs, due to the Sinc function in Equation (20) limiting the effecting spectral bandwidth. This can be seen as a convolution of the original image with a Rect function given by the Fourier transform of the Sinc function [54].

3.3.2. Lateral Motion Effects

Transverse motion will degrade the lateral resolution leading to a blurring of the image in the direction of travel [55] and a reduction in SNR [54]. For a continuous wave (CW) source, the SNR degradation $(SNR_{decrease})$ due to lateral motion is given by Equation (21) [54]:

$$SNR_{decrease} \cong -5log_{10}\left(1 + 0.5\frac{\Delta x^2}{w_0^2}\right), \tag{21}$$

where x is the scanning distance during the camera integration time and w_0 denotes FWHM of the beam profile. The normalised displacement is defined as $\Delta x^2/w_0^2$. For pulsed illumination, Δx is replaced by (T_{pulse}/T_{camera}). Δx, where T_{pulse} is the pulse width in time and T_{camera} is the integration time of the camera for a single data point capture. To combat the lateral motion artefacts, the spectra capture speed can be increased, the motion speed decreased, or a pulsed or stroboscopic instead of CW light source can be used. [54,56,57]. Additionally, pulsed illumination can also offer SNR improvements for measurement on the fly over CW sources [58].

The irradiance distribution profile for a Gaussian beam can be found using Equation (22) whilst stationary. However, once there is a transverse movement between the probe and surface, then the irradiance distribution profile for a fixed exposure period (T) can be calculated by Equation (23) [54]:

$$g(x,y,z) \approx \frac{4ln2}{\pi w_0^2} exp\left[\frac{-4ln2(x^2+y^2)}{w_0^2}\right], \tag{22}$$

$$G(x,y) = \frac{1}{T}\int_{-T/2}^{T/2} g(x+v_x t, y+vyt)dt, \tag{23}$$

where x and y are locations along the beam profile, w_0 is the FWHM of the beam profile and v_x and v_y represent constant velocities along the transverse x and y-axis respectively. The impact on transverse resolution in the x-axis for instance can be found by solving for $G(x, y = 0)$ and then locating the FWHM of the calculated irradiance profile.

3.4. Dispersion Compensation

Many manufacturing conditions do not offer an ideal environment to operate within, with machining processes including various liquids for lubrication and for flushing of debris. OCT has already shown that measurement errors due to chromatic dispersion in optically dense materials occur, resulting from the frequency dependence of propagation constants for different mediums [59]. The result of unbalanced dispersion within an LCI system is a reduction in signal peak intensity and a broadening of its Point Spread Function (PSF) shown in literature to be dependent on refractive index, central wavelength, bandwidth and propagation distance [59–61]. LCI is very tolerant of chromatic dispersion if the amount of dispersion occurring in the reference and sample arms are equal. This can be achieved through insertion of artefacts of varying thickness and refractive index into the reference arm of the spectrometer [60] or through using a common-path setup. Software based compensation is also achievable offering a more flexible method of compensation [59,62,63].

The propagation constant for a material can be described by a Taylor series expansion shown by Equation (24) with the expansion made around the light sources central frequency k_0 [63]:

$$\theta(k) = \theta(k_0) + \frac{\partial\theta(k)}{\partial k}(k_0 - k) + \frac{1}{2!}\cdot\frac{\partial^2\theta(k)}{\partial k^2}(k_0-k)^2 + ... + \frac{1}{n!}\cdot\frac{\partial^n\theta(k)}{\partial k^n}(k_0-k)^n. \tag{24}$$

The terms in Equation (24) relate to the zeroth, first and second orders of dispersion which respectively represent: constant offset, group velocity, both of which are not related to dispersive broadening and group-delay dispersion per unit length, though higher order terms may be present.

In practice, the frequency dependent dispersion variation can be compensated for by multiplying the obtained interferometric signal by a phase correction term such that $I_{corrected}(k) = I(k) \times e^{(-i\theta(k))}$ and can be obtained during post-processing [46]. This is achieved through isolation of the coherence function of a strong reflector, followed by performing an inverse Fourier transform on the masked signal and then taking the arctangent of the imaginary component divided by the real component. By performing a linear fit and then identifying the difference through subtraction, an error term can be achieved indicating how much wavenumbers are out of phase with one another. This can then be used to form a phase adjustment value $e^{(-i\theta(k))}$. Using the knowledge that, for a properly dispersion compensated system the signal peak intensity will be at its maximum [60], an iterative procedure can be carried out until the maximum peak value is achieved.

3.5. Design Exemplar

This section contains a walk-through for component specification dependent on an exemplar set requirement list as outlined in the design framework shown in Figure 4:

- Axial resolution: 5 µm
- Operating range: 2.5 mm
- Measurement speed: 200 mm/s
- Lateral resolution: 20 µm

Axial resolution: The operating mediums absorption spectrum, degree of scattering present in the measurement and the required axial resolution will determine the potential range of λ_0. It is also worth noting that some bandwidth regions have been more developed than others in terms of product availability and pricing. If a central wavelength of 1 µm is chosen, then the corresponding bandwidth for a set axial resolution can be calculated using Equation (17)—for this example, $\Delta\lambda \approx 88$ nm.

Operating range: For a required operating range and light source bandwidth, the minimum number of pixels sampling the wavelength range can be found using Equation (13) and solving for N_s (i.e., $N_s = z_{max}\frac{4\Delta\lambda}{\lambda_0^2}$). For the given example, this would come to N_s = 880 pixels.

Measurement speed: The measurement speed is limited by the linescan rate of the camera. However, signal strength will impact the possibility of operating at such speeds which requires a knowledge of sample reflectivity and potential scattering or absorption of the operating medium; the amount of signal fall-off will also impact this. As shown by Equation (23), when relative transverse motion between the probe and sample occurs for a set exposure time, there will also be a reduction in the transverse resolution.

Lateral resolution: To achieve a lateral resolution of 20 µm at a translational speed of 200 mm/s with a linescan rate of 100 µs, the maximum static FWHM of the beam profile should be ~6.8 µm from Equation (23), if the linescan rate is halved to 50 µs, then the static FWHM should be ~18.7 µm. The static FWHM of the beam profile is dependent on the focusing optics used and will only be applicable over a certain axial field of view as described by Equations (18) and (19), respectively. The use of objective lenses can be employed to extend the field of view [64].

4. Signal Processing Algorithms

Signal processing is increasingly playing a key role in the analysis of optical measurements due to the enhanced power of microprocessors. There are many approaches to improving signal quality of LCI systems presented in literature; however, these again remain dispersed across the literature. This section will provide a concise overview of the key analysis procedures along with highlighting practical application signal issues which the user will encounter during real-life measurement scenarios. A signal processing framework for the steps required to go from raw detected signal to spatial distance value is illustrated in Figure 7. Here, the solid bounding boxed elements represent essential parts of post processing and the dashed bounding boxed elements represent procedures that will enhance the signal quality.

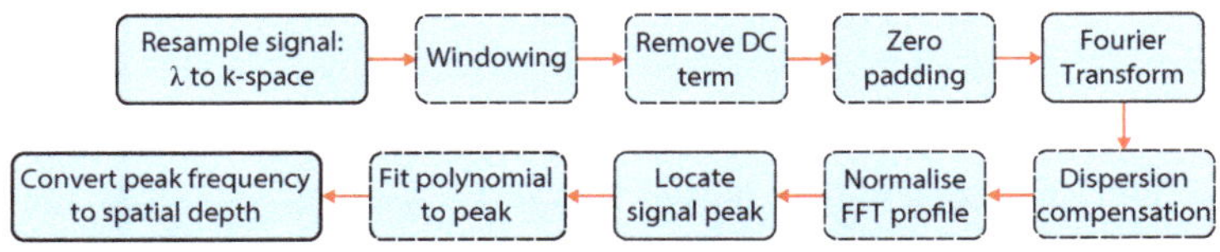

Figure 7. Framework for LCI signal data processing and distance offset retrieval, outlining required procedures in solid bounding boxes along with processing steps which can improve the signal quality in dashed bounding boxes.

4.1. Resampling Captured Interferogram

The majority of spectrometers utilises diffractive components to uniformly sample the signal by wavelength. However, FFT algorithms are most efficient when the data points are evenly spaced within the frequency domain [65]. Due to the relationship between wavenumber and wavelength, (λ), ($k = 2\pi/\lambda$), the data will be non-uniformly spaced by wavenumber, requiring a re-sampling process before a conventional FFT can be effectively applied. If the dataset is left as linear in λ before the FFT is applied, then severe deterioration of axial resolution and SNR will be present in the signal, with an OPD dependent broadening of the PSF function leading to a reduced operational limit due to signal fall-off. This phenomena can be seen in Figure 8a.

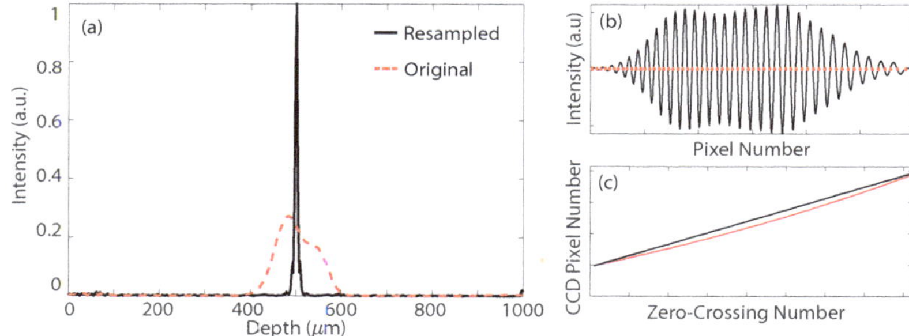

Figure 8. (**a**) a comparison between FFT of a resampled spectrum and FFT of the original spectrum; (**b**) a sample signal spectra with the DC component removed, the red dots show the zero crossing (ZC) points on the curve; (**c**) the red line shows the nonlinearly spaced ZC locations for a raw spectrum, and the black line shows the ZC points spacing once a resampling procedure has been undertaken.

It is possible to avoid this requirement for re-sampling of the data by using a Non-uniform Direct Fourier Transform (NuDFT) instead of an FFT algorithm. However, the NuDFT is more computationally intensive than other alternatives for re-sampling. Since computational time is particularly important for high-speed imaging, the FFT is preferred. Re-sampling approaches can be categorised into two methods. The first method uses a special hardware configuration which samples the returned signal directly to be linear in k [66,67]. However, this also adds additional cost, design complexity and is not easily translated between systems. The second method utilises software to take the signal acquired nonlinearly in k and re-sample it during post processing through the use of a λ to k-mapping function [68]. However, software-based methods do not perform as well for reflectors close to the maximum achievable depth (Nyquist limit) due to high frequency fringes in the interferogram that are poorly sampled [65].

The approach taken in this work is similar to that presented for automatic calibration [68], relying on the knowledge that, for a singular reflector and whilst operating in a common-path configuration negating the requirement for dispersion compensation, the interference fringes will be evenly spaced in k-space. By using zero-crossing detection shown in Figure 8b, fringe spacing as a function of pixel number is obtained from which a mapping relationship shown by the red line in Figure 8c between linear in λ and linear in k-space can be created; this allows interpolation of the data using a B-spline interpolation to create an evenly sampled dataset in k-space before performing an FFT.

To increase the accuracy of the zero-crossing detection, a zero-filling procedure can be followed which addresses the issue of errors at greater depths because of the increased shoulders around the peaks [14,65]. This involves taking the FFT, Fourier shifting the result and then zero padding the start and end of the array to increase the data-array length by a factor M, followed by an inverse FFT back to the spectral data. These spectral data, which was M times larger than the original array, have a greater datapoint density hence a more accurate resampling procedure is possible. In practice, this gives an increase in signal magnitude and a reduced PSF of the signal peak.

4.2. Windowing Signal

Applying a windowing function to the spectral interferogram can improve the sensitivity of FFT spectral-analysis techniques and reduce spectral leakage and side-lobes amplitude, resulting in a greater SNR. There are various types of windowing functions available, and the appropriateness of a window function depends on the frequency content of the measured signal. For example, if measuring a multi-layered reflecting sample results in signal components close to one another, then a window function with a narrow main lobe should be used to prevent spectral resolution degradation. Various windows should be compared against one another to find the best fit for the measurement situation. Table 2 shows four commonly employed windowing functions applied to an LCI signal demonstrating an increase in PSF FWHM compared with the original unwindowed signal but also a substantial increase in SNR.

Table 2. Comparison of windowing functions applied to LCI signal, showing change in PSF and SNR from unwindowed data.

Window Function	Hann	Hamming	Taylor	Blackman
Δ PSF FWHM (μm)	0.81	0.7	0.45	1.23
Δ SNR (dB)	39.09	33.92	22.13	38.54

4.3. Fixed Pattern Noise Removal

Fixed pattern noise refers to items in the captured signal that are separate to the interference modulation pattern which contains the depth information as a response to OPD between the reference reflected signal and the signal reflected from the measured sample [46]. For example, variations in response of pixels in the CCD camera or spurious etalons in the interferometer due to back-reflectance within the interferometer or stray rays and the DC signal term. There have been various software methods to remove these unwanted signal additions suggested in literature with the most common techniques being high-pass filtering, subtracting the measured reference signal [69] and subtracting the mid-point between the upper and lower signal envelopes [12].

4.4. Zero-Padding

A zero-padding methodology can be applied to the captured spectrum to increase datapoints in the Fourier domain achieved by joining an array of zero values to the end of the spectrum array. This does not add any frequency components to the signal but when an FFT is performed the number of frequency bins that will be compared with the signal will be increased, thus improving the datapoint density of the Fourier transform signal. This can be seen in Figure 9a,b showing the FFT peak of a resampled spectrum without and with zero-padding before Fourier-transforming, respectively. However, as array size is increased naturally so will the computational time. This increased datapoint density can improve polynomial fitting to the FFT peak as shown in Figure 9c.

Figure 9d shows the resolution of the interferometer when processing the raw data by detecting the location based on the maximum intensity detected peak, this value is calculated as being 2.17 μm according to the number of bins assigned by the 2048 CCD pixel array in the spectrometer. This resolution can be improved through increasing the number of bins as shown through zero-padding in Figure 9b, or by fitting a polynomial to the datapoints surrounding the maximum peak location and extracting the vertex. In practice, a mixture of zero-padding and polynomial fitting can improve the system's ability to resolve depth values.

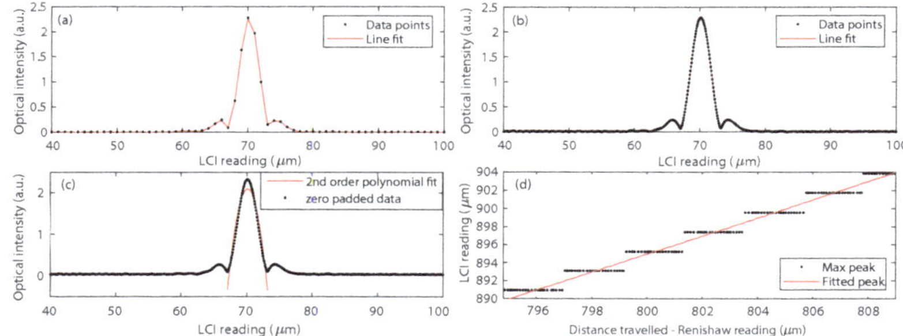

Figure 9. (**a**) FFT of original dataset, (**b**) FFT of zero-padded dataset, (**c**) improving resolution through fitting a 2nd order polynomial to the FFT peak, (**d**) comparing maximum peak detection versus a polynomial peak fitting against OPD.

4.5. Locate Signal Peak(s)

The signal peak(s) which have the largest magnitude will correspond to regions from the sample that the light signal has been backreflected or backscattered from. Thus, these frequencies correspond to the depth values at areas of interest on the sample. A straightforward method of finding the peak(s) within a signal is through use of the MATLAB function "findpeaks". The resolution of the measurement can be increased by fitting a 2nd order polynomial to datapoints surrounding the peak location relating to the measured sample offset distance as shown in Figure 9b. The obtained peak frequency can then be converted to a spatial depth as described in Section 2.2.

5. Conclusions

This work has identified a growing interest around the application of developments in OCT configurations of LCI to the industrial domain for in situ measurement. However, due to the wealth of knowledge accrued over the past 30 years of progress in the field, an in-depth structured collation of theory with an industrial application mindset was required.

As a result, this work for the first time presents an end-to-end holistic framework for the development of OCT configured LCI systems, linking the journey of light reflected from the object surface to the definition of the m measurand. This includes the breakdown of achieved signal contributions, practical operating limits from measurement range due to sampling rate and sensitivity fall-off, lateral resolution variation across the region of measurement, motion effects on SNR/introduction of spurious artefacts and dispersion mismatch compensation.

In addition to identification of factors impacting the measurement signal, a simple practical approach to component selection resulting from critical operating qualifiers is also given and finally a comprehensive step by step dataflow approach to digital signal processing techniques for optimising the quality and robustness of achieved system measurements.

Some general limitations of OCT configured spectral domain LCI measurements which still require improvement include: achieving a high lateral resolution over a large work volume, difficulty in measuring surfaces with large gradients, limited penetration depth through scattering mediums, relatively small operating range, and signal strength dependent on scatterers reflectively, impacted by motion blur and suffering from sensitivity fall-off.

This work will equip the reader with the necessary underlying theoretical and practical information to apply the developments that have been made in OCT/ LCI to industrial applications for in process metrology.

Author Contributions: Conceptualization, T.H., L.J., J.P., P.K.; methodology, T.H.; software, T.H.; validation, T.H.; formal analysis, T.H.; investigation, T.H.; resources, J.P., P.K.; data curation, T.H.; writing—original draft preparation, T.H.; writing—review and editing, T.H., L.J., J.P., P.K.; supervision, L.J., J.P., P.K.; funding acquisition, P.K., J.P. All authors have read and agreed to the published version of the manuscript.

Funding: The authors wish to acknowledge support of the UK Engineering Physical Science and Research Council (EPSRC) for funding this work (Grant Nos. EP/M020746/1, EP/L01498X/1).

Conflicts of Interest: The authors declare no conflict of interest.

References

1. Schwenke, H.; Knapp, W.; Haitjema, H.; Weckenmann, A.; Schmitt, R.; Delbressine, F. Geometric error measurement and compensation of machines—An update. *CIRP Ann.* **2008**, *57*, 660–675, [CrossRef]
2. Uhlmann, E.; Mullany, B.; Biermann, D.; Rajurkar, K.P.; Hausotte, T.; Brinksmeier, E. Process chains for high-precision components with micro-scale features. *CIRP Ann.* **2016**, *65*, 549–572, [CrossRef]
3. Altintas, Y.; Aslan, D. Integration of virtual and online machining process control and monitoring. *CIRP Ann.* **2017**, *66*, 349–352, [CrossRef]
4. Li, Z.; Wang, G.; He, G. Surface quality monitoring based on time-frequency features of acoustic emission signals in end milling Inconel-718. *Int. J. Adv. Manuf. Technol.* **2018**, *96*, 2725–2733, [CrossRef]
5. Fang, F.Z.; Zhang, X.D.; Weckenmann, A.; Zhang, G.X.; Evans, C. Manufacturing and measurement of freeform optics. *CIRP Ann.* **2013**, *62*, 823–846, [CrossRef]
6. Gao, W.; Haitjema, H.; Fang, F.Z.; Leach, R.K.; Cheung, C.F.; Savio, E.; Linares, J.M. On-machine and in-process surface metrology for precision manufacturing. *CIRP Ann.* **2019**, *68*, 843–866, [CrossRef]
7. Huang, D.; Swanson, E.A.; Lin, P.; Schuman, J.S.; Stinson, W.G.; Chang, W.; Hee, M.R.; Flotte, T.; Gregory, K.; Puliafito, C.A.; et al. Optical coherence tomography. *Science* **1991**, *254*, 1178–1181. [CrossRef]
8. Kirtane, T.S.; Wagh, M.S. Endoscopic optical coherence tomography (OCT): Advances in gastrointestinal imaging. *Gastroenterol. Res. Pract.* **2014**, *2014*, 1–7. [CrossRef]
9. Schmitt, J.M. Optical Coherence Tomography (OCT): A Review. *IEEE J. Sel. Top. Quantum Electron.* **1999**, *5*, 1205–1215. [CrossRef]
10. Targowski, P.; Iwanicka, M. Optical Coherence Tomography: Its role in the non-invasive structural examination and conservation of cultural heritage objects—A review. *Appl. Phys. A Mater. Sci. Process.* **2012**, *106*, 265–277, [CrossRef]
11. Lin, H.; Zhang, Z.; Markl, D.; Zeitler, J.A.; Shen, Y. A Review of the Applications of OCT for Analysing Pharmaceutical Film Coatings. *Appl. Sci.* **2018**, *8*, 2700, [CrossRef]
12. Hovell, T.; Matharu, R.S.; Petzing, J.N.; Justham, L.; Kinnell, P. Lensless fiber-deployed low-coherence interferometer for in situ measurements in nonideal environments. *Opt. Eng.* **2020**, *59*, 1–11, [CrossRef]
13. Hovell, T.; Petzing, J.; Justham, L.; Kinnell, P. in situ measurement of electrochemical jet machining using low coherence interferometry. In Proceedings of the Euspen's 20th International Conference & Exhibition, Geneva, Switzerland, 8–12 June 2020.
14. Choma, M.A.; Sarunic, M.V.; Yang, C.H.; Izatt, J.A. Sensitivity advantage of swept source and Fourier domain optical coherence tomography. *Opt. Express* **2003**, *11*, 2183–2189. [CrossRef]
15. Bosselmann, T.; Ulrich, R. High-Accuracy Position-Sensing with Fiber-Coupled White-Light Interferometers. In Proceedings of the 2nd International Conference on Optical Fiber Sensors, Stuttgart, Germany, 5–7 September 1984; [CrossRef]
16. Takada, K.; Yokohama, I.; Chida, K.; Noda, J. New measurement system for fault location in optical waveguide devices based on an interferometric technique. *Appl. Opt.* **1987**, *26*, 1603–1606, [CrossRef]
17. Youngquist, R.C.; Carr, S.; Davies, D.E.N. Optical coherence-domain reflectometry: A new optical evaluation technique. *Opt. Lett.* **1987**, *12*, 158-160, [CrossRef]
18. Gilgen, H.H.; Novak, R.P.; Salathe, R.P.; Hodel, W.; Beaud, P. Submillimeter Optical Reflectometry. *J. Light. Technol.* **1989**, *7*, 1225–1233, [CrossRef]
19. Liang, H.; Cucu, R.; Dobre, G.M.; Jackson, D.A.; Pedro, J. Application of OCT to Examination of Easel Paintings. In Proceedings of the Second European Workshop on Optical Fibre Sensors, Santander, Spain, 9–11 June 2004.

20. Targowski, P.; Gora, M.; Wojtkowski, M. Optical Coherence Tomography for Artwork Diagnostics. *Laser Chem.* **2006**, *2006*, [CrossRef]
21. Kononenko, V.V.; Konov, V.I.; Pimenov, S.M.; Volkow, P.V.; Goryunov, V.A.; Ivanov, V.V.; Novikov, M.A.; Markelov, V.A.; Tertysnik, A.D.; Ustavshikov, S.S. Control of laser machining of poly crystalline diamond plates by the method of low-coherence optical interferometry. *Quantum Electron.* **2005**, *35*, 622–626. [CrossRef]
22. Wiesner, M.; Ihlemann, J.; Muller, H.H.; Lankeanu, E.; Huttmann, G. Optical coherence tomography for process control of laser machining. *Rev. Sci. Instrum.* **2010**, *81*, 033705. [CrossRef]
23. Webster, P.J.L.; Wright, L.G.; Mortimer, K.D.; Leung, B.Y.; Yu, J.X.Z.; Fraser, J.M. Automatic real-time guidance of laser machining with inline coherent imaging. *J. Laser Appl.* **2011**, *23*, 022001. [CrossRef]
24. Purtonen, T.; Kalliosaari, A.; Salminen, A. Monitoring and adaptive control of laser processes. *Phys. Procedia* **2014**, *56*, 1218–1231. [CrossRef]
25. Kononenko, V.V.; Bushuev, E.V.; Zavedeev, E.V.; Volkov, P.V.; Luk'yanov, A.Y.; Konov, V.I. Low-coherence interferometry as a tool for monitoring laser micro- and nanoprocessing of diamond surfaces. *Quantum Electron.* **2017**, *47*, 1012–1016, [CrossRef]
26. Wiesauer, K.; Pircher, M.; Goetzinger, E.; Bauer, S.; Engelke, R.; Ahrens, G.; Gruetzner, G.; Hitzenberger, C.K.; Stifter, D. En-face scanning optical coherence tomography with ultra-high resolution for material investigation. *Opt. Express* **2005**, *13*, 1015–1024. [CrossRef] [PubMed]
27. Liu, P.; Groves, R.M.; Benedictus, R. Signal processing in optical coherence tomography for aerospace material characterization. *Opt. Eng.* **2013**, *53*, 033201. [CrossRef]
28. Liu, P.; Groves, R.M.; Benedictus, R. 3D monitoring of delamination growth in a wind turbine blade composite using optical coherence tomography. *NDT E Int.* **2014**, *64*, 52–58. [CrossRef]
29. Kirillin, M.Y.; Alarousu, E.; Fabritius, T.; Myllylä, R.; Priezzhev, A.V. Visualization of paper structure by optical coherence tomography: Monte Carlo simulations and experimental study. *J. Eur. Opt. Soc.* **2007**, *2*, 07031. [CrossRef]
30. Tuukka, P.; Jakub, C.; Erkki, A.; Risto, M. Optical coherence tomography as an accurate inspection and quality evaluation technique in paper industry. *Opt. Rev.* **2010**, *17*, 218–222, [CrossRef]
31. You, J.; Kim, S. Optical inspection of complex patterns of microelectronics products. *CIRP Ann. Manuf. Technol.* **2008**, *57*, 505–508. [CrossRef]
32. Czajkowski, J.; Prykäri, T.; Alarousu, E.; Palosaari, J.; Myllylä, R. Optical Coherence Tomography as a Method of Quality Inspection for Printed Electronics Products. *Opt. Rev.* **2010**, *17*, 257–262. [CrossRef]
33. Majumdar, A.; Huang, H. Development of an in-fiber white-light interferometric distance sensor for absolute measurement of arbitrary small distances. *Appl. Opt.* **2008**, *47*, 2821–2828. [CrossRef]
34. Donato, A.D.; Pietrangelo, T.; Anzellotti, A.; Monti, T.; Morini, A.; Farina, M. Infrared imaging in liquid through an extrinsic optical microcavity. *Opt. Lett.* **2013**, *38*, 5094–5097. [CrossRef] [PubMed]
35. Tang, D.; Gao, F.; Jiang, X. On-line surface inspection using cylindrical lens-based spectral domain low-coherence interferometry. *Appl. Opt.* **2014**, *53*, 5510–5516, [CrossRef] [PubMed]
36. Park, H.M.; Jung, H.W.; Joo, K.N. Dual low coherence scanning interferometry for rapid large step height and thickness measurements. *Opt. Express* **2016**, *24*, 28625–28632, [CrossRef] [PubMed]
37. Park, H.M.; Joo, K.N. High-speed combined NIR low-coherence interferometry for wafer metrology. *Appl. Opt.* **2017**, *56*, 8592–8597, [CrossRef]
38. Krauter, J.; Osten, W. Nondestructive surface profiling of hidden MEMS using an infrared low-coherence interferometric microscope. *Surf. Topogr. Metrol. Prop.* **2018**, *6*, 015005, [CrossRef]
39. Cheymol, G.; Brichard, B.; Villard, J.F. Fiber optics for metrology in nuclear research reactors-Applications to dimensional measurements. *IEEE Trans. Nucl. Sci.* **2011**, *58*, 1895–1902. [CrossRef]
40. Amaral, M.M.; Raele, M.P.; Caly, J.P.; Samad, R.E.; Nilson, D.; Freitas, A.Z. Roughness measurement methodology according to DIN 4768 using optical coherence tomography (OCT). In Proceedings of the Modeling Aspects in Optical Metrology II, Munich, Germany, 15–16 June 2009. [CrossRef]
41. Quinsat, Y.; Tournier, C. In situ non-contact measurements of surface roughness. *Precis. Eng.* **2012**, *36*, 97–103. [CrossRef]
42. Koller, D.M.; Hannesschläger, G.; Leitner, M.; Khinast, J.G. Non-destructive analysis of tablet coatings with optical coherence tomography. *Eur. J. Pharm. Sci.* **2011**, *44*, 142–148. [CrossRef]

43. Meemon, P.; Yao, J.; sung Lee, K.; Thompson, K.P.; Ponting, M.; Baer, E.; Rolland, J.P. Optical Coherence Tomography Enabling Non Destructive Metrology of Layered Polymeric GRIN Material. *Sci. Rep.* **2013**, *3*, 1709. [CrossRef]
44. Karim, F.; Bora, T.; Chaudhari, M.B.; Habib, K.; Mohammed, W.S.; Dutta, J. Optical fiber-based sensor for in situ monitoring of cadmium sulfide thin-film growth. *Opt. Lett.* **2013**, *38*, 5385–5388. [CrossRef]
45. Brezinski, M.E. *Optical Coherence Tomography: Principles and Applications*; Elsevier: Amsterdam, The Netherlands, 2006.
46. Drexler, W.; Fujimoto, J.G. *Optical Coherence Tomography Technology and Applications*; Springer: Berlin/Heidelberg, Germany, 2014; Volume 1.
47. Leitgeb, R.; Hitzenberger, C.K.; Fercher, A.F. Performance of fourier domain vs. time domain optical coherence tomography. *Opt. Express* **2003**, *11*, 889–894. [CrossRef] [PubMed]
48. Yun, S.H.; Tearney, G.J.; Bouma, B.E.; Park, B.H.; de Boer, J.F. High-speed spectral-domain optical coherence tomography at 1.3 μm wavelength. *Opt. Express* **2003**, *11*, 3598–3604. [CrossRef] [PubMed]
49. Tan, K.M.; Mazilu, M.; Chow, T.H.; Lee, W.M.; Taguchi, K.; Ng, B.K.; Sibbett, W.; Herrington, C.S.; Brown, C.T.A.; Dholakia, K. In-fiber common-path optical coherence tomography using a conical-tip fiber. *Opt. Express* **2009**, *17*, 2375–2384, [CrossRef] [PubMed]
50. Xu, Y.; Chng, X.K.B.; Adie, S.G.; Boppart, S.A.; Carney, P.S. Multifocal interferometric synthetic aperture microscopy. *Opt. Express* **2014**, *22*, 16606–16618, [CrossRef]
51. Ralston, T.S.; Marks, D.L.; Carney, P.S.; Boppart, S.A. Real-time interferometric synthetic aperture microscopy. *Nat. Phys.* **2007**, *3*, 129–134. [CrossRef]
52. Mason, J.H.; Davies, M.E.; Bagnaninchi, P.O. Blur resolved OCT: full-range interferometric synthetic aperture microscopy through dispersion encoding. *Opt. Express* **2020**, *28*, 3879–3894, [CrossRef]
53. Wiley, C.A. Pulsed Doppler Radar Methods and Apparatus. U.S. Patent 3196436, 13 August 1954.
54. Yun, S.H.; Tearney, G.J.; de Boer, J.F.; Bouma, B.E. Motion artifacts in optical coherence tomography with frequency-domain ranging. *Opt. Express* **2004**, *12*, 2977–2998. [CrossRef]
55. Markl, D.; Hannesschläger, G.; Buchsbaum, A.; Sacher, S.; Khinast, J.G.; Leitner, M. In-line quality control of moving objects by means of spectral-domain OCT. *Opt. Lasers Eng.* **2014**, *59*, 1–10, [CrossRef]
56. You, J.W.; Chen, T.C.; Mujat, M.; Park, B.H.; de Boer, J.F. Pulsed illumination spectral-domain optical coherence tomography for human retinal imaging. *Opt. Express* **2006**, *14*, 6739–6748, [CrossRef]
57. Moneron, G.; Boccara, A.C.; Dubois, A. Stroboscopic ultrahigh-resolution full-field optical coherence tomography. *Opt. Lett.* **2005**, *30*, 1351–1353. [CrossRef]
58. Drexler, W.; Morgner, U.; Kärtner, F.X.; Pitris, C.; Boppart, S.A.; Li, X.D.; Ippen, E.P.; Fujimoto, J.G. In vivo ultrahigh-resolution optical coherence tomography. *Opt. Lett.* **1999**, *24*, 1221–1223. [CrossRef] [PubMed]
59. Wojtkowski, M.; Srinivasan, V.J.; Ko, T.H.; Fujimoto, J.G.; Kowalczyk, A.; Duker, J.S. Ultrahigh-resolution, high-speed, Fourier domain optical coherence tomography and methods for dispersion compensation. *Opt. Express* **2004**, *12*, 2404–2422. [CrossRef]
60. Hitzenberger, C.K.; Baumgartner, A.; Drexler, W.; Fercher, A.F. Dispersion effects in partial coherence interferometry: implications for intraocular ranging. *J. Biomed. Opt.* **1999**, *4*, 144–152. [CrossRef] [PubMed]
61. Hillman, T.R.; Sampson, D.D. The effect of water dispersion and absorption on axial resolution in ultrahigh-resolution optical coherence tomography. *Opt. Express* **2005**, *13*, 1860–1874, [CrossRef]
62. Marks, D.L.; Oldenburg, A.L.; Reynolds, J.J.; Boppart, S.A. Autofocus algorithm for dispersion correction in optical coherence tomography. *Appl. Opt.* **2003**, *42*, 3038–3046. [CrossRef] [PubMed]
63. Cense, B.; Nassif, N.; Chen, T.C.; Pierce, M.C.; Yun, S.H.; Park, B.H.; Bouma, B.E.; Tearney, G.J.; de Boer, J.F. Ultrahigh-resolution high-speed retinal imaging using spectral-domain optical coherence tomography. *Opt. Express* **2004**, *12*, 2435–2447. [CrossRef] [PubMed]
64. Song, S.; Xu, J.; Wang, R.K. Long-range and wide field of view optical coherence tomography for in vivo 3D imaging of large volume object based on akinetic programmable swept source. *Biomed. Opt. Express* **2016**, *7*, 4734–4748, [CrossRef] [PubMed]
65. Dorrer, C.; Belabas, N.; Likforman, J.P.; Joffre, M. Spectral resolution and sampling issues in Fourier-transform spectral interferometry. *J. Opt. Soc. Am. B* **2000**, *17*, 1795–1802, [CrossRef]
66. Hu, Z.; Rollins, A.M. Fourier domain optical coherence tomography with a linear-in-wavenumber spectrometer. *Opt. Lett.* **2007**, *32*, 3525–3527, [CrossRef]

67. Payne, A.; Podoleanu, A.G. Direct electronic linearization for camera-based spectral domain optical coherence tomography. *Opt. Lett.* **2012**, *37*, 2424–2426, [CrossRef]
68. Liu, X.; Balicki, M.; Taylor, R.H.; Kang, J.U. Towards automatic calibration of Fourier-Domain OCT for robot-assisted vitreoretinal surgery. *Opt. Express* **2010**, *18*, 24331–24343, [CrossRef] [PubMed]
69. Nassif, N.; Cense, B.; Park, B.H.; Pierce, M.C.; Yun, S.H.; Bouma, B.E.; Tearney, G.J.; Chen, T.C.; de Boer, J.F. In vivo high-resolution video-rate spectral-domain optical coherence tomography of the human retina and optic nerve. *Opt. Express* **2004**, *12*, 367–376. [CrossRef] [PubMed]

Article

Detecting and Measuring Defects in Wafer Die Using GAN and YOLOv3

Ssu-Han Chen [1,2,*], Chih-Hsiang Kang [1,2] and Der-Baau Perng [3,4]

1 Department of Industrial Engineering and Management, Ming Chi University of Technology, New Taipei City 24301, Taiwan; C.H.Kang@o365.mcut.edu.tw
2 Center for Artificial Intelligence & Data Science, Ming Chi University of Technology, New Taipei City 24301, Taiwan
3 Department of M-Commerce and Multimedia Applications, Asia University, Taichung 41354, Taiwan; perng@asia.edu.tw
4 Department of Business Management, Asia University, Taichung 41354, Taiwan
* Correspondence: ssuhanchen@mail.mcut.edu.tw

Received: 16 October 2020; Accepted: 3 December 2020; Published: 5 December 2020

Featured Application: Die defect detection and measurement.

Abstract: This research used deep learning methods to develop a set of algorithms to detect die particle defects. Generative adversarial network (GAN) generated natural and realistic images, which improved the ability of you only look once version 3 (YOLOv3) to detect die defects. Then defects were measured based on the bounding boxes predicted by YOLOv3, which potentially provided the criteria for die quality sorting. The pseudo defective images generated by GAN from the real defective images were used as the training image set. The results obtained after training with the combination of the real and pseudo defective images were 7.33% higher in testing average precision (AP) and more accurate by one decimal place in testing coordinate error than after training with the real images alone. The GAN can enhance the diversity of defects, which improves the versatility of YOLOv3 somewhat. In summary, the method of combining GAN and YOLOv3 employed in this study creates a feature-free algorithm that does not require a massive collection of defective samples and does not require additional annotation of pseudo defects. The proposed method is feasible and advantageous for cases that deal with various kinds of die patterns.

Keywords: wafer die; defect detection; generative adversarial network (GAN); you only look once version 3 (YOLOv3)

---

1. Introduction

Wafer is the major material for making integrated circuits (ICs), and it plays an indispensable role in electronic products. The upstream of the semiconductor industry are IC design companies and silicon wafer manufacturing companies. IC design companies design circuit diagrams according to customer needs, while silicon wafer manufacturing companies use polysilicon as the raw material for silicon wafers. The primary task of IC manufacturing companies in the midstream is to transfer the circuit diagrams to wafers. The completed wafer is then sent to the downstream IC packaging and testing companies for packaging and testing the functions of ICs, concluding the whole manufacturing process.

With the continuous evolution of wafer manufacturing technology, wafer sizes have become larger and the patterns on the die have become more diverse. In order to inspect surface defects in the dies of a wafer, automated optical inspection (AOI), mainly using one or more optical imagery charge-coupled devices (CCDs), has gradually replaced traditional manual visual inspection (VI).

The current die AOI inspection methods of die inspection can be divided into design-rule checking [1,2], neural network [3–5], golden template method [6,7], and golden template self-generating method [8,9]. Design-rule checking is also known as the "knowledge database method". Mital and Teoh [1] and Tobin et al. [2] found the geometric features and textural features of the components on the dies and stored them in the knowledge library, and then compared the features to determine whether they are defective or not. Neural network methods learn the mapping relationship between die features and defect classes throughout learning their synaptic weights. Su et al. [3] cut the die images into several windows of fixed size, and extracted the corresponding average gray value as features, and then established the back-propagation neural network (BPNN), the radial basis function neural network (RBFNN), and the learning vector quantization neural network (LVQNN) models. Chang et al. [4] used the kmeans to distinguish the P-electrode, the light-emitting area, and the background in the LED die. Then they extracted the geometric features and the textural features from the P-electrode and the light-emitting area, and distinguished the defects using the LVQNN. Timm and Barth [5] used radially encoded features to measure the discontinuity around the P-electrode of the LED die. Since there were relatively few defects, an anomaly detection method such as the one-class support vector machine (SVM) could obtain extremely high classification accuracy. However, the geometric features and appearance of various die images are quite different. It is difficult to find effective handcrafted features and establish criteria. Golden template method is also called the "image-to-image reference method". This method compares the golden template with the image to be inspected, and the areas with significant differences are regarded as the ones where the defects may occur. Chou et al. [6] used the golden template method to highlight defects and measure the size, shape, location, and color of the defects, and then distinguished the types of defects based on design-rule checking. Zhang et al. [7] cut the boundary between the pads and the dies before establishing a golden template to highlight defects, and then determined the defect types based on features such as location, number of objects, and area of objects. This type of golden template method usually uses pixel-wise difference to identify the difference. Before performing the pixel-wise difference, the image to be inspected must be carefully aligned with the golden template. To alleviate alignment problem, golden template self-generating methods were introduced. Guan et al. [8] used the characteristics of repetitive patterns of the dies in the wafer to self-generate a defect-free template to highlight the defects. Liu et al. [9] used a discrete wavelet transform (DWT) to extract the standard image from three defective IC chip images, and then used the difference between the standard image and the defect image to highlight the defects. Liu et al. [9] also used the image gray-scale matching method to reduce the impact of different brightness on the detection results. This method avoided the alignment problem, but was limited by the need to zoom out to capture the die's repetitive pattern when shooting the image. Since the image resolution was reduced, small defects became difficult to detect. However, the defect inspection algorithms used in AOI systems often need to be highly customized, highly accurate alignments, and rely heavily on experts to design hand-crafted features. As a result, existing algorithms described above cannot be applied to different product types [1–9]. A cruel truth is that die AOI inspection machines are often idle in practice when the production line is changed.

To overcome the above problem, deep learning methods have been introduced into die defect classification in recent years. Cheon et al. [10] proposed a four-layer convolutional neural network (CNN) architecture based on stacked convolution and max-pooling to classify five types of die defects. They also used the k nearest neighbor (KNN) in the 3D autoencoder output space to detect unknown defects that were very different from the existing types. Lin et al. [11] designed a six-layer CNN to classify LED chip defects, and used class activation mapping (CAM) to create a heat map corresponding to the analyzed image to locate the defective area. Chen et al. [12] constructed a CNN network based on the separable convolutional and bottleneck for six times to classify four types of die defects. From the existing literature, it finds that deep learning methods are shown not to require any feature extraction process, to be able to avoid shift, rotation, exposure, and so on, which is very powerful and has attracted global attention. However, most studies focused on how to use deep learning methods to solve the problem of die defect classification, while relatively few focused on how to use those to solve the problem of die defect detection. The latter is the focus of the present study and makes a contribution to introduce an object detection method, you only look once version 3 (YOLOv3), to solve the die defect detection problem. The YOLOv3 model is able to predict the center coordinate, width, and height of each bounding box where the defect is located, and the confidence that each bounding box contains a defect. There is no need to rely on experts for feature engineering and have a certain degree of invariance to interference such as translation and rotation, which are attractive characteristics for companies that face constant changes die patterns. In addition, since the particle defects embedded on the dies are very tiny and some of the defects are dense, YOLOv3 uses DarkNet53 as the backbone and introduces multiscale detection, which is able to detect defects of different sizes on the extracted feature maps. In this way it can effectively detect tiny and dense defects, ensuring the quality of the die.

Moreover, there is another issue regarding defective samples collection and annotation in the factory environment. Operators do not have much time to collect various appearances of different kinds of defects. Since the collection of defect images is time-consuming, recent research on generating pseudo defective images with GAN has attracted attention. Chen et al. [12] uses affine transformation and naïve generative adversarial networks (GAN) to tackle the problem of having unbalanced quantities of defect-free and defective images. They expanded the number of defective images that enhanced the classifier's generalization ability. Tsai et al. [13] applied cycle-consistent adversarial network (CycleGAN) to generate the saw-mark defect in heterogeneous solar wafer images and to solve the unbalanced classification problem arising in manufacturing inspection. Their experiment showed that the CNN's classification accuracy rates of GAN-based data augmentation were better than those of doing over-sampling and assigning higher class-weights on minor classes. In addition to the research related to defect classification, the GAN-based method was also applied to the field of defect detection. Yang et al. [14] introduced an image generating process for welded joints, which was based on the affine transformation and CycleGAN. Then the YOLOv3 model was used for the welding head detection, with a better average precision (AP) of 91.02% than faster region-based convolutional neural network (Faster RCNN). Tian et al. [15] used CycleGAN to augment images of healthy apples and apples with anthrax, thereby increasing the number of images and enriching its content. After that, YOLOv3-dense was used to test the apple for anthracnose. Experiments showed that their model performs at an AP of 95.57%. This method could also be applied to the detection of apple surface diseases in orchards. However, some welded joints on metals or anthraces on apples generated by CycleGAN are not expected output [14,15]. After augmenting the data using GAN, rich pseudo images are obtained. Although the appearance diversity of defects increases, no corresponding annotation files are provided [14,15], as operators have no time for time-consuming annotation work. Besides, GAN is currently unable to form specific structures, generative images that are not only blurry but also incorrectly colored. These undesirable generative images should be manually deleted. Cooperating GAN with an automatic annotation method is another contribution of this study, so that the data pairs are available for training the YOLOv3 die defect detection model. This research uses a series of pre- and post- digital image processing (DIP) techniques to reduce the generation loading of GAN and to develop an autoannotation procedure for pseudo defective images. The DIP techniques not only help to generate realistic pseudo defects but also save the time needed for annotating defective pseudo images.

This paper is composed of four parts. Following this introductory section, which has summarized the literature on die inspection and presents the contributions of this study, is the second section that describes the hardware architecture for capturing images, and the methodology. It also introduces GAN, the automatic annotation mechanism, and the modified application of YOLOv3. The third section presents the experimental results. A spot-checking process helps us to determine the YOLOv3 as the base model. The hyperparameters to be used in the GAN + YOLOv3 mechanism are derived, based on the design of the experiment (DOE). The defect detection results are reported, compared, and analyzed. The final part is the conclusion.

2. Research Method

The overall research process of this study is shown in Figure 1. First, we captured the images through the image-capturing system. We examined the die image structure and composition, and the appearance, characteristics, and specification of the particles. The next step was to separate the image set into training, validation, and testing sets. We manually marked the fine particles embedded on the surface of the die through an annotation tool to create the annotation file corresponding to each image. Defect-free dies do not need to be annotated and are not included in the training process. In order to increase the diversity of defects, the study created pseudo particle defects with the help of GAN's automatic generation ability. The study also automatically generated an annotation file corresponding to each pseudo defective image through the connected component labeling (CCL) [16]. The next step was to feed real and pseudo defective images to the YOLOv3 model for learning. Finally, we measured size of defects. Details of the research procedure are explained in the following sub-sections.

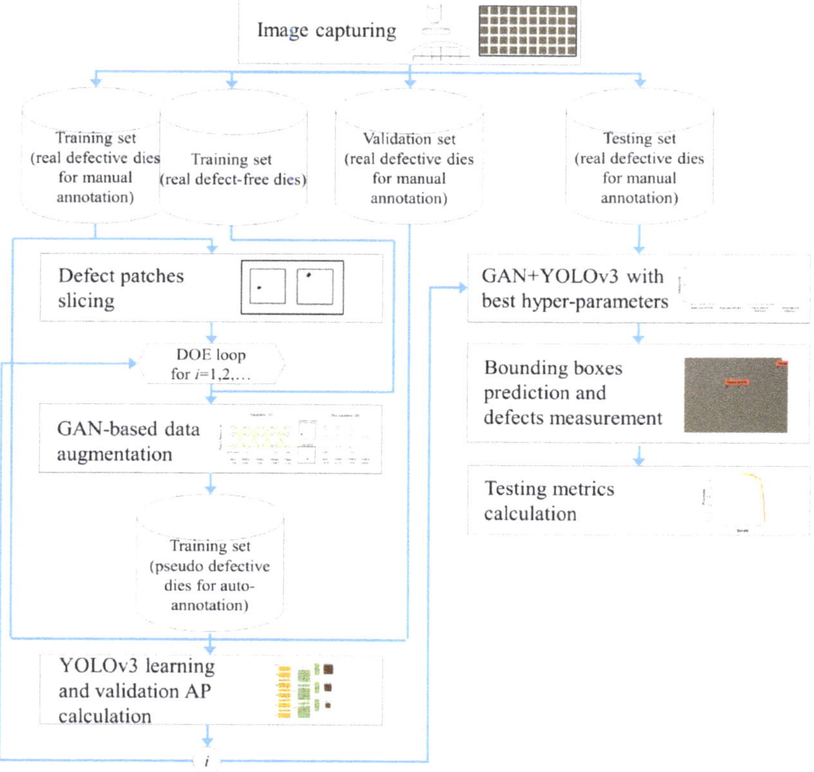

Figure 1. The research process.

2.1. Hardware Structure and Composition of the Die

In order to retrieve the surface images of the dies on the wafer, the study used the image-capturing system shown in Figure 2a. The CCD in this system was Hitachi KP-FD202GV, and the resolution was a 1620 × 1220 color image. The lens is an Olympus Lens with an optical magnification of 5×, a working distance of 19.6 mm, and a resolution of 3.36 μm, coupled with a lighting source with a 12 V/100 W coaxial yellow ring halogen lamp. The front-illuminated light source emphasizes the surface characteristics of the inspection object. During the shooting process, the researchers used the

XY axis motion controller to capture the image of each die to be inspected with an S-shaped scanning path. By lowering the requirements for positioning precision, one image could contain multiple dies, as shown in Figure 2b. However, only the die pattern at the center of the image was intact, called region of interesting (ROI), and the eight neighboring dies had only partial patterns.

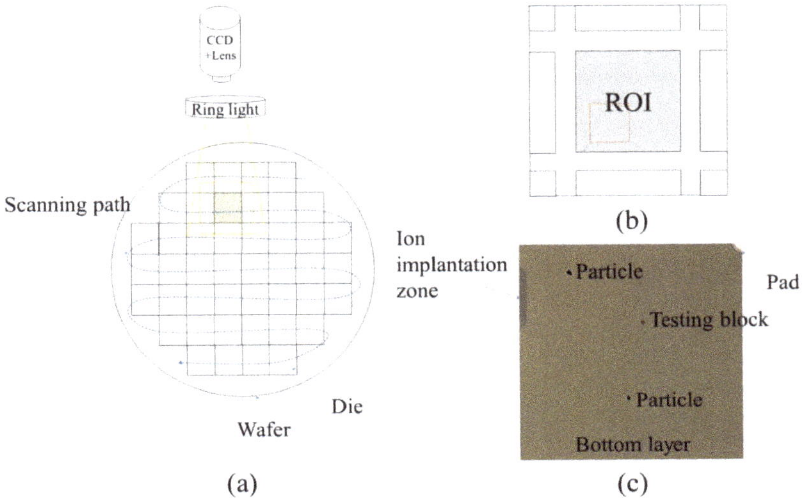

Figure 2. The image-capturing system and composition of the die. (**a**) The scanning path and field of view (**b**) the ROI die and (**c**) the appearance of die.

The appearance of the image of the die surface in this research is shown in Figure 2c. In compliance with a confidentiality agreement with the case company, images displayed in this paper are only part of the die, and the images have been flipped and discolored before presentation. The die was composed of the pad, the ion implantation zone, the bottom layer and the testing block. The pad was mainly used for electrical testing to ensure the function of the die. The bottom layer was a protective layer protected by a thin film, which could protect the components from chemical reactions, moisture, corrosion, pollution, etc. The testing block was used by the foundry customers to perform special tests. During the manufacturing process, the wafer might contaminate the surface of the die due to particle residues, which could result in defective products. These particles appeared at random positions, and they might be seen on the surface of the entire die. The shape of the particle was irregular, either large or small, and sometimes dense clusters occurred. The testing block on the die was a dark rectangular pattern with an appearance similar to that of particle defects, which increased the difficulty of defect detection.

2.2. Manually Annotating Defects

After building the die image set, we needed a corresponding annotation set before the model could be trained. LabelImg was used as an annotating tool by the researchers to manually annotate the locations and names of the defects in the image one by one. These annotation messages were stored in the XML format, and the filename was the same as the filename of the annotated image, except for the filename extension. As shown in Figure 2b, only the central die in the image was the ROI. The traditional method might have been to design an algorithm to perform ROI image segmentation before proceeding to subsequent actions. However, since this study would use the object detection method in deep learning, the preprocess of extracting ROI could be omitted. As long as the researchers focused on framing the defects on the ROI die when annotating, later the algorithm would naturally ignore the defects on the eight neighboring dies when detecting defects.

2.3. Defect Data Augmentation by GAN and Their Autoannotation

As proposed by Goodfellow et al. [17], GAN has a wide range of applications, such as fashion, advertising, science, and games. Since the images for defect detection are usually captured in a stable environment, each image is roughly the same regardless of location or color. Therefore, the general traditional data augmentation method is not necessarily applicable. The case company does not have much time for engineers to collect a huge image set, let alone an additional time-consuming annotation work for a large number of image sets. To overcome this difficulty, this study took advantage of the powerful generative capabilities of GAN to create richer types of defects. The basic idea of GAN is shown in Appendix A.1 of Appendix A.

When we directly input a set of die images into the GAN model, however, we found that its objective function value fluctuated during the iteration and converged only with difficulty. Additionally, we also found that the GAN model could not generate high-resolution pseudo images effectively. It could only generate the approximate outline of the die, but the details could not be identified. Consequently, a strategy of simply generating the particle defects was adopted.

The detailed process is shown in Figure 3. We used the defect coordinate position, and the information of length and width that the annotator has previously noted in the real image, which helped to cut out the patch that indicates the area of each particle defect. Otsu binarization [18] was then used to eliminate the background in the patch as far as possible, retaining the original appearance of the particle defects, and the defects were attached to a white background image with the same size as the GAN input image. As shown in the bottom left of Figure 3, GAN is composed of two networks: a generator and a discriminator. During the first iteration, the generator generated poor pseudo images and the discriminator distinguished them from real images easily. During the second iteration, the quality of pseudo images generated by the generator was improved, which fooled the underlying discriminator. With the rise of the ability of the discriminator, real images and pseudo images can be recognized, which will also drive the improvement of the generator. The training process of adversarial learning between the two networks was continued and a generative model similar to the real image distribution was created. As the learning object became simpler, the objective function converged rapidly, and generated more realistic pseudo particle defects. Finally, we embedded the pseudo particle defects into the defect-free dies to create a generative pseudo image set.

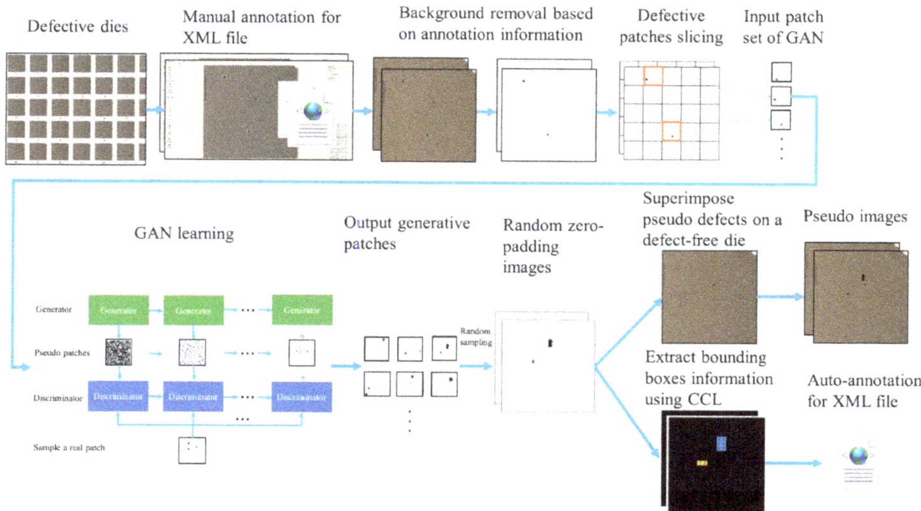

Figure 3. The flowchart of GAN-based data argumentation and auto-annotation.

Although we used GAN for data augmentation to increase the diversity of defects, the annotation files of these pseudo defective images were not generated. In line with the previous literature, an additional manual method was adopted to annotate the pseudo defect images [14,15]. In order to save time when annotating the pseudo defective images, DIP techniques were used to automatically annotate the pseudo particles as shown in the bottom right of Figure 3. The CCL algorithm [16] scanned the image from left to right and top to bottom. If the gray values between adjacent pixels were found to be similar during scanning, they were labeled with a same index. Each pseudo particle defect would be regarded as a blob, and information of its minimum bounding box could also be registered. Then the XML annotation file of the pseudo defective image could be output, which reduced the time spent in annotating the pseudo defective images.

2.4. Defect Detection and Measurement Using YOLOv3

This research used YOLOv3 [19] as the basis for die defect detection and measurement. The basic idea of YOLOv3 is shown in A.2 of Appendix A. The YOLOv3 model is a one-stage method, end-to-end training process that can be realized using a single network. The inference can predict the center coordinate, width and height of each bounding box where the defect is located, and the confidence that each bounding box contains a defect.

After YOLOv3 output the predicted bounding boxes, the study would further measure the defects of the corresponding patches and sort the quality of a die, as shown in Figure 4. The process included Ostu binarization [18], the estimation of the bounding ellipse, and the calculation of the major and the minor axis. The process was able to potentially assist to conduct the sorting of the dies in accordance with the quality specifications of the customers. For example, there were three classes of die products: an excellent die had no particle defect after inspection; a qualified die had particle defects with the major axis length between 50 and 149 μm and the minor axis length less than 20 μm; an unqualified die had particle defects that exceeded the quality specification.

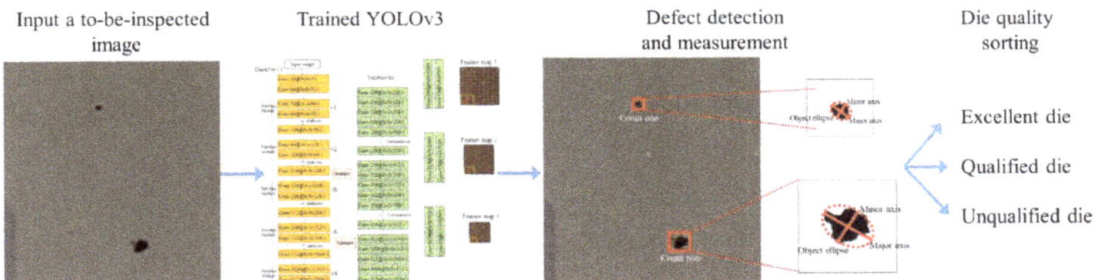

Figure 4. Die defect detection and measurement.

3. Analysis and Results of the Experiments

This study collected 669 die images, 198 defect-free images, and 471 defective images with particles. Since training an object detection model needs only defect samples of the "object", this research randomly selected 300 defective images as the training image set. The remaining defective images were used as testing image sets to evaluate the inference performance of the model. In addition, on the production line, the appearance of defects is multifaceted. It is not possible to produce distinctively different defect appearances if we only depend upon the jitter mechanism of YOLOv3 itself. In addition to the images generated by the jitter mechanism, this research also applied the GAN to generate images of pseudo die defects.

3.1. Spot-Checking Experiment

The spot-checking experiment gets a quick assessment of different models on a custom dataset. Researchers are able to know which type of models is suitable at picking out the structure of the dataset. In order to demonstrate the performance of the object detection models for the detection of particle defects on the dies, this study compared the YOLOv3, the Faster RCNN [20], and the single shot multibox detector (SSD) [21]. After all the training processes were completed, the validation AP was used to evaluate the performance of the models on defective images. As shown in second column of Table 1, there were significant gaps of validation APs between YOLOv3 and other two models. In practice, the inference speed of a model was always concerned. The frames per second (FPS) was adopted here to evaluate the inference speed of the models as shown in the last column of Table 1. We found that the inference speed of SSD was fastest, followed by the YOLOv3 and lastly by faster RCNN. Even though the FPS of YOLOv3 was not the fastest, it was enough to be applied to the production line. The spot-checking experiment indicates that the YOLOv3 was the best model at learning the structure in the dataset so we could focus the attention to optimize it.

Table 1. Spot-checking comparison using different evaluation metrics.

Evaluation Metrics Methods	Validation AP	FPS
YOLOv3	81.59%	13.447
Faster RCNN	14.00%	2.783
SSD	57.56%	23.852

3.2. Hyperparameter Sensitivity Experiment

The hyperparameters of a model are related to the flexibility and potential of its learning, and directly influence the degree of the generalization when the model makes inferences. Since training a deep learning model often takes a long time, it is extremely inefficient to find the optimal hyperparameter combination manually for a deep learning model. Based on DOE, the research analyzed the validation AP with various hyperparameter combinations. It endeavored to identify the key hyperparameter combinations that affect the AP of die defect detection, which provides the basis for improving the AP of the model in a reasonable way. The DOE of this research includes four factors, and each factor has three levels.

- Input image size of GAN: The quality of the image generated by the GAN is not only influenced by the background complexity of the input image, but also by the size of the input image. If the size is too small, the GAN cannot generate detailed defects; if the size is too large, the GAN would weaken during the generation process. This research set three levels of the factor to be 28 × 28, 64 × 64, and 96 × 96, with 28 × 28 as the default value, and 96 × 96 the upper limit of the defect size.
- Fold size of GAN image augmentation: This study proposed an image augmentation technology based on the GAN. Defect patches were generated from the adversarial learning process. Then the patches were pasted onto the defect-free die image to generate pseudo defect images. Compared with the original defective die images in Figure 5(a1,a2), the defects in Figure 5(b1,b2) were naturally embedded in the die image. This process generated various shapes, sizes and numbers of defects, and increased the quantity of training images and the diversity of defects, allowing YOLOv3 to learn a richer appearance of defects, and improving the performance of model training. In this study, the factors were set to three levels: 1, 1.5, and 2. The 1 meant that the GAN image augmentation mechanism was turned off; 1.5 and 2 meant that the original number of defective die images was multiplied by 1.5 and 2 respectively as the pseudo defective die images for later training.
- Upper limit of the input image size: There are only convolutional layers in YOLOv3, and its input image size is unrestricted. However, in order to strengthen the robustness of model inference, YOLOv3 adopted a multiscale training strategy. During the training process, the size of the input image was changed after a certain number of iterations. YOLOv3 could define the upper limit of the size range of the input image during the training process. In addition, because the minimum input feature map in Yoloblock employed downsampling 32 times, the upper limit of the image

size range must also be a multiple of 32. In the experimental design, the factor was set to three levels: 416 × 416, 480 × 480, and 544 × 544. 416 × 416 was the default value of YOLOv3.

- Degree of jitter: In addition to the pseudo images generated by the GAN, the YOLOv3 also had its own data augmentation program, called the degree of jitter. It could flip, zoom, crop, and perform HSV contrast conversion of the input image to augment the images and suppress overfitting. This research set the factor to three levels: 0, 0.15, and 0.3. The 0.3 was the default value of YOLOv3, and 0 indicated that the jitter was turned off.

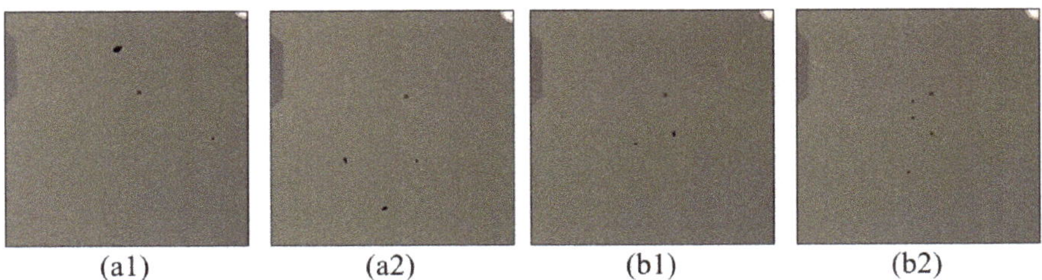

Figure 5. Comparison of defect images. (**a1**,**a2**) Original defects and (**b1**,**b2**) pseudo defects.

Next, this study removed 20% from the training image set to be used as the validation image set (not including any pseudo defective images). After conducting 3^4 DOE, the main effect plots of the validation AP for all the factor and level combinations were drawn as shown in Figure 6. Using the criterion "the larger the better" (LTB) for validation AP, the researchers selected the hyperparameter combination: the input image size and augmentation fold size of the GAN were 64 × 64 and 2, and the upper limit of the input image size and jitter degree of YOLOv3 were 416 × 416 and 0.3.

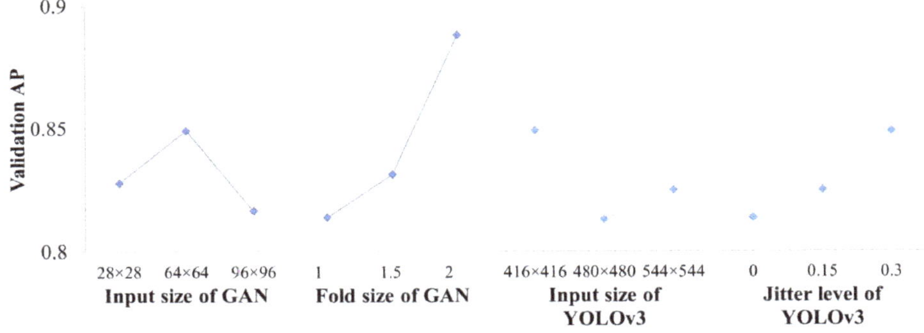

Figure 6. The main effect plot of 3^4 design of the experiment (DOE).

3.3. Results of Die Defect Detection and Measurement

After deciding the hyperparameters of the GAN + YOLOv3 model and training the model, defect detection and measurement of the remaining test images were performed. The pipeline of the testing process first inferred the predicted bounding boxes of the defects through YOLOv3. Then, the major and the minor axis of the defects were measured for the content inside the bounding box. After the inference was completed, different evaluation metrics were used to measure the generalization ability of the proposed algorithm.

The testing AP was used to measure the performance of the predicted bounding boxes: after the testing image set was inferred by the object detection method, the predicted bounding boxes were compared with the ground truth boxes, and the average of the maximum precision values calculated when recall ≥ 0, 0.1, ..., 1.0. Coordinate prediction error was used to measure the accuracy of the coordinate prediction: after the testing image set was inferred by the object detection method, the closeness of the center coordinates, length, and width of the predicted bounding boxes were compared with those of the ground truth boxes, which could be calculated through the first two items in Equation (A2) of Appendix A.

Figure 7 demonstrates the patches of defect detection results. When inputting the die image of the product, as shown in Figure 7(a1–a4), the model precisely box-bounds the corresponding particles, as in Figure 7(b1–b4). The testing blocks in Figure 7(b1–b4) are not falsely box-bounded. YOLOv3 is able to discriminate between irregular-shaped particles and rectangular testing blocks. The model effectively detected particle defects on the surface of the die, and even very small defects could be detected successfully.

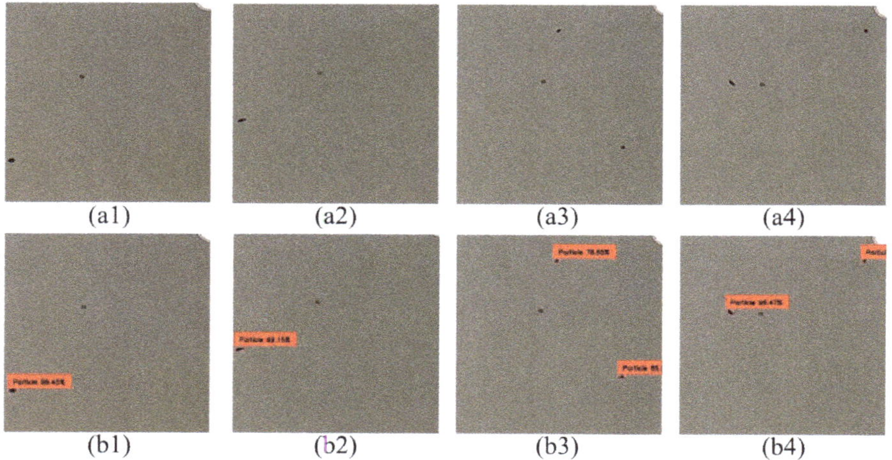

Figure 7. Defect detection results by GAN + YOLOv3. (**a1–a4**) Images to be inspected and (**b1–b4**) detection results.

In order to further demonstrate the performance improvement of the GAN-based image augmentation technology for the detection of particle defects on the dies, this research also constructed the YOLOv3 model, the GAN + YOLOv3 model (augmenting 1.5 times the training sample), the GAN + YOLOv3 model (augmenting 2 times the training sample), and the CycleGAN + YOLOv3 model (augmenting 1.5 times the training sample). After the training of the four models was completed, the study used the testing AP and the testing coordinate prediction error to evaluate the models on testing images.

Before calculating AP, the precision–recall (PR) curve of each model was drawn as shown in Figure 8. The pseudo defective die image generated by the GAN could work along well with the true defective die image to train the YOLOv3. In addition, when GAN helped to increase the training image by about 1.5 times, the PR curve tended to converge as shown in Figure 8, and the corresponding testing AP jumped from 81.39% to more than 88%, an increase of about 7% as shown in the second column of Table 2. As indicated by the results of the testing coordinate prediction error indicator, the coordinates and length and width of the predicted bounding boxes and the ground truth boxes, were very close to the ground truth labels. Even without the help of the GAN, the bounding box error predicted by the naïve YOLOv3 model was below three decimal places. After adding the GAN, the testing coordinate prediction error could be reduced to a level below four decimal places as shown in the last column of Table 2. This experiment shows that the pseudo defect images generated by the GAN play an important role in enriching the diversity of defects, which helps to improve the efficacy and versatility of the model. Beside, we also compared the CycleGAN + YOLOv3 with proposed GAN + YOLOv3. The corresponding result is shown in the last row of Table 2. It clearly found that its testing AP and coordinate prediction error were not satisfied. The main reason is the appearance of the particle patches generated by CycleGAN was far from the real particle. Not only the area of defect was large, but also the edge of defect was not smooth.

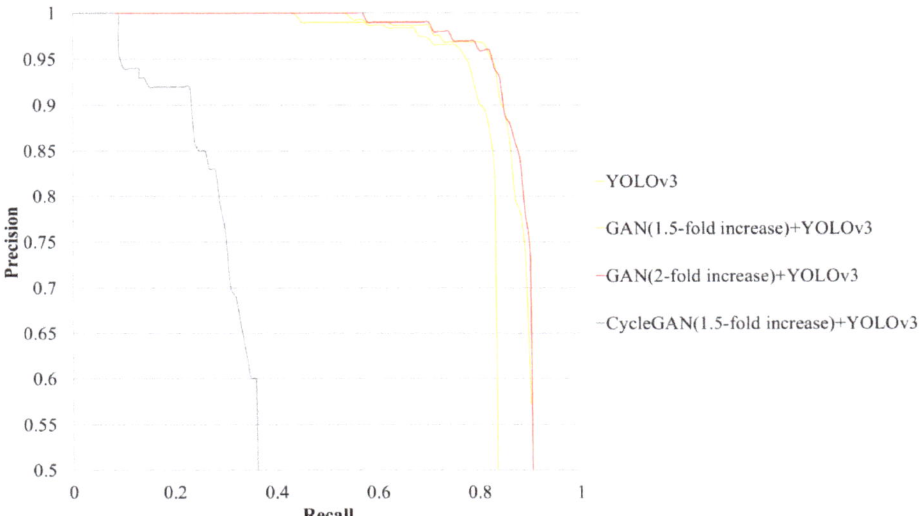

Figure 8. Precision–recall (PR) curve for each method.

Table 2. Comparison of methods by different evaluation metrics.

Evaluation Metrics Methods	Testing AP	Testing Coordinate Prediction Error
YOLOv3	81.39%	1.6456×10^{-3}
GAN + YOLOv3 (1.5-fold increase)	88.12%	2.9662×10^{-4}
GAN + YOLOv3 (2-fold increase)	88.72%	1.5851×10^{-4}
CycleGAN + YOLOv3 (1.5-fold increase)	33.14%	6.4880×10^{-1}

4. Conclusions

Defect sample collection, defect annotation, and feature engineering have always been the most time-consuming tasks in defect detection. To address this issue, this research integrated the technology of generative pseudo defective samples (using GAN), automatic pseudo defect annotation (using DIP), and automatic feature extraction (using YOLOv3). The methods proposed in this study need not to rely on experts for feature engineering and did not need bulk defect samples. Massive defect annotations are not required, either. Users need only to prepare a few defect image sets, manually annotate them, and complete the model training before conducting the inferences. This means that the method has great potential for application in various die patterns, where appearances are changeable and complex. In addition, the experimental results show that after the addition of the GAN mechanism, both the overall detection precision of the predicted bounding box and the measurement accuracy of quality classification were improved. This indicates that the pseudo defect images generated by the GAN help enrich the diversity of the training data set, which to some extent improved the versatility of the model.

If sematic segmentation methods make a breakthrough in the inference speed in the future, it may be possible to combine the GAN and sematic segmentation methods to perform defect segmentation. The annotation process of the object segmentation model captures the outline of the defect in the image, rather than simply annotating the rectangular bounding box, as happens in the object detection model. Therefore, the annotation does not contain the background and does not consider the angle. The annotation may include other rectangles near the defects. In this way, the process of removing the background and the process of extracting blobs from the predicted bounding box can be omitted, and the efficiency of model inference can be improved.

Author Contributions: Conceptualization, S.-H.C. and D.-B.P.; methodology, S.-H.C.; software, C.-H.K.; validation, S.-H.C., C.-H.K. and D.-B.P.; formal analysis, C.-H.K.; data curation, D.-B.P.; writing—original draft preparation, S.-H.C.; writing—review and editing, D.-B.P.; funding acquisition, S.-H.C. All authors have read and agreed to the published version of the manuscript.

Funding: This research was funded by the Ministry of Science and Technology, Taiwan, grant number MOST 108-2221-E-131-006-MY2 and The APC was funded by Ming Chi University of Technology.

Acknowledgments: We are grateful to three anonymous reviewers for comments. The authors acknowledge all participants, Yuan-Shuo Chang, Yu-Hsin Yen and Hsin-Chi Chang, for their domain knowledge and annotation support of the study.

Conflicts of Interest: The authors declare no conflict of interest.

Appendix A

We described the technical sections related to GAN and YOLOv3 in Appendix A.1 and Appendix A.2 respectively, so that the readers can focus on the important messages in the main texts.

Appendix A.1. Descriptions of the GAN

The network structure of GAN is shown in Figure A1 [17]. GAN is composed of two networks: a generator (G) and a discriminator (D). The generator is a four-layer neural network of regression. By obtaining the distribution of the real image and inputting the noise source (**z**) to the generator, it produces a pseudo image similar to the real one. The discriminator is a three-layer neural network of the binary classifier, which is responsible for evaluating the authenticity of the pseudo image. With the rise of the ability of the discriminator, real images, and simulated images can be recognized, which will also drive the improvement of the generator. Subsequently the generator generates pseudo images closer to real ones for the discriminator to distinguish. Finally, the generator can generate pseudo images very similar to real ones. Between the two networks described above, the training process of adversarial learning is continued, interactive learning is realized, and a generative model similar to the real image distribution is created. The loss function of the model is shown in Equation (A1):

$$\min_{\mathrm{G}} \max_{\mathrm{D}} V(\mathrm{D}, \mathrm{G}) = \mathrm{E}_{\mathbf{x} \sim P_{data}(\mathbf{x})}[\log \mathrm{D}(\mathbf{x})] + \mathrm{E}_{\mathbf{z} \sim P_{\mathbf{z}}(\mathbf{z})}[\log(1 - \mathrm{D}(\mathrm{G}(\mathbf{z})))] \quad \text{(A1)}$$

where **x** is an image from the real data distribution P_{data}; **z** is a noise vector sampled from a uniform distribution or a normal distribution $P_{\mathbf{z}}$; and E represents the expectation of real data and that of noise.

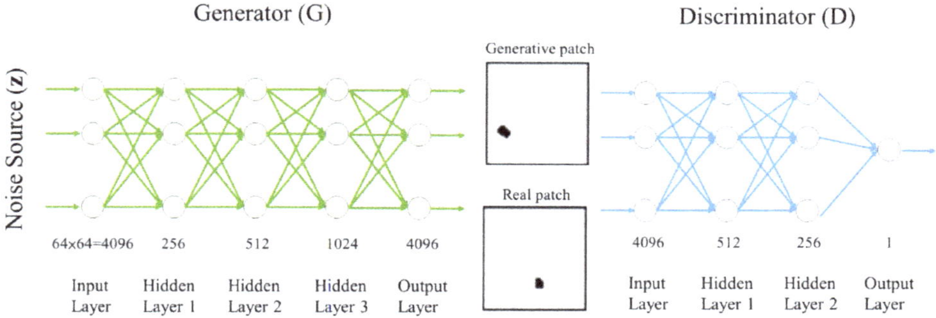

Figure A1. Architecture of the GAN network.

Appendix A.2. Descriptions of the YOLOv3

The network structure of YOLOv3 is shown in Figure A2. With DarkNet53 as its backbone, the model used a series of 1 × 1 and 3 × 3 convolution layers for feature extraction without a pooling layer and a fully connected layer. YOLOv3 introduced a residual block, which adds the corresponding dimensions of the input and the output feature maps to control the magnitude of the gradient propagate and alleviate the vanishing gradient problem faced by the deep network. In addition, the feature pyramid network (FPN) structure was used for multiscale detection effects. After the input image passed through DarkNet53, the feature map generated by Yoloblock was taken for two purposes. The first was to generate feature map 1 with a size of 13 × 13 after passing through the 3 × 3 and the 1 × 1 convolution layer. The second use was to add an upsampling layer after passing through the 1 × 1 convolutional layer, and to splice it with the output result of the intermediate layer of the DarkNet53 network, which generated feature map 2 with a size of 26 × 26. After the same loop, feature map 3 with a size of 52 × 52 was generated. 13 × 13, 26 × 26, and 52 × 52 are the number of grid cells of the output feature maps for each scale. The depth of these output feature maps, was set as $B \times 5$, where B represented the number of predicting bounding boxes for each scale, set here as 3. The number 5 represented the 5 levels of x, y, w, h, and confidence that must be predicted for each bounding box, where x and y represent the shift levels between the predicted bounding box center and the upper left corner of the grid cell; w and h represent the ratio of the width and height of the predicted bounding box to the width and height of the entire image; confidence represents the confidence value of the defect. The depth of the feature map output for traditional YOLOv3 needs to include the probability of the predicted bounding box of each grid cell for C categories, so its depth should be $B \times (5 + C)$. However, this research only sought to classify the single problem of particle defects, so the prediction of C could be omitted.

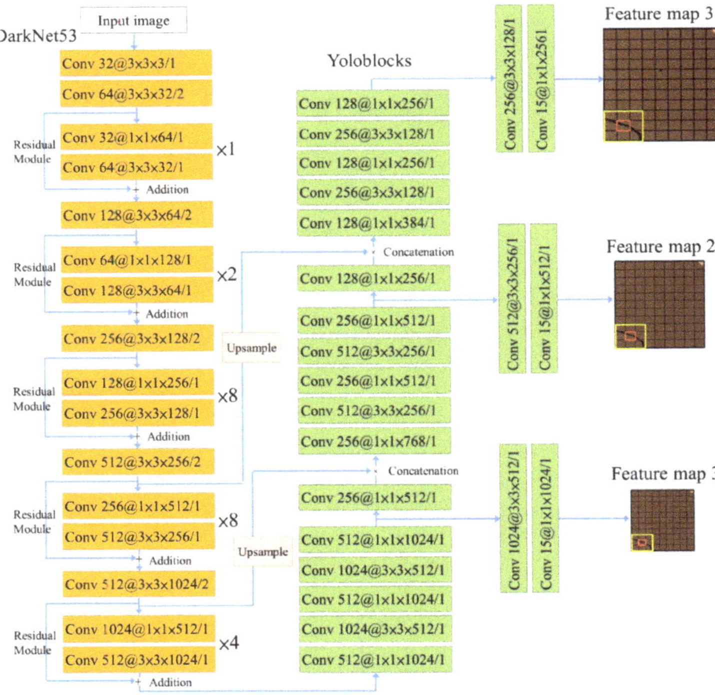

Figure A2. Architecture of the YOLOv3 network.

During training, YOLOv3 used the modified loss function and the back-propagation algorithm to learn the weights, as shown in Equation (A2). The loss function of YOLOv3 was originally composed of three parts, namely coordinate prediction error, intersection over union (IoU) error, and classification error [19]. However, since this research focused on the single classification problem, the classification error could be omitted.

$$\begin{aligned}\text{Loss} = {} & \lambda_{coord} \sum_{i=0}^{S^2} \sum_{j=0}^{B} \mathbf{I}_{ij}^{obj} [(x_i - \hat{x}_i)^2 + (y_i - \hat{y}_i)^2] + \lambda_{coord} \sum_{i=0}^{S^2} \sum_{j=0}^{B} \mathbf{I}_{ij}^{obj} [\left(\sqrt{w_i} - \sqrt{\hat{w}_i}\right)^2 + \left(\sqrt{h_i} - \sqrt{\hat{h}_i}\right)^2] \\ & + \sum_{i=0}^{S^2} \sum_{j=0}^{B} \mathbf{I}_{ij}^{obj} (C_i - \hat{C}_i)^2 + \lambda_{noobj} \sum_{i=0}^{S^2} \sum_{j=0}^{B} \mathbf{I}_{ij}^{noobj} (C_i - \hat{C}_i)^2\end{aligned} \tag{A2}$$

The first two terms in Equation (A2) represent the coordinate prediction error, and λ_{coord} was the weight hyperparameter given in advance. Since the number of grid cells that did not contain objects far exceeded the number of grid cell that contained objects, the loss of confidence without objects would be great. In order to reduce the impact of this problem on the network, it was generally assumed to be 5. I_{ij}^{obj} described the predicted bounding box j of the grid cell i had an indicator function containing objects. The $\hat{x}_i$, $\hat{y}_i$, $\hat{w}_i$, and $\hat{h}_i$ represent the central coordinates and the length and width of the ith predicted bounding box and the x_i, y_i, w_i, and, h_i represented those of the ith ground truth box.

In addition, the last two terms in Equation (A2) represented the IoU error, where λ_{noobj} was the weight hyperparameter given in advance, which generally defaulted to 0.5. I_{ij}^{noobj} represented that the predicted bounding box j of the grid cell i did not contain the indicator function of the object. The $\hat{C}_i$ represented the ith predicted value of confidence; the C_i referred to whether the ith ground truth box contained object.

References

1. Mital, D.; Teoh, E. Computer based wafer inspection system. In Proceedings of the Proceedings IECON '91: 1991 International Conference on Industrial Electronics, Control and Instrumentation, Kobe, Japan, 28 October–1 November 1991; Institute of Electrical and Electronics Engineers (IEEE): New York, NY, USA, 2002; pp. 2497–2503.
2. Tobin, K.W.J.; Karnowski, T.P.; Lakhani, F. Integrated applications of inspection data in the semiconductor manufacturing environment. In *Metrology-based Control for Micro-Manufacturing*; Tobin, K.W.J., Lakhani, F., Eds.; SPIE: Bellingham, WA, USA, 2001; Volume 4275, pp. 31–40.
3. Su, C.-T.; Yang, T.; Ke, C.-M. A neural-network approach for semiconductor wafer post-sawing inspection. *IEEE Trans. Semicond. Manuf.* **2002**, *15*, 260–266. [CrossRef]
4. Chang, C.-Y.; Chang, C.-H.; Li, C.-H.; Jeng, M. Learning Vector Quantization Neural Networks for LED Wafer Defect Inspection. In Proceedings of the Second International Conference on Innovative Computing, Informatio and Control (ICICIC 2007), Kumamoto, Japan, 5–7 September 2007; Institute of Electrical and Electronics Engineers (IEEE): New York, NY, USA; p. 229.
5. Timm, F.; Barth, E. Novelty detection for the inspection of light-emitting diodes. *Expert Syst. Appl.* **2012**, *39*, 3413–3422. [CrossRef]
6. Chou, P.B.; Rao, A.R.; Sturzenbecker, M.C.; Wu, F.Y.; Brecher, V.H. Automatic defect classification for semiconductor manufacturing. *Mach. Vis. Appl.* **1997**, *9*, 201–214. [CrossRef]
7. Zhang, J.-M.; Lin, R.-M.; Wang, M.-J.J. The development of an automatic post-sawing inspection system using computer vision techniques. *Comput. Ind.* **1999**, *40*, 51–60. [CrossRef]
8. Guan, S.-U.; Xie, P.; Li, H. A golden-block-based self-refining scheme for repetitive patterned wafer inspections. *Mach. Vis. Appl.* **2003**, *13*, 314–321. [CrossRef]
9. Liu, H.; Zhou, W.; Kuang, Q.; Cao, L.; Gao, B. Defect detection of IC wafer based on two-dimension wavelet transform. *Microelectron. J.* **2010**, *41*, 171–177. [CrossRef]
10. Cheon, S.; Lee, H.; Kim, C.O.; Lee, S.H. Convolutional Neural Network for Wafer Surface Defect Classification and the Detection of Unknown Defect Class. *IEEE Trans. Semicond. Manuf.* **2019**, *32*, 163–170. [CrossRef]

11. Lin, H.; Li, B.; Wang, X.; Shu, Y.; Niu, S. Automated defect inspection of LED chip using deep convolutional neural network. *J. Intell. Manuf.* **2019**, *30*, 2525–2534. [CrossRef]
12. Chen, X.; Chen, J.; Han, X.-G.; Zhao, C.; Zhang, D.; Zhu, K.; Su, Y. A Light-Weighted CNN Model for Wafer Structural Defect Detection. *IEEE Access* **2020**, *8*, 24006–24018. [CrossRef]
13. Tsai, D.-M.; Fan, M.; Huang, Y.-Q.; Chiu, W.-Y. Saw-Mark Defect Detection in Heterogeneous Solar Wafer Images using GAN-based Training Samples Generation and CNN Classification. In Proceedings of the 14th International Joint Conference on Computer Vision, Imaging and Computer Graphics Theory and Applications, Prague, Czech Republic, 25–27 February 2019; pp. 234–240.
14. Yang, L.; Liu, Y.; Peng, J. An Automatic Detection and Identification Method of Welded Joints Based on Deep Neural Network. *IEEE Access* **2019**, *7*, 164952–164961. [CrossRef]
15. Tian, Y.; Yang, G.; Wang, Z.; Li, E.; Liang, Z. Detection of Apple Lesions in Orchards Based on Deep Learning Methods of CycleGAN and YOLOV3-Dense. *J. Sensors* **2019**, *2019*, 1–13. [CrossRef]
16. Dillencourt, M.B.; Samet, H.; Tamminen, M. A general approach to connected-component labeling for arbitrary image representations. *J. ACM* **1992**, *39*, 253–280. [CrossRef]
17. Goodfellow, I.; Pouget, A.J.; Mirza, M.; Xu, B.; Warde, -F.D.; Ozair, S.; Courville, A.; Bengio, Y. Generative adversarial nets. In *Advances in Neural Information Processing Systems*; Ghahramani, Z., Welling, M., Cortes, C., Lawrence, N., Weinberger, K.Q., Eds.; Curran Associates, Inc.: Red Hook, NY, USA, 2014; Volume 27, pp. 2672–2680.
18. Otsu, N. A Threshold Selection Method from Gray-Level Histograms. *IEEE Trans. Syst. Man, Cybern.* **1979**, *9*, 62–66. [CrossRef]
19. Redmon, J.; Farhadi, A. Yolov3: An Incremental Improvement. 2018. Available online: https://pjreddie.com/media/files/papers/YOLOv3.pdf (accessed on 27 November 2020).
20. Ren, S.; He, K.; Girshick, R.; Sun, J. Faster r-cnn: Towards realtime object detection with region proposal networks. *IEEE Trans. Pattern Anal. Mach. Intell.* **2015**, *39*, 1137–1149. [CrossRef] [PubMed]
21. Liu, W.; Anguelov, D.; Erhan, D.; Szegedy, C.; Reed, S.; Fu, C.Y.; Berg, A.C. SSD: Single shot multibox detector. Computer Vision—ECCV 2016, Proceedings of The 14th European Conference on Computer Vision, Amsterdam, The Netherlands, 8–16 October 2016; Leibe, B., Matas, J., Sebe, N., Welling, M., Eds.; Springer: Cham, Germany, 2016; pp. 21–37.

 *applied sciences*

Article

Dynamic Pad Surface Metrology Monitoring by Swing-Arm Chromatic Confocal System

Chao-Chang A. Chen *, Jen-Chieh Li, Wei-Cheng Liao, Yong-Jie Ciou and Chun-Chen Chen

Department of Mechanical Engineering, National Taiwan University of Science and Technology, Taipei 106, Taiwan; d10603010@mail.ntust.edu.tw (J.-C.L.); M10703221@mail.ntust.edu.tw (W.-C.L.); josephciou@mail.ntust.edu.tw (Y.-J.C.); ccchen@tlhome.com.tw (C.-C.C.)

* Correspondence: artchen@mail.ntust.edu.tw; Tel.: +886-2-2733-3141 (ext. 1193)

Abstract: This study aims to develop a dynamic pad monitoring system (DPMS) for measuring the surface topography of polishing pad. Chemical mechanical planarization/polishing (CMP) is a vital process in semiconductor manufacturing. The process is applied to assure the substrate wafer or thin film on wafer that has reached the required planarization after deposition for lithographic processing of the desired structures of devices. Surface properties of polishing pad have a huge influence on the material removal rate (MRR) and quality of wafer surface by CMP process. A DPMS has been developed to analyze the performance level of polishing pad for CMP. A chromatic confocal sensor is attached on a designed fixture arm to acquire pad topography data. By swing-arm motion with continuous data acquisition, the surface topography information of pad can be gathered dynamically. Measuring data are analyzed with a designed FFT filter to remove mechanical vibration and disturbance. Then the pad surface profile and groove depth can be calculated, which the pad's index PU (pad uniformity) and PELI (pad effective lifetime index) are developed to evaluate the pad's performance level. Finally, 50 rounds of CMP experiments have been executed to investigate the correlations of MRR and surface roughness of as-CMP wafer with pad performance. Results of this study can be used to monitor the pad dressing process and CMP parameter evaluation for production of IC devices.

Keywords: pad dressing; dynamic measurement; CMP; pad uniformity; pad lifetime

Citation: Chen, C.A.; Li, J.-C.; Liao, W.-C.; Ciou, Y.-J.; Chen, C.-C. Dynamic Pad Surface Metrology Monitoring by Swing-Arm Chromatic Confocal System. *Appl. Sci.* **2021**, *11*, 179. https://dx.doi.org/10.3390/app11010179

Received: 31 October 2020
Accepted: 23 December 2020
Published: 27 December 2020

Publisher's Note: MDPI stays neutral with regard to jurisdictional claims in published maps and institutional affiliations.

1. Introduction

Chemical-mechanical planarization/polishing (CMP) has been known as a key process for global and local planarization in IC fabrication. Because of the urgent demands for conducting linewidth of IC device downsizing to nanometers, the stability and availability of CMP process have become critically significant [1,2] for high volume production. The polishing pad used in CMP process is one of the most important consumables for affecting CMP process output [3]. The material removal rate (MRR) and planarization ability of the process are determined by the structure and material properties of polishing pads [4]. The slurry contains abrasive particles on the wafer surface for removal of the passivated layer after chemical activation. Currently, a CMP tool is not capable of fully monitoring the polishing pad on-line. Usually it only measures the groove depth and pad thickness or by empirical analysis [5–7]. Some efficiency indicators of pad performance can be established with measuring the surface topography, such as roughness and bearing area ratio so that the polishing pad could be efficiently utilized [8–10]. The asperities and profile of pad are associated with MRR and final quality of as-CMP wafer. The asperities and groove depth of pad are gradually worn with CMP processes. The pad conditioning or dressing process is necessary to restore the pad surface, but the profile and groove depth are changed with the numbers of conditioning. As the pad topography will effectively influence the MRR and polishing results, different kinds of measuring methods are developed to monitor the change of the pad surface [11–13]. Nowadays, the methods of analysis of pad topography are mostly for static and partial area [5]. Additionally, judging the efficiency of polishing pad and lifetime are based on groove depth and thickness. Better methods are those that establish a system which can achieve bigger area scanning and extract pad surface topography easily and faster. Accurate results can be obtained using optical microscope, but the polishing pad needs to be cut and comprehensive measurement cannot be achieved. The pad profiler can achieve a scan of a full pad, but still requires a high measurement time and the mechanical parts are easily affected by environmental pollution. Thus, there is not yet available for dynamic measurement of pad topography before and after CMP or pad conditioning processes.

This study aims to develop a dynamic pad monitoring system (DPMS) for measuring the surface topography of polishing pad. A swing-arm type conditioner is widely used in modern polishing machines. In this study, a chromatic confocal sensor is attached on a designed fixture arm to acquire data from pad topography. Topography of total working area of the polishing pad is detected by rotation of polishing pad and arm motion. Because the mechanism is fixed on the swing arm, the entire area of polishing pad can be scanned and it can effectively reduce the measuring time. The DPMS then can provide a performance index of polishing pad to maximize the utilization of the polishing pad in a relatively shorter time.

2. Design and Configuration of DPMS

2.1. System Description

The DPMS is shown in Figure 1 with rotating the platen and the swing-arm motion on a CMP tool. The monitoring mechanism is based on a concentric circle as shown in Figure 1a. The surface topography is built by height information from a chromatic confocal sensor. This developed DPMS is divided into motion module and measuring module. The system is designed to ensure the movement does not have any interference with the space constraints of the polishing tool. Experimental set-up is shown in Figure 1b. A chromatic confocal measurement sensor with STIL MG140/CL3 sensor is used in the measuring module. The chromatic confocal measurement is mainly done through the multi-wavelength white dispersion lens and spectrometer design. Different wavelengths of light are focused to different height positions, and through the pinhole design, unfocused wavelengths are blocked out, by avoiding entering inside the spectrometer, while controlling the hole size to control the measurement accuracy.

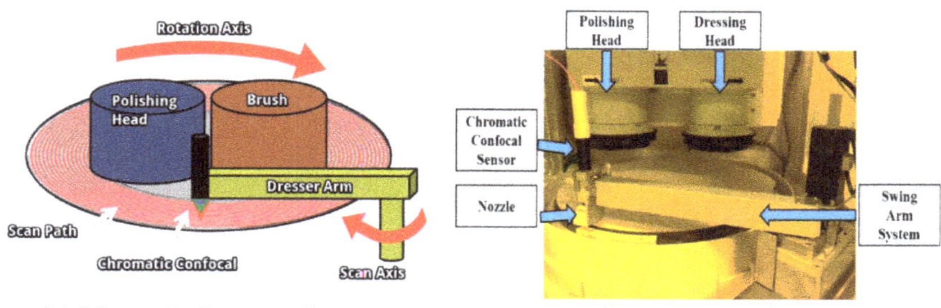

(**a**) Schematic diagram of system (**b**) experimental setup

Figure 1. Dynamic swing-arm chromatic confocal system.

With dynamic measurement by the swing-arm confocal system, the raw data contains surface non-uniformity, groove depth, surface height change or roughness [13]. The system disturbance from motor vibration or electrical noise needs to be considered for pre-processing of analysis. A filter based on FFT method is used and the filtered signals from specific frequencies are identified before the experiments [14–16]. Then, metrology data can be obtained from the signal measured from the system by reducing the disturbance of external environment.

2.2. Mathematical Model of Scanning Locus

The main purpose of the system is to obtain the total surface information of polishing pads. Because the system is set up on the polishing machine with a rotating platen, the sensor's scanning locus is combined with platen rotating and swing motion of the dressing arm [17]. The scanning locus of height sensor needs to be calculated with the actual position of the sensor, as the thickness of the pad changes with the radius during the CMP process. After combining the sensor's location and measurement data, the distribution of height and groove depth of pad are shown in the results. The motion locus is expressed as a spiral line; the diagram is shown in Figure 2.

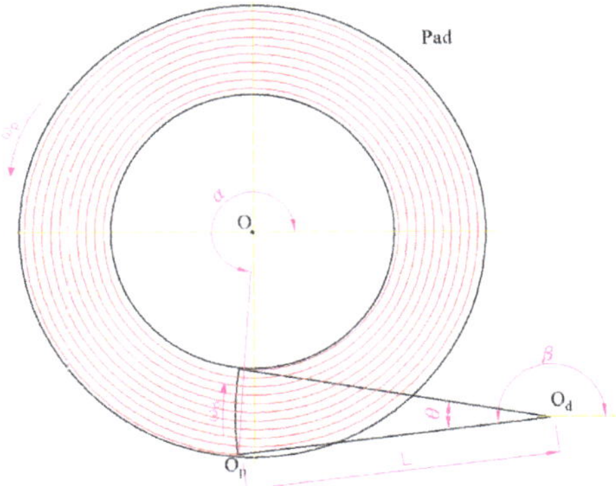

Figure 2. Schematic of scanning locus of confocal sensor.

The equation of locus can be expressed as:

$$\left[\frac{X(t)}{Y(t)}\right] = D\left[\frac{coscos\ (\alpha - \omega \times t)}{sinsin\ (\alpha - \omega \times t)}\right] \tag{1}$$

$$D = \sqrt{(d_x + L(\beta - \omega_d \times t))^{\ 2} + (d_y + L(\beta - \omega_d \times t))^{\ 2}} \tag{2}$$

where D means the distance between pad center and sensor location, d_x *and* d_y are the distance between pad center and arm's rotating center, L is the length of swing arm, α *and* β mean the initial angle of pad and swing arm, ω *and* ω_d mean the rotating speed of pad and swing arm.

2.3. Signal Processing and Filter Design

The measured data from the confocal sensor can be analyzed in three parts including vibration signal, rotation signal, and height data. Since the disturbance signal will couple with the real surface height data, the real surface features are separated by the signal processing. The rotation signal and system vibration can be separated by FFT method in this system. The rotating speed is defined in the beginning of the experiments. The vibration from the structure can be filtered by determining the mechanical frequency of the swing arm. The working frequency can be analyzed by rotating the arm independently on a ceramic platen. The frequency data are shown in Figure 3. The IIR filter is used to eliminate the influences of vibration and disturbances. Then the measured signal is analyzed and presented as Figure 4a. The comparison of processed and original signals is shown in Figure 4b,c. The original height data are combined with arm tilting, pad waviness, and pad. As the signal is analyzed by the designed algorithm, the measurement signal can be extracted and calculated.

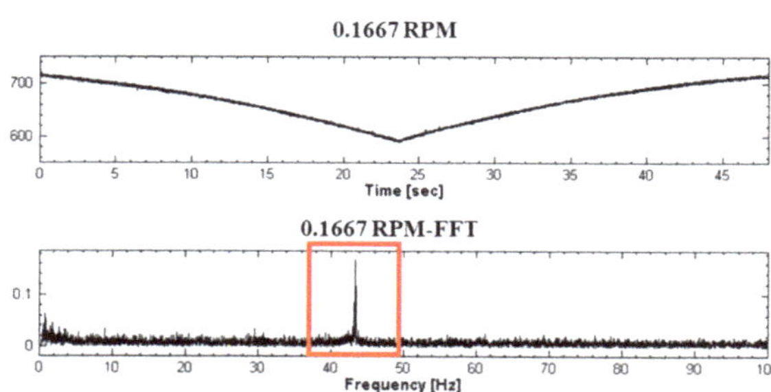

Figure 3. Working frequency of swing arm.

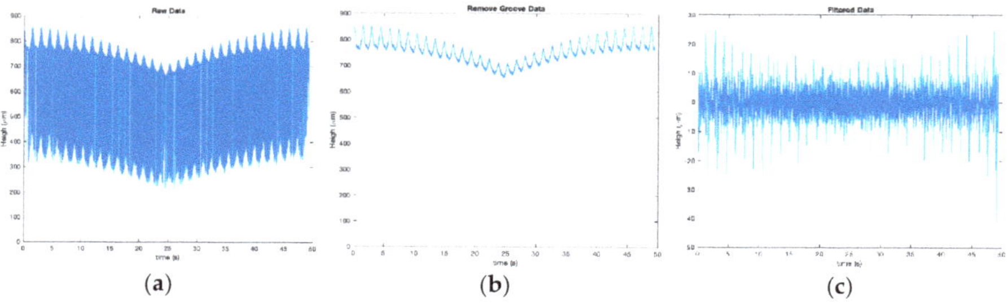

Figure 4. Measurement data of pad surface. (**a**) Raw data combined with arm tilting, pad waviness and asperities. (**b**) Extraction of pad surface profile. (**c**) Extraction pad asperities.

3. Experimental Method and Parameters

In CMP experiments, a HAMAI HS-720C polishing machine is attached with the confocal system to achieve in situ monitoring. An IC1000 polyurethane pad with x-y type groove is implemented in the experiments and its characteristics are shown in Table 1. A Kinik 3EA-3 diamond dresser is adopted in this experiments as shown in Figure 5, which has grit size around 100 ± 15 μm and 40–60 μm height. Related experimental parameters of pad conditioning and CMP are listed in Tables 2–4. Some 3 × 40 mm^2 Cu blanket substrates are used in the CMP experiments with 50 rounds. The pad surface is measured between each polishing process. The MRR and roughness of Cu substrates are measured to compare with the change of the pad performance index. With the swing-arm monitoring system, the measurement of pad surface has been accomplished without taking pad off-line or pausing the CMP process. Each measurement data has been completed within 50 s during the conditioning process for arm swinging from outer edge to inner position with sampling rate as 1 kHz. Changes of pad thickness and groove depth can be measured before and after each round of CMP process. Then the correlations of wafer quality and pad performance index can be analyzed and discussed.

Table 1. Characteristics of IC1000 pad.

Pad	IC1000
Thickness (mm)	1.36
Grid size (mm)	7.5
Groove width (mm)	0.5
Groove depth (mm)	0.436
Hardness (Shore D)	53
Compressibility (%)	5%
Recovery (%)	76%

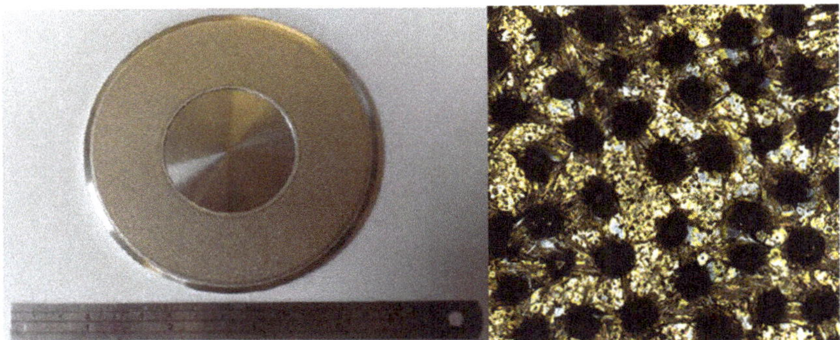

Figure 5. Kinik 3EA-3 diamond dresser.

Table 2. Break-in parameters of pad dressing.

Pressure	6.89 kPa (1 psi)
Pad speed	80 rpm
Dresser head speed	70 rpm
Slurry	DIW
Slurry feed rate	200 mL/min
Conditioning time	12 min

Table 3. Diamond conditioning parameters of pad.

Pressure	13.79 kPa (2 psi)
Pad speed	80 rpm
Dresser head speed	70 rpm
Slurry	DIW
Slurry feed rate	200 mL/min
Conditioning time	1.5 min

Table 4. Chemical mechanical planarization/polishing (CMP) parameters.

Pressure	20.68 kPa (3 psi)
Pad speed	80 rpm
Dresser head speed	70 rpm
Slurry	C8902
Slurry feed rate	200 mL/min
Conditioning time	1 min

4. Results and Discussion

4.1. Measuring Points Allocating and Processing

Figure 6 shows the spiral locus obtained by a confocal sensor in different rotating speeds of pad platen. Figure 7 shows the results to allocate the height data to pad surface by placing target symbols on the ceramic platen of polishing machine. By calibrating the location of measuring data with angle feedback signals from motor driver, the outlook shape of the target symbols can be displayed clearly. With the calibration data, the surface profile can be mapped into the corresponding location.

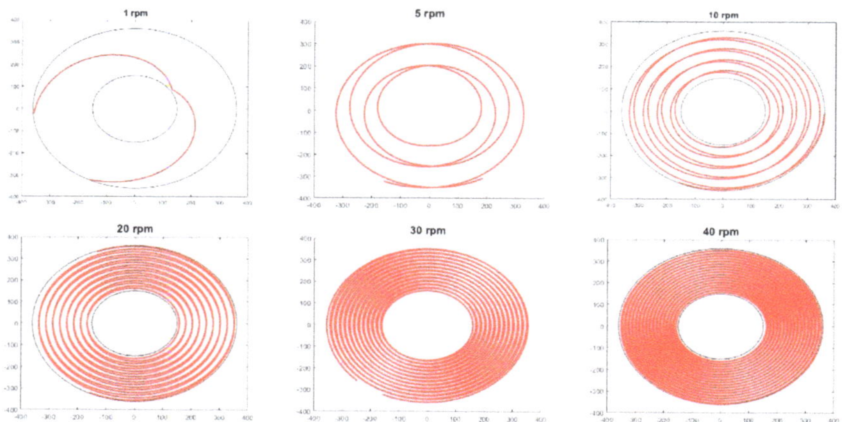

Figure 6. Diagram of scanning locus by different pad speed.

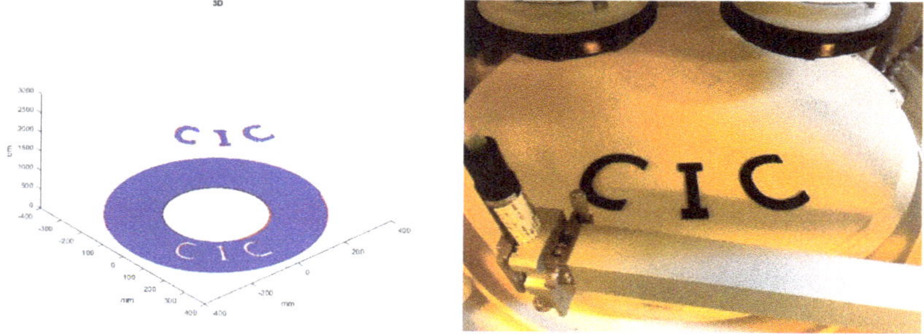

Figure 7. Re−allocating the measurement data into the surface.

4.2. Metrological Parameters

Two major indexes are investigated in this study, which are PU (pad non-uniformity) and PELI (pad effective life time index). PU shows the pad profile and the wear condition during the conditioning process. PELI shows the remaining lifetime of pad by evaluating the change of groove depth. The definition of PU is as Equation (3), where T_{max} *and* T_{min} mean the maximum and minimum value of measuring height data. T_{pad} means the original pad thickness without conditioning. The PU represents the difference of pad profile, whose value varies with the initial setup or within the dressing process. The PELI defines the available lifetime of pad by the remaining pad groove depth. The definition of PELI shows as Equation (4) and H_g and H_{g*} mean the groove depth before and after conditioning process.

$$PU = \frac{T_{max} - T_{min}}{2 \times T_{pad}} \times 100\% \quad (3)$$

$$PELI = \frac{H_{g*}}{H_g} \times 100\% \quad (4)$$

To examine the system metrological parameters, 50 measurements were tested for specific rotation speeds. The detailed experimental parameters are listed as Table 5. The results are shown in Figure 8. For the total swing time for each measurement is 50 s, so the length of scanning locus on pad will increase with the faster rotation speed. The result of PU is larger than others when rotation speed is 1 rpm. Lower rotation speed will reduce the scanning area, so the PU value will become unstable. The PELI value is stable because the groove depth is more evenly distributed on the pad surface. The standard deviation of

PU in each rotation speed except 1 rpm is less than 1.2%. the overall SD of experiments is 0.48%. SD of groove depth in each rotation speed is under 3 μm, and the overall SD of rotation speed from 1 rpm to 100 rpm is 5.58 μm.

Table 5. Parameters of metrological experiments.

Platen Rotation Speed	1, 5, 10, 20, 30, 40, 50, 60, 70, 80, 90, 100 rpm
Swing Arm Rotation Speed	1°/s
Sampling Rate	1000 Hz
Measurement times	50 times
Scan time	50 s

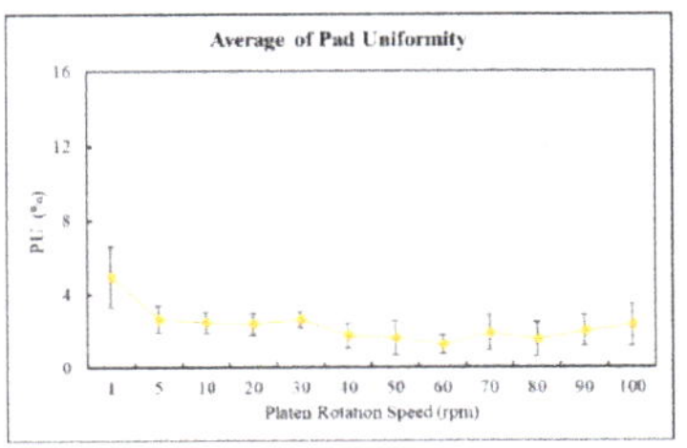

(**a**) Average pad non-uniformity (PU) in different rotation speed.

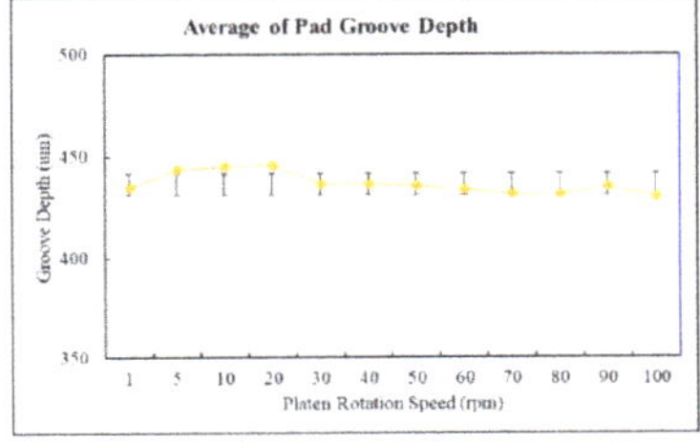

(**b**) Average pad effective lifetime index (PELI) in different rotation speed.

Figure 8. Results of metrological experiments.

4.3. PU and PELI in CMP Experiments

Figure 9 shows the change of PELI (pad effective lifetime index) and PU (pad non-uniformity) value in CMP experiments. The PELI decreases to 30.5% after 50 rounds of CMP experiments and PU increases to 61.9% in the same time. Figure 10 shows the re-mapped pad surface profile from the measuring data. After 50 rounds of CMP process with conditioning between each polishing, a dish-type pf pad profile is measured. The dish-type shape is formed because of the difference of relative speed with pad radius. The pad cutting rate (PCR) is higher at inner area due to the higher relative speed. The scanning electron microscope(SEM) photos of cross-section of pad after 50 rounds of tests are shown as Figure 11 to verify the change of the groove depth during CMP process. Figure 11a shows the locations to take the SEM pictures. The pad's area is separated into 6 ring sections from outer ring to inner ring. Figure 11b is the cross-section on groove of new IC1000 pad. Figure 11c–h shows the wear of pad's groove from outer area to pad center.

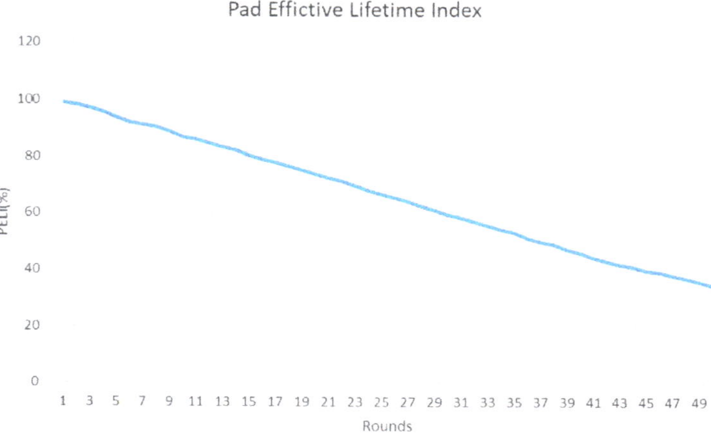

(**a**) Change of PELI in 50 rounds of experiments

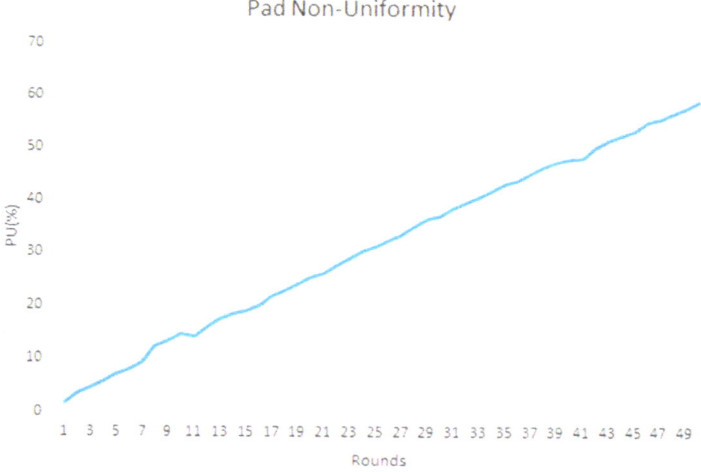

(**b**) Change of PU in 50 rounds of experiments

Figure 9. Results of PELI and PU for 50 rounds of CMP experiments.

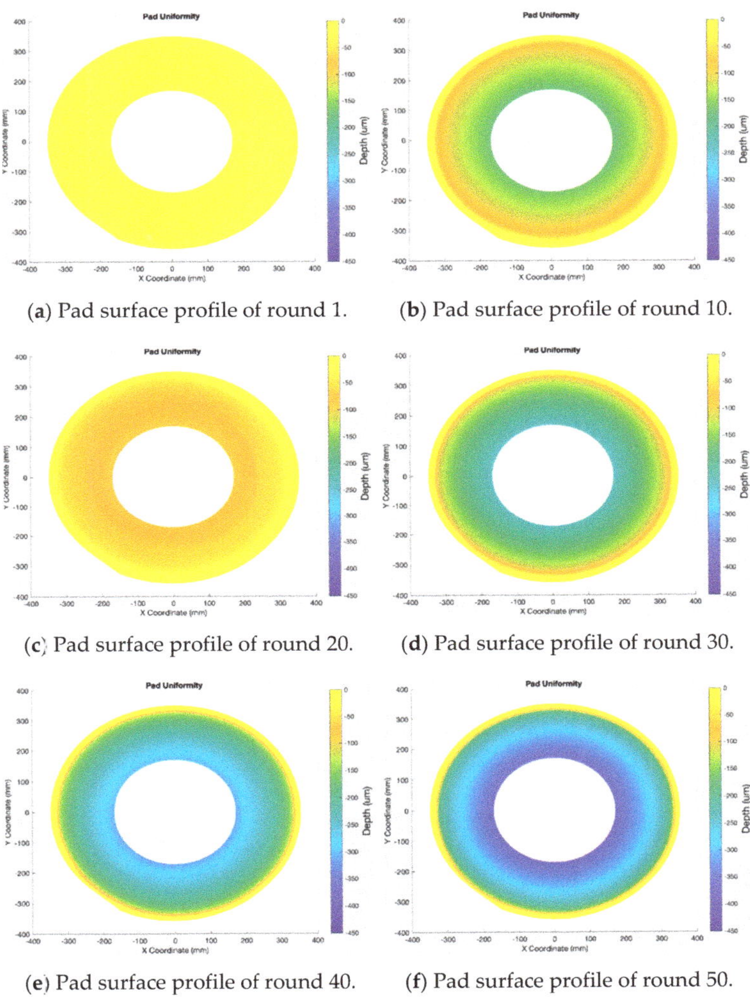

(a) Pad surface profile of round 1. (b) Pad surface profile of round 10.

(c) Pad surface profile of round 20. (d) Pad surface profile of round 30.

(e) Pad surface profile of round 40. (f) Pad surface profile of round 50.

Figure 10. Re−mapped pad surface profile from round 1 to round 50.

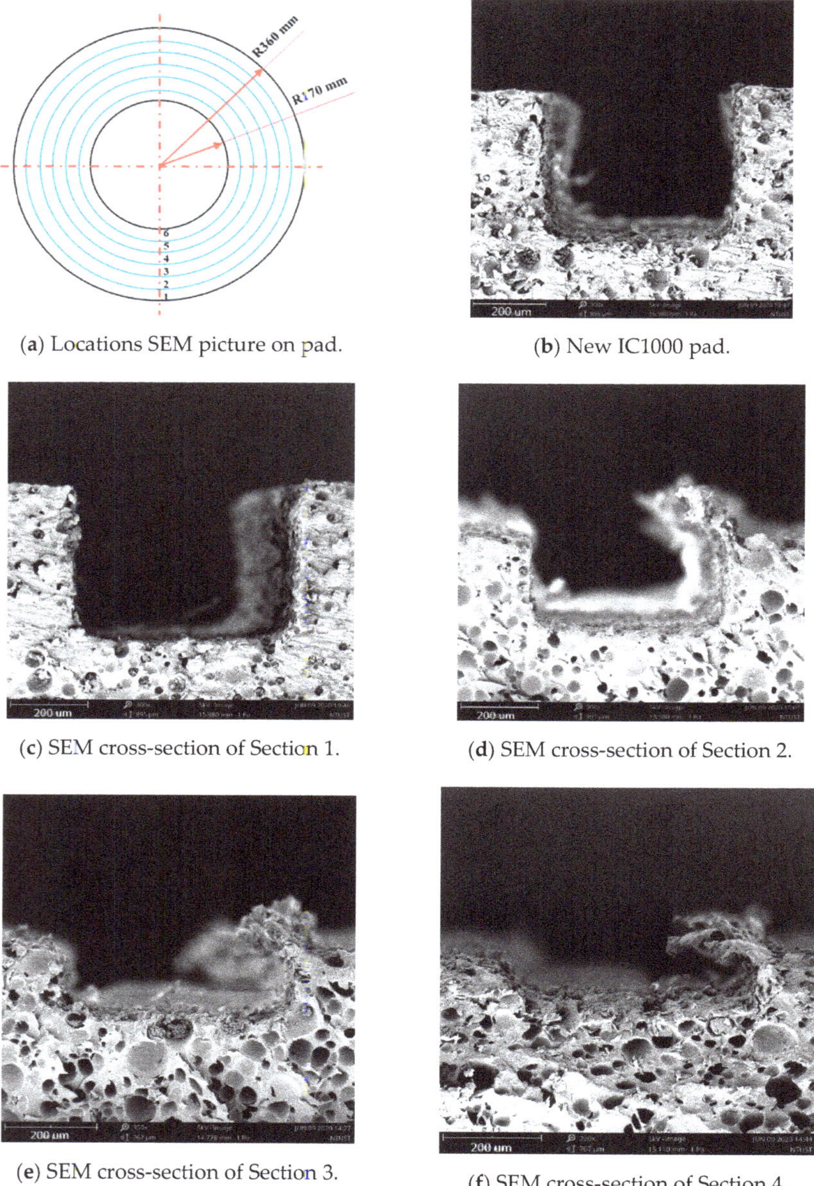

(a) Locations SEM picture on pad.

(b) New IC1000 pad.

(c) SEM cross-section of Section 1.

(d) SEM cross-section of Section 2.

(e) SEM cross-section of Section 3.

(f) SEM cross-section of Section 4.

Figure 11. *Cont.*

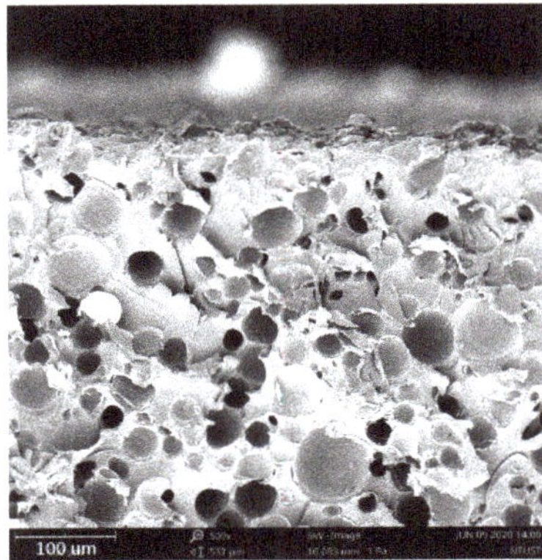

(g) SEM cross-section of Section 5.

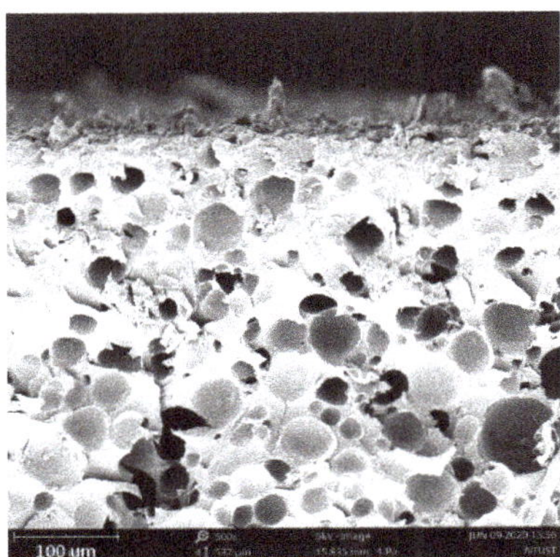

(h) SEM cross-section of Section 6.

Figure 11. Cross-section SEM photo of pad.

4.4. Correlations of CMP Results with Pad Performance Index

After 50 rounds of CMP tests with totally 150 CuB substrates, the MRR of CMP and surface roughness S_a are presented in Figures 12 and 13. The average MRR is 602.97 nm/min and the average S_a is 3.496 nm. The MRR is 710 nm/min and reaches a maximum 762.5 nm/min in the third round of CMP experiment. From experimental results, the PELI and PU remain the same during the first three rounds of tests. The MRR decreases under average value after 25 rounds of tests and S_a of wafer increases over the average line after 31 rounds of test, but Sa value has a trend to increase around 25 rounds of test.

The PELI refers to the remaining groove depth of pad, which represents the ability to store and transfer slurry during CMP process. The effective groove depth can be used to refresh and spread the slurry into the surface between pad asperities and CMP area of Cu film. With the wearing of pad thickness or reducing pad groove depth, the MRR of CMP keeps decreasing. When PELI is smaller than 65%, i.e., the groove depth is less than 280 μm, the MRR of CMP becomes unstable. The MRR is 525 nm/min as PELI is between 35~50%. The MRR of CMP decreases 35% as the PELI of pad is over 70%.

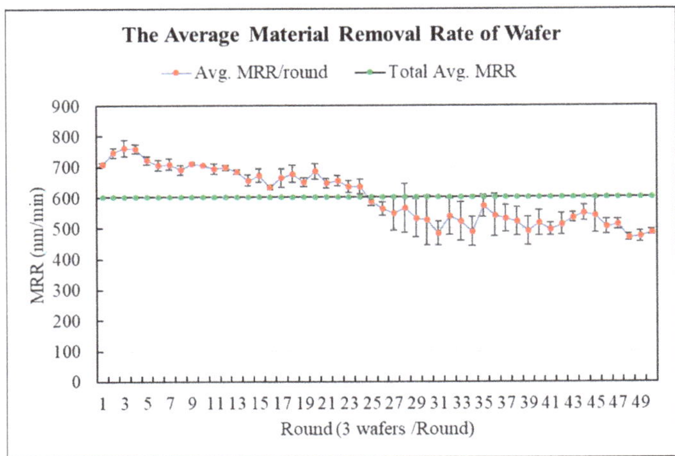

Figure 12. Average MRR of CuB wafer of 50 rounds of test.

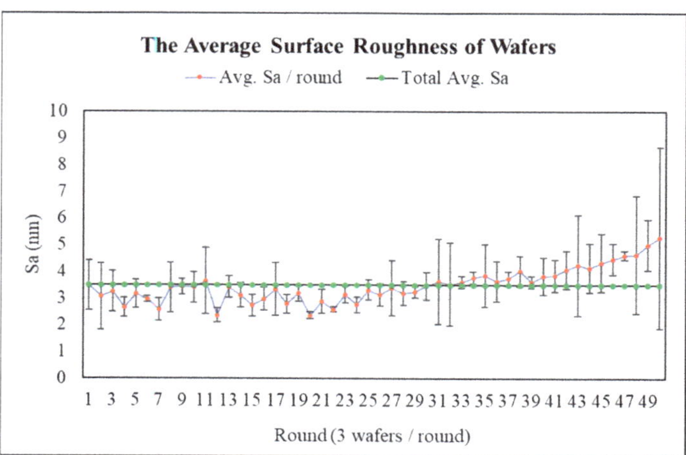

Figure 13. Average Sa of CuB wafer of 50 rounds of test.

Figures 14 and 15 show the correlations between MRR of CMP with PELI and PU of polishing pad. The MRR of CMP has a high correlation as 0.94 with PELI and −0.94 with PU. The high correlation factor shows the MRR of CMP is highly influenced by the PELI of pad.

Consequently, Figures 16 and 17 show the correlations between wafer S_a with PU and PELI. The correlations are obtained as 0.74 and −0.74. The wafer S_a keeps in the same level in the first 25 rounds of the test, and then raises with continuous tests. The correlations are obtained 0.93 and −0.91 by calculating only the last 25 rounds of the CMP tests, where the PELI is below 66.8%. The wafer S_a is significantly affected when pad is conditioned after rounds of processes. The correlations of each performance index and wafer quality are shown in Table 6.

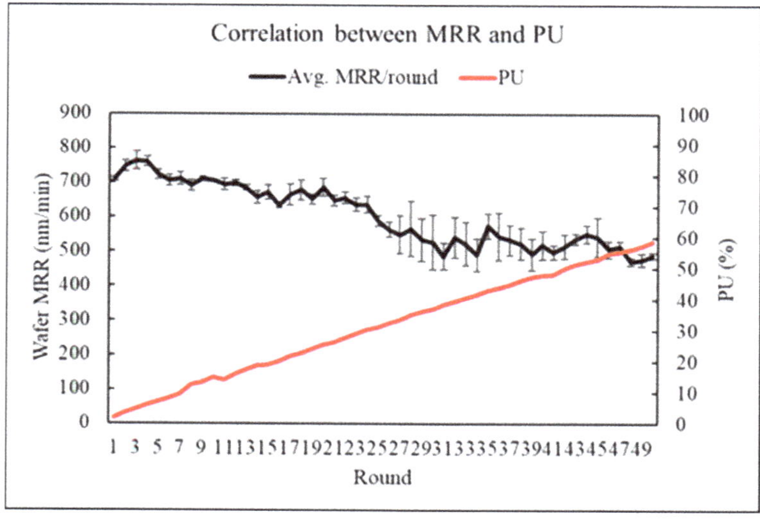

Figure 14. The correlations between MRR and PU.

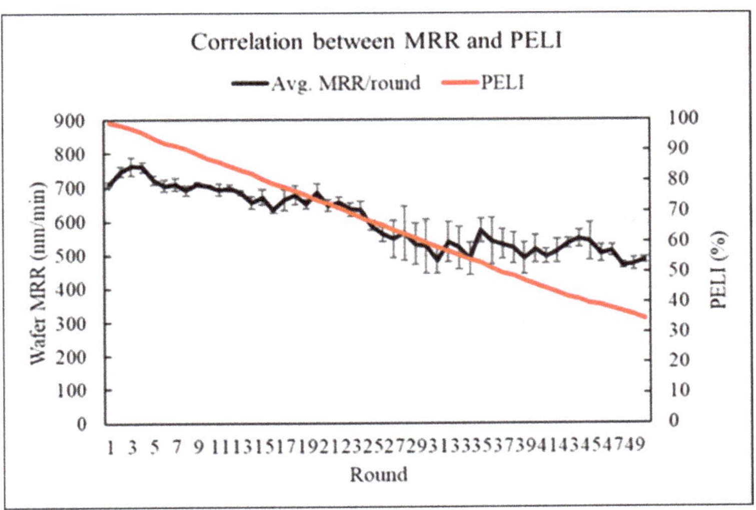

Figure 15. The correlations between MRR and PELI.

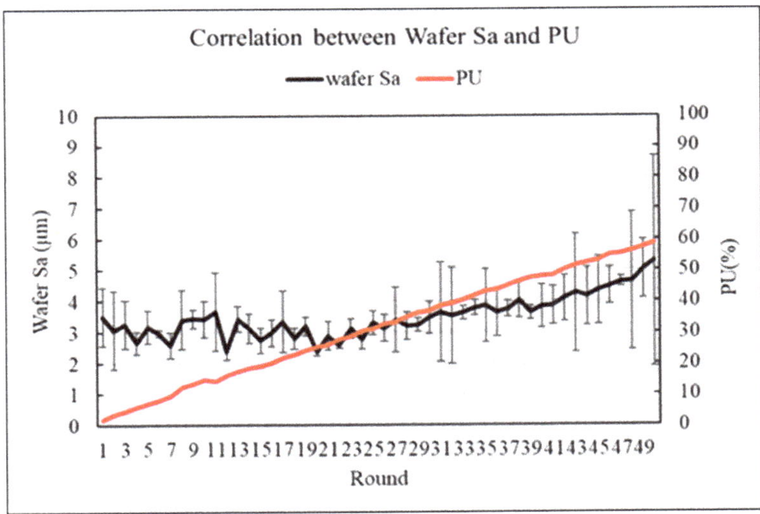

Figure 16. The correlations between wafer Sa and PU.

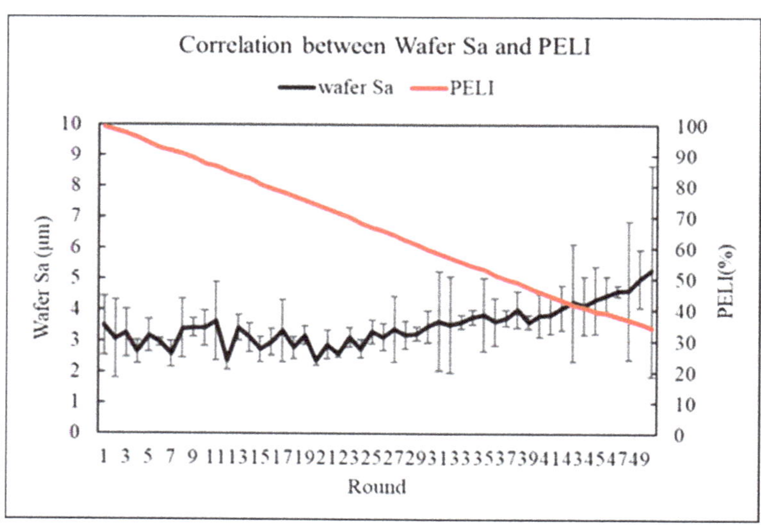

Figure 17. The correlations between wafer Sa and PELI.

Table 6. Correlations of Wafer and pad index.

Wafer/Pad Index	MRR(nm/min)	Sa(µm)	Sa(µm) Last 25 rounds
PU(%)	−0.94	0.76	0.93
PELI(%)	0.94	−0.76	−0.91

5. Conclusions

This study has developed and completed a dynamic pad measurement system (DPMS) of surface topography for chemical mechanical polishing/planarization (CMP) process of IC fabrication. The integration of a chromatic confocal measurement probe into a dressing arm in CMP tool can be used for in-process acquiring pad topography for accessing the pad performance index. The measuring time can be minimized with the motion of the pad conditioning arm and not affecting the CMP throughput. Two major indexes of PU and PELI are presented to identify the status and performance level of the pad during the CMP process. Relationship of wafer quality and pad performance index are discussed by 50 rounds of CMP experiments. The change of PELI and PU are obvious, the wear of pad can be observed by the SEM cross-section photos. The PELI starts from 99.2% and ends with 34.61%, in which the groove is almost gone in the inner part of pad. The PU is 1.9% to 58.7% from start to end. The PU and PELI have high correlations, −0.94 and 0.94, with wafer MRR. Considering wafer Sa remains in the early stage of the experiments, the PU and PELI also highly correlated with wafer Sa for calculating the late stage of the experiments of 0.93 and −0.91. The MRR is changing with the wear of pad during CMP experiments, and the wafer S_a is affected by pad profile when the pad cutting rate (PCR) increases to a certain level. In this study, the S_a value will be highly correlated when PELI is below 66.8%. The MRR drops by 64% and wafer S_a raises 35% with the PELI decreasing by 64.6% and PU increasing by 56.8%.

Results of the study show that the developed DPMS can monitor the change of pad surface profile, which are significantly correlated with wafer quality by CMP. Experimental results can be used positively to predict pad life time for in-process process control of the CMP process.

Author Contributions: Conceptualization, C.-C.A.C. and J.-C.L.; Methodology, J.-C.L.; Software, J.-C.L., W.-C.L.,Y.-J.C. and C.-C.C.; Validation, J.-C.L., W.-C.L. and Y.-J.C.; Formal Analysis, C.-C.C.; Investigation, J.-C.L. and C.C; Data Curation, J.-C.L.; Writing—Original Draft Preparation, J.-C.L.; Writing—Review & Editing, J.-C.L.; Project Administration, C.-C.A.C.; Funding Acquisition, C.-C.A.C. All authors have read and agreed to the published version of the manuscript.

Funding: Authors appreciate for the financial funding of the research project collaborated with Ta Liang Technology Co., Ltd. supported by the Industrial Value Creation Program (grant number: 108-EC-17-A-05-S3-054) from the Academia by the Ministry of Economic Affairs (MEA), Taiwan.

Institutional Review Board Statement: Not applicable.

Informed Consent Statement: Not applicable.

Data Availability Statement: Data sharing not applicable.

Conflicts of Interest: The authors declare no conflict of interest.

References

1. McGrath, J.; Davis, C. Polishing pad surface characterisation in chemical mechanical planarisation. *J. Mater. Process. Technol.* **2004**, *154*, 666–673. [CrossRef]
2. Lee, E.-S.; Cha, J.-W.; Kim, S.-H. Evaluation of the wafer polishing pad capacity and lifetime in the machining of reliable elevations. *Int. J. Mach. Tools Manuf.* **2013**, *66*, 82–94. [CrossRef]
3. Hooper, B.; Byrne, G.; Galligan, S. Pad conditioning in chemical mechanical polishing. *J. Mater. Process. Technol.* **2002**, *123*, 107–113. [CrossRef]
4. Park, B.; Lee, H.; Park, K.; Kim, H.; Jeong, H. Pad roughness variation and its effect on material removal profile in ceria-based CMP slurry. *J. Mater. Process. Technol.* **2008**, *203*, 287–292. [CrossRef]
5. Choi, W.J.; Jung, S.P.; Shin, J.G.; Yang, D.; Lee, B.-H. Characterization of wet pad surface in chemical mechanical polishing (CMP) process with full-field optical coherence tomography (FF-OCT). *Opt. Express* **2011**, *19*, 13343–13350. [CrossRef] [PubMed]
6. Del Monaco, S.; Calderone, F.; Fritah, M.; Le Tiec, T.; Laurent, A. Chemical Mechanical Planarization (CMP) In-Situ pad groove monitor through Fault Detection and Classification (FDC) system. In Proceedings of the ICPT 2012-International Conference on Planarization/CMP Technology, Grenoble, France, 15–17 October 2012.
7. Elledge, J.B. In-Situ Chemical-Mechanical Planarization Pad Metrology Using Ultrasonic Imaging. U.S. Patent 7,306,506, 11 December 2007.
8. Chen, C.-C.A.; Wang, P.-K. Study on Bearing Area Ratio Analysis of CMP Pad Topography Measured by Confocal Laser Scanning System. In Proceedings of the International Conference on Planarization/CMP Technology, Hsinchu, Taiwan, 29 October 2013; pp. 114–117.
9. Lee, H.; Zhuang, Y.; Sugiyama, M.; Seike, Y.; Takaoka, M.; Miyachi, K.; Nishiguchi, T.; Kojima, H.; Philipossian, A. Pad flattening ratio, coefficient of friction and removal rate analysis during silicon dioxide chemical mechanical planarization. *Thin Solid Films* **2010**, *518*, 1994–2000. [CrossRef]
10. Matsumura, Y.; Hirao, T.; Kinoshita, M. Analysis of Pad Surface Roughness on Copper Chemical Mechanical Planarization. *Jpn. J. Appl. Phys.* **2008**, *47*, 2083–2086. [CrossRef]
11. Khanna, A.J.; Jawali, P.; Redfield, D.; Kakireddy, R.; Chockalingam, A.; Benvegnu, D.; Yang, M.; Rozo, S.; Fung, J.; Cornejo, M.; et al. Methodology for pad conditioning sweep optimization for advanced nodes. *Microelectron. Eng.* **2019**, *216*, 111101. [CrossRef]
12. Lee, H.; Lee, S. Investigation of pad wear in CMP with swing-arm conditioning and uniformity of material removal. *Precis. Eng.* **2017**, *49*, 85–91. [CrossRef]
13. Lee, S.; Kim, H.; Dornfeld, D. Development of a CMP pad with controlled micro features for improved performance. In Proceedings of the ISSM 2005, IEEE International Symposium on Semiconductor Manufacturing, San Jose, CA, USA, 13–15 September 2005; pp. 173–176. [CrossRef]
14. Huang, B.; Kunoth, A. An optimization based empirical mode decomposition scheme. *J. Comput. Appl. Math.* **2013**, *240*, 174–183. [CrossRef]
15. Huang, N.E.; Shen, Z.; Long, S.R.; Wu, M.C.; Shih, H.H.; Zheng, Q.; Yen, N.-C.; Tung, C.C.; Liu, H.H. The empirical mode decomposition and the Hilbert spectrum for nonlinear and non-stationary time series analysis. *Proc. R. Soc. A Math. Phys. Eng. Sci.* **1998**, *454*, 903–995. [CrossRef]
16. Lee, S.; Jeong, S.; Park, K.; Kim, H.; Jeong, H. Kinematical Modeling of Pad Profile Variation during Conditioning in Chemical Mechanical Polishing. *Jpn. J. Appl. Phys.* **2009**, *48*, 126502. [CrossRef]
17. Chen, C.-C.A.; Pham, Q.-P. Study on diamond dressing for non-uniformity of pad surface topography in CMP process. *Int. J. Adv. Manuf. Technol.* **2017**, *91*, 3573–3582. [CrossRef]

MDPI
St. Alban-Anlage 66
4052 Basel
Switzerland
Tel. +41 61 683 77 34
Fax +41 61 302 89 18
www.mdpi.com

Applied Sciences Editorial Office
E-mail: applsci@mdpi.com
www.mdpi.com/journal/applsci

www.ingramcontent.com/pod-product-compliance
Lightning Source LLC
LaVergne TN
LVHW071614170726
843515LV00009B/2394

* 9 7 8 3 0 3 6 5 2 9 8 6 8 *